THE BIRDS OF
Pennsylvania

THE BIRDS OF *Pennsylvania*

Gerald M. McWilliams
Naturalist

Daniel W. Brauning
Ornithologist, Pennsylvania Game Commission

WITH A FOREWORD BY

Kenn Kaufman
Field Editor, *Audubon Magazine*

COMSTOCK PUBLISHING ASSOCIATES a division of

CORNELL UNIVERSITY PRESS | ITHACA AND LONDON

First published 2000 by Cornell University Press

Printed in the United States of America

Cornell University Press strives to use environmentally responsible suppliers and materials to the fullest extent possible in the publishing of its books. Such materials include vegetable-based, low-VOC inks and acid-free papers that are recycled, totally chlorine-free, or partly composed of nonwood fibers. Books that bear the logo of the FSC (Forest Stewardship Council) use paper taken from forests that have been inspected and certified as meeting the highest standards for environmental and social responsibility. For further information, visit our web site at www.cornellpress.cornell.edu.

Library of Congress Cataloging-in-Publication Data

McWilliams, Gerald M.
The birds of Pennsylvania / Gerald M. McWilliams and Daniel W. Brauning : with a foreword by Kenn Kaufman.
p. cm.
Includes index.
ISBN 0-8014-3643-5 (cloth)
1. Birds—Pennsylvania. 2. Birds—Pennsylvania—Geographical distribution I. Brauning, Daniel W. II. Title.
QL684.P4M36 1999
598′.09748—dc21 99-38164

Cloth printing 10 9 8 7 6 5 4 3 2 1

I am indebted to my wife, Linda, who spent many long hours with me in the preparation of this book, who was patient with my impatience during the long evenings of writing, and who managed to endure nearly a decade with a bird-obsessive writer; and to my sons, Nathan and Aaron, who grew into men during the time it took to write this book. I will always be grateful for their understanding in my absence.

Jerry McWilliams

I would like to express particular appreciation to my wife, Marcia, for her enduring patience during the preparation of this book and for sharing with me a fascination with creation. Also, to my sons, Timothy, Andrew, and David, for their understanding when Dad was glued to his desk. My parents, Marilyn and Wayne Brauning, deserve my gratitude for giving me the freedom to pursue this hobby and unlikely career and for supporting me in my enjoyment of birds.

Soli Deo Gloria

Daniel Brauning

CONTENTS

MAPS

PHOTOGRAPHS

FOREWORD

The state of Pennsylvania looms large on the map of eastern North America and larger still in the history of bird study on this continent. The young Alexander Wilson, off the boat from Scotland in 1794, was walking to Philadelphia when the sight of a Red-headed Woodpecker captured his imagination and changed his life. Young John James Audubon, newly arrived from France in the early 1800s, neglected his family business at Mill Grove, Pennsylvania, so that he could watch the Eastern Phoebes and other birds there. Wilson and Audubon, of course, went on to become the two major founders of American ornithology.

Pennsylvania has held many other distinctions in the avian realm. Its Academy of Natural Sciences in Philadelphia, the oldest natural history museum in the Americas, is still a leader in ornithology, as is the Carnegie Museum in Pittsburgh. Hawk Mountain, north of Reading, was one of the first private bird sanctuaries and probably the birthplace of the current worldwide craze in hawk watching. The state has also given us many great ornithologists, from Spencer Baird to Ted Parker.

Yet I would say that Pennsylvania's stellar position in ornithology is eclipsed by its importance to the birds themselves. Not just because it is a big state, stretching from the Great Lakes almost to the Atlantic coast, and not just because vast numbers of birds migrate over or through this region. The state is also significant as a major reserve for nesting birds. In recent decades, we have seen widespread declines among many of our migratory forest species, largely owing to loss of habitat. The extensive forests of Pennsylvania, many of them on protected public lands, have become more and more essential to maintaining healthy populations of Scarlet Tanagers, Wood Thrushes, and many other species. This key importance makes Pennsylvania the Keystone State in an avian sense as well as in U.S. history.

In light of the great significance of Pennsylvania to birding and birds, it might seem strange that no comprehensive state bird book has appeared for more than a century—but perhaps its significance is the explanation. The sheer scope of the task, the amount of information to be summarized, makes this much more of a challenge than describing a smaller or less-studied state.

Undaunted by the challenge, however, were Gerald McWilliams and Daniel Brauning, outstanding experts on bird distribution. Their complementary backgrounds made them the ideal combination to write this book. McWilliams is a fine all-around naturalist who has focused on Erie County, on the lakeshore in the northwestern corner of the state. For years now, it has been well-nigh impossible for any bird to fly across northwestern Pennsylvania without intersecting with Jerry's binocular field. His knowledge of migration timing in the state is unequaled, and for this book he also worked out the first detailed accounts of winter distribution. Dan Brauning has lived in southeastern and central Pennsylvania and has done extensive field studies all over the state, first as coordinator of the Breeding Bird Atlas and later in his work on bird conservation projects. No one knows more than he about the state's nesting birds or their habitats. In addition, Dan's statewide contacts made it possible to draw on data from scores of other observers. Between the two of them, McWilliams and Brauning command a formidable understanding of Pennsylvania bird life.

With this book, they make this understanding available to all of us. The depth of information here is remarkable. The status of every species in every season is presented in clear, thorough terms, and habitat preferences are described in detail. The sections on the historical status of each bird are outstanding, illuminating and explaining the long-term and recent changes in our avifauna. This book will be absolutely indispensable for anyone who wants to understand the bird life of Pennsylvania or of any of the surrounding states, whether for recreation, for research, or for conservation work.

When I was living west of Philadelphia in the 1980s, I often wondered if the woodpeckers and phoebes I watched might have been direct descendants of the ones that had inspired Wilson and Audubon almost two centuries earlier. I also wondered about the status and distribution of various bird species around the state and wished there were a good book on the subject. Now there is an outstanding book to fill that need. I congratulate McWilliams and Brauning on an impressive achievement and heartily recommend this volume to anyone who cares about birds.

KENN KAUFMAN

PREFACE

Few things in the natural world have inspired naturalists more than birds. Early pioneers of bird study in North America, such as John James Audubon and Alexander Wilson, crossed Pennsylvania many times but left few written accounts of their studies there. The first systematic list of birds of Pennsylvania was not published until 1845. In 1890, Benjamin Warren's extensive report on the birds of Pennsylvania marked a new era of ornithological history and inspired other observers to study birds. Warren's was the first and, until now, the only thorough statewide reference of the state's birds. Over 100 years have passed since a comprehensive book covering the birds of Pennsylvania has been published. W. E. Clyde Todd's ambitious work on the birds of western Pennsylvania is still recognized as the most important study ever made of the birds of that region. Earle Poole and Merrill Wood published annotated checklists that brought the state list up to date but fell short of providing a comprehensive reference to the state's birds. Poole's comprehensive work on the birds of Pennsylvania unfortunately was never published. The unpublished manuscript, completed in the early 1960s, was a keystone reference and serves to define the current period.

Until the early part of the twentieth century most bird study was undertaken by professional ornithologists. The information accumulated over decades was acquired primarily by collecting bird skins, nests, and eggs. By 1934, with the publication of the first Peterson Field Guide, the shotgun was largely replaced by binoculars. With the use of new descriptive field guides and site guides, it became easier to find and identify birds in the field. The collecting of specimens was no longer necessary to document most records in Pennsylvania. That method was replaced by the camera and detailed written field notes, augmented by binoculars and spotting scopes that came to be widely used for bird study by amateurs and professionals. What was once considered a primarily professional occupation became a popular pastime among people of all ages and professions.

Much has changed in the ensuing years. Bird watching, now referred to by most people as birding, has become increasingly popular in Pennsylvania and across the nation. Recordings of bird songs and the calls of most species are now available, helping to sharpen identification skills. The American Birding Association is a national organization devoted to field identification and bird distribution. Its magazine, *Birding*, contains many articles with photographs and valuable tips on difficult identifications. Books on where to go to find birds are available for many states and regions, including Pennsylvania. Other tools that play a major part in today's expanded knowledge of Pennsylvania birds are the many regional telephone hot lines and the Internet, which frequently update the presence of rare and interesting birds. The growing interest among the state's birders put over 2000 people into the field during the Breeding Bird Atlas project and continues to add substantially to our knowledge of our birds, providing a new foundation of information. Recent species splits, name changes, and changes in taxonomic sequence also accelerated, creating a need for an up-to-date book on the birds of Pennsylvania.

Pennsylvania Birds, Poole's bird manuscript (ca. 1960), and the *Atlas of Breeding Birds in Pennsylvania* (Brauning 1992), as well as several local publications are the keystones of this book. We began the research for the manuscript in the fall of 1990. Our goal was to create a comprehensive modern reference work, with a historical review, on the status and distribution of the birds of Pennsylvania. The word *modern* is defined in this book as the period from about 1960 to 1998. This book serves to extend our knowledge not only of the status and distribution of the birds in Pennsylvania but also of the status and distribution of the birds of the eastern United States. It is intended to be used by all audiences from novices to advanced birders and ornithologists.

Physiographic regions are clearly described, and maps assist the reader in locating sites where most of the data for this book were collected. A brief summary of the biology of each family of birds is followed by the species accounts. Most species accounts open with a general note followed by a brief description of the habitat and then information on the migration, breeding, and wintering status and distribution. The distribution and abundance of every bird have not been precisely defined for all regions or in all seasons in Pennsylvania. Breeding distributions have been documented over most of the state, but little is yet known about the status and distribution of birds that winter here. However, more information is presented on the state's wintering birds in this book than in any other publication to date. The account closes with a summary of the history of that species in Pennsylvania prior to 1960.

Some species are illustrated with a photograph, or their breeding distributions are mapped on the basis of Breeding Bird Survey data. The material in this book includes the most recent information, but there is still much to be learned about the birds of Pennsylvania. Even though nine years of research have gone into this publication, we do not claim to have included every published or unpublished record. Nevertheless, readers should find that the *The Birds of Pennsylvania* is a valuable source of information for anyone interested in learning more about the status and distribution of the state's birds.

ACKNOWLEDGMENTS

Grants were provided by the McLean Contributionship to subsidize the publication and by the Wild Resource Conservation Fund to support the research for this book. We also thank the Pennsylvania Game Commission for providing logistical support and for allowing Dan Brauning time to work on the manuscript.

This book includes the observations of thousands of birders from across Pennsylvania who made the effort to document their sightings and contribute to our collective knowledge of birds. Their contributions are foundational. The following people collected, compiled, and contributed to this book an immense amount of data on county winter bird distributions (data were received from all but 10 counties). Their hard work and dedication made it possible to make accurate assessments of the distribution of wintering birds in Pennsylvania for the first time. They also contributed rare and unusual bird sightings on a countywide scale (data were received from all but 20 counties). Thanks to all the others, far too many to list, who also assisted in the making of this book.

County	Contributors
Adams	Art and Eleanor Kennell
Allegheny	Ted Floyd, Paul Hess, Joyce Hoffmann
Armstrong	Margaret and Roger Higbee
Beaver	Dr. John Cruzan
Bedford	Janet Shafer
Berks	Rudy Keller
Blair	Stan Kotala
Bradford	Trudy Gerlach
Bucks	Ron French, Ken Kitson
Butler	Paul Hess
Cambria	Gloria Lamer, Georgette Syster
Cameron	Bill Hendrickson
Carbon	Rick Wiltraut
Centre	John and Becky Peplinski, Paul and Glenna Schwalbe
Chester	Phyllis Hurlock
Clarion	Margaret Buckwalter
Clearfield	Jocelynn Smrekar
Clinton	Paul and Glenna Schwalbe
Crawford	Jim Barker
Cumberland	Jane Earle, Ramsay Koury
Delaware	Al Guarante, John Miller, Nick Pulcinella
Elk	Linda Christenson
Erie	Jean Stull Cunningham
Forest	April Walters
Franklin	Ken Gabler
Greene	Ralph Bell
Huntingdon	Greg Grove
Indiana	Margaret and Roger Higbee
Juniata	Linda Whitesel
Lancaster	Eric Witmer
Lawrence	Barbara Dean
Lehigh	Bernie Morris
Luzerne	Mark Blauer
Lycoming	Wesley Egli, Paul and Glenna Schwalbe, Stanley Stahl
McKean	John Dzemyan
Mercer	Marty McKay
Mifflin	Margaret Kenepp
Montgomery	Gary Freed
Montour	Allen Schweinsberg
Northampton	Rick Wiltraut
Northumberland	Allen Schweinsberg
Perry	Deuane Hoffman
Philadelphia	Ed Fingerhood, John Miller
Potter	David Hauber

Schuylkill	Dan Knarr
Snyder	Allen Schweinsberg
Somerset	Ruth and Glenn Sager
Sullivan	Nick Kerlin
Susquehanna	Dr. Jerry Skinner
Tioga	Joanna Stickler
Union	Allen Schweinsberg
Venango	Russ States
Warren	Ted Grisez
Washington	Margaret and Roger Higbee, Dr. Roy Ickes
Wayne	Joe and Voni Strasser
Wyoming	William Reid
York	Al Spiese

A special thanks to the following people for their technical assistance and contributions: Nancy Clupper, Jean Stull Cunningham, Eileen James, Linda McWilliams, Charles Murray, and Dr. Kenneth Parkes. The following institutions contributed data or provided access to their collections: the Academy of Natural Sciences of Philadelphia, the American Museum of Natural History, the Breeding Bird Census, the Carnegie Museum of Natural History, the Cleveland Museum of Natural History, the Cornell Laboratory of Ornithology Nest Card Record Program, the Delaware Museum of Natural History, the North Museum of Lancaster, the Pennsylvania Society for Ornithology Special Area Project, the Pennsylvania State Museum of Natural History, the U.S. Geological Survey Breeding Bird Survey, the U.S. Geological Survey Bird Banding Laboratory, the United States National Museum (Smithsonian), and the Western Foundation of Vertebrate Zoology.

We are especially indebted to the following people, who assisted with their guidance and extensive knowledge of the birds of Pennsylvania and contributed many hours of their time to review the manuscript: David Cutler, John P. Dunn, Jon Dunn, Douglas Gross, Franklin and Barbara Haas, George Hall, Margaret and Roger Higbee, Robert Leberman, Robert Mulvihill, Glenna and Paul Schwalbe, Phillips Street, William Reid, and Rick Wiltraut.

Thanks are also due to Chandler Robbins, wildlife biologist for the U.S. Department of the Interior for the careful and thorough review of the final manuscript; to Peter Prescott, science editor at Cornell University Press for his encouragement and guidance; and to the Cornell production staff for their assistance with technical matters.

All of the photographs in this book depict birds that are rarely seen in Pennsylvania. Every photograph was taken in Pennsylvania. Some of these photographs have been published elsewhere, but many have not. We are thankful to the following persons, who permitted the use of their valuable originals for this book: Alan Brady, Dave Darney, Mike Fialkovich, Franklin Haas, Ed Kwater, Randy Miller, Edwin Johnson, Walter Shaffer, Robert Schutsky, Paul Schwalbe, and Rick Wiltraut.

Part I. Introduction

Historical Perspectives on Bird Populations and Habitats

Pennsylvania, as one of the 13 original United States and among the first to be settled by European colonists, has been molded by intensive human activities for over three centuries. This chapter attempts to describe the broad-scale changes in habitats and resulting populations of birds within this Commonwealth since about 1600. It briefly addresses the changing state of knowledge of bird life but does not attempt to duplicate the summary of ornithological activity provided by Fingerhood (Brauning 1992).

Native American populations had been living in semipermanent villages and farming and hunting the valleys and mountains for many centuries when Henry Hudson first entered the Delaware Bay in 1609 (Muller 1989). Fire was the predominate tool used to clear and open the otherwise contiguous forest, and some fires burned out of control over large areas. The resulting open habitats, as well as beaver impoundments and blowdowns, provided habitat for birds associated with brushy habitat and, possibly, grassy areas. Wetland habitats also provided breaks in the contiguous forests. The deciduous forest had reached its climax, covering much of the state in mature stands. Pennsylvania was predominately a rough, unbroken wilderness covered by a mixture of deciduous and coniferous forests at the time of first European settlement.

Early Dutch and Swedish colonists vied for control of North America; settlements were limited primarily to the banks of the lower Delaware River until the English claimed rights to America in 1664. William Penn was granted the charter to the Commonwealth in 1681. Through much of the 1700s he negotiated purchases from the Native Americans of progressively western portions of Pennsylvania. Little was recorded of the bird life during this era other than anecdotes of explorers and settlers. References usually were limited to the larger birds that served as food. Penn wrote, in part to attract settlers, of the abundance of wildlife in the new lands, citing "turkey . . . pheasants, heath birds, pigeons, and partridges in abundance" and also mentioned "great numbers of swan, white and grey goose, brand, ducks, snipe and curloe [*sic*]" (fide Thomas 1876). Life was hard, and few people took the time to document natural history observations. Early records also were impaired by a lack of systematic nomenclature. Linnaeus's system of Latin names was not developed until 1758 (Gill 1990). Many birds were first named by Linnaeus during the final years of the eighteenth century (AOU 1998). Peter Kalm (1753–1761), a student of Linnaeus, included references to a few birds in Pennsylvania, mostly game species, in the three-volume record of his travels in North America.

At the time that the state's first systematic list of birds was produced by William Bartram, in 1791, all but a part of what we now know as Erie County had been purchased from Native Americans and given local government. Pennsylvania, and Philadelphia in particular, was the cultural and scientific capital of the newly formed colonies, and a strong agricultural community had been established east of the mountains. Early naturalists (e.g., Audubon, Barton, Bartram, and Wilson) working in Pennsylvania at the end of the 1700s and in the early 1800s began the state's bird list. Their published records in some cases identified arrival dates of migrants and distinguished permanent residents. These prominent naturalists contributed to the naming of many North American birds on the basis of specimens collected around Philadelphia, but the basic movements and patterns were just being determined.

The early ornithological publications, written a few decades after the Revolutionary War, already reflected the influence of European settlement. Birds of open grasslands, such as Upland Sandpiper and Eastern Meadowlark, appear in the earliest regional lists, as do birds of the forest, such as the Brown Creeper and Pileated Woodpecker (Barton 1799). Most of the birds in these earliest records, with the obvious exception of the extinct Carolina Parakeet, Passenger Pigeon, and Heath Hen, still are characteristic of Pennsylvania 200 years later. Bird populations in the vast untouched mountains were rarely mentioned in these earliest ornithological records. Todd (1940) lamented the absence of such records. Differences between late eighteenth and early nineteenth century bird lists and modern knowledge of bird populations remain enigmatic problems. Implications of dramatically different seasonal occurrence provide a tantalizing suggestion of very different patterns for at least some species. For example, "Pigeon Hawks" (i.e., Merlins) were recorded as permanent residents (Barton 1799). But such contrasts with current status are the exception, and some cases may be explained by an obvious error of identification or seasonal occurrence. Also, the question will remain unanswered whether grassland-associated species expanded into the colonists' agricultural fields or were present, in smaller numbers, in natural openings and in openings created by Native Americans.

With the final purchases of land from the Native Americans and the Land Act of 1792, settlers spread rapidly west across Pennsylvania. By 1820, the boundaries of what is now known as Pennsylvania were drawn and a low-density population of settlers was forging a rugged living from the land across much of the state. Title to much of Pennsylvania was claimed within a remarkably short period of time. Agriculture dominated the southeastern counties, and Pennsylvania was the national leader in the production of grains and cattle.

It was not until nearly all of the state had been settled that Baird published the first regional bird list in 1845. He systematically described seasonal occurrence and abundance of the birds of Cumberland County in a way that became a model for many similar works over the following decades. His work reflected the influence of agriculture on the landscape. Many grassland species (e.g., Upland Sandpiper, Vesper Sparrow, and Dickcissel) were listed as abundant summer residents. Forest dwellers, from warblers to woodpeckers, also remained abundant, at least in the mountains. Differences between Baird's list and modern records are many, but generally the differences are a matter of degree rather than of substance. He failed to find the Black-throated Green Warbler in Cumberland County in 1845, but he included the Black-throated Blue Warbler, which was absent from subsequent lists until it recolonized the mountains during the final years of the twentieth century.

Industrial development and a new surge of immigration promoted incredible changes through the middle decades of the nineteenth century. Canals and railroads provided large-scale transportation to remote areas, and mining changed the face of the coal regions. The state experienced its most rapid industrial growth between 1860 and 1900. The state's human population grew from 2 million to about 6 million during that same period (Zelinksy 1989). The lumber boom altered the landscape. Timber operations moved from southern and riverside forests into the heart of the mountains and made Pennsylvania the nation's largest producer of timber by 1880 (Muller 1989). In the midst of this industrial boom was a surge of county bird publications and the first regional treatises. Annotated lists following Baird's 1845 model were published for several southern counties, such as Bucks, Delaware, Westmoreland, and Chester. Turnbull's 1869 and Gentry's 1876 and 1877 volumes provided broad descriptions of the birds of eastern Pennsylvania. Several key publications of bird life from this period make frequent reference to forest birds in the "remnant" old forest stands (e.g., Todd 1893; Cope 1902). These provide only hints of the status of birds that once dominated the state.

It was at this apex of resource exploitation that the first and, until now, the only thorough statewide report of Pennsylvania's birds was published. Warren produced a first edition of his *Report on the Birds of Pennsylvania* in 1888 and a more thorough edition in 1890. By then much of the wildness of Penn's Woods had been extinguished. The wolf had been extirpated, and reports of mountain lions were unverified. The vast clouds of Passenger Pigeons had been reduced to a few roaming flocks, and the Heath Hen was a distant memory. Those losses were the legacy of an era of reckless exploitation. Bounties that had been established on birds of prey pushed many hawks and owls from settled areas into remote regions. A hatred of bird-eating hawks was seen even in the writings of the first state ornithologist, George Sutton (1928a), who, for example, called the Sharp-shinned and Cooper's hawks "our most objectionable birds." Unregulated and commercial hunting had dramatically reduced most populations of game mammals and birds, notably turkey and waterfowl (Todd 1940). The boom of industrialization and mining reduced water quality and, consequently, reduced nesting populations of wetland birds.

Farming followed lumbering, and by the end of the 1800s Pennsylvania was only about 25% forested (Muller 1989). A radically different landscape was presented to native and migrant birds than had been here two centuries earlier. Todd (1940) said that "the stands of original forest are now reduced to a few pitiful remnants." The loss of the original forests dramatically affected some species, such as Pileated Woodpecker, which became restricted to the remote mountains (Rhoads 1899). Although a species' abundance was sometimes described in terms of how many a gunner could kill in a day, systematic lists were being produced from the far corners of the state, such as the Poconos (e.g., Carter 1904) and the central mountains (e.g., Cope 1898). The change in bird life had been "largely caused by the cutting away of the original primeval forest" (Cope 1902). A common thread was the decline of the "Canadian" element of birds, species associated with northern, boreal forests such as the Swainson's Thrush, Winter Wren, Olive-sided Flycatcher, and Blackburnian, Canada, and Magnolia warblers and others. Southern, "Carolinian" species, such as the Chipping Sparrow, Yellow-breasted Chat, and others, were described to be moving north onto the borders of remaining forest patches (Cope 1898). Also, a recent study shows that old-growth stands support much higher densities of forest birds than did nearby second-growth forests (Haney and Schaadt 1996). Pennsylvania's bird communities were in rapid transition.

Although these forest-associated birds were dramatically affected by the massive timbering activity, almost all survived to the present time by adapting to the regrowing forests. Todd (1893) considered many forest birds, such as Black-throated Blue and Black-and-white warblers, to be abundant or common in the second growth, as well as in the remaining stands of original forest, of Indiana and Clearfield counties. The adaptability of such birds was clear as they passed through a serious population bottleneck at the dawn of the twentieth century.

An outcry over the loss of game species sponsored the formation of the Game Commission and new hunting regulations at the end of the 1800s. Unregulated killing of birds spurred formation of chapters of the Audubon Society in the 1890s and federal regulations, in the form of the Migratory Bird Treaty Act of 1918. Hunting regulations, even moratoriums, were placed on dwindled stocks, and market hunting was outlawed (Kosack 1995), setting the stage for the restoration of many of the more prominent species and avoiding extinction for others. Scientific inquiry was achieving new levels of organization with the formation of the American Ornithologists' Union in 1883. Newly organized ornithological organizations (e.g., the Delaware Valley Ornithology Club in Philadelphia in 1890) provided new details of bird life in regional publications such as *Cassinia* and the *Cardinal* in the early 1900s.

Agricultural acreage reached a peak of about 20 million acres, about two-thirds of our land area, in 1900. Farming with horses and mules and extensive pasturing acreage provided new landscapes for open-country birds. Grassland species that had colonized the southern counties with the settlers expanded with the growing agriculture to become nearly statewide, abundant species. The meadowlark, for example, was considered common in cleared areas of Potter, Clinton, and McKean counties (Cope 1902; Keim 1905), areas that had not long before been extensive forest. The populations of introduced birds exploded in this human-altered landscape. Some, such as the House Sparrow and the European Starling, were pests; others, such as the Ring-necked Pheasant, were game birds. Attempts to establish certain other exotic game birds, such as Chukars and Sharp-tailed Grouse, failed.

As the settlers pushed west, the poor rocky soils of Pennsylvania's mountain regions could not compete with the newly broken sod of the Midwestern farms being settled in the early 1900s. Human populations peaked in north-central counties during lumbering booms, but growing needs for industrial labor drew people from the rugged rural areas into the cities. As farms were abandoned and forests regenerated, an explosion of shrub-associated bird populations was observed. Sutton (1928a) commented on the dramatic northward expansion of shrubland species from the south, including the Northern Cardinal, Lark Sparrow, and Northern Mockingbird. Northerly birds of brushy habitats, such as the Chestnut-sided Warbler and Alder Flycatcher, also were more widely distributed than previously reported. At the same time, regulations on hunting were being successful. By the 1920s the Wood Duck was coming back from the brink of extirpation and turkey populations were being restored by trap-and-transfer programs.

The flooding of much of the 10,000-acre Pymatuning Swamp in 1932 and 1933 was just the most dramatic example of the losses of vegetated wetlands (Tiner 1987). Sutton (1928b) called it a "wonderful sanctuary" and it was unequaled anywhere in Pennsylvania, but much of the boreal element was lost with the rising water level (Trimble 1940). Various waterfowl, including diving ducks, swans, and others exploited the newly formed lakes, some establishing short-lived breeding populations. Many smaller wetlands, not named or cataloged, were drained, paved, or flooded during the nineteenth and twentieth centuries, resulting in a loss of 56% of the state's wetlands (Thorne et al. 1995). Severe reductions of wetland habitat in the southeastern corner of the state, notably at Tinicum, went largely undocumented during the nineteenth century but continued well into the twentieth century, resulting in ongoing declines of bitterns, shorebirds, and many other wetland species.

Birds of agriculture, from pheasants to meadowlarks, continued at high levels through the 1960s while agricultural set-aside programs encouraged farmers to idle fields even though farms were being abandoned and acreage declined (Klinger et al. 1998). Farming became increasingly industrialized after World War II. New equipment made larger fields more practical, reducing hedgerows. New cultivars enabled more-frequent cutting schedules, resulting in mowing more frequently than the time required by most birds to produce a brood. The conversion of pasture and hay land to high-intensity row crops rendered fields unattractive to grassland species, and new chemical pesticides, herbicides, and fertilizers increased crop yields by reducing pests and weeds, making farm fields inhospitable to wildlife. Certain pesticides, the chlorinated hydrocarbons, had tremendous impacts on already depleted populations of birds of prey. The outcry against the use of DDT contributed, in part, to a greater awareness of conservation issues during the 1960s. These changes, and the continued loss of agricultural acreage to development, contributed to grassland bird populations' experiencing the largest declines of any habitat group during the last half of the twentieth century (Sauer et al. 1997).

At the dawn of a new millennium, Pennsylvania more closely resembles the landscape at the time William Penn signed the Charter than it has for 150 years. Forested land covers about 59% of the state (Alerich 1993), the highest since the Civil War. That forest is also, on average, more mature than it has been since the great old forests were first felled. More widely enforced protection of migratory birds also has contributed to the return of some species, such as large targets like the Pileated Woodpecker, which now come into close contact with human populations in maturing timber of suburbs and urban parks. All but a fraction of Penn's woods are still young, for a forest, but huge tracts of the central and northern mountains support the vast majority of breeding and migrant species that were found during Bartram's or Audubon's day. A comprehensive approach to conservation has been organized through the Pennsylvania Biological Survey, begun in 1979 (Genoways and Brenner 1985) and reflected in their biodiversity conservation strategy (Thorne et al. 1995). Conservation challenges remain, but Pennsylvania remains a stronghold of birds in the northeast.

Seasonal Calendar of Bird Life

Birds' ability to fly sets them apart from most vertebrates and enables them to travel great distances in relatively short periods of time. As a result, the types and numbers of birds change dramatically in Pennsylvania through the year. This seasonal calendar of bird life in Pennsylvania gives the reader an overview of the birds that inhabit the state.

The low ebb of bird life in Pennsylvania occurs during January. During this season, when most ponds and lakes are frozen, most migrants move south or concentrate around open water. Permanent residents, such as chickadees and migrant juncos, are the predominant birds of the winter woodlands, particularly in the north. Mid-January usually marks the beginning of "true" winter, when only the hardiest birds survive. Birds become more concentrated as winter progresses and food becomes harder to find. This is the time of year with the least movement. Waterbirds look for open large flowing rivers, warm-water discharges, or other sources where water remains open. Passerines look for feeding stations, fruit-bearing trees, and weedy fields. Many birds do not remain during harsh winters, particularly when snow cover is heavy. Half-hardy resident species such as the Carolina Wren are especially vulnerable to extreme winter conditions, but populations usually recover within a few years. Ground-dwelling birds such as Wild Turkeys and Northern Bobwhites may have difficulty finding food because of heavy snow cover.

The first bird song and the first northbound migration may start in late January. Great Horned Owls, Horned Larks, Northern Cardinals, Song Sparrows, Tufted Titmice, and Brown Creepers are among the first to be heard. The lengthening of daylight, in combination with encouragement from warm southwest winds, triggers migration. Species that winter in Pennsylvania (Horned Larks and Red-winged Blackbirds, for example) are among the first to move north.

By late February and early March, after the first spring thaw, Tundra Swans, Canada Geese, American Black Ducks, and Northern Pintails begin their northerly flight across the state. The deeper water of lakes and rivers attracts transient diving ducks such as Common Mergansers, Common Goldeneyes, Canvasbacks, and scaup. Great Horned Owls are nesting in February. The flight song of American Woodcock, mixed with the chorus of spring peepers, can be heard on warm evenings in February. Huge flocks of Common Grackles, Red-winged Blackbirds, and Brown-headed Cowbirds blacken the trees and fields by mid-March. A few raptors that migrate early, such as Northern Harriers and Red-tailed Hawks, drift northward searching fields for rodents. Eastern Phoebes and perhaps an early Tree Swallow arrive by mid-March, and the first waves of robins move north. Waterfowl migration peaks by mid-March, and by early April the majority of Blue-winged Teal and Northern Shovelers pass. Hawk migration is well under way by this time and is especially evident along the south shore of Lake Erie, where hundreds of hawks of up to 10 species may be counted in a day, especially when a storm front approaches from the southwest. Red-tailed Hawks, Bald Eagles, and Peregrine Falcons begin nesting in late March. More passerines, especially sparrows, arrive by early to mid-April. Among the flocks of sparrows, an observer should be able to find Eastern Towhees; Chipping, Field, Fox, Swamp and abundant White-throated sparrows and Dark-eyed Juncos. Swallows return, and during days of frigid temperatures all six of the species in the state can be seen feeding together over ponds and lakes. Winter residents such as Snowy Owls, Northern Shrikes, and American Tree Sparrows depart by mid-April. Nesting begins for many permanent residents, and others, such as early migrants, set up territory and are in full song. Migrant shorebirds such as Greater and Lesser yellowlegs, Pectoral Sandpipers, and Solitary Sandpipers can be found resting and feeding along the edges of pools and on exposed mud bars left by receding water from spring floods. By late April, as temperatures rise and daylight hours continue to lengthen, a few warblers arrive to liven the woods with their song. Among those to be expected in late April are Nashville, Yellow, Black-throated Green, Yellow-throated, Pine, and Black-and-white warblers. Occasionally some herons that breed south of the state, such as Snowy Egret, Little Blue and Tricolored herons, and Glossy Ibis, wander north of their normal breeding range into Pennsylvania.

The greatest number and species of birds may be found during May. By early May, migration for most long-distance migrants is well under way and peaks some time between May 10th and 18th. In some locations, more than 150 species have been recorded in 24 hours within this eight-day period, including large numbers of flycatchers, wrens, thrushes, vireos, warblers, tanagers, grosbeaks, and orioles. Migrant numbers rapidly decrease in the last few days of May, and almost all breeding species are on territory. Among the birds still migrating in early to mid-June are Semipalmated Sandpipers, Sanderlings, Olive-sided and Yellow-bellied flycatchers, Gray-cheeked Thrushes, and Blackpoll Warblers. The vast majority of species are busy at nest building and egg laying by the end of May.

June is the peak month for nesting in Pennsylvania, and young of permanent residents are already out of the nest. Stragglers of some species not known to breed in the state, such as Common Loons and Great Black-backed Gulls, continue here through the summer. Sometime between June 18th and 28th, migration almost ceases. When shorebirds appear at this time, it is difficult to determine whether they are northbound or southbound. The last few days in June or the first days of July mark the beginning of fall migration, when the first adult Lesser Yellowlegs, Least Sandpipers, and Short-billed Dowitchers appear. Singing slows considerably by mid-July, and many species have fledged young. Birds that raise more than one brood begin another nest in late June or July, and the late-nesting American Goldfinches begin to nest. By late July, herons have completed their nesting, and southern species wander across Pennsylvania borders, especially in the southeast.

Bird song all but ends in August, and the nesting season is usually over for most species. Passerines may be difficult to find at this time, when they are molting and preparing for migration and winter. Many of the adult passerines that nested to our north begin their journey south over Pennsylvania in August. Most of the adults of many species of shorebirds have already migrated south of Pennsylvania by early August, about the time their young begin to arrive in the state. Thousands of swallows, preparing for their trip to their wintering grounds, can be seen lined up side by side on utility lines and on bare branches of trees. Late August marks the peak migration time for Common Nighthawks, when evening flights may be spectacular. It is also the period when Eastern Kingbirds congregate and are most conspicuous.

September is the month of fall migration, when the greatest number of birds may be found. Birds migrating south lack the urgency to reach their wintering grounds that northbound birds exhibit when migrating to their breeding grounds, so migration tends to be less compacted in fall than in spring. Some species of birds stretch their migration period out over several weeks or even months. Most warblers seen in early September are immatures. Hawk migration begins, and the first migrants are usually observed along the mountain ridges. Ospreys and Bald Eagles are usually the first to move, often as early as mid-August. Their greatest numbers are seen from early to late September. At the arrival of the season's first cold front, usually between the second and third week of September, thousands of Broad-winged Hawks move south with the northerly winds. They are most easily observed kettling along hawk lookouts, such as the famous Hawk Mountain Sanctuary. Kinglets arrive by late September as do White-throated Sparrows, which can outnumber all other species at this time.

Although fewer hawks are observed in October, there is a greater diversity of species. This is the best time to see Peregrine Falcons and all three species of accipiters. Puddle ducks, such as Wood Ducks and American Wigeons, are most common at this time. Most flycatchers, vireos, and warblers have passed to the south by the middle of the month. Several species of sparrows migrate through the state in October: Fox Sparrows, White-crowned Sparrows, and Lapland Longspurs, for example. The most common sparrows seen in October are the White-throated Sparrow and Dark-eyed Junco. Tens of thousands of blackbirds once again blacken the skies and fields. If strong north winds are accompanied by freezing rain or snow in late October, loons and grebes appear. Winter residents such as American Tree Sparrows arrive from the far north at this time. By late October, the appearance of Pine Siskins and Evening Grosbeaks may signal the beginning of a winter invasion of northern finches.

Winter residents and late migrants such as diving ducks return to Pennsylvania in November from northern breeding grounds. Severe storms may force migrating flocks of sea ducks such as scoters and Oldsquaws down onto inland lakes, but they leave as soon as the weather has cleared. Gulls return to inland lakes and rivers, where many remain until forced south by freezing water. Snowy Owls, Northern Shrikes, and, during invasion years, crossbills may wander into Pennsylvania in November from farther north. The first flocks of Snow Buntings are usually seen in November. Most Neotropical migrants have gone, but a few stragglers will remain into early winter.

December usually marks the beginning of winter weather. The last of the raptors, such as Rough-legged Hawks and Golden Eagles, pass through. Fall migration has nearly ended, and the birds that appear at this time usually attempt to winter until forced south by freezing water or scarcity of food. Some individuals that normally would have migrated south by this time may remain because of illness or injury. The Christmas Bird Count period brings may observers into the field, resulting in reports of many birds found during this season.

Physiographic Regions

In the species accounts, the locations of sightings are given by county; Map 1 shows the boundaries of Pennsylvania's counties. In addition, the state has been divided into the eight physiographic regions shown on Map 2. The regions are based on the Physiographic Provinces of a Pennsylvania map produced by the Commonwealth of the Pennsylvania Department of Environmental Resources, Office of Resources Management, Bureau of Topographic and Geologic Survey. For simplicity, we have slightly modified the names and boundaries of these regions. The names used in this work are Coastal Plain, Piedmont, Ridge and Valley, Glaciated Northeast, High Plateau, Southwest, Glaciated Northwest, and Lake Erie Shore. These areas are primarily defined by topographic features and represent sufficiently distinctive bird populations to serve as the primary geographic reference points for this book. We occasionally refer in the species accounts to several sections within these regions; the Blue Ridge Mountains, the Poconos, and the Allegheny Mountains. The Coastal Plain includes a narrow strip of floodplain bordering the Delaware River in

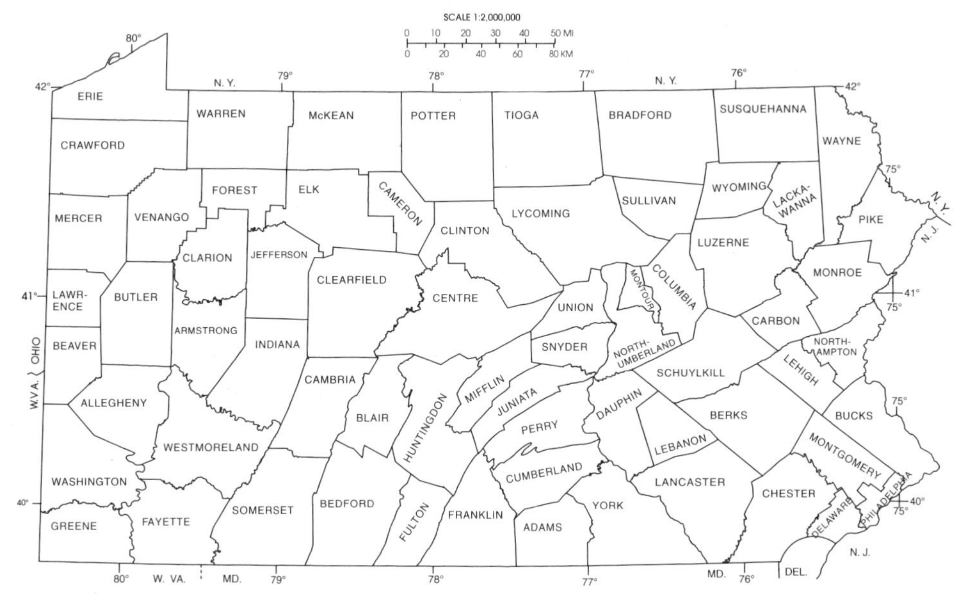

1. Pennsylvania counties

the southeastern corner of the state. The Piedmont is an extensive area of gentle rolling hills and valleys carved out by the lower Susquehanna and Delaware rivers. The Ridge and Valley dominates the landscape east of the Allegheny Front. Its major river valleys drain to the Atlantic Ocean and flow into the Juniata, Potomac, and West Branch of the Susquehanna. The easternmost ridge of the Ridge and Valley region stands about 1000 feet above the Great Valley. The Great Valley, technically the first valley of the Ridge and Valley, is treated with the Piedmont. The Blue Ridge Mountains lie in the Piedmont but are most like the Ridge and Valley. The Poconos lie between the Glaciated Northeast and the Ridge and Valley but are most like the High Plateau. To the northeast of the Ridge and Valley lies the Glaciated Northeast, an area affected by glaciers that left behind natural wetlands and numerous lakes. The Allegheny Mountains are a southern extension of the High Plateau.

Pennsylvania's dominant physiographic feature is the Appalachian Mountain system, which passes through the state from a northeast to a southwest direction and separates the state into two distinctive parts. The eastern boundary of the High Plateau is known as the Allegheny Front. The Appalachian Plateau comprises the majority of the state west and north of the Allegheny Front. It reaches its highest points just behind the Front and gradually slopes downward to the north and west. The Allegheny Plateau has been divided into three regions to reflect geological and ornithological patterns: the High Plateau, Glaciated Northeast and Glaciated Northwest. The High Plateau includes the most mountainous and wooded portions of the state; the Southwest includes portions of the Allegheny, Ohio, Monongahela, and Youghiogheny river systems; and the Glaciated Northwest contains most of our natural wetlands. Along the south shore of Lake Erie is a narrow strip that forms part of the original basin of

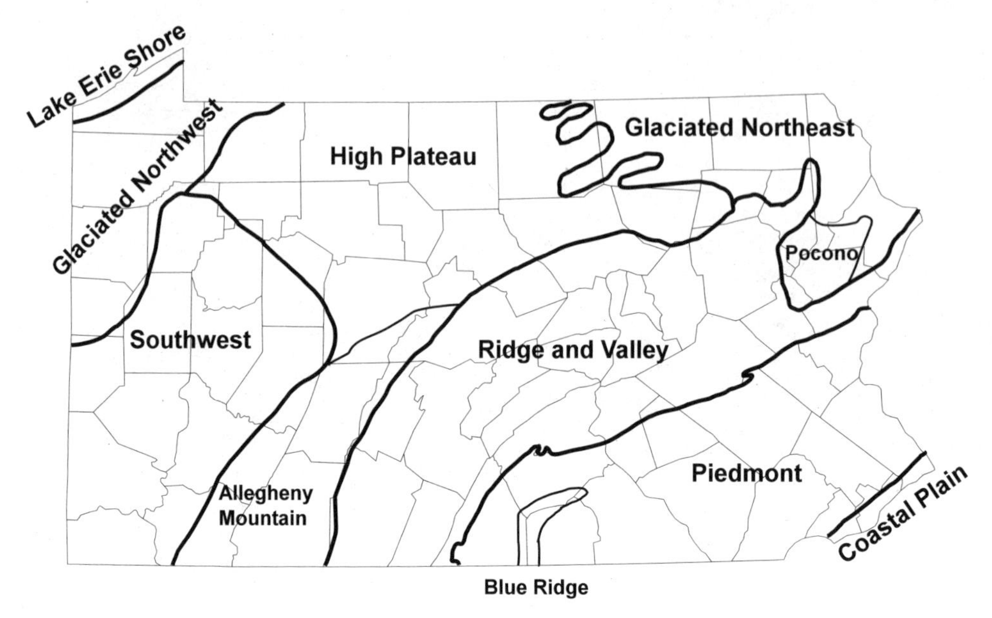

2. Pennsylvania physiographic regions

Lake Erie. All of the streams in this region flow into Lake Erie.

Even though about 59% of Pennsylvania is forested, many diverse habitats make up the landscape, including large slow-moving rivers, fast-flowing streams, wetlands, tidal marshes, sandy beaches, and grasslands. In the twentieth century, manmade lakes have been created for the purpose of flood control and recreation. These add to the natural lakes that were formed in the northwestern and northeastern corners of the state during periods of glaciation. Strip mining and agriculture have also played important roles in human and bird life in recent years. Modern agricultural techniques have had a negative impact on many grassland species. However, with the reclamation of strip mines, some of the habitat that was lost to agriculture was restored as habitat for grassland species. Wetlands have suffered as a result of the draining of swamps and marshes to accommodate urban expansion and economic development.

The physiographic regions are described below in further detail. In addition, sites of particular ornithological significance are briefly described for each region. These sites are among the best known or most frequently visited sites and have consistently contributed to our knowledge of bird life in Pennsylvania. Many are located along roads, in parks, or in areas close to major cities. Some of these sites are representative of the bird communities in the region. The two sites that attract more species of birds than anywhere else in Pennsylvania are Presque Isle State Park in the northwestern corner and John Heinz National Wildlife Refuge at Tinicum in the southeastern corner of the state.

Coastal Plain

The Coastal Plain, a narrow band in the southeastern corner of the state, is defined by the tidal portions of the Delaware River floodplain. Nearly all of the Coastal Plain region is urbanized. The Coastal Plain historically contained extensive tidal and freshwater wetlands, but only remnants remain at undeveloped sites along the Delaware River and at the John Heinz National Wildlife Refuge at Tinicum, where a remnant area of protected tidal marsh is surrounded by urbanization. This region is best known for the tidal marshes and impoundments that attract a wide variety of herons, shorebirds, and waterfowl. Very little forested habitat remains within this region.

The John Heinz National Wildlife Refuge at Tinicum is located in Philadelphia and Delaware counties, about a mile from the Philadelphia International Airport. The area has been a sanctuary since 1955 and has been under federal management since 1972. It protects 205 acres of freshwater tidal marsh in Pennsylvania. It has been diked, dredged, and filled to create a wide variety of habitats. Nearly 300 bird species have been recorded here. Formerly, the area included extensive tidal mudflats that attracted thousands of shorebirds and herons in migration and was an important wintering site for waterfowl. This area is famous for southern waders such as White and Glossy ibises and Little Blue and Tricolored herons. The area is still an important location for breeding marsh birds, both migrating and wintering waterfowl, and shorebirds. The area's boundaries have changed significantly over time. Much of the open tidal flats have disappeared. Historical accounts referred to a variety of sites as Tinicum, including the Philadelphia sewage treatment plant, wetlands near Fort Mifflin, and sometimes the airport area. Even today, the refuge and the surrounding areas are referred to as Tinicum.

Piedmont

The Piedmont is a vast expanse of rolling hills and valleys. It contains much of the state's farmland and has been settled the longest of the state's regions. It has experienced intense urban and suburban developmental pressure and has the least wooded acreage in the state aside from the Coastal Plain region but still is the best area for woodland species such as Kentucky Warbler. The proximity to the ocean provides relatively mild winter weather. This allows more species such as Chipping, Field, and Savannah sparrows to winter here and in the Coastal Plain region than anywhere else in the state. This region covers much of the southeastern corner of the state and extends northward to the Kittatinny Ridge, the south ridge of the Appalachian Mountains.

The lower Susquehanna River of Lancaster County is the best studied area of the Susquehanna drainage. Pennsylvania's border lies only about 15 miles from the outflow into Chesapeake Bay. The river's north-to-south bearing also contributes to making this a significant migratory route. Anywhere along the river may provide excellent birding, but several sites stand out as premier places to observe birds. Washington Boro, located about 2 miles south of Columbia in Lancaster County along Route 441, was, until recently, known for the large heron rookery located on one of the large islands. Thousands of Cattle Egrets nested on the island, and many herons, egrets, and occasionally ibises roosted there. Though the herons have disappeared from the island, the mudflats connecting the islands, known as the Conejohela Flats, are an important staging area for shorebirds and gulls. Farther downriver, the water slows as it is dammed at Safe Harbor and at Holtwood. Both of these areas may attract large numbers of gulls and waterfowl, especially Tundra Swan. Muddy Run Reservoir, located about 2 miles east of Holtwood Dam, is well known for the thousands of staging waterfowl during migration. This area also hosts large Turkey and Black vulture roosts and several Bald Eagle nests. The widest portion of the Susquehanna

River, Conowingo Pond, is a reservoir formed by the Conowingo Dam in Maryland. It attracts thousands of gulls and waterfowl, especially in winter near the warm-water discharge of the Peach Bottom Atomic Power Station. Between the power station and the Maryland border, the Philadelphia Electric Company supports a major winter roosting site of up to two dozen Bald Eagles. This is best viewed from the York County side of the river.

Middle Creek Wildlife Management Area is located along the Lebanon-Lancaster county line about a mile south of Kleinfeltersville. This area was purchased by the Pennsylvania Game Commission for the protection and management of wildlife. It comprises about 5000 acres consisting of patches of forest, a 400-acre shallow lake, several small ponds and impoundments, and about 1300 acres of farmland. The fields support nesting populations of grassland birds. Waterfowl are the main attraction for birders at this site. It is most famous for the huge flocks of Tundra Swans, Canada Geese, many species of puddle ducks and, recently, Snow Geese.

Several Piedmont lakes have been important wintering sites for ducks, geese, and gulls and important staging sites for shorebirds. These lakes include Lake Marburg at Codorus State Park in southwestern York County, Octoraro Reservoir in Lancaster County, Blue Marsh Lake and Lake Ontelaunee in Berks County, Struble Lake in Chester County, and Marsh Creek Lake at Marsh Creek State Park in Chester County. Others include Green Lane Reservoir in northern Montgomery County; Lake Galena at Peace Valley Park in central Bucks County and Lake Nockamixon at Nockamixon State Park in north-central Bucks County; and Minsi Lake in northern Northampton County.

The northernmost extension of the Blue Ridge Mountains is found in southern Cumberland County, the western half of Adams County, and eastern Franklin County. It is referred to as South Mountain and is treated as a section in this volume. This section rises over 1000 feet above the Piedmont. The birds of the Blue Ridge are most like those in the Ridge and Valley region. The Blue Ridge Mountains have the highest reporting rate of any state section for species such as Great Crested Flycatcher, White-breasted Nuthatch, and Worm-eating Warbler.

Ridge and Valley

Of all of the regions in the state the Ridge and Valley is the most sharply defined geologically. The southeastern border of this region, variously named the Kittatinny Mountain in the east and Blue Mountain to the west, is nearly unbroken from the Maryland border in southwestern Franklin County to northeastern Northampton County on the New Jersey border. The world's most famous hawk watch lookout is found along this ridge. The western and northern edge of this region is defined for most of its length by the Allegheny Front. The often parallel-running ridges within this region rise to 2700 feet above sea level and are heavily forested, whereas the river valleys generally are farmed and settled. The Susquehanna River valley provides lower elevations in broad valleys that create a wide corridor for southern species to wander north into this region. The higher elevations are home to nesting northern species, but few birds remain here in winter.

The world's first sanctuary for the birds of prey, Hawk Mountain Sanctuary, is located along the Kittatinny Ridge about a mile east of Drehersville on the Schuylkill and Berks county line. It was designated a hawk sanctuary in the fall of 1934 and encompasses over 2300 acres. Thousands of people visit this sanctuary yearly to witness the spectacular fall hawk flights. Nearly every species of hawk that has been recorded in Pennsylvania has been seen here. It is probably best known for the Broad-winged Hawk migration in September, when on occasion over 20,000 birds have been recorded in a single day. The lookouts can also be an excellent place to see migrating loons and flocks of swans, geese, and finches. The sanctuary has recorded over 240 species. Other hawk watches along the Kittatinny Ridge have received less fame than Hawk Mountain but still have contributed extensive information to our knowledge of the fall raptor migration. These include Bake Oven Knob on the Lehigh-Carbon county line; Baer Rocks located about 1.5 miles southwest of Bake Oven Knob; Little Gap Raptor Research Station on the Carbon-Northampton county line; and Delaware Water Gap in eastern Monroe County on the Delaware River. Waggoner's Gap is located on Blue Mountain in Cumberland County on the Perry County line. West of the Kittatinny Ridge there are several other hawk lookouts that have contributed extensive data of fall raptor flights. Two examples are the Pulpit, located on Tuscarora Mountain east of McConnellsburg along the Franklin-Fulton county line and Bald Eagle Mountain, located west of State College in Centre County, which is noted for the high daily counts of migrant Golden Eagles.

Bald Eagle State Park in Centre County, Raystown Reservoir in Huntingdon County, Montour Preserve in Montour County, Rose Valley Lake in Lycoming County, and Penn Forest and Wild Creek reservoirs, and Beltzville State Park in Carbon County are of ornithological importance. All of these sites have attracted numbers of loons, grebes, herons, waterfowl, shorebirds, gulls, and terns. Access areas along the West Branch of the Susquehanna River where it passes through Lycoming County and Clinton County and the main branch of the Susquehanna River have been sites of significant waterbird records as well. Some of the largest breeding populations of Cliff Swallows, Golden-winged, Prairie, and Worm-eating warblers in the state are also found in the Ridge and Valley.

Glaciated Northeast

The Glaciated Northeast is in the northeastern corner of the state. It extends from eastern portions of Tioga County east to Pike County and south to northern Wyoming County, northern and southern Lackawanna County, and most of Monroe County. It contains natural ponds and lakes carved out by glaciers. The region is heavily forested, and human habitation is sparse but growing. The North Branch Susquehanna River creates a low-elevation corridor for southern species to wander north into this region. The high elevation Poconos contain most of the state's native spruce trees. Many northern species of passerines breed in this region. The numerous lakes and the Delaware River attract waterfowl and wintering Bald Eagles.

Despite the numerous ponds and lakes in this region, the area is inadequately studied. The 10-mile-long and 1-mile-wide Lake Wallenpaupack is one of the few lakes in this area that has been noted to attract loons, grebes, and ducks during migration. Pocono Lake Preserve, located in western Monroe County, supports many breeding species that are usually associated with Canadian-zone forests, such as Red-breasted Nuthatch, Golden-crowned Kinglet, Nashville Warbler, Yellow-rumped Warbler, and White-throated Sparrow.

High Plateau

The High Plateau lies between the glaciated corners of the state and extends from the New York state line southward in a narrow band through Clearfield, Cambria, eastern Westmoreland and Fayette counties, and all of Somerset

County. The High Plateau is the largest region in the state as well as the most remote and mountainous. Elevations range from about 800 feet in the river valleys to over 3000 feet in the southern mountains. It is interesting to note that the elevation in Pennsylvania plays a smaller role in the distribution of northern breeding species in Pennsylvania than in states north and south of Pennsylvania (Brauning 1992). An exception is the Blue Ridge Mountains; some northern species breed here and nowhere else in the southeastern part of the state. Because of the inaccessibility and low human population of the northern portion of this region, the northern-tier counties remain the least known of the state's avian regions. More species of warblers are found breeding in this region than anywhere else in the state. In winter this region is the coldest and harbors the fewest birds. The Allegheny Mountains contain the state's highest mountains, rising to a peak of 3200 feet at Mt. Davis. Yet its low valleys support more wintering species of birds than elsewhere in the region, probably because of this section's more southerly latitude.

Powdermill Nature Reserve, located about 3 miles south of Rector in the Allegheny Mountain section of Westmoreland County, is a 2200-acre reserve and research station of the Carnegie Museum of Natural History, Pittsburgh. The area consists of fields, second-growth deciduous forest, and several small ponds. At a banding station established at the reserve in 1961 mist nets are used to capture birds for study. It has continued to be the largest and most important banding site for passerines in the state and is a center for educational programs. Banding returns from birds banded at this site have come from throughout the United States, Canada, and Central and South America.

Glendale Lake in northern Cambria County, Donegal Lake in southeastern Westmoreland County, and Lake Somerset in northwestern Somerset County are three reservoirs that regularly support resting and feeding loons, grebes, waterfowl, gulls, and terns during migration and occasionally shorebirds when low lake levels expose shoreline mudflats. In Warren County at the western edge of this region lies Kinzua Reservoir, created by the damming of the Allegheny River about 7 miles east of the city of Warren. The area along the Allegheny River between Warren and the southern part of the reservoir has attracted numbers of migrant and wintering waterfowl. Many species of birds that have been found in winter in this area have not been recorded anywhere else in the High Plateau. Resident Bald Eagles can be seen along the river and over the reservoir. Several huge colonies of Cliff Swallows nest on bridges and manmade structures around the dam.

Southwest

The Southwest region encompasses the entire southwestern corner of the state. It extends north to the Glaciated Northwest in Venango County and east to the High Plateau region. Most of Allegheny County in the central portion of the Southwest contains the city of Pittsburgh and surrounding suburbs. Most of the area south of the city of Pittsburgh is made up of small towns, fields, and forested, rolling hills with elevations to about 1600 feet. Several southern species, such as Carolina Chickadee, Summer Tanager, and formerly Bachman's Sparrow, reach their northernmost breeding range in the southwestern corner of Pennsylvania. Strip mining has altered much of the land of this region, some of which has been reclaimed into grasslands. The stronghold of the state's Henslow's Sparrow population is in this region.

The Piney Tract is an example of reclaimed strip mines that has received considerable ornithological attention. It was discovered in the early 1980s as a nesting site for several species of grassland sparrows such as Grasshopper, Henslow's, Vesper, and Savannah sparrows. Many other reclaimed strip mines in neighboring counties have been explored and yield similar species lists, including Upland Sandpipers and Short-eared Owls.

Lake Arthur at Moraine State Park in west-central Butler County, Crooked Creek Lake at Crooked Creek Lake Park in south-central Armstrong County, and Yellow Creek Lake at Yellow Creek State Park in central Indiana County are reservoirs of interest. These reservoirs have produced numerous records of loons, grebes, and ducks during migration in recent years, at least partly because of increased coverage by birders. Vagrant southern herons have also made appearances along their weedy shorelines. During years of drought, receding lake levels form mudflats that attract shorebirds, gulls, and terns. These species can occasionally be found resting on the swimming beaches when lake levels are normal. Increased use of these reservoirs for human recreational purposes continues to reduce the number of waterfowl that feed and rest there during migration.

Glaciated Northwest

The Glaciated Northeast and Glaciated Northwest, together, encompass more than 50% of Pennsylvania's wetlands. Most of the state's marshes and swamps, as well as many lakes, occur in the Glaciated Northwest. The region extends from approximately 6 miles from Lake Erie south through all of Crawford, Mercer, and Lawrence counties and east to northeastern Warren County. Many of the state's waterfowl and threatened wetland bird nesting records come from this region. No other region in the state experiences such extreme weather conditions, especially in late fall and early winter before Lake Erie freezes over. Cold air passing over the warm water of Lake Erie condenses and falls in the form of snow when the air reaches the cooler higher elevations of the Glaciated Northwest, forming a distinct snowbelt. The heavy snow here, as in the High Plateau, prevents all but the hardiest birds from surviving.

Pymatuning Reservoir was formed by the damming of the Shenango River near the town of Jamestown in Mercer County in 1935. The lake follows the creek in a northerly direction about 12 miles, with the western half of the lake in the state of Ohio. The lake then turns abruptly to the east-southeast another 6 miles and averages about 2.5 miles wide. Water levels of the portion from Ford Island southeast are maintained separately, and the area is protected as a sanctuary. The lake has been an important staging area for geese and swans since it was formed. Several species of ducks, such as Redhead and Ring-necked Duck, not previously known to nest in Pennsylvania, bred here shortly after the lake was flooded. Bald Eagles have nested here since 1937. When water levels are low, this area becomes an important fall staging area for shorebirds. Aside from Presque Isle State Park and Pymatuning Reservoir, many of Todd's (1940) records of herons and waterfowl in northwestern Pennsylvania came from the largest natural lake in the state, Conneaut Lake in Crawford County. The vast marshes of Conneaut Marsh in Crawford County, which is the outlet of Conneaut Lake, stretch for about 11 miles before emptying into French Creek. This area is famous for nesting marsh birds such as Pied-billed Grebe, American Bittern, rails, and Common Moorhens, as well as several species of ducks, Bald Eagle, and the Prothonotary Warbler. Shenango Reservoir, created by the damming of the Shenango River in the west-central portion of Mercer County, is best known for the shorebirds the mudflats attract in late summer when water levels are low during years of drought.

Lake Erie Shore

The Lake Erie Shore is a narrow band that runs in a northeast to southwest direction along the coast of Lake Erie and extends southward to meet the Glaciated Northwest region at an elevation of approximately 700 or 800 feet above sea level. The immediate shoreline and Presque Isle peninsula are ecologically unique in Pennsylvania. This is the only site in the state where Piping Plover and Common Tern have nested. Lake Erie greatly influences the region's weather, keeping it relatively cool in summer and warm in early winter before the lake freezes. Lake Erie acts as a barrier, interrupting northbound bird migration and concentrating birds along the shoreline. This phenomenon is especially evident on Presque Isle, where birds funnel onto the peninsula in their attempt to continue northerly over land. Southbound birds find Presque Isle to be the first landfall after crossing the lake. Here they collect to rest and feed before continuing their journey. Southbound waterfowl stream past within a couple of miles offshore, much as waterfowl do when migrating along the Atlantic coastline.

Presque Isle State Park is a narrow lowlying sandspit that extends about 7 miles in a northeastward direction into Lake Erie. The base of this peninsula is connected to the mainland of Erie County about 4 miles from the center of the city of Erie. Not only the bird life but also the plant life of this area have been studied and written about for over a century. Over 320 species of birds have been recorded in the park area. No other single site in Pennsylvania concentrates such large numbers of passerines during spring migration. For many species, average daily highs at this park exceed average daily highs at all other sites in the state by several times. The sandy beaches and the lake create habitats similar to an ocean, attracting birds such as Piping Plover and Purple Sandpiper, the latter never recorded anywhere else in Pennsylvania. Presque Isle Bay draws thousands of diving ducks and gulls in migration and through the winter. Gull Point, the eastern tip of the peninsula, was established as a bird sanctuary in 1927 and is famous for the large number of shorebird species that rest here on their flights to and from the nesting grounds of the far north. The recent interest in this area of ecological importance as a resting place for shorebirds has prompted park officials to close this area to all human activity during the migration and nesting periods.

Bird Sites

Throughout the twentieth century ornithologists and birders have discovered many sites that attract a wide diversity of species and substantial numbers of birds because of their geographic location or the habitats they contain. Unique land formations such as mountain ridges, valleys, lakes and wetlands formed by glaciers, manmade impoundments, and rivers have created ideal habitats for concentrating birds in many areas of the state. Map 3 identifies the areas that have contributed substantial ornithological information that formed the foundation of this book. Many of these sites have been visited by ornithologists, birders, and naturalists for more than 150 years. For information on directions to many of these sites, when to visit, and what birds to expect, refer to the following recently published local, regional, and statewide sources.

Estabrooks, B. ed. 1996. Birding in western Pennsylvania. Audubon Society of Western Pennsylvania, Pittsburgh.

Ford, P. 1995. Birder's guide to Pennsylvania. Gulf Publishing, Houston.

Freeland, D.B. 1975. Where to find birds in western Pennsylvania. Typecraft Press, Pittsburgh.

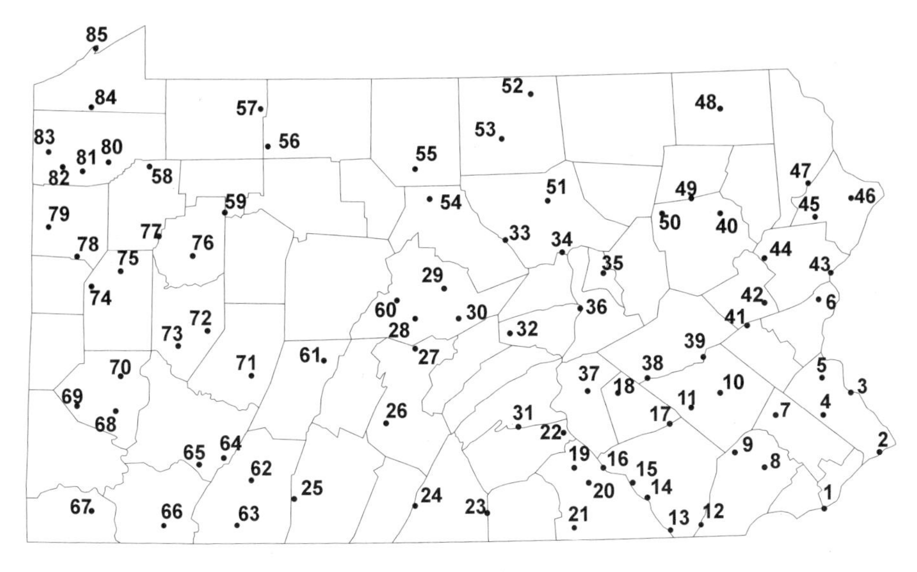

3. Locations of selected bird sites

Coastal Plain
1. John Heinz National Wildlife Refuge (Tinicum), Delaware and Philadelphia counties
2. Delaware River at Pennsbury Manor, Bucks County

Piedmont
3. Delaware River at New Hope, Bucks County
4. Peace Valley Park, Bucks County
5. Nockamixon State Park, Bucks County
6. Minsi Lake, Northampton County
7. Green Lane Reservoir, Montgomery County
8. Marsh Creek State Park, Chester County
9. Struble Lake, Chester County
10. Lake Ontelaunee, Berks County
11. Blue Marsh Lake, Berks County
12. Octoraro Reservoir, Lancaster County
13. Muddy Run Reservoir and Conowingo Reservoir, Lancaster County
14. Susquehanna River at Safe Harbor, Lancaster County
15. Susquehanna River at Washington Boro (Conejohela Flats), Lancaster County
16. Susquehanna River at Brunner's Island, York County
17. Middle Creek Wildlife Management Area, Lancaster and Lebanon counties
18. Memorial Lake State Park, Lebanon County
19. Gifford Pinchot State Park, York County
20. Rocky Ridge hawk lookout, York County
21. Codorus State Park, York County
22. Susquehanna River at West Fairview, Cumberland County
23. South Mountain, Adams, Franklin, and Cumberland counties

Ridge and Valley
24. Tuscarora Mountain hawk lookout, Fulton County
25. Allegheny Front hawk lookout and Shawnee State Park, Bedford County
26. Raystown Lake, Huntingdon County
27. Tussey Mountain hawk lookout, Huntingdon County
28. Bald Eagle Moutain hawk lookout and Scotia Barrens, Centre County
29. Bald Eagle State Park, Centre County
30. Colyer Lake, Centre County
31. Waggoner's Gap hawk lookout, Cumberland County
32. Snyder-Middlesworth and Tall Timbers Natural Areas, Snyder County
33. West Branch Susquehanna River, Clinton and Lycoming counties
34. Allenwood State Game Lands 252, Lycoming and Union counties
35. Montour Preserve and Pennsylvania Power and Light Company Fly Ash Basin, Montour County
36. Confluence of the North and West Branches of the Susquehanna River, Sunbury, Northumberland County
37. Clark's Valley, State Game Lands 210 and 211, Dauphin County
38. Route 183 hawk lookout, Berks and Schuylkill counties
39. Hawk Mountain Sanctuary, Berks and Schuylkill counties
40. Susquehanna River, Exeter area, Luzerne County
41. Baer Rocks and Bake Oven Knob hawk lookouts, Carbon and Lehigh counties
42. Beltzville State Park and Penn Forest and Wild Creek reservoirs, Carbon County
43. Delaware Water Gap hawk lookout, Monroe County

Glaciated Northeast
44. Pocono Lake and surrounding areas, Tobyhanna State Park, Monroe County
45. Promised Land State Park and Bruce Lake Natural Area, Pike County
46. Shohola Lake, State Game Lands 180, Pike County
47. Lake Wallenpaupack, Pike and Wayne counties
48. Woodbourne Sanctuary, Susquehanna County

High Plateau
49. Harveys Lake, Luzerne County
50. Ricketts Glen State Park, Luzerne County
51. Rose Valley Lake, Lycoming County
52. Hammond Lake, Tioga County
53. Pine Creek Gorge, Colton Point State Park, Tioga County
54. Tamarack Swamp Natural Area, Clinton County
55. Susquehannock State Forest, Potter County
56. Tionesta Natural and Scenic Area and Allegheny National Forest, McKean and Warren counties
57. Allegheny River including Kinzua Reservoir, Warren County
58. Oil Creek State Park, Venango County
59. Cook Forest State Park, Clarion, Forest, and Jefferson counties
60. Black Moshannon State Park, Centre County
61. Prince Gallitzin State Park, Cambria County
62. Lake Somerset, Somerset County
63. Mt. Davis, Somerset County
64. Laurel Highlands, Linn Run State Park, and Powdermill Nature Reserve, Westmoreland County
65. Donegal Lake, Westmoreland County
66. Ohiopyle State Park, Fayette County

Southwest
67. Clarksville area, Greene County
68. Point State Park, Allegheny County
69. Imperial reclaimed strip mine, Allegheny County
70. Parks within the city and surrounding areas of Pittsburgh (e. g., Fricke and Shenley parks), Allegheny County
71. Yellow Creek State Park, Indiana County
72. Keystone State Park, Armstrong County
73. Crooked Creek Lake Park, Armstrong County
74. Moraine State Park, Butler County
75. "The Glades," State Game Lands 95, Butler County
76. Piney Tract reclaimed strip mine, Clarion County
77. Kahle Lake, Venango and Clarion counties

Glaciated Northwest
78. Pennsy Swamp, State Game Lands 284, Mercer County
79. Shenango Reservoir, Mercer County
80. Erie National Wildlife Refuge, Crawford County
81. Conneaut Marsh, Crawford County
82. Conneaut Lake, Crawford County
83. Pymatuning area and Hartstown marsh, Crawford County
84. Edinboro Lake, Erie County

Lake Erie Shore
85. Presque Isle State Park, Erie County

Kitson, K. 1998. Birds of Bucks County. Bucks County Audubon Society, New Hope, Pa.
Leberman, R.C. 1976. The birds of the Ligonier Valley. Special Publication No. 3. Carnegie Museum of Natural History, Pittsburgh.
Leberman, R.C. 1988. The birds of western Pennsylvania and adjacent regions. Special Publication No. 13. Carnegie Museum of Natural History, Pittsburgh.
Morrin, H.B., ed. com. chair. 1991. A guide to the birds of Lancaster County, Pennsylvania. Lancaster County Bird Club.
Morris, B.L., R.E. Wiltraut, and F.E. Brock. 1984. Birds of the Lehigh Valley area. Lehigh Valley Audubon Society, Emmaus, Pa.
Schweinsberg, A.R. 1988. Birds of Central Susquehanna Valley. Privately published.
Stull, J., J.A. Stull, and G.M. McWilliams. 1985. Birds of Erie County, Pennsylvania including Presque Isle. Allegheny Press, Elgin, Pa.
Uhrich, W.D., ed. 1997. A century of bird life in Berks County, Pennsylvania. Reading Public Museum, Reading, Pa.
Wood, M. 1979. Birds of Pennsylvania, when and where to find them. Pennsylvania State Univ., University Park.
Wood, M. 1983. Birds of central Pennsylvania, 3rd ed. Pennsylvania State Univ., University Park.

Overview and Format of the Species Accounts

A total of 429 species of birds are listed in the species accounts. Of this total, about 200 species have been known to breed in the state, and 30 species are listed as hypothetical. A species is listed as hypothetical if it occurred in the state but documentation no longer exists; if it is believed to have occurred in the state but has not been accepted to the official list of birds of Pennsylvania, 1995; or if it escaped from captivity. In the species accounts, brackets around the common name identify a species as hypothetical. A photograph, a specimen, a taped recording, or a satisfactorily written description constitutes documentation.

A summary of the status and distribution provided in the AOU check-list is included for each species of wild bird recorded within the boundaries of Pennsylvania from earliest historical records until October 1998. English and scientific names, as well as the new taxonomic sequence of species, are based upon the classification and nomenclature of the seventh edition of The American Ornithologists' Union's Check-list of North American Birds (AOU 1998). Each account is divided into categories to assist the reader in locating specific data. The categories used are General status; Habitat; Seasonal status and distribution, which includes the sections Spring, Summer, Breeding (where applicable), Fall, and Winter; and History or Comments (or both). The categories used may vary from species to species depending on the relevance to a species' status and distribution in the state. Below is an explanation of the contents in each category.

Format

General Status

For each species, we describe the North American breeding and wintering range, and, for some species, the worldwide breeding and wintering range. Unless cited otherwise, ranges are based on the seventh edition of the AOU's *Check-list of North American Birds* (1998). In this general status category we include a general overview of the species' seasonal status in Pennsylvania (see Abundance and Frequency Categories, p. 23, for an explanation of terms). Here you will also find the general distribution in Pennsylvania for each species. Distributions of some species are described by physiographic region (see Physiographic Regions for explanation). Interesting anecdotes have also been included for some species.

Habitat

The environment in which the species is typically observed when in Pennsylvania is described here. Any seasonal difference in habitat is noted as well. The habitat is not listed for species that have occurred in Pennsylvania fewer than 10 times since 1960.

Seasonal Status and Distribution

An attempt has been made in the seasonal status and distribution category to define the species' distribution, frequency of occurrence, and abundance for each of four seasons and in each physiographic region where sufficient data permit. The seasons are defined on the basis of the arrival and departure times for each species, not by climatic seasons. When a species has been recorded only a few times within a season, dates, locations, and numbers are listed and referenced. For rare species for which there are too many records to cite each individual occurrence, the counties are listed. Most records are presented in the species accounts by location following the order of the physiographic regions listed in the introduction. In some species accounts records are presented chronologically when dates are the primary focus. Below is an explanation of these seasons.

Spring

Spring is defined as the time when northbound migrants are passing through the state. The spring migration period may extend from about the third week of February to about the first week of June, including most of the very early arrivals and late spring departures of most species.

Breeding

The breeding category is presented only for those species that have nested or have shown overwhelming evidence that nesting occurs within the boundaries of Pennsylvania. The breeding status and distribution of the species are described here, and a limited description of nesting characteristics is given (see Background on Breeding Information Sources).

Summer

When there is no evidence of nesting in Pennsylvania, summer occurrences are placed in this category: used primarily for identifying the time between spring and fall when little or no migration occurs. The summer season is arbitrarily defined, and onset and end of this season may vary from family to family or from species to species. Records placed in this section include birds that have wandered here from areas outside their normal breeding range or for some reason have failed to migrate out of the state. For most species, summer begins at the end of spring migration—generally the first or second week of June. The end of the summer season may be from the third or fourth week of July to the third or fourth week of August, depending on the species. A Canvasback in the third week of August, for example, is more likely to be a summer visitor than an early fall migrant because the species normally does not begin migrating through the state before the first week of November. However, a Bay-breasted Warbler in the third week of August would be typical for a first fall migrant. For most species of shorebirds the summer season is not used, because spring and fall migration often come within only a few days of each other or even overlap.

Fall

Fall is defined as the period when southbound migrants are passing through the state. Active fall banding stations in Pennsylvania, such as Powdermill Nature Reserve and Presque Isle State Park, have added significant information about arrival and departure times of passerines. Banding information used in this section came primarily from these two banding stations. Some banding recoveries have been listed in the Comments section for species that may be especially interesting to the reader. Fall migration may be difficult to discern for many species of birds, especially those that also breed within the state. For some shorebirds, fall migration may begin as early as the last week of June or the first week of July. Many passerines begin migrating south the first or second week of August, whereas many species of diving ducks are not seen until late October. Therefore, the beginning of fall migration is determined by each species' typical date of departure. During July and August there may be some overlap in the summer (nesting of residents) and fall seasons. All birds seen in September or later are listed in the fall season. Some authors have chosen the end of November as the cut-off period of fall and the beginning of winter, but many birds are still migrating through the state in December or early January. So, for the purposes of this book, we have chosen to extend the fall season to the first or second week of January, when the ground and most lakes and slow-moving streams have frozen. After this time only the hardiest species, or "true" winter residents, remain.

Winter

The period labeled winter may span only four to six weeks. It is the harshest time of year, when the wild food crop may be nearly depleted or inaccessible under snow. Most birds that are present from the second or third week of January to the second or third week of February are not migrating. There are some exceptions. The occasional very early spring arrival of migrants during mild winters or warm spells (as early as the fourth week of January or the first week of February) is noted in those species' accounts. Most birds present after the third week of February are placed in the spring season.

History

The history category summarizes, for most species, the status, distribution, and trends primarily before 1960 in Pennsylvania. In this book, *history* is defined as those years before Earl L. Poole completed, but unfortunately never published, his comprehensive manuscript on the birds of Pennsylvania (about 1960). Another "book," *Pennsylvania Birds* (Poole 1964), was but an outline of the data gathered for the major publication. There were few people in historical times with the skills necessary to correctly identify birds, especially birds in the field, so for accuracy and authenticity the historical data used in this book were taken only from published records reviewed by qualified ornithologists and from Poole's unpublished manuscript. No comprehensive statewide bird book has been published since, so the information in his manuscript has been invaluable in the preparation of this work. Since the late 1950s and early 1960s, there has been a dramatic increase in amateur involvement in serious ornithological studies, especially quantitative studies, which evolved into a growing list of local and regional guides. Even nationwide serious cooperative studies began in the late 1950s, with the Operation Recovery program and establishment of the first three bird migration observatories: Manomet, Mass., Long Point, Ontario, and Point Reyes, Calif. In 1960, the Christmas Bird Count reached 10,000 participants for the first time. The popular Golden field guide *The Birds of North America* was published in 1966. The Powdermill banding operations began in the early 1960s and continue today; Powdermill is one of the largest banding stations in the Appalachian Mountains. It could be argued that recent or modern ornithology began in the late nineteenth or early twentieth century with the first major conservation efforts, including the formation of the first Audubon Societies or the publication of the first field guide in 1935. However, because many of the events listed above took place in the late 1950s and early 1960s, it seems that ornithology in Pennsylvania moved into a new era during that time.

Comments

Anecdotes, specific facts, theories or comments by the authors that may not apply in other categories in the species accounts are placed in the Comments section. Questions as to the validity of records are mentioned in this category. Field-identifiable subspecies are also listed and discussed in the Comments category.

Journal References

The following journals provided the majority of the bird sight records referenced in this book.

American Birds. Journal of the National Audubon Society. 1971–1973. Replaced *Audubon Field Notes* with vol. 25 in 1971; became *National Audubon Society Field Notes* after 1993.

Audubon Field Notes. Journal of the National Audubon Society in collaboration with the U.S. Fish and Wildlife Service. 1947–1970. Became *American Birds* with vol. 25.

The Auk. Journal of the American Ornithologists' Union. 1884–1996.

Cassinia. 1891–1960. Journal of ornithology of eastern Pennsylvania, southern New Jersey, and Delaware. Delaware Valley Ornithological Club.

Pennsylvania Birds. 1987–1998. A quarterly journal, became a journal of the Pennsylvania Society for Ornithology in 1992.

Background on Breeding Information Sources

The Breeding Bird Survey

The Breeding Bird Survey (BBS) provides the most important standardized data on breeding bird populations nationwide. These data are cited frequently in the breeding section of species accounts to describe relative abundance and population trends. The U.S. Fish and Wildlife Service (USFWS) has coordinated the BBS program since 1966, using volunteers to run more than 3000 routes across the United States and Canada each year. Within each state and Canadian province, survey routes are distributed randomly within physiographic provinces. Volunteers survey a route comprising 50 stops at half-mile intervals, once a year on a June morning (Robbins et al. 1986). All birds seen or heard in three minutes are counted at each stop. In Pennsylvania, 122 routes distributed across the state have been surveyed since 1966 (Map 4).

Breeding population trends mentioned in species accounts are based on analysis of BBS routes developed by USFWS statisticians with data through 1996 (Sauer et al. 1997). The analysis compensates for change of observers between years. Trends computed since the inception of the survey, a 30-year period, are identified as "long-term" and those since 1980 are considered "short-term." We cite only statistically significant trends ($p < 0.10$) and list statewide results unless a specific physiographic region is identified. These analyses are now available on the Internet at the Breeding Bird Survey's home page (www.mbr-pwrc.usgs.gov/bbs/).

Relative abundance maps are based on the average number of birds counted on 313 routes surveyed from 1985 through 1994 in Pennsylvania and neighboring states. A 10-year average was used to smooth the variability among years and observers. Routes from neighboring states were included to provide continuous patterns across the state border. We generated maps with Winsufer (Golden Software, Incorporated), using analytical methods and mapping procedures similar to those of Price et al. (1995) and those used in the Birds of North America series (A. Poole and Gill 1992–1995) by the American Ornithologists' Union and the Academy of Natural Sciences in Philadel-

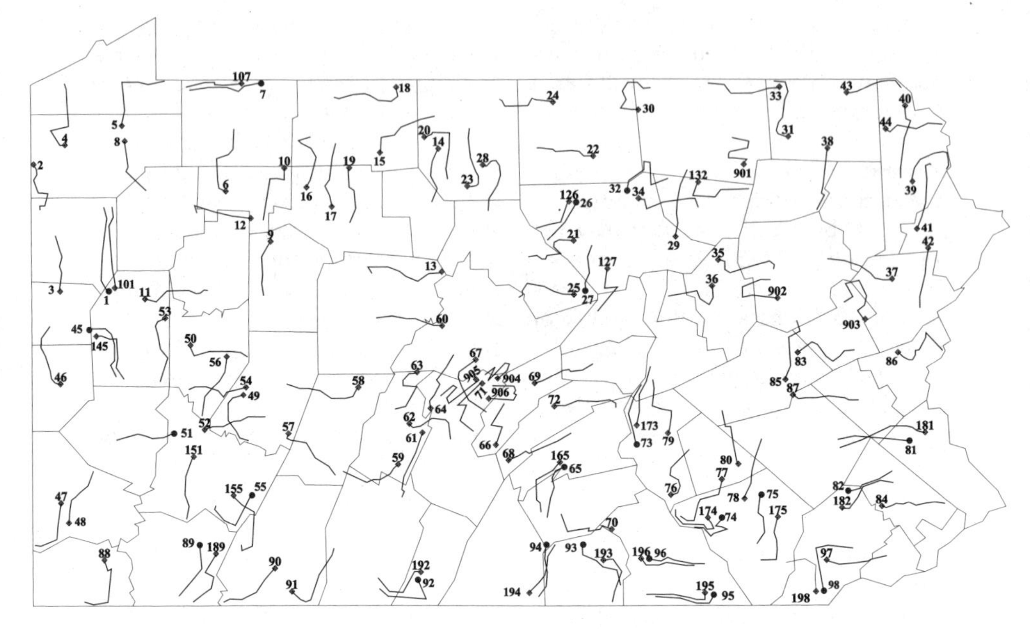

4. Breeding Bird Survey routes. Inactive routes are designated by a dot; active routes are designated by a diamond. Not to scale. (Original map provided by Keith Pardieck, Paxtuent Wildlife Research Center.)

phia. Maps were generated using the Kriging method. The influence of individual routes was reduced by generating each mapping grid point using 10 neighboring routes and error factors typically equivalent to the square root of the mean relative abundance. A few routes were eliminated because of data quality issues. Maps received minor editing to reflect other biases and were sent to Cornell University Press as camera-ready figures. We include maps only for species that occurred on more than two-thirds of Pennsylvania's routes at an average rate of at least 1.0 bird per route statewide. Significant biases of the daytime breeding bird survey include variability among observers and the restriction of routes to roadsides. The three-minute survey period also limits the detection of some species, notably those that occur at low densities (such as raptors, woodpeckers, and herons). Maps for such species could be misleading. As a result, BBS maps were used for only 40 of the state's approximately 200 breeding species.

The maps provide another level of insight into the distribution of widespread species. They complement the distribution information provided by Pennsylvania's Breeding Bird Atlas (Brauning 1992) and illustrate broad-scale patterns of abundance. The Breeding Bird Atlas project, based at the Academy of Natural Sciences of Philadelphia, was a statewide project completed between 1983 and 1989 and published in 1992 (Brauning 1992). It provided a detailed assessment of the summer distribution of Pennsylvania's breeding birds by documenting nesting evidence in 10-square-mile "blocks." Four thousand nine hundred twenty-eight atlas blocks were surveyed. The relative abundance maps reflect both the availability of habitat and the species' density within that habitat but not necessarily the number of birds that may be seen at any given location. Detectability differences among species influence relative abundance values, so specific values are not necessarily comparable among species. Three abundance intervals are presented in most maps. Abundance categories were 1, 3, 9, 30, and 100, unless these contour levels failed to distinguish patterns. Maps are not included for species in which there was insufficient variation in abundance across the state.

Summary of Nest Placement and Timing

The description and placement of nests were summarized from historical egg-collection records, historical authors—particularly Todd (1940) and Warren (1890)—and recent nest and egg books—notably H. H. Harrison (1975) and C. Harrison (1978). The nesting season was defined by egg dates obtained from diverse sources, notably historical egg collections currently housed at the Vertebrate Museum of Natural History, the Carnegie Museum of Natural History, the Delaware Museum of Natural History, Bart Snyder's personal collection, and other lesser collections. These historical sources were merged with the Cornell Nest Record Card program and with the dates of nests that had eggs reported to the Breeding Bird Atlas project. The range of egg dates described in the accounts as "most eggs sets" reflects all but the extreme 5% each of earliest and latest egg dates.

Abundance and Frequency Categories

Terms to describe abundance and frequency have been used in varying ways in many publications. Our terminology for relative abundance and frequency is defined here.

Relative abundance refers to the number of individuals likely to be encountered at the particular season and place. The numerical ranges given below are guides to the approximate number of individual birds an experienced observer could see or hear in a single day or season.

Abundant	100 or more per day
Common	26 to 99 per day
Fairly common	6 to 25 per day
Uncommon	up to 5 per day, up to 25 per season
Rare	1 to 5 per day, up to 5 in a season

Frequency of occurrence reflects the number of times a species has been recorded over a period of years.

Regular	recorded nearly every year
Irregular	not recorded every year, but at least once every two or three years
Casual	three or more years between records, but recorded at least 10 times ever
Accidental	recorded fewer than 10 times or not recorded in the past 10 years

A *record* is defined in this book as documentation of a sighting of one or more birds even if the exact number of birds is not known. A *vagrant* is a bird which is outside its normal breeding and wintering range and migration corridor. A bird is frequently referred to as a *visitor* when it appears and stays usually only a few days or less within an area. A bird becomes listed as a *resident* when it remains for several days or more in an area. Some species of birds listed as residents spend their entire lives within an area.

Documentation of Bird Sightings

The Pennsylvania Ornithological Records Committee (PORC) was organized in 1989. The PORC is a subcommittee of the Ornithological Technical Committee (OTC) of the Pennsylvania Biological Survey. The OTC is a member of the Pennsylvania Biological Survey. It serves as an official advisory group of the Pennsylvania Game Commission (PGC) and selects the birds of special concern. The purpose of PORC is to determine the authenticity of rare or unusual bird sightings and to maintain the official check-list of the birds of Pennsylvania. The PORC has established standards for collecting and submitting quality field data and provides a means by which sight records can gain acceptance as credible scientific data. It maintains a file on past bird records and a photographic records library.

The word *accepted* used throughout the species accounts indicates that the documentation was voted on and accepted as a credible record by the PORC. *Pending review* is used to indicate that the sighting is pending a PORC decision.

Abbreviations

Place Name and Title Abbreviations

AB	*American Birds*
AFN	*Audubon Field Notes*
ANSP	Academy of Natural Sciences of Philadelphia
ASWPB	*Audubon Society of Western Pennsylvania Bulletin*
BBA	Breeding Bird Atlas
BBS	Breeding Bird Survey
BESP	Bald Eagle State Park
BNA	Birds of North America series
CBC	Christmas Bird Count
CCC	Civilian Conservation Corp
CMNH	Carnegie Museum of Natural History
HMS	Hawk Mountain Sanctuary
JHNWR	John Heinz National Wildlife Refuge (Tinicum)
MCWMA	Middle Creek Wildlife Management Area
MSP	Moraine State Park
NASFN	*National Audubon Society Field Notes*
OTC	Ornithological Technical Committee
PB	*Pennsylvania Birds*
PGC	Pennsylvania Game Commission
PISP	Presque Isle State Park
PNR	Powdermill Nature Reserve
PORC	Pennsylvania Ornithological Records Committee
SGL	State Game Lands
USNM	United States National Museum (Smithsonian Institution)
VIREO	Visual Resources for Ornithology, ANSP
WB	*The Wilson Bulletin*
YCSP	Yellow Creek State Park

Other Abbrevations of Frequently Used Words

The first three letters of each month are used throughout the species accounts.

Co.	County
cos.	counties
pers. comm.	personal communication
pers. obs.	personal observation
pers. recs.	personal records
unpbl. ms.	unpublished manuscript

Part II. Species Accounts

Order Gaviiformes
Family Gaviidae: Loons

Three of the world's five species of loons have been recorded in Pennsylvania. Red-throated and Common loons are seen here on migration between their breeding grounds in the northern U.S. and Canada and wintering grounds in the Atlantic Ocean. Pacific Loons winter along the Pacific coast and are casual visitors in eastern North America. Loons are rather large birds, well equipped for life on the water and usually found on lakes or on large slow-moving rivers. The body is long and narrow, the neck short, and the head long and tapered with a dagger-like bill. Loons have short legs and four toes on each foot, with the front three completely connected by webs. Because the legs are attached far back on the body and the wings are short, loons have difficulty walking on land. They are excellent swimmers and divers. They feed mostly on fish and are physiologically adapted for spending considerable periods of time under water. When feeding, they usually dive with head thrust forward. When escaping danger, loons usually dive or sink rather than fly. They are powerful swimmers, often swimming under water great distances before surfacing. Their takeoff is labored and space-consuming, but, once airborne, they fly swiftly and directly. All loons fly with their neck and legs extended and drooped, giving them a hunchbacked profile. When trying to move on land, they usually hold their body nearly vertical and shuffle along, make short hops, or just push themselves along on their breast.

Red-throated Loon *Gavia stellata*

General status: A few Red-throated Loons pass through Pennsylvania in spring and fall every year, usually coinciding with the migration of Common Loons. Red-throated Loons are uncommon to rare migrants in both seasons, flying to and from their breeding grounds in arctic Canada and their wintering grounds along the Pacific and Atlantic coasts. Very few are in breeding plumage during the brief period in spring when they are present in the state. They are occasionally forced down onto inland lakes during the passage of storms. Red-throated Loons are often seen in migration along the lower Delaware River,[1] more rarely along mountain ridges such as at HMS,[2] Baer Rocks,[3] and regularly over Lake Erie.[4] In summer and winter, Red-throated Loons are accidental in Pennsylvania.

Habitat: Red-throated Loons are usually found on the deeper portions of large lakes and slow-moving rivers.

Seasonal status and distribution

Spring: Usually single individuals or small groups are observed in spring resting on lakes or large rivers, especially after heavy precipitation in conjunction with southerly winds. Red-throated Loons have been recorded from late Mar (rarely) or early Apr to mid-May. They rarely remain until early Jun.

Summer: There have been three summer records of Red-throated Loons: 1 bird on 18 Aug 1974 at Montour Preserve (Schweinsberg 1988), another bird on 17 Jun 1988 in Lake Erie near Lake City,[5] and the most recent on Quemahoning Reservoir, Somerset Co., on 9 and 11 Jun 1997.[6]

Fall: Migration usually begins in late Oct, sometimes mid-Oct or early Nov during the passage of cold fronts. Most birds are seen in early to mid-Nov, during peak Common Loon migration. An unusually high count of 40 birds at Lake Carey in Wyoming Co. on 23 Nov 1991 followed a major rainstorm.[7] The fall of 1997 was exceptional at PISP, when over 200 Red-throated Loons were counted migrating over Lake Erie.[4] The normal average fall tally for this area is eight birds. By mid-Dec most birds have departed, but lingering loons are occasionally seen until mid-Jan.

Winter: Red-throated Loons have been recorded in winter at least five times. They have been recorded in Jan and Feb on the Delaware River in Philadelphia Co.[8] There was one bird at Codorus State Park on 21 Jan 1998,[9] and one was found on the Susquehanna River at Lock Haven in Clinton Co. on 28 Jan 1972.[10] A bird was on the Allegheny River between Irvine and Starbrick in Warren Co. from 12 to 14 Feb 1967.[11]

History: Poole (1964) listed Red-throated Loons as rare and irregular transients prior to 1964. Two historic summer records exist in Pennsylvania, one bird from 21 Jul to 2 Aug 1936 on the West Chester Reservoir (Conway 1940) and two birds in nonbreeding plumage at Oneida Lake in Butler Co. during Jun 1933 (Todd 1940). Several historic records of this species have been in winter; one bird was grounded by an ice storm in Philadelphia on 29 Jan 1951,[12] and two birds were recorded in Warren on 12 Feb 1904 (Todd 1940). A bird was reported on the Shenango River at Greenville in Mercer Co. on 20 Feb 1936 (Todd 1940).

Comments: The regularity of the Red-throated Loon in modern times may be the result of the increased number of impoundments that have been created, especially in the High Plateau, in the Ridge and Valley, and in the Piedmont away from the Delaware and Susquehanna rivers.

[1] N. Pulcinella, pers. comm.
[2] Paxton, R.O., K.C. Richards, and D.A. Cutler. 1979. Hudson-Delaware region. AB 34:144.
[3] Paxton, R.O., W.J. Boyle Jr., and D.A. Cutler. 1986. Hudson-Delaware region. AB 41:64.
[4] G.M. McWilliams, pers. obs.
[5] Hall, G.A. 1988. Appalachian region. AB 42: 1286.
[6] Tilly, J. 1997. Local notes: Somerset County. PB 10:104.
[7] Paxton, R.O., W.J. Boyle Jr., and D.A. Cutler. 1992. Hudson-Delaware region. AB 46:67.
[8] J.C. Miller, pers. comm.
[9] Spiese, A. 1998. Local notes: York County. PB 12:27.
[10] P. Schwalbe and G. Schwalbe, pers. comm.
[11] T. Grisez, pers. comm.
[12] Potter, J.K., and J.J. Murray. 1951. Central Atlantic Coast region. AFN 5:196.

Pacific Loon *Gavia pacifica*

General status: Formerly conspecific with the Arctic Loon, the Pacific Loon recently has been recognized as a distinct species. In North America, Pacific Loons breed from the Arctic coast of Alaska east across northern Canada to Northwestern Quebec. They winter along the Pacific coast to West Mexico. Migrants are casual east to the Great Lakes region and to the Atlantic and Gulf coasts. Three accepted records of this species are listed for Pennsylvania.

Seasonal status and distribution

Spring: On 29 Apr 1996, a bird in breeding plumage was seen and photographed by J. Horn with a flock of 37 Common Loons at Green Lane Reservoir in Montgomery Co.[1]

Summer: Quite unusual was the discovery by B. Fisher and F. Fisher of a Pacific Loon in breeding plumage. It was observed from 13 to 15 Jun 1998 at Long Arm Dam in York Co., where it was documented with a photograph.[2]

Fall: A Pacific Loon in juvenile plumage was found by G. M. McWilliams on 25 Nov 1992 in Presque Isle Bay, Erie Co. south of Big Pond. It was seen feeding very close to Common Loons, but usually was alone. It remained until 2 Dec 1992 and was seen by many observers.[3] The loon was documented with several photographs, which are in the files of the PORC and VIREO.

Comments: A bird believed to be of this species was seen in the company of 25 Common Loons on the Susquehanna River at Marietta in Lancaster Co. on 30 Oct 1993.[4] Another individual described as a Pacific Loon was found at Penn Forest Reservoir in Carbon Co. on 12 May 1990 (Street and Wiltraut 1996). However, descriptions of both birds did not eliminate Arctic Loon. A Pacific Loon observed on the lower Susquehanna River at Peach Bottom in Lancaster Co. on 8 Nov 1997 is currently under review.[5]

[1] Freed, G.L. 1996. Local notes: Montgomery County. PB 10:103.
[2] Spiese, A. 1998. Local notes: York County. PB 12:83.
[3] McWilliams, G.M. 1993. First Pennsylvania record of Pacific Loon. PB 6:144–145.
[4] Pulcinella, N. 1994. Rare bird reports. PB 7:133.
[5] Hoppes, J. 1998. Local notes: Lancaster County. PB 11:239.

Common Loon *Gavia immer*

General status: Few sounds of the far north are as enchanting or as mysterious as the laughing call of the Common Loon. Common Loons breed across Canada and New England states and winter primarily along the Atlantic and Pacific coasts. These loons are common to abundant regular migrants along ridges of central Pennsylvania and over Lake Erie. In other areas of the state, Common Loons are uncommon to fairly common regular migrants. Frequently, some are grounded during severe storms, landing on ice-covered roadways, parking lots, and fields. Unless they are taken to open water where they have plenty of space to take off, they usually die. Common Loons are regular nonbreeding summer residents or visitors on undisturbed lakes. They are irregular in winter over most of the state.

Habitat: Common Loons are found on large lakes, slow-moving rivers, and occasionally small lakes and ponds.

Seasonal status and distribution

Spring: Their typical migration period is from mid- to late Mar until mid-May. Migration apparently has no peak time. Storms occasionally force large numbers down on inland lakes, such as 250 at Lake Ontelaunee on 15 Apr 1962 (Poole 1964) and 450 on Penn Forest Reservoir on 21 Apr 1992 (Street and Wiltraut 1996). Stragglers continue to pass through until mid-Jun.

Summer: Common Loons are rare in Pennsylvania after spring migration. They have been found throughout the state. Most birds recorded in summer are immatures; adults in breeding plumage are recorded in summer, but nests, eggs, or downy young have not been found since the 1950s. The closest breeding population is in the Adirondack Mountains of New York (Andrle and Carroll 1988).

Fall: Some birds begin to appear in early Oct. Most fall migrants are seen after major cold fronts during Nov. After several days of southerly winds, a wind shift from the north usually produces loose migrating flocks comprising 50 or more birds, especially over Lake Erie or along the mountain ridges. A passage of several hundred birds in a single morning during a Nov cold front is not unusual.[1] One-day high fall counts include 746 flying past the hawk-watch lookout on Tuscarora Mountain on 12 Nov 1977,[2] 488 or more birds flying overhead at Hawk Mountain on 13 Nov 1977,[3] 450 at Pymatuning Lake on 19 Nov 1978, and 435 at Quemahoning Dam on 17 Nov 1978.[4] Approximately 700 birds settled down on Yellow Creek Lake in Indiana Co. on 23 Nov 1996,[5] and an impressive 1077 passed PISP in the first two hours of daylight on the morning of 11 Dec 1993, a late date for that number of birds.[6] Numbers usually decline by early Dec, with birds lingering to late Jan.

Winter: Common Loons are regular winter residents only in Delaware Co., where they are rare.[7] In the High Plateau they are accidental. In that region they have been recorded wintering on the Allegheny River in Warren Co.,[8] in the Mt. Jewitt area in McKean Co. in mild winters,[9] and at Quemahoning Dam in Somerset Co.[10]

History: The number of birds recorded during migration has changed very little historically. There is some evidence of nesting: loons were reported nesting on Long Pond in Monroe Co. in 1908;[11] a nest with eggs was found at Pocono Lake in Monroe Co. around 1946 (Street 1954); and a nest with two eggs was found on a small island on Pocono Lake on 17 Jul 1955 (Street 1954).

[1] G.M. McWilliams, pers. obs.
[2] Hall, G.A. 1978. Appalachian region. AB 32:203.
[3] Buckley, P.A., R.O. Paxton, and D.A. Cutler. 1981. Hudson-Delaware region. AB 32:183.
[4] Hall, G.A. 1979. Appalachian region. AB 33:176.
[5] M. Higbee and R. Higbee, pers. comm.
[6] G.M. McWilliams, pers. obs.
[7] N. Pulcinella, pers. comm.
[8] Hall, G.A. 1991. Appalachian region. AB 45:272.
[9] J.P. Dzemyan, pers. comm.
[10] R. Sager and G. Sager, pers. comm.
[11] Harlow, R.C. 1908. Breeding of the loon in Pennsylvania. Auk 25:471.

Order Podicipediformes
Family Podicipedidae: Grebes

Grebes are waterbirds that resemble loons, but they are smaller, and the body is shorter and rounder. Their tails are vestigial, and their legs are attached far back on the body, making them rather awkward on land. Unlike loons, which have webbed toes, grebes have lobed toes. Their bills vary in size and general shape from short, stubby, and chickenlike to long, narrow and daggerlike. They are excellent divers and usually escape danger by diving rather than flying. They are able to slowly lower themselves in water by altering their specific gravity until only their head or bill is above the surface. Their takeoff is labored and space-consuming, but, once airborne, they fly swiftly and directly, usually low over the water. Occasionally, storm-grounded Horned and Red-necked grebes have mistaken ice-covered roadways, parking lots, or fields for water and have been unable to take off again once landed. Without human intervention they usually die. Three species regularly occur as migrants, and one breeds in Pennsylvania.

Pied-billed Grebe *Podilymbus podiceps*

General status: These rather secretive grebes are fairly common, regular migrants throughout Pennsylvania. Pied-billed Grebes are easily overlooked, in part because of their habit of sinking out of sight in the water when approached and emerging in the concealment of vegetation. Pied-billed Grebes breed across southern Canada and throughout the U.S. and winter over most of their breeding range. They breed in the state and are regular winter residents except in the Ridge and Valley, where they are casual. They are accidental in winter in the High Plateau.

Habitat: Pied-billed Grebes are found on open shallow ponds, lakes, slow-moving streams and rivers and marshes in migration and in winter. During the breeding season they prefer extensive marshes and are found in small lakes, ponds, and beaver dams with emergent and aquatic vegetation.

Seasonal status and distribution

Spring: Birds may appear soon after ice has melted in early Mar. Their typical migration period is from mid-Mar to mid-May. They are most often seen singly or in small loose groups.

Breeding: Pied-billed Grebes are uncommon to fairly common regular breeders in large marshes of glaciated northwestern counties, in such localities as Conneaut Marsh and Hartstown Swamp, Crawford Co. They are rare and irregular elsewhere statewide, with some summer residents not breeding. Summer records have come from most counties, with confirmed breeding in at least 22 widely scattered counties. Since the 1960s, these grebes have declined as breeders in the southeastern counties. A notable exception is Glen Morgan Lake, Berks Co., where a sizable breeding population was found in 1995[1] and as many as 118 adults and young were seen in Jul 1996.[2] The grebes sometimes occupy newly established wetland habitat for a few years in response to a flush in fertility, but they do not persist in smaller wetlands. They are highly vocal through the breeding season, in contrast to the rest of the year. Eggs are placed in a floating nest of vegetation anchored to stems. Egg dates range from 9 May to 23 Jul. Eggs are pale blue or green when first laid, turning brown when incubation begins. Adults often approach their nests under water and surface upon reaching the nest. Two broods a year have been documented,[3] but one is typical. Small young, sometimes observed riding on a parent's back, have been reported as late as 15 Aug (Grimm 1952).

Fall: Migrants usually arrive in late Aug and remain until ice begins to form. Numbers are usually larger in fall than in spring. Occasionally, loose flocks of 80 or more birds may be found.

Winter: Pied-billed Grebes are rare throughout their winter range in Pennsylvania. They may be found where water is open, especially on large slow-moving rivers. Rarely are more than one or two birds found at any site. They have been recorded in the Glaciated Northeast only in Wyoming Co. at Lake Carey.[4] In the High Plateau, Pied-billed Grebes were found on eight occasions between 1966 and 1993 along the Allegheny River in or near Warren.[5] In the Allegheny Mountains they have been observed in Somerset[6] and Westmoreland[7] cos. An unusually high count of 22 birds was found at PISP on 22 Jan 1987.[8]

History: Pied-billed Grebes have had a variable breeding history. The earliest records indicate they were common migrants but rare breeders in Pennsylvania, nesting irregularly in Pymatuning Swamp before the lake was impounded (Sutton 1928b). They became very common breeders after Pymatuning Lake was created in 1935 (Grimm 1952). Numerous nesting records were reported during the 1930s and 1940s in the southeastern counties, including some in the Philadelphia area. Poole (1964) still regarded them as "most numerous" in the southeast and northwest, but Miller and Price (1959) suggested that they were rare breeders, indicating there were "a few nesting records" at Tinicum.

Comments: Pied-billed Grebes are listed in Pennsylvania by the OTC as Candidate–Rare on the Species of Special Concern list.

[1] Keller, R. 1995. Local notes: Berks County. PB 9:89.
[2] Keller, R. 1996. Local notes: Berks County. PB 10:161.
[3] Miller, R.F. 1942. The Pied-billed Grebe: A breeding bird of the Philadelphia region. Cassinia 32:22–34.
[4] W. Reid, pers. comm.
[5] T. Grisez, pers. comm.
[6] Hall, G.A. 1990. Appalachian region. AB 44:267.
[7] Leberman, R., and R. Mulvihill. 1991. County reports—January through March 1991: Westmoreland County. PB 5:51.
[8] Hall, G.A. 1987. Appalachian region. AB 41:281.

Horned Grebe ***Podiceps auritus***

General status: Aptly named for the hornlike tufts of feathers on their heads, Horned Grebes breed across most of Canada and winter along both coasts. In Pennsylvania they are fairly common to common regular migrants. They are regular winter residents, except in the High Plateau and the Glaciated Northeast, where nearly all winter records are from birds grounded by storms. Horned Grebes are casual nonbreeding summer residents or visitors. Migrants can become abundant on lakes and rivers throughout the state in spring and on Lake Erie in fall. Horned Grebes are more widely reported as being grounded by storms and sudden freezes than any other species of waterbird in Pennsylvania. They also account for most storm-related avian mortalities.

Habitat: Horned Grebes are usually found on ponds, lakes, and slow-moving rivers in migration and winter. They prefer more open and deeper water with less vegetation than do Pied-billed Grebes.

Seasonal status and distribution

Spring: In early Mar, rarely late Feb, Horned Grebes arrive singly or in small flocks. Occasionally, flocks containing hundreds of birds can be found at rest on lakes. Their typical migration period is from the first week of Mar to the first week of May. The majority of birds arrive, often overnight, during the last few days of Mar or the first few days of Apr. Although large numbers are expected during spring migration, the 1700 forced down by an ice storm on 9 Apr 1979 on the Susquehanna River at Lock Haven in Clinton Co. was remarkable.[1] Other notable numbers of spring migrants were 1600 at Bald Eagle State Park in Centre Co. on 5 Apr 1973[2] and 400 that arrived on the unusual date of 15 Feb 1991 in Presque Isle Bay.[3] The majority of birds depart within a day or two of their arrival.

Summer: Fewer than a dozen modern records of Horned Grebes are in summer; most are of single birds that appeared in mid- to late summer. They have been recorded in the Piedmont, Ridge and Valley, Southwest, and Lake Erie Shore regions from Adams, Butler, Chester, Erie, Franklin, Lancaster, Lycoming, and Snyder cos.

Fall: Horned Grebes are uncommon in fall across the state. In the Glaciated Northwest they can be common to abundant, with a high count of over 900 birds at Pymatuning Lake on 26 Nov 1978.[4] They are fairly common to abundant on Lake Erie, where numbers vary from year to year. Individuals or small groups usually can be observed migrating along the Lake Erie shore, especially with northerly winds. Hundreds may be seen passing by PISP in a single day, and in some years, hundreds of birds stage on Lake Erie or Presque Isle Bay for several weeks.[5] Whereas in spring the flocks are tightly packed, Horned Grebes are usually scattered or in loose flocks in fall. Horned Grebes trickle through during Oct, with single birds arriving as early as the first week of Sep. Their typical migration period is from late Oct to mid-Dec. Peak migration usually occurs during the first half of Nov. Most birds are chased out by the freezing of smaller bodies of water in mid- to late Jan. Probably one of the most memorable groundings in recent years occurred on the night of 6–7 Jan 1979, when a sleet storm over much of southwestern Pennsylvania grounded large numbers of Horned Grebes. Hundreds were found at State College, at least 20 in the Ligonier Valley, 23 in Greene Co., and 40 in Fayette Co.[6]

Winter: Birds grounded by storms or sudden freezes can appear almost anywhere and at any time during winter in the state. Many individuals have been retrieved in an area of only a few square miles after some groundings. Birds that attempt to winter remain on large lakes as long as water is open, but most move to large slow-moving rivers. In the High Plateau, other than those grounded by storms, Horned Grebes have been recorded wintering only along or near the Allegheny River in Warren Co. Five records, from 1966 to 1985, occurred between Warren and Kinzua Dam.[7]

[1] Hall, G.A. 1979. Appalachian region. AB 33:770.
[2] Hall, G.A. 1973. Appalachian region. AB 27:772.

[3] Hall, G.A. 1991. Appalachian region. AB 45:272.
[4] Hall, G.A. 1979. Appalachian region. AB 33:176.
[5] G.M. McWilliams, pers. obs.
[6] Hall, G.A. 1979. Appalachian region. AB 33:281.
[7] T. Grisez, pers. comm.

Red-necked Grebe ***Podiceps grisegena***

General status: Red-necked Grebes are the largest of the regularly occurring grebes found in Pennsylvania. In North America, they breed across western and southern Canada east to Quebec and in the northern states east to Wisconsin. Most winter along the Pacific and Atlantic coasts. These grebes are regular rare migrants in spring and fall in Pennsylvania and are casual winter visitors. Most winter records are of birds downed after severe storms or sudden freezes.

Habitat: Red-necked Grebes are found on lakes and slow-moving rivers. They usually prefer deeper water than Pied-billed and Horned grebes.

Seasonal status and distribution

Spring: Most records are in spring. In some years, first spring migrants arrive as early as the second week of Feb if there is an early thaw, but they are not expected until mid-Mar. Individuals or small groups may appear at any time

Red-necked Grebe at Washington Crossing State Park, Bucks County, February 1994. (Photo: Franklin C. Haas)

thereafter until early May. Usually fewer than five birds are reported at a time, but some notable numbers have been reported. There were 48 Red-necked Grebes counted along the Delaware River during the spring of 1959 (Poole 1964). One of the largest flights ever in the state occurred during the spring of 1994, perhaps from the freezing of the Great Lakes: a high count of 26 birds was seen on the Susquehanna River in Lancaster Co. from Long Level to Marietta on 20 Feb;[1] 21 along the Allegheny River in Armstrong Co. between Kittanning and Rosston on 17 Feb;[2] and 35 in Erie Co. at Presque Isle Bay on 31 Mar.[3]

Summer: A bird found along the Susquehanna River in Dauphin and Northumberland cos. on 15 and 20 Jul 1994[4] provides the only summer record. This record followed a spring invasion of Red-necked Grebes.

Fall: Except along the Lake Erie Shore, where they have become regular in fall, Red-necked Grebes are rarer in fall than in spring. Their typical migration period is from late Oct, rarely mid-Oct, to early Dec. In most years, singles can be seen migrating over Lake Erie at this season during the passage of a cold front.[5] Early migrant records outside the normal season include 1 bird on 10 Aug 1992 at Sheppard Myer's Dam,[6] 3 on 28 Aug 1993 at Codorus State Park in York Co.[6,7] and 1 bird on 10 Sep 1962 at the Churchville Reservoir.[8]

Winter: Like Horned Grebes, most Red-necked Grebes found between mid-Jan and mid-Feb are storm-grounded and emaciated. Birds in healthy condition usually do not linger more than a few days. Storm-related birds have been sighted along the lower Susquehanna River in Lancaster and York cos. and along the West Branch of the Susquehanna River in Centre and Clinton cos. Others have been reported during winter in Erie, Montour, Northampton, Potter, Union, and Warren cos.

[1] Heller, J. 1994. Notes from the field: Lancaster County. PB 8:41.
[2] Higbee, M., and R. Higbee. 1994. Notes from the field: Armstrong County. PB 8:36.
[3] McWilliams, G.M. 1994. Notes from the field: Erie County. PB 8:38.
[4] Haas, F.C., and B.M. Haas, eds. 1994. Rare and unusual bird reports. PB 8:163.
[5] G.M. McWilliams, pers. obs.
[6] A. Spiese, pers. comm.
[7] Haas, F.C., and B.M. Haas, eds. 1993. Rare and unusual bird reports. PB 7:103.
[8] Scott, F.R., and D.A. Cutler. 1963. Middle Atlantic Coast region. AFN 17:19.

Eared Grebe *Podiceps nigricollis*

General status: Eared Grebes breed from southwestern Canada south to Texas and California, and they winter across most of the West and east across the southern U.S. to Louisiana and north along the Atlantic coast to Maryland. They were not reported in Pennsylvania before 1959. Thereafter, they have been infrequently reported and are casual or accidental in the state. In the last 15 years, Eared Grebes have become more regular, though they still are quite rare. Most records are from the Lake Erie Shore at PISP, where they have occurred regularly since 1990. They often associate with Horned Grebes, especially in fall. Reliable records have come from Butler, Cambria, Chester, Erie, Luzerne, Lycoming, Montgomery, Perry, Philadelphia, Venango, and Westmoreland cos. Photographs of Eared Grebes from Pennsylvania are in the files of PORC and VIREO.

Habitat: Most sightings have been on lakes. The habitat preferences of Eared Grebes appear to be similar to those of Horned Grebes, and Eared Grebes are likely to occur where Horned Grebes are found.

Seasonal status and distribution

Spring: The first Eared Grebe in Pennsylvania was identified by J. G. Stull in

Eared Grebe at Honeybrook, Chester County, October 1994, (Photo: Franklin C. Haas)

Presque Isle Bay on 28 Mar 1959.[1] Earliest arrivals since have been recorded about the third week of Mar, with some birds appearing as late as mid-May. Most grebes are either in full breeding plumage or at least molting into breeding plumage. All spring sightings have been of just one or two birds.

Fall: They are not expected before the last week of Oct but have been reported as early as the second week of Sep.[2] Most sightings are in Nov. Some have lingered to the first week of Jan at PISP.[3]

Comments: Eared Grebes frequently have been confused with nonbreeding-plumaged Horned Grebes. However, more published material on identification of Eared Grebes in nonbreeding plumage and more birders carefully examining grebes have been significant factors in the increased number of correctly identified birds in recent years.

[1] Stull, J.H., and J.G. Stull. 1959. Field notes. Sandpiper 1:51.
[2] Hall, G.A. 1988. Appalachian region. AB 42:73.
[3] Hall, G.A. 1991. Appalachian region. AB 45:272.

[Western/Clark's Grebe] *Aechmophorus occidentalis/clarkii*

General status: This genus of grebes is listed as hypothetical in Pennsylvania. The Western/Clark's Grebe complex breeds from south-central Canada south to northern New Mexico and east to Wisconsin. They winter primarily along the Pacific coast south into Mexico and east to western Texas.

Comments: There have been several sight records, mostly before the genus was separated into two species. At this writing, the occurrence of Western or Clark's grebes has not been confirmed with a photograph or specimen in Pennsylvania, and written field notes have not conclusively identified either species. A specimen in the collection of St. Vincent College, collected on the Greensburg Reservoir southeast of Latrobe some time between 1898 and 1906 (Todd 1940), was apparently lost in a fire at the college a few decades ago. A Western Grebe shot very close to the Pennsylvania border at Youngstown in Ohio on 28–30 Oct 1913 (Peterjohn 1989) is in the collection at the Carnegie Museum (Todd 1940).

Order Procellariiformes
Family Procellariidae: Shearwaters and Petrels

Birds of the open ocean, shearwaters and petrels, rarely come to land except to nest on sea cliffs and offshore islands or when carried inland by hurricanes. They primarily differ from gulls, which they closely resemble, by having bills covered with horny plates separated by sutures and both nostrils encased in well-formed tubes (hence the name tubenoses). Only four species of this worldwide family have been confirmed in Pennsylvania and on only four occasions. All but two birds were found grounded by the passage of powerful hurricanes in 1989 and 1996 that carried them inland from the Atlantic Ocean. Even when taken alive to a rehabilitator, they rarely survive for more than a few days. Other species of shearwaters have been reported in the state, but they are listed as hypothetical because they could not be identified or the specimens were lost.

Northern Fulmar ***Fulmarus glacialis***

General status: In North America the circumpolar Northern Fulmar nests along coastal Alaska and in the North Atlantic in colonies from Baffin Island and south to Newfoundland. It winters in the Pacific and North Atlantic oceans and is listed as accidental in Pennsylvania, where it has been recorded only once.

Seasonal status and distribution

Fall: The circumstances surrounding this sighting were as unusual as the occurrence itself. The fulmar was seen well by J. Peplinski and B. Peplinski for several minutes flying over the rain-soaked crowd at the Penn State–Notre Dame football game at State College, Centre Co. on 16 Nov 1985.[1] An excellent detailed description was written, and the observers made sketches.

[1] Hall, G.A. 1986. Appalachian region. AB 40:112.

Kermadec Petrel ***Pterodroma neglecta***

General status: Kermadec Petrels breed in the South Pacific and range at sea across the South Pacific. A sighting of this pelagic bird has been recorded on motion picture film in Pennsylvania. Since the discovery of this bird in 1961 the film of the petrel has been reviewed by many seabird experts with debate as to whether it was a Herald or a Kermadec petrel. The PORC has not been able to review the film. However, the acceptance of this species in Pennsylvania is in accordance to the AOU (1998) recent decision listing it as a Kermadec Petrel.

Seasonal status and distribution

Fall: On 3 Oct 1959, three days after a tropical hurricane dissipated in south-central Pennsylvania, several observers watched a petrel circling over the lookout at Hawk Mountain. Light conditions were not favorable, but a reasonably satisfactory motion picture film was obtained. The bird was initially identified as a dark morph *P. neglecta* by Dr. Robert C. Murphy.[1]

[1] Heintzelman, D.S. 1961. Kermadec Petrel in Pennsylvania. WB 73:262–267.

Black-capped Petrel ***Pterodroma hasitata***

General status: Black-capped Petrels breed on Hispaniola and range at sea in the Carribean and western Atlantic Ocean north regularly to North Carolina. They are occasionally carried inland by hurricanes and have been recorded in Ontario, Pennsylvania, New York, Massachusetts, Virginia, Kentucky, Ohio, and Florida. Accidental in Pennsylvania, they have been recorded here on only two occasions, both during the passage of hurricanes. All but one of these birds were found in or near the path of the eye.

Seasonal status and distribution

Fall: During the passage of Hurricane Hugo on 22 Sep 1989, five dead or moribund birds were found. Three were found on 23 Sep; one moribund bird was in a parking lot in Nicholson, Wyoming Co.; another on the west bank of the Youghiogheny River in Buena Vista, Allegheny Co.; and one on Rockwood

Avenue, Oil City, Venango Co. in a yard. The other two birds were found the next day in Venango Co., one at the park office of Oil Creek State Park, and the other near Venango Manor County Home, 2 miles west of Franklin, Sugarcreek Township. All birds died within a day or two of their discovery.[1] Several of these specimens were deposited in the CMNH and ANSP. During the passage of Hurricane Fran in Sep 1996, three birds were found in the state. One was picked up at the Butler Co. Airport on 7 or 8 Sep and sent to Florida for rehabilitation, but later died. Two were picked up in Somerset Co. near Mt. Davis on 7 Sep and were taken to a local rehabilitation center, where one escaped and the other died.[2]

[1] States, R. 1989. Rare birds in Venango Co. PB 3:124–125.
[2] Hess, P. 1996. Local notes: Butler County. PB 10:162.

Cory's Shearwater *Calonectris diomedea*

General status: The closest breeding populations to North America are on islands in the eastern North Atlantic Ocean in the Azores, on Berlenga Island off Portugal, in the Madeira, Salvage, and Canary islands. They range widely over the Atlantic Ocean west to North America. This species has been recorded once in Pennsylvania.

Seasonal status and distribution

Fall: A Cory's Shearwater was recorded by P. Schwalbe and G. Schwalbe at Williamsport in Lycoming Co. on 9 Sep 1996. It was picked up, still alive, in a parking lot during the passage of Hurricane Fran. The bird later died, was measured and photographed, and the specimen was sent to the ANSP.[1]

Comments: This bird was determined to belong to the subspecies *C. d. borealis*,[2] which breeds in the Azores, Madeira, Canary, and Berlenga islands (Harrison 1983).

[1] W. Egli. 1996. Local notes: Lycoming County. PB 10:166.
[2] Schwalbe, P., and G. Schwalbe. 1996. Cory's subspecies determined. PB 10:206.

[Greater Shearwater] *Puffinus gravis*

General status: This seabird breeds on Tristan da Cunha, Gough Island, and in the Falkland Islands in the South Atlantic Ocean and disperses north in the western Atlantic Ocean to Greenland. It is listed as hypothetical in Pennsylvania.

Comments: Greater Shearwaters were reported twice in Pennsylvania. A bird reportedly was found dead in Chester Co. prior to 1881 (Poole, unpbl. ms.), and there is one probable sight record from Chester Co. 7 Sep 1974.[1]

[1] Scott, F.R., and D.A. Cutler. 1974. Middle Atlantic Coast region. AB 29:35.

[Audubon's Shearwater] *Puffinus lherminieri*

General status: Widespread in tropical oceans, Audubon's Shearwaters nest closest to North America on islands in the Carribean and disperse north in the Atlantic Ocean to Massachusetts. This species is listed as hypothetical in Pennsylvania because there is no written description, photograph, or specimen.

Comments: One bird was reported on the Delaware River near Tinicum, Philadelphia on 2 Aug 1953.[1]

[1] Potter, J.K., and J.J. Murray. 1954. Middle Atlantic Coast region. AFN 8:10.

Family Hydrobatidae: Storm-Petrels

Storm-petrels are smaller than shearwaters and petrels and have shorter wings. Their bills are similar, except that rather than having two tubes over their nostrils, storm-petrels have only one. Unlike shearwaters and petrels, which soar on flat wings with shallow rapid wing beats, storm-petrels are nearly always flapping their wings and flitting about the water surface in swallowlike fashion. They spend most of their lives over the open ocean far from shore and come ashore only to nest on offshore islands and when carried inland by hurricanes. Leach's Storm-Petrel is the only species in this family that has been confirmed in Pennsylvania. Most birds have been carried into the state during the passage of hurricanes. There has been no report from Pennsylvania since 1955.

[Wilson's Storm-Petrel] ***Oceanites oceanicus***

General status: Wilson's Storm-Petrels breed on islands in the Southern Hemisphere, nesting closest to North America on islands off southern South America. They are common transequatorial migrants in all oceans (Harrison 1983). This species is listed as hypothetical in Pennsylvania.

Comments: A bird believed to be of this species was collected on the Greensburg Reservoir near Latrobe sometime between 1898 and 1906. The bird was mounted and was received by Reverend Maximilian Duman, O.S.B., curator of the collection at St. Vincent College, where the specimen was housed (Todd 1940). Apparently the specimen was lost in a fire at the college.

Leach's Storm-Petrel ***Oceanodroma leucorhoa***

General status: Most records of Leach's Storm-Petrel are of birds blown inland after hurricanes. They breed along the North Pacific and Atlantic coasts and offshore islands. The closest breeding colonies to Pennsylvania are in the Gulf of St. Lawrence, Newfoundland, coastal Maine, and Massachusetts. This species is listed as accidental in Pennsylvania. All records are historical. Warren (1890) stated that "in September, 1879, [I] had a specimen presented to me by the late Dr. George Martin of West Chester, who had picked it up in his yard in an exhausted and dying condition." Witmer Stone (1894) mentioned a specimen taken at Tinicum on 18 Dec 1890. After a severe storm on 23 and 24 Aug 1933, many were seen and others found in a weakened condition as far inland as the Susquehanna River, where several were collected (Frey 1943). Three collected on the Lackawanna River at Scranton were deposited in the Everhart Museum (Poole, unpbl. ms.). A specimen found along the Wissahickon Creek on 25 Aug was placed in the ANSP and, on 24 Aug, at least 12 birds were seen on Lake Ontelaunee, five on the Angelica Dam, and one on the Schuylkill River near Reading. Three picked up in exhausted condition in Reading, West Reading, and Hamburg were placed in the collection at the Reading Public Museum. Five birds were seen on 14 Aug 1955, the day after a hurricane, at the Conejohela Flats at Washington Boro. One that was captured and photographed later died and was mounted and displayed in the Franklin and Marshall College Museum (Poole, unpbl. ms.). Leach's Storm-Petrels are included on the official list of the birds of Pennsylvania on the basis of a specimen collected at Reading on 25 Aug 1933.

Band-rumped Storm-Petrel ***Oceanodroma castro***

General status: This species breeds in the tropical Pacific and Atlantic oceans and disperses north in the Atlantic Ocean to Massachusetts. One recent record exists for Pennsylvania.

Seasonal status and distribution: On 24 Feb 1998, a beachwalker picked up the remains of a petrel along Manchester Beach along Lake Erie in Erie Co. Even though the only identifiable remains were the wings, a few rectrices, the legs and feet, the petrel was identified by Dr. Kenneth Parkes at the CMNH and Dr. David Lee at the North Carolina State Museum of Natural Sciences as this species.[1]

Comments: Supposedly a bird identified as this species was found dead along one of the streets of Chambersburg in Franklin Co. The specimen was brought to Frank S. Flack on 15 Apr 1912, where it remained in his collection (Todd 1940). "The example in question agrees well with authentic specimens in the collection of the Carnegie Museum" stated Todd (1940). The whereabouts of the specimen is not known.

[1] McWilliams, G.M. 1998. A Band-rumped Storm-Petrel *(Oceanodroma castro)* found on the shore of Lake Erie in Pennsylvania. PB 12: 127–128.

Order Pelecaniformes
Family Phaethontidae: Tropicbirds

Tropicbirds are seabirds found in tropical and subtropical oceans. They are rarely observed away from the open ocean or offshore islands, where they nest, except when they are carried inland by hurricanes. Tropicbirds are gull-like in appearance and are mainly white. The bill slopes slightly, gently tapers to a point, and lacks the hooked tip of gulls. Tropicbirds have extremely long central tail projections that are at least as long as their body. The White-tailed Tropicbird has been reported once from Pennsylvania.

White-tailed Tropicbird *Phaethon lepturus*

General status: The closest breeding range of the White-tailed Tropicbird to eastern North America is from Bermuda south to the Carribean Islands. Two specimens have been collected in Pennsylvania.

Seasonal status and distribution

Fall: Two White-tailed Tropicbirds were collected on 16 Oct 1954 after Hurricane Hazel. The first tropicbird was captured alive at Coldsmith's Stables, near Gettysburg, by C. H. Wolford. According to Chandler S. Robbins, the bird was a female and was placed in the USNM. The second was a female taken at Nanticoke in Luzerne Co. by D. H. Christian. This specimen was also placed in the USNM (Poole, unpbl. ms.). Both were verified by a curator of the USNM.

Comments: After the same hurricane, birds were reported from Staunton, Virginia, and in western New York.[1]

[1] Griscom, L. 1955. A summary of the fall migration. AFN 9:4.

Family Sulidae: Gannets

Gannets are large marine birds usually not found far from the coast. They have long narrow wings and a long tapering bill and tail. When in flight they frequently interrupt their flapping with short glides. They feed by making spectacular dives, often from great heights, into the water to catch fish. Fishermen use flocks of feeding gannets to locate schools of fish. In Pennsylvania, gannets are very rare vagrants. Apparently they simply wander into the state rather than being storm-assisted, as are most other seabirds.

Northern Gannet *Morus bassanus*

General status: Northern Gannets breed in the Gulf of St. Lawrence, off Quebec, in Labrador, Newfoundland, and off New Brunswick. They occasionally wander inland and have been recorded in New England and the Great Lakes. They are accidental in Pennsylvania, where most records are from the Delaware River. There have been five sightings since 1960.

Seasonal status and distribution

Spring: An immature was seen by J. Carrol and J. Devlin on 3–4 May 1961 and was found dead on 5 May 1961 at Tinicum.[1] One was at Tinicum on 4 May 1967 (J.C. Miller 1970).

Fall: An injured bird was found by Gillespie at Chadds Ford in Delaware Co. 15 Oct 1975.[2] The only record of Northern Gannet outside southeastern Pennsylvania occurred during the fall. A first-year bird was identified by S. Stull and G. M. McWilliams on 29 Nov 1981 at PISP.[3] It was seen by many observers until 25 Dec 1981,[4] occasionally observed diving into the shallow water of Presque Isle Bay from heights of over 100 feet, usually at an angle of 45 degrees or less to the surface of the water.[5] This bird is documented with a photograph in the files of the PORC.[6]

Winter: An injured bird was picked up by J. Northwood at Montgomeryville, Montgomery Co., on 27 Jan 1962.[7]

History: Apparently the first report of Northern Gannet in Pennsylvania was of a bird that broke its wing, probably after flying into wires. It was picked up during the early part of Jul 1921, location unknown, and the specimen was mounted.[8] Another was found at Secane, Delaware Co., on 22 Nov 1922 (Stone 1937). An individual was killed by a hunter in Glenolden on 21 Nov 1932, and the specimen was placed in the ANSP.[9]

Comments: Peterjohn (1989) stated that young gannets wander up the St. Lawrence River to Lake Erie, the most likely route for the bird that appeared at PISP.

[1] Scott, F.R., and D.A. Cutler. 1961. Middle Atlantic Coast region. AFN 15:397.

[2] Grantham, J. 1976. General notes. Cassinia 56:27.

[3] Hall, G.A. 1982. Appalachian region. AB 36:176.

[4] Hall, G.A. 1982. Appalachian region. AB 36:293.

[5] G.M. McWilliams, pers. obs.

[6] Inadvertently omitted in publication, but listed, on the official list in the files of PORC.

[7] Scott, F.R., and D.A. Cutler. 1962. Central Atlantic Coast region. AFN 16:316.

[8] Green, H.T. 1926. Gannet in Bucks Co., Pennsylvania. Auk 43:363.

[9] Worth, C.B. 1933–1937. Field notes. Cassinia 30:8.

Family Pelecanidae: Pelicans

Pelicans are large, heavy water birds with a massive flat bill. They have lower mandibles with large naked pouches that are connected with the throat and are capable of great distention. They have a short tail and short legs with fully webbed toes. Their wings may span up to 8 or 9 feet. Pelicans are powerful flyers and have little difficulty in taking off despite their short legs and heavy body. Two species are found in North America, and both are rare vagrants in Pennsylvania. American White Pelicans prefer inland freshwater, and Brown Pelicans, birds of the coast, prefer saltwater. Both species use their pouches as dip nets for catching fish. American White Pelicans feed on or near the surface of the water, whereas Brown Pelicans usually feed by plunge-diving from the air. Flocks of pelicans often feed in unison. When one or two pelicans dip their heads under water the rest of the flock often dip their heads at the same time. When in flight, flocks frequently alternately flap their wings and glide in unison.

American White Pelican *Pelecanus erythrorhynchos*

General status: One of the largest species in North America, these huge black-and-white birds cannot be mistaken for other North American species. Their breeding range is from central Canada south and locally throughout the West. They winter from California south along the coast of Mexico and along the Gulf coast to peninsular Florida. In Pennsylvania, they are casual visitors in spring and fall and accidental in summer and winter. They are rare in all seasons; most records are of single birds.

Habitat: They are usually found on large lakes and rivers.

American White Pelican at Presque Isle State Park, Erie County, May 1995. (Photo: Gerald M. McWilliams)

Seasonal status and distribution

Spring: There have been at least 10 records of American White Pelican in recent years, with spring records from Armstrong, Crawford, Dauphin, Elk, Erie, Indiana, and Lancaster cos. Spring dates range from 2 Apr to 21 May.

Summer: They have been recorded twice during summer in recent years. One was discovered on the Delaware River at the mouth of Lackawaxen River in Pike Co. on 26 Jun 1986. Possibly the same bird made a brief visit at the Peace Valley Nature Center in Bucks Co. on 29 Jun 1986.[1]

Fall: There are five recent fall records. On 14 Aug 1992, a bird was found at Fort Mifflin in Philadelphia Co. and stayed to 3 Oct 1992.[2] Two were discovered at the Conejohela Flats, Lancaster Co., on 22 Aug 1993.[3] An exceptional concentration of 21 pelicans was at Montoursville in Lycoming Co. on 22 Sep 1996,[4] and three that were seen at Montour Preserve in Montour Co. from 20 Sep to 2 Oct 1996 were photographed.[5] One very late bird appeared on 5 Dec and remained until 8 Dec 1982 at Pymatuning in Crawford Co.[6]

Winter: An American White Pelican on the Ohio River in Allegheny Co. on 9 Jan 1995 remained to 13 Feb 1995.[7,8]

History: American White Pelicans have always been rare visitors in Pennsylvania. The earliest record from the state comes from Conneaut Lake in Crawford Co., where one was shot on 1 Oct 1861 (Sutton 1928b). Two shot on Conneaut Lake on 15 Oct 1898 were mounted and placed at the CMNH (Sutton 1928b). Warren (1890) stated that three or four birds were reported on the Susquehanna River at Keating, Clinton Co., sometime around 1885. One was seen on the dam at Moselem Springs in Berks Co. about 1892 (Poole, unpbl. ms.). A small flock was seen on the Susquehanna River about 1905 (Frey 1943). This species was not recorded again until 5 Oct 1926, when one was captured alive at Overview in Cumberland Co. One was shot out of a flock of five near Shippensburg on 7 Oct 1926. Apparently the latter bird was sent to the State Museum at Harrisburg (Poole, unpbl. ms.). Other birds collected include one from Kurtz's Dam, Valley Township in Chester Co., which was mounted and placed in the Reading Museum in 1928 (Poole, unpbl. ms.). They were reported in 1935 when a bird was found on 19 Jun and remained to 23 Jun at Pymatuning Swamp (Todd 1940). Another bird summered in the same location until 23 Oct 1956.[9]

[1] Paxton, R.O., W.J. Boyle Jr., and D.A. Cutler. 1986. Hudson-Delaware region. AB 40:1183.
[2] Haas, F.C., and B.M. Haas, eds. 1992. Rare and unusual bird reports. PB 6:128.
[3] Haas, F.C., and B.M. Haas, eds. 1993. Rare and unusual bird reports. PB 7:103.
[4] Egli, W. 1996. Local notes: Lycoming County. PB 10:166.
[5] Brauning, D.W. 1997. Local notes: Montour County. PB 10:228.
[6] Hall, G.A. 1983. Appalachian region. AB 37:301.
[7] Floyd, T. 1995. Local notes: Allegheny County. PB 9:31.
[8] Kwater, E. 1997. An unusual pelican in Beaver County. PB 11:55.
[9] Brooks, M. 1957. Appalachian region. AFN 11:345.

Brown Pelican *Pelecanus occidentalis*

General status: Brown Pelicans are resident along the west coast of the U.S. and Mexico. In the eastern U.S., Brown Pelicans are residents of the Atlantic and Gulf coasts. They began a northward expansion along the Atlantic Coast in the mid-1980s. Since then, they have been reported from many inland areas of the U.S. Prior to about the early 1980s this saltwater species was regarded as an escapee and was never expected to occur in Pennsylvania as a wild vagrant. Only one record has been accepted in the state.

Seasonal status and distribution

Summer and fall: A Brown Pelican at Lake Leboeuf in Erie Co. on 5 Nov 1981 was captured by S. Stull and later proved to be an escapee.[1] One was on a farm pond near Wind Gap in Northampton Co. on 23 Jul 1984 reported by W. Reid. It was believed to be a wild bird, but the record was not confirmed.[2] Brown Pelicans were advancing northward along the Atlantic Coast in record numbers in the 1980s.[3] By the late 1980s and early 1990s their northward movement had taken them to New Jersey in numbers, and they were beginning to appear in several interior states far from the coast. On 22 Jul 1991 one was observed flying along the western Lake Erie shore of Pennsylvania near the Ohio state line.[4] Unfortunately the only credible description of this bird came from Ashtubula, Ohio. Local residents along the lake reported to have seen at least one Brown Pelican during the summer for two or three years before this sighting. On 13 Jul 1992 an adult discovered by J.G. Stull, J. H. Stull, and L. McWilliams in Presque Isle Bay remained until about noon the next day. It was photographed by a boater for the first confirmed Pennsylvania record.[5] At about this time, other vagrant Brown Pelicans were appearing in other parts of the U.S., such as Pottawatomie Co., Kansas.[6] Since 1992 there has been no other report in Pennsylvania, and Brown Pelicans have retreated from the earlier northern advancements along the Atlantic coast.

[1] G.M. McWilliams, pers. obs.
[2] Paxton, R.O., W.J. Boyle Jr., and D.A. Cutler. 1986. Hudson-Delaware region. AB 38:1002.
[3] Roberson, D. 1982. The changing seasons. AB 36:950.
[4] Hall, G.A. 1991. Appalachian region. AB 45: 1114–1115.
[5] Hall, G.A. 1992. Appalachian region. AB 46: 1134.
[6] Grzybowski, J.A. 1992. Southern Great Plains region. AB 46:1151.

Family Phalacrocoracidae: Cormorants

Cormorants are large all-black to nearly all-black waterbirds. They have a long neck, stiff tail, and a long narrow body. Their short legs are attached far back on the body. Cormorants have a long slender bill with a strong hook on the end of the upper mandible. They differ from other birds in several ways: all four toes are connected by webs; adults have no external nostrils and breathe through the mouth; and they have a small naked throat pouch. Cormorants are usually seen perched on rocks, buoys, or posts or in dead trees over or along water. They stand very erect, often with their wings held partially open. In flight they are usually seen in flocks formed in lines or in V formation. They feed primarily on fish caught by diving from the surface, and they frequently swim with just their neck out of the water, especially when feeding. Cormorants are good indicators of water quality. They have increased dramatically in recent years, especially since the implementation of stricter pollution laws in the Great Lakes. Two species occur in Pennsylvania: the Great Cormorant and the Double-crested Cormorant. As elsewhere in North America, both species have been rapidly expanding in the state. The Double-crested Cormorant has recently been confirmed nesting.

Great Cormorant ***Phalacrocorax carbo***

General status: Great Cormorants are typically found in saltwater. They breed in northeastern North America from the north shore of the Gulf of St. Lawrence in Quebec south along the coast to Maine. They winter within their breeding range and regularly south to South Carolina. Recently they have entered Pennsylvania waters via the Delaware and Susquehanna rivers. They were first reported in the state in Bucks Co. on the Delaware River after 1942.[1] Since 1983, Great Cormorants have been seen annually but are rare to locally fairly common from fall through spring on the Delaware. They are often seen in company with their smaller relatives, Double-crested Cormorants. Great Cormorants are regular winter residents only on the lower Delaware River. Expanding northward, they are casual visitors on the lower Susquehanna River and accidental in the Piedmont away from the Delaware and Susquehanna rivers.

Habitat: Great Cormorants are most frequently perched over water on rocks, pilings, or buoys along large rivers, especially the lower Delaware, and rarely on lakes in the Coastal Plain and Piedmont.

Seasonal status and distribution

Spring: In the spring, birds begin to wander up the Delaware River. They have been seen flying past hawk lookouts as far north along the Delaware as Morgan Hill in Northhampton Co. Most have moved out of the state by late Apr, and some linger into late May. Two records of Great Cormorants have occurred far from the Delaware and Susquehanna rivers: an immature was at Carr's Recreational Park at Glen Morgan Lake in Berks Co. on 28 Apr 1996,[2] and one was at Green Lane Reservoir in Montgomery Co. from 3 to 24 May 1997.[3]

Summer: The only summer record was at Glen Morgan Lake in Berks Co. This individual was discovered on 28 Apr 1996 and remained until 19 Jul 1996.[4]

Fall: Great Cormorants were first confirmed in the state on 17 Dec 1983, when five were photographed off Andalusia in Bucks Co.[5] They begin wandering into Pennsylvania sometimes as early as late Aug. The greatest number of cormorants arrives in late Nov or early Dec as singles or small groups composed mostly of immatures. On the Delaware River they range as far north as Northampton Co. Three fall records are away from the Delaware or Susquehanna river. Three were found at Beltzville State Park on 20 Dec 1992;[6] one was at Nockamixon State Park in Bucks Co. on 18 Dec 1996;[7] and one was at Green Lane Reservoir in Montgomery Co. on 22 Oct 1997.[8] Great Cormorants have not been reported north of the Conejohela Flats on the Susquehanna River in Lancaster Co.

Winter: The only area where this species is known to winter regularly is the lower Delaware River from Chester Co. north to Trenton Falls in Bucks Co. The number of wintering birds has increased in recent years. A count of 39 birds was made on an early winter bird count at Marcus Hook in Delaware Co. on 11 Jan 1997.[9] An individual at Nockamixon State Park in Bucks Co. from 22 to 28 Jan 1989[10] is the only winter record to date away from the Delaware River.

[1] K. Kitson, pers. comm.
[2] Keller, R. 1996. Local notes: Berks County. PB 10:97.
[3] Freed, G.L. 1997. Local notes: Montgomery County. PB 11:98.
[4] Keller, R. 1996. Local notes: Berks County. PB 10:161.
[5] Paxton, R.O., W.J. Boyle Jr., and D.A. Cutler. 1984. Hudson-Delaware region. AB 38:181.
[6] Haas, F.C., and B.M. Haas, eds. 1993. Rare and unusual bird reports. PB 6:179.
[7] Haas, F.C., and B.M. Haas, eds. 1997. Birds of note—October through December 1996. PB 10:220.
[8] Haas, F.C., and B.M. Haas, eds. 1998. Birds of note—October through December 1997. PB 11:230.
[9] Haas, F.C., and B.M. Haas, eds. 1997. Birds of note—January through March 1997. PB 11:22.
[10] French, R. 1989. County Reports—January through March 1989: Bucks County. PB 3:138.

Double-crested Cormorant *Phalacrocorax auritus*

General status: Double-crested Cormorants have steadily increased in Pennsylvania and throughout the eastern Great Lakes. They breed locally from Alaska across southern Canada and the U.S. They winter primarily along the Pacific and Atlantic coasts and Gulf states. After pesticides, such as DDT, were banned, Double-crested Cormorant populations began recovering, especially in the early 1980s. By 1984, daily flocks of up to 200 could be seen migrating along the Delaware River.[1] By the mid-1990s, up to 1000 birds could be seen staging at PISP.[2] With the increase of birds at traditional sites, they began to expand and are now regular on lakes and rivers where they never occurred before. They are uncommon to fairly common and locally abundant regular migrants. Double-crested Cormorants are regular summer visitors or residents in the state. They recently nested on the Susquehanna River. In winter, they are local but regular along the lower Delaware River and casual to accidental elsewhere in the state.

Habitat: They are usually found on rivers and lakes and are often seen perched over water on buoys, rocks, stumps, snags, or trees. They also roost on exposed mudflats and sandbars.

Seasonal status and distribution

Spring: Double-crested Cormorants are usually more common in spring than in fall. Migration may begin as early as mid-Mar. Their typical migration period is from late Mar to about the third week of May. Birds can regularly be seen migrating, often in large flocks, along large rivers, especially the Delaware and Susquehanna and over Lake Erie. Numbers usually peak around mid- to late Apr, but sometimes sizable flocks remain into May.

Breeding: They have been reported in summer from all sections of the state except the Glaciated Northeast and the lower Southwest. Cormorants seen in summer are usually immatures, single, or in small flocks. Nonbreeding cormorants are most common along the lower Delaware and Susquehanna rivers. A flock of 35 on the lower Delaware River in mid-Jun 1979 was an unusually large concentration for this month.[3] Nesting was first confirmed in the state on 1 Jul 1996, when a single nest was found on Wade Island in the Susquehanna River near Harrisburg. On 12 Jul 1996 three young were seen in the nest about 35 feet above the ground on the eastern side of the island.[4] Two pairs were observed at the same site in 1997.

Fall: Fewer are seen in areas where they are found during spring migration, but they can be observed in flocks flying in formation along the Ridge and Valley. Their typical migration period is from mid- to late Aug until early Dec. Peak migration occurs during the first half of Oct, often accompanying a cold front. A few birds linger to late Dec or early Jan, depending on the availability of open water.

Winter: Double-crested Cormorants are usually very rare in the state after mid-Jan, especially on inland lakes and smaller rivers. They have successfully wintered on the lower Delaware and at Lake Erie during mild winters. Other winter sightings have been reported from Allegheny, Bucks, Crawford, Lancaster, Lycoming, and York cos.

History: During the nineteenth century, cormorants were regarded as rather rare. Warren (1890) did not mention their occurrence east of Erie. Stone (1894) regarded them as "occasional on the lower Delaware River." Since then they have been steadily increasing. Flocks of up to 100 birds were reported flying north near West Chester on 12 Mar 1936, and 25 passed over Hawk Mountain on 8 Oct 1937 (Poole, unpbl. ms.). Populations of cormorants and other fish-eating species declined rapidly during the 1950s when pesticides affected nesting success (Peterjohn 1989).

[1] Boyle, W.J., Jr., R.O. Paxton, and D.A. Cutler. 1984. Hudson-Delaware region. AB 38:890.

[2] Hall, G.A. 1992. Appalachian region. AB 46:421.

[3] Richards, K.C., R.O. Paxton, and D.A. Cutler. 1979. Hudson-Delaware region. AB 33:848.

[4] McConaughy, M. 1996. First Double-crested Cormorant nesting in Pennsylvania. PB 10:151.

Family Anhingidae: Darters

Darters are very similar to cormorants except that they have a slimmer neck, longer tail, and unhooked heronlike bill. They are often referred to as snake birds because of the snakelike appearance of their neck and their habit of swimming with just their head and neck out of the water. Darters feed by swimming under water and use their long sharply pointed bill for spearing fish. They often soar in circles high overhead with neck extended and tail fanned out. The Anhinga is the only species in this family that is found in North America. It has only recently been added to the official list of birds of Pennsylvania.

Anhinga *Anhinga anhinga*

General status: In the U.S., Anhingas breed from Texas east across the Gulf States to coastal North Carolina. They are rare postbreeding wanderers well north of the breeding range. Two records have been accepted in Pennsylvania.

Seasonal status and distribution

Spring: N. Pulcinella saw two males and a female Anhinga soaring high overhead at Swarthmore in Delaware Co. on 18 Apr 1993.[1] A. Mirabella saw a male Anhinga sitting on a beaver lodge in Montgomery Co. on 15 May 1996.[2]

Comments: Two recent reports are currently under review. On 6 May 1997 one was reported flying south in Upper Gwynedd in Montgomery Co.,[3] and three were observed soaring over Media in Delaware Co. on 25 May 1998.[4]

[1] Haas, F.C., and B.M. Haas, eds. 1993. Rare and unusual bird reports. PB 7:59.

[2] Freed, G.L. 1996. Local notes: Montgomery County. PB 10:103.

[3] Freed, G.L. 1997. Local notes: Montgomery County. PB 11:98.
[4] Pulcinella, N. 1998. Rare bird reports. PB 12:62.

Family Fregatidae: Frigatebirds

Found throughout the world over tropical seas, frigatebirds are powerful flyers capable of sustaining their position high in the air even in strong wind. They are mostly black and have extremely long narrow wings that in flight are usually sharply bent back at the wrist. The tail is very long and deeply forked. Frigatebirds have a short neck with a throat pouch that can be greatly distended and a long narrow bill with both mandibles sharply hooked at the tip. Frigatebirds are pirates of the sea; they frequently pursue gulls and terns, forcing them to regurgitate their food and then stealing it. The name Man-o'-war-bird was given to them by seafarers who named them after a swift, powerful warship. Most inland North American records are of birds displaced by hurricanes. The only accepted frigatebird record at this writing in Pennsylvania came from Hurricane Flossie in 1956. However, some rather convincing, though unsubstantiated, sightings of frigatebirds have been reported in the state by casual observers during the passage of recent hurricanes.

Magnificent Frigatebird *Fregata magnificens*

General status: This large high-soaring species breeds from the Florida Keys south and is frequently displaced by hurricanes. Magnificent Frigatebirds have been observed as far inland as the upper Mississippi River Valley after the passage of hurricanes. There have been several recent unconfirmed sightings of frigatebirds in Pennsylvania. The species is included on the list of birds of Pennsylvania on the basis of a historical record of a bird collected in 1956.

History: A Magnificent Frigatebird was shot near New Kensington in Westmoreland Co. after Hurricane Flossie on 3 Oct 1956 (Poole 1964). The specimen is at the CMNH.[1]

[1] K.C. Parkes, pers. comm.

Order Ciconiiformes
Family Ardeidae: Bitterns and Herons

Bitterns and herons are long-legged wading birds with a long daggerlike bill, long neck, short tail, and rather short rounded wings. The head and back of most species are decorated with long wispy plumes that are especially noticeable during the breeding season. All birds in this family fly with their neck drawn back into their shoulders. Bitterns inhabit marshes and are more secretive than herons. They are more frequently heard than seen and have suffered from the reduction of wetlands throughout the state. Herons are usually found roosting in trees or skulking in shallow water along edges of ponds, lakes, streams, and rivers, where they search for fish, frogs, and snakes. Eleven species of this family are represented in Pennsylvania. Nine of those breed within the state, and two, the Tricolored Heron and Little Blue Heron, are only visitors from areas south of Pennsylvania. Many species of herons nest in colonies, often in swamps or on river islands. They build loose shallow nests in trees, from only a few feet off the ground to over 100 feet up. A single tree may hold many nests. Other species in this family are solitary nesters, or they nest in small loose colonies using the ground, cattails, brush, or small trees over or along water. Bitterns and herons lay from three to seven white, bluish green, or buff eggs. Only Great Blue Herons and Black-crowned Night-Herons regularly winter in the state.

American Bittern *Botaurus lentiginosus*

General status: American Bitterns are secretive, more often heard giving their territorial pumping notes than seen.

When disturbed, they usually fly only a short distance before dropping out of sight in the vegetation. American Bitterns breed from southern Canada south to California, Texas, and Georgia. They winter along the Pacific and Atlantic coasts and the southern U.S. In Pennsylvania they are uncommon to rare regular migrants and are regular breeding residents only in Crawford Co. in scattered large wetlands. They are casual in winter. This species has suffered greatly from the loss of wetland habitat.

Habitat: American Bitterns prefer large cattail marshes interspersed with pockets of open water. They are occasionally found around small bodies of water, streams, and rivers bordered with cattails, tall marsh grasses, or sedges, especially during migration.

Seasonal status and distribution

Spring: Their typical migration period is from early Apr to mid-May. Migration usually peaks during the second and fourth weeks of Apr. An interesting phenomenon has been witnessed annually since 1990 at PISP, where birds are seen leaving the marshes at dusk and flying north. Usually from 10 to 15 birds daily can be seen leaving the marshes during the last half of Apr, but 25 birds were counted from one spot on the evening of 25 Apr 1996. They were observed flying out of the marshes and circling to gain altitude before flying north over Lake Erie.[1]

Breeding: American Bitterns are regular breeders only in the largest wetlands in Crawford Co., where they are uncommon. They are frequently seen or heard from various access points to Conneaut Marsh, Crawford Co. These bitterns are local, rare, and irregular in summer elsewhere; some are nonbreeders. Breeding Bird Atlas observations came from 25 counties, but since 1980 breeding has been confirmed in only eight counties: Crawford, Delaware, Erie, Lackawanna, Lawrence, Monroe, Potter, and Sullivan.[2] These large bitterns are secretive in tall aquatic vegetation, but the loud distinctive song may be heard at a distance of over a mile across a marsh or open water. Three to five olive-buff eggs are laid in a nest, a foot-high mound of reeds. A nest once was found in an upland field adjoining wetlands (Grube 1959). Egg dates range widely, 8 May to 3 Jul, with most in early May while migrants are still passing through the state. Only one brood per year is expected.[3] Most sites previously occupied (e.g., Tinicum, Delaware Co., and Long Pond, Monroe Co.) no longer support regular nesting, reflecting a population decline and retraction of range. Populations are declining across the northeastern U.S.[4]

Fall: It is very difficult to determine the fall migration of American Bitterns. Separating lingering birds found in traditional nesting sites from actual migrants is difficult. Where they are not known to nest, they have been recorded from Aug through early Jan. American Bitterns are rarely seen perched in the open or in trees. However, Schweinsburg (1988) reported flushing a bird from an evergreen along Bucknell University's golf course on 7 Dec 1973.

Winter: Winter records include one captured by a dog in Butler on 31 Jan 1966[5] and a sighting in Huntingdon Co. on 10 Feb 1967.[6] Wood (1983) mentioned one wintering in central Pennsylvania in 1976–1977.

History: Frey (1943) cited an early spring date on 9 Mar in Cumberland Co. Trimble (1940) thought that the flooding of Pymatuning Reservoir "greatly extended the area" for bitterns and stated that many nests were found. Before 1960 they were found in many marshes in Crawford Co. (W.C. Grimm 1952) and in larger marshes statewide (Poole

1964). American Bitterns have declined precipitously during the twentieth century. Many nesting locations since the early 1900s have been abandoned or are only irregularly occupied. Nesting was confirmed or suspected in at least 17 counties during the twentieth century (BBA; Poole, unpbl. ms.). Loss of wetlands reduced the nesting incidence, particularly in the Coastal Plain and the Piedmont.

Comments: American Bitterns were listed as Threatened in Pennsylvania from 1979 (Genoways and Brenner 1985) until 1997, when their status was changed to Endangered by the PGC.

[1] G.M. McWilliams, pers. obs.
[2] Leberman, R.C. 1992. American Bittern *(Botaurus lentiginosus)*. BBA (Brauning 1992): 46–47.
[3] Gibbs, J.P., F.A. Reid, and S.M. Melvin. 1992. American Bittern *(Botaurus lentiginosus)*. BNA (A. Poole and Gill 1992), no. 17.
[4] Gibbs, J.P., and S.M. Melvin. 1992. American Bittern. *In* Migratory nongame birds of management concern in the Northeast. U.S. Dept. of Interior, Fish and Wildlife Service. Newton Corner, Mass.
[5] Hall, G.A. 1966. Appalachian region. AFN 20: 422.
[6] Hall, G.A. 1967. Appalachian Region. AFN 21: 419.

Least Bittern ***Ixobrychus exilis***

General status: The smallest member in this family, Least Bitterns are quite secretive and predominantly nocturnal. Early in the breeding season they are vocal and visible, but as summer progresses they become increasingly more difficult to find. To escape danger, they often run through vegetation rather than fly to safety. Least Bitterns breed in southern California and the eastern half of the U.S. In Pennsylvania they are locally uncommon to rare in the Coastal Plain, Glaciated Northwest, and Lake Erie Shore. Like American Bitterns, they have suffered from loss of wetlands.

Least Bittern at Presque Isle State Park, Erie County, May 1990. (Photo: Ed Kwater)

Habitat: They prefer large emergent and brushy wetlands interspersed with pockets of open water. Least Bitterns can also be found around small bodies of water, streams, or rivers bordered with cattails, tall marsh grasses, or sedges. Unlike American Bitterns, which spend much of their time on the ground, Least Bitterns prefer to climb about emergent vegetation, tolerating much deeper water than the former. They have been seen foraging in buttonbush that surrounds some of the marshes on PISP.

Seasonal status and distribution

Spring: They arrive soon after the new emergent vegetation is high enough to conceal them, which may be as early as mid-Apr. They are more active and frequently are heard and seen from May through early Jun than at other times of year.

Breeding: Least Bitterns are locally uncommon breeders in the Tinicum area in Philadelphia Co., PISP in Erie Co., and larger emergent wetlands in the Glaciated Northwest. They are casual in suitable habitat elsewhere, except in the High Plateau, where they are accidental. Least Bitterns are declining in areas with the historically largest populations, notably Tinicum. In recent years, only a few pairs have been reported in Tinicum, where as many as 27 nests were reported in 1958 (J.C. Miller and Price 1959). Similar declines were suggested in the Pymatuning area, where habitat has not been significantly reduced. Egg dates range from 19 May to 22 Jul, with most found by 22 Jun (Harlow 1918). Nests, supporting four to five pale blue to greenish eggs, are constructed above the water level in dense stands of cattails, shrubs, or sedges (H.H. Harrison 1975). Under good conditions they may nest semicolonially.

Fall: The fall migration period is difficult to determine because these birds are rarely found after the breeding season. Most seen in fall probably have lingered on the breeding grounds. They are usually not found after late Sep or early Oct, but some have lingered to the first week of Dec.

History: Historically Least Bitterns nested in Berks, Bucks, Butler, Centre, Chester, Crawford, Delaware, Erie, Huntingdon, Mercer, and Philadelphia cos. Formerly these bitterns were considered by Harlow (1918) to be very common in the Delaware marshes as far north as Bucks Co. Severe loss of tidal marsh to development reduced populations of this and many other wetland species.

Comments: Least Bitterns were listed as Threatened in Pennsylvania from 1979 (Genoways and Brenner 1985) until 1997, when they were downgraded to Endangered by the OTC.

Great Blue Heron ***Ardea herodias***

General status: Great Blue Herons are the largest and most familiar herons in Pennsylvania. Although regular and fairly common to common statewide, they are uncommon to rare in remote wooded sections of the state. They breed across southern Canada and throughout the U.S. and winter over most of their breeding range. In Pennsylvania they breed locally over most of the state. Great Blue Herons become less common by Jun and then increase in number during postbreeding dispersal in mid- to late summer. They are regular winter residents over most of the state.

Habitat: Great Blue Herons use a wide range of aquatic habitats including marshes, lakes, ponds, and rivers in all seasons. Nests are sometimes placed far from water.

Seasonal status and distribution

Spring: Their typical migration period is from mid-Mar until mid- to late May.

Loose flocks containing 10 or more birds can sometimes be seen migrating high overhead statewide in Apr.

Breeding: Great Blue Herons regularly nest in colonies or singly in scattered locations across northern Pennsylvania and locally in the southern half of the state. Birds range many (more than 12) miles from known nests to forage at a variety of wet habitats, but many individuals are found far from where any nesting activity is known and are believed to be nonbreeders. They are fairly common, found singly or in small groups while foraging along any water body in all counties. Colonies may be found in forested hillsides far from water, particularly outside the glaciated regions. Most colonies larger than 40 nests occur in the glaciated northern corners of the state, often in a wetland or along a stream. The largest colonies have been near Greenville, Mercer Co., with average nest counts of 200 at Barrows and 190 at Brucker (1985–1995). The Brucker site was abandoned in 1997.[1] Approximately 120 nesting colonies were documented during the 1990s, but additional small colonies and single nests are believed to be overlooked and widespread, and not all colonies are active in a given year. Most colonies support fewer than 15 nests. Some colonies may be occupied for many decades, but others have been abandoned from disturbance or for unknown reasons. Nests are usually high in a tall tree or snag. Nesting begins in late Mar to mid-Apr, with most nests containing four eggs. By mid-Jul, postbreeding vagrants and fledglings bolster numbers. The BBS shows a population increase in Pennsylvania of 4.5% per year since 1966.

Fall: Fall migration is not well defined. Fewer birds are observed as fall progresses, and the number of herons continues to decrease as winter approaches.

Winter: Great Blue Herons are uncommon to rare over most of the state; abundance varies from year to year, depending upon availability of open water. They are absent from most mountains away from major rivers. They may become concentrated at limited food sources, such as fish hatcheries, especially during severe freezes.

History: Great Blue Herons historically nested across the state, but direct persecution and cutting of nest trees reduced their numbers. Warren (1890) mentioned the destruction of nesting colonies for the bird's feathers. Colonies became scarce and restricted to the northern counties by the 1910s (Harlow 1913). Great Blue Herons were protected in 1923 (Kosack 1995), but the nesting population continued to decline and colonies became restricted primarily to the northwestern counties (Poole 1964). The recovery since 1960 has been dramatic. Counts of 234 active nests in 1989 at the Brucker colony and 216 nests in 1990 at Barrows document the largest Great Blue Heron colonies in the state's history.[1]

Comments: Rather unusual was the occurrence of a white-morph Great Blue Heron at Pymatuning Reservoir in Crawford Co. This form, found chiefly in the Florida Keys, was formerly known as the Great White Heron. One was observed with a Great Egret, along Ford Island at Pymatuning on 29 Jul 1961. The bird was described as large and pure white with yellowish legs and bill. The bird was seen again by a different observer on 9 Aug 1961 in about the same area. Both observers were employees of the PGC. An adult female white morph was observed at Pymatuning Lake on 14 May 1938 for several days before it was collected and placed in the Ford Island museum at Pymatuning.[2] Another bird was also found in Oct 1938 at the same location. B. L. Oudette believed that there may have been a third bird as well (Todd 1940). Until recently, Great Blue Herons have been considered a problem at fish hatcheries, where they readily take fish from holding units. The practice of shooting birds at hatcheries has been unsuccessful at solving the problem. Construction of covers over the rearing ponds of many aquaculture facilities has been successful in saving not only production fish but also the lives of the herons and their allies.[3]

[1] Brauning, D.W. 1996. Colonial birds. Annual report. Pennsylvania Game Commission, Harrisburg.
[2] Leberman, R.C. 1961. Field notes. Sandpiper 4:15.
[3] G.M. McWilliams, pers. obs.

Great Egret ***Ardea alba***

General status: In North America Great Egrets breed in the West from Oregon south through California. In the East they breed locally in the northern part of their range from southern Ontario and the New England states south to Florida. They winter from coastal Oregon and Maryland south. Great Egrets are the most common and largest of the four white herons that regularly occur in Pennsylvania. They are uncommon to rare regular visitors throughout the state where suitable habitat exists. Great Egrets may be fairly common locally during migration and postbreeding dispersal. This species breeds locally along the lower Delaware and Susquehanna rivers. Great Egrets are accidental in winter in southeastern Pennsylvania.

Habitat: They inhabit the edges of marshes, ponds, lakes, and slow-moving streams and rivers.

Seasonal status and distribution

Spring: They appear in the Coastal Plain north to the Ridge and Valley as early as the first week of Mar. They are not expected in the High Plateau and the Lake Erie Shore until the first week of Apr. Great Egrets are irregular and very rare any other time of year in the High Plateau and Lake Erie Shore. The peak of the northbound migrants occurs from about mid-Apr to the first week of May. A notable concentration of over 100 birds was at Enola in Dauphin Co. on 14 Apr 1992.[1] Migration is usually complete by the last week of May, with stragglers into the first week of Jun.

Breeding: Great Egrets are regular and fairly common in early summer only at Tinicum and along the lower Susquehanna River. A colony of up to 18 nests occupied Mud Island (near Fort Mifflin) in Philadelphia Co. in 1989 with Black-crowned Night-Herons (PGC file data). That colony was abandoned in 1991 after dredge spoil was deposited around the nest trees. Birds were observed building nests at Tinicum in 1997.[2] Nests are active from late Apr through Jul. The only known breeding site in the mid-1990s was at Wade Island, in the Susquehanna River at Harrisburg. It supported an average of 87 active nests (1986–1997) and a high of 161 nests in 1992,[3] with generally increasing numbers. At Rookery Island, Lancaster Co., up to three nests were observed in 1987 and 1988 in a large Cattle Egret colony that was abandoned in 1988.

Fall: Postbreeding dispersal is evident by the second week of Jul statewide. Up to 150–200 birds regularly congregate at roosts during the evening in southeastern locations such as Darby Creek and Tinicum.[4] At the Conejohela Flats or below Holtwood Dam, 15 to 40 birds is a typical midsummer count (Morrin et al. 1991). An unusually high fall count of

30 birds, for the Glaciated Northwest, was found on 3 Oct 1965 at Pymatuning.[5] Migrants occasionally linger as late as the last week of Dec or first week of Jan.

Winter: Great Egrets stay in marshes from Tinicum south into Delaware Co. along the Delaware River, but they leave once the water freezes. Five birds remained at Tinicum to the end of Jan 1980.[6] Birds have attempted to spend the winter in Lehigh Co., where a bird remained from Dec 1960 to Feb 1961 (Morris et al. 1984), and a frostbitten immature was picked up on 14 Jan 1982 in Lehigh Co. and later released.[7] In Lancaster Co., Great Egrets appeared for one day, 29 Jan 1995.[8]

History: Many early bird lists considered Great Egrets fall vagrants. In a likely reference to this species, Baird (1845) listed the "White Crane" as a breeder in Cumberland Co. A later list simply stated that the species was occasionally seen on the Susquehanna River in Lancaster Co. (Libhart 1869). Numbers of vagrants increased during the middle decades of the twentieth century after populations were decimated by the millinery trade in the late 1800s and early 1900s. An invasion of Great Egrets was noted on 5 Aug 1933, when Todd (1940) saw 25 birds at one time at Hartstown. The first documented nest in Pennsylvania was in Delaware Co. at Tinicum in 1957 (J.C. Miller and Price 1959).

Comments: Great Egrets were listed as State Threatened in 1990, and were downgraded to Endangered by the PGC in 1999.

[1] Rannels, S. 1992. County reports—April through June 1992: Dauphin County. PB 6:77.
[2] T. Floyd, pers. comm.
[3] Brauning, D.W. 1996. Colonial birds. Annual report. Pennsylvania Game Commission, Harrisburg.
[4] N. Pulcinella, pers. comm.
[5] Hall, G.A. 1966. Appalachian region. AFN 20:42.
[6] Richards, K.C., R.O. Paxton, and D.A. Cutler. 1980. Hudson-Delaware region. AB 34:257.
[7] Paxton, R.O., W.J. Boyle Jr., and D.A. Cutler. 1982. Hudson-Delaware region. AB 36:277.
[8] Haas, F.C., and B.M. Haas, eds. 1995. Seasonal occurrence tables—January through March 1995. PB 9:42.

Snowy Egret *Egretta thula*

General status: Small, delicate white herons, Snowy Egrets are easily separated from the other white herons found in the state by their very erratic feeding behavior. They move briskly through shallow water, constantly picking at the surface for food that they have stirred up from the bottom. They breed in the West in California and locally in the Rocky Mountain states. In the eastern U.S. Snowy Egrets breed primarily in the Gulf states and along the Atlantic coast north locally to Maine. They winter in the southern part of their breeding range. In Pennsylvania, they are uncommon to rare regular spring visitors or residents in the Coastal Plain and Piedmont. They become fairly common locally in these regions after postbreeding dispersal. Snowy Egrets have bred in the state on rare occasions. In the Ridge and Valley, they are irregular visitors. West of the Allegheny Front, Snowy Egrets are casual to accidental and have been recorded in Bradford, Butler, Cambria, Crawford, Erie, Fayette, Indiana, McKean, Somerset, Tioga, and Westmoreland cos.

Habitat: Snowy Egrets inhabit the edges of marshes, ponds, lakes, and slow-moving streams and rivers.

Seasonal status and distribution

Spring: The usual migration period is from the first week of Apr until the first week of Jun. Most sightings of Snowy Egrets occur from the third week of Apr to the second week of May. Small groups or individuals may wander as far north as Bradford Co. along the Susquehanna River or west of the Ridge and Valley. Birds west of the Ridge and Valley may be vagrants from the Ohio Valley rather than from coastal populations.

Breeding: Snowy Egrets are casual nonbreeding visitors at Tinicum and in the Piedmont along the Susquehanna River. They are accidental elsewhere until postbreeding dispersal in Jul. They were confirmed nesting in Pennsylvania on only three occasions: on Rookery Island, Lancaster Co., in 1975 and 1987, and at Tinicum in 1978.[1] They probably nested between 1975 and 1987 at Rookery Island, when nest searches were not conducted, and possibly in 1985 at Tinicum. Summer sightings along the lower Delaware and Susquehanna rivers suggest the potential for this species to reestablish a nesting outpost in an established heron colony. A pair was present on an island in the Susquehanna River at Wyoming in Luzerne Co. to at least 27 Jun 1964 and may have nested.[2]

Fall: Postbreeding Snowy Egrets usually appear around the second week of Jul. Sites used in this season are similar to those in spring. High concentrations may be found at this season in the marshes of Tinicum and Darby Creek[3] and along the Susquehanna River. A high count of 59 birds was recorded on 6 Aug 1980 on the Conejohela Flats in Lancaster Co. (Morrin et al. 1991). There were 20 birds on Brunner's Island in York Co. on 10 Aug 1988;[4] 19 were at West Fairview in Cumberland Co. on 1 Aug 1991;[5] and 16 were found between Berwick and Pittston, Luzerne Co., in late Aug 1983.[6] Most birds retreat by mid-Sep; stragglers remain to the fourth week of Oct.

History: Some of the earliest bird lists from southeastern Pennsylvania include this species, listing it as frequent in autumn in Delaware Co. (Cassin 1862) and rare in Lancaster Co. (Libhart 1869). Warren (1890) considered Snowy Egrets to be less common than Great Egrets, as they are in recent times.

[1] Schutsky, R.M. 1992. Snowy Egret *(Egretta thula)*. BBA (Brauning 1992):54–55.

[2] W. Reid, pers. comm.

[3] N. Pulcinella, pers. comm.

[4] Spiese, A. 1988. County reports—July through September 1988: York County. PB 2:117.

[5] Hoffman, D. 1991. County reports—July through September 1991: Cumberland County. PB 5:128.

[6] Paxton, R.O., W.J. Boyle Jr., and D.A. Cutler. 1984. Hudson-Delaware region. AB 38:182.

Little Blue Heron *Egretta caerulea*

General status: Little Blue Herons breed primarily along the Gulf states and north along the Atlantic coast to Maine. They winter in the southern portion of their breeding range. Little Blue Herons disperse widely after the nesting season, across the U.S. and southeastern Canada from their southeastern U.S. breeding range. Both the white immatures and the blue-gray adults wander into Pennsylvania. Little Blue Herons are rare regular visitors to the Coastal Plain and the Piedmont, especially along the Delaware and Susquehanna rivers. They are casual visitors in the Ridge and Valley and west of the High Plateau. Little Blue Herons are accidental elsewhere in the state. Most sightings of these herons are of single birds.

Habitat: Little Blue Herons are found along the edges of marshes, ponds, lakes, and slow-moving rivers.

Seasonal status and distribution

Spring: East of the Allegheny Front fewer birds are seen in spring than in fall. A bird found on 2 Apr 1981 at Wyalusing in Bradford Co. is the only spring record from the Glaciated Northeast.[1] The single record in the High Plateau was near Powdermill in Westmoreland Co. on 21 May 1983.[2] Most records west of the Allegheny Front are at this season, and the majority reported are adults. Little Blue Herons have been recorded as early as the first week of Apr. Most birds are not seen until the third week of Apr through the third week of May. Stragglers may remain into Jun.

Summer: Immature or nonbreeding adults occasionally linger through Jun. Some birds appear in early Jul. A nestling was reported, but not confirmed, in a nest on Rookery Island on the lower Susquehanna River during the summer of 1975; nesting again was suspected, but not confirmed, at this location in 1986 (Morrin et al. 1991).

Fall: Postbreeding dispersal begins when southern birds wander up the Delaware and Susquehanna rivers and onto inland lakes and ponds in late Jul. Most records east of the Allegheny Front are at this season. Nearly all sightings are of immature birds, and the majority are found in the Coastal Plain and Piedmont. Little Blue Herons retreat by the second week of Sep. Stragglers may remain to the third week of Nov.

History: Little Blue Herons were considered rare or casual by Warren (1890) and Stone (1894), but by 1920 they were more frequently seen, often in large concentrations. Beck (1924) considered them the most common of the southern herons along the lower Susquehanna River. Beck wrote of seeing upwards of 150 Little Blue Herons feeding at Washington Boro in Aug 1920. They have always been rare west of the Appalachian Mountains, but there was a notable movement of birds from Ohio into western Pennsylvania when at least 19 birds were reported during the period between 11 Jul and 16 Aug 1932 (Todd 1940). G. M. Sutton pointed out that the supposed records of Snowy Egret listed by Warren were probably based on young individuals of Little Blue Herons (Todd 1940).

Comments: It is likely that many first-year (all-white) birds were misidentified as Snowy Egrets, especially historically. All Little Blue Herons west of the Allegheny Front may be strays from the Ohio Valley rather than from coastal populations.

[1] Paxton, R.O., W.J. Boyle Jr., and D.A. Cutler. 1981. Hudson-Delaware region. AB 35:805.

[2] Hall, G.A. 1983. Appalachian region. AB 37:869.

Tricolored Heron *Egretta tricolor*

General status: Tricolored Herons breed primarily along the Gulf coast and Atlantic coast north to southern Maine. They winter in the southern part of their breeding range. Tricolored Herons are the rarest of the regularly occurring herons found in Pennsylvania. They are rare irregular visitors to the Coastal Plain and Piedmont, mostly during the postbreeding season. Tricolored Herons are accidental elsewhere in the state, except at PISP and the Ridge and Valley, where they are casual in spring. In the Glaciated Northeast there is no record of this species. Most sightings of Tricolored Herons are of single birds, but as many as eight have been reported.[1]

Habitat: Tricolored Herons are found along the edges of marshes, ponds, lakes, and slow-moving streams and rivers.

Seasonal status and distribution

Spring: In the Ridge and Valley they have been recorded three times: one bird over Montandon Marsh in Northumberland Co. on 16 May 1976 (Schweinsberg 1988), one at Scranton in Lackawanna Co. on 6 Jun 1982,[2] and one in Montour Co. on 16 Apr 1993.[3] Nearly all records west of the Allegheny Front are from this season. Records outside the Lake Erie Shore include: one bird at Conneaut Marsh near Geneva in Craw-

Tricolored Heron at Clintonville, Venango County, August 1993. (Photo: Walter Shaffer)

ford Co. on 25 and 30 May 1959 (Poole, unpbl. ms.), one at Deer Lake in Westmoreland Co. on 18 May 1977,[4] one in Greene Co. 23 May 1985,[5] one at Beaver Run Reservoir in Westmoreland Co. on 28 Apr 1994,[6] and one near New Castle in Lawrence Co. on 11 May 1996.[7] Statewide, birds have been reported from early Apr to late May, and stragglers have occurred to early Jun.

Summer: At least two records of a summering individual have been reported: one at Tinicum in 1964[8] and one at Camp Hill in Cumberland Co. in 1979.[9]

Fall: Most records of Tricolored Herons east of the Allegheny Front are in fall. Three fall records are west of the Allegheny Front: one at Prince Gallitzin State Park in Cambria Co. on 6 Sep 1987,[10] one at YCSP in Indiana Co. on 2 Oct 1987,[11] and an immature near Clintonville in Venango Co. from 8 to 12 Aug 1993.[12] Postbreeding dispersal begins with birds arriving in mid- to late Jul. Tricolored Herons leave the state earlier than other species of southern herons. By late Sep, Tricolored Herons have moved south of the state. One at Tinicum on 13 Oct 1973 was a rather late date for this species.[13]

History: Poole (unpbl. ms.) reported that this species has always been considered the rarest of the southern herons that occur in eastern Pennsylvania. He reported that they had been recorded only about 11 times in the 70 years prior to 1960.

[1] Boyle, W.J., Jr., R.O. Paxton, and D.A. Cutler. 1980. Hudson-Delaware region. AB 34:879.
[2] Paxton, R.O., W.J. Boyle Jr., and D.A. Cutler. 1982. Hudson-Delaware region. AB 36:959.
[3] Haas, F.C., and B.M. Haas, eds. 1993. Rare and unusual bird reports. PB 7:59.
[4] F.C. Haas and B.M. Haas, pers. comm.
[5] Hall, G.A. 1985. Appalachian region. AB 39:299.
[6] Leberman, R.C. 1994. Notes from the field: Westmoreland County. PB 8:104.
[7] Haas, F.C., and B.M. Haas, eds. 1996. Birds of note—April through June 1996. PB 10:90.
[8] Scott, F.R., and D.A. Cutler. 1964. Middle Atlantic Coast region. AFN 18:501.
[9] Richards, K.C., R.O. Paxton, and D.A. Cutler. 1979. Hudson-Delaware region. AB 33:848.
[10] Hall, G.A. 1988. Appalachian Region. AB 42:74.
[11] M. Higbee and R. Higbee, pers. comm.
[12] Haas, F.C., and B.M. Haas, eds. 1993. Rare and unusual bird reports. PB 7:103.
[13] Scott, F.R., and D.A. Cutler. 1974. Middle Atlantic Coast region. AB 28:33.

[Reddish Egret] ***Egretta rufescens***

General status: Reddish Egrets breed along the Gulf coast of Texas, Louisiana, and Alabama, and at the southern tip of Florida. They winter in most of their breeding range. Reddish Egrets rarely wander inland away from saltwater. In the east, away from the coast, they have been recorded in northwestern South Carolina, Kentucky, New York, and Massachusetts. In Pennsylvania this species is listed as hypothetical.

Comments: On 9 May 1953 a dark-phase bird was seen at PISP in Erie Co. and remained at this site until the following day.[1] It was seen by many observers, including several familiar with the species.[2] However, no photograph or written description has been found to document the sighting.

[1] Mehner, J.F. 1954. Reddish Egret and White Pelicans in northwestern Pennsylvania. WB 66: 70–71.

[2] Sundell, R. 1954. Additions to the ornithology of western Pennsylvania since 1940. M.S. thesis; Allegheny College.

Cattle Egret ***Bubulcus ibis***

General status: Over the last two centuries Cattle Egrets have spread throughout the world from their original home in Africa. They arrived in the U.S. in the 1940s and now breed across the southern and eastern half of the country. In winter they remain throughout their southern breeding range. Cattle Egrets made their first appearance in Pennsylvania on 13 May 1956, when a single bird appeared at Tinicum (Morrin et al. 1991). Since that time, Cattle Egrets infiltrated the southeastern portion of the state and have been observed in all geographic regions. By 1960 they had reached the Lake Erie Shore (Stull et al. 1985). They quickly became established as a breeding species in the Piedmont by the mid-1970s, but the population suddenly began to decline in the early 1980s and they have not nested here since 1987. Since the early 1990s Cattle Egrets have become uncommon to rare regular visitors and irregular summer residents in the Coastal Plain and Piedmont. They are irregular to casual visitors west of the Allegheny Front, where they have been recorded from the counties of Allegheny, Bradford, Clarion, Crawford, Erie, Fayette, Greene, Indiana, Luzerne, Potter, Somerset, Tioga, Warren, and Westmoreland.

Habitat: They are found along the edges of ponds and lakes. Cattle Egrets are also found in pastures and wet fields, where they feed on terrestrial organisms that are stirred by grazing animals.

Seasonal status and distribution

Spring: Since the abandonment of the colony at Washington Boro, Cattle Egrets have become irregular in spring in southeastern Pennsylvania. In the Coastal Plain and Piedmont, Cattle Egrets have arrived as early as early Mar, but they are usually not expected until early or mid-Apr. In the remainder of the state, most are not recorded before late Apr. The greatest numbers of birds found in spring have usually departed by late May, but occasionally nonbreeding birds will remain through the summer.

Breeding: Cattle Egrets were first found nesting in the state in 1975, when 772 nests were found on an island in the Susquehanna River near Washington Boro. A peak of 7580 birds was counted in Aug 1981 at this island, known as Rookery Island. Cattle Egret numbers rapidly declined thereafter until 1988, when the rookery was abandoned for unknown reasons (Morrin et al. 1991).

Fall: As do other southern herons, Cattle Egrets begin dispersing from the breeding grounds in mid-Jul, with most postbreeding birds wandering north along the Delaware and Susquehanna drainage systems. Most birds have departed from the state by mid-Sep, but a few may linger well into Nov. A rather late individual was found on 11 Dec 1979 at State College in Centre Co.[1]

[1] Hall, G.A. 1980. Appalachian region. AB 34:272.

Green Heron ***Butorides virescens***

General status: These small herons are rather secretive and are usually found singly or in pairs. They are usually observed hunched quietly along the edge of a wooded pond or perched partially concealed on a low branch above the water. Green Herons breed across most of the U.S. and they winter from coastal South Carolina south through Florida and west through the Gulf states to southern California. They are fairly common regular migrants in Pennsylvania, except in the High Plateau and in mountainous sections, where they are uncommon and local. They are widespread breeders in the state.

Habitat: Green Herons prefer swamps, ponds, lakes, streams, and rivers bordered by woods.

Seasonal status and distribution

Spring: Birds arrive in early Apr in the Piedmont and Coastal Plain, but they may appear as early as the first week of Mar. They arrive later in the remainder of the state, usually not until mid- to late Apr. Green Herons usually are found only singly or in small groups, even when migration is well under way.

Breeding: Green Herons are uncommon to fairly common statewide in association with a wide variety of wetlands. Nesting has been confirmed in all counties, but least frequently in the extensively forested High Plateau. Nests are placed in dense shrubs or low in trees, often near water or some distance from water. Green Herons are adaptable and sometimes nest near people. The flimsy nests may be solitary or in small colonies but are not often with other species in Pennsylvania. Populations appear stable, with no significant changes in numbers on BBS routes. Nests with eggs have been found from 13 May to 13 Jun.

Fall: Fall migration is not well defined, but an increase in the number of birds has been noted in early Sep in the Coastal Plain.[1] A noticeable decline in sightings begins in mid- to late Sep, and by late Oct most birds have departed from the state. A few stragglers occasionally remain well into Dec. Extreme dates are 27 Dec 1965 in West Chester[2] and 28 Dec 1987 at MCWMA (Morrin et al. 1991). An unusually late bird found on 2 Jan 1983 along Penns Creek near New Berlin remained for six days (Schweinsberg 1988).

Winter: The only true winter record was of a bird at Glade Bridge on the Allegheny River at Warren in Warren Co. on 24 Jan 1993.[3]

History: Poole's (unpbl. ms.) comment that it is "undoubtedly our commonest and most widely distributed heron" reflects the general distribution enjoyed by this species. Warren (1890) mentioned a nesting colony of 25 to 30 birds and indicated that Green Herons will nest in company with Great Blue Herons and Black-crowned Night-Herons.

[1] N. Pulcinella, pers. comm.
[2] Scott, F.R., and D.A. Cutler. 1966. Middle Atlantic Coast region. AFN 20:407.
[3] T. Grisez, pers. comm.

Black-crowned Night-Heron ***Nycticorax nycticorax***

General status: Nocturnal in habits, Black-crowned Night-Herons are usually observed during the day roosting in trees or in shrubs near water. They can also be seen at dusk or dawn, flying to or from feeding sites. Black-crowned Night-Herons breed locally across southern Canada and the U.S. and winter primarily across their southern breeding range and north along the east and west coasts. Migrants are uncommon to rare at most sites but may be fairly common locally. In the Glaciated Northeast and the High Plateau, these birds are rarely reported at any season. They nest locally in fewer than eight colonies in the state. The stronghold of nesting colonies is the lower Susquehanna Valley, but the number of active nests has been steadily declining in the state. Wintering birds regularly occur only along the lower Delaware River.

Habitat: They are found near water in cattails or brush, particularly around marshes, lakes, and ponds but also along rivers. They generally roost during the day in trees or shrubs.

Seasonal status and distribution

Spring: Singles or small groups first appear in early to mid-Mar in the southeast and by early to mid-Apr in the northwestern portion of the state. Most spring records come from the Delaware and Susquehanna rivers and along Lake Erie. Migrant groups, usually numbering fewer than 10 birds, may consist of both adults and subadults. Migration usually peaks around mid- to late Apr and has ended by the second or third week of May. Stragglers remain into the first week of Jun.

Breeding: Black-crowned Night-Herons are irregular and uncommon away from breeding colonies, which are located at Tinicum in Philadelphia, in the Piedmont, along the North Branch Susquehanna River in Luzerne and Lackawanna cos., and at Pymatuning Swamp, Crawford Co. About 14 colonies have been active since the 1980s statewide, but not more than eight colonies were active in any given year. The two largest colonies were on islands in the southern Susquehanna River: Rookery Island, Lancaster Co., and Wade Island, Dauphin Co. Rookery Island, the state's largest colony for the species, supported 456 nests in 1985 but was abandoned during the 1988 nesting season (Morrin et al. 1991). Wade Island supported an annual average of 225 Black-crowned Night-Herons from 1985 to 1994, with the highest count in 1990 (344 nests) and a low of 99 nests in 1997.[1] Colonies of eight to 100 nests have been found in woodlots, pine plantations, and even suburban yards in the Piedmont, often some distance from water. The average total state nesting population, 1990–1997, was about 475 nests per year, but numbers are declining. The flimsy nests are close together, most often in tops of small trees, 20 to 40 feet above ground. Eggs are pale blue to greenish and unmarked. Clutches are completed from the last week of Apr through May, with young in the nest to mid-Jul.

Fall: Away from breeding colonies and a few other concentration points in the Piedmont, Black-crowned Night-Herons are much less common at this season than in spring. The beginning of fall migration is not well defined. Individuals appear away from breeding sites from the second week of Aug to late Oct, with stragglers to late Dec or early Jan.

Winter: The only area in the state where Black-crowned Night-Herons regularly winter is the Coastal Plain, at Tinicum, although they are uncommon there. They have been recorded in winter in the Lehigh Valley at Dorney Pond and Bethlehem in Northampton Co. (Morris et al. 1984), in Lancaster Co. (Morrin et al. 1991), and along the Ohio River in Allegheny Co.[2] They have also been recorded in winter in Indiana Co.[3] and at PISP in Erie Co. (Stull et al. 1985).

History: Black-crowned Night-Herons were considered by Warren (1890) to be the next most common herons after Green Herons, although his assessment may reflect a southeastern bias that overlooked the statewide prominence of Great Blue Herons. Colonies were always concentrated in the Piedmont, but development pressures and disturbance eliminated many historical sites. A notable exception was the colony established in the Pymatuning Refuge, which grew from 30 nests in 1934 to about 90 in 1938 (Trimble 1940). The colony at Rookery Island was active from 1950 to 1988 (Morrin et al. 1991). The reasons for its abandonment were not determined.

Comments: Birds congregate at unprotected fish-farming facilities from spring through fall. More than 100 were reported during summer at Huntsdale Fish Hatchery, although few remained after netting was installed in 1995. As many as 80 drowned at a private hatchery in Lebanon Co. in 1994, and up to 10 individuals were shot in 1995 under a depredation permit.[1] The OTC listed Black-crowned Night-Herons as Candidate–At Risk in 1997.

[1] Brauning, D.W. 1995. Colonial birds. Annual report. Pennsylvania Game Commission, Harrisburg.

[2] Hall, G.A. 1977. Appalachian region. AB 31:331.

[3] M. Higbee and R. Higbee, pers. comm.

Yellow-crowned Night-Heron *Nyctanassa violacea*

General status: Yellow-crowned Night-Herons breed locally across the eastern U.S. (except inland Maine) south through the Appalachian Mountains. They winter along the Atlantic coast from South Carolina south and west along the Gulf coast. Yellow-crowned Night-Herons are regular visitors or residents in the Piedmont, where they breed locally. In the Coastal Plain, they are rare irregular visitors. They are casual to accidental elsewhere in the state. Most records away from breeding sites are of individuals or pairs. Birds have also been reported from Armstrong, Butler, Centre, Crawford, Erie, Greene, Huntingdon, Luzerne, Snyder, Union, Warren, Washington, and Westmoreland cos.

Habitat: They are found along lakes and ponds, in marshes, and along streams and rivers. In the lower Susquehanna Valley they prefer rivers bordered by sycamores (Morrin et al. 1991). Several sightings at the Lake Erie Shore have been of birds roosting in conifers.[1]

Seasonal status and distribution

Spring: Birds usually appear in early to mid-Apr in the southeast. Most records west of the Allegheny Front are seen during this season from mid- to late Apr into late May.

Breeding: Nesting Yellow-crowned Night-Herons are limited to the Piedmont where they are regular. Most nonbreeding records are along tributaries of the Susquehanna River in Cumberland and Lancaster cos. and some parts of York Co. They are accidental in Chester and Montgomery cos. and some parts of York Co. Single nests or loose colonies are regular along the Conodoquinet Creek at West Fairview near its confluence with the Susquehanna River and in Camp Hill, Cumberland Co., where nests have been recorded since the early 1950s.[2] Nests are also found along the Conestoga and Little Conestoga creeks near Brownstown and Lancaster City, Lancaster Co., where they are often found within suburban or commercial developments. Surveys in the 1990s counted not more than 8–12 nests in any year.[3] Ten nests were counted in 1987 on a small island in the Susquehanna River near the Governor's Mansion,[4] but not after 1990. Nesting was documented near Rocky Hill, Chester Co., in 1973.[5] The loose stick nests containing three to four eggs are established by mid-May in a tree, usually 30–40 feet high over water. Some young may remain in the nest until late Jul.

Fall: Yellow-crowned Night-Herons depart sooner than Black-crowned Night-Herons, usually in late Aug or early Sep. By late Sep most have left the state, but they have been recorded to mid-Oct. A very late bird was seen on a CBC on 21 Dec 1996 in Harrisburg.[6]

History: Turnbull (1869) listed Yellow-crowned Night-Herons as stragglers in the Philadelphia area. Warren (1890) stated that no Pennsylvania records had been reported in 20 years. In western Pennsylvania this species was recorded only once before 1940; an adult was taken along Ten Mile Creek near Waynesburg in Greene Co. (Todd 1940). Several historical records suggested possible breeding (e.g., Burns 1919). Yellow-crowned Night-Herons were persecuted along with other wading birds at the beginning of the twentieth century and were considered accidental in the state before the 1940s, when breeding was first confirmed by C. Platt near Ambler, Montgomery Co. (Poole, unpbl. ms.). Nesting began along various creeks in Lancaster Co. in the 1950s (Morrin et al. 1991). Colonization in Pennsylvania corresponded with a significant expansion of the species northward after 1925. The greatest expansion took place from 1945 to 1955, when new state breeding records ranged from Ohio to Massachusetts.[7] This expansion may simply have been a recolonization of the species' historical, unrecorded range.

Comments: Immature Black-crowned Night-Herons look very similar to immature Yellow-crowned Night-Herons and may occasionally be misidentified, especially in late summer and early fall, when young birds outnumber adults. Yellow-crowned Night-Herons feed predominantly on crawfish. They were listed as State Threatened in 1990 because of their limited population and restricted range. The PGC downgraded their status to Endangered in 1999 because of their small population.

[1] G.M. McWilliams, pers. obs.
[2] Potter, J. 1951. Middle Atlantic Coast region. AFN 5:284.
[3] Brauning, D.W. 1996. Colonial birds. Annual report. Pennsylvania Game Commission, Harrisburg.
[4] Schutsky, R.M. 1992. Yellow-crowned Night-Heron *(Nyctanassa violacea)*. BBA (Brauning 1992):62–63.
[5] Scott, F.R., and D.A. Cutler. 1973. Middle Atlantic Coast region. AB 27:755.
[6] Haas, F.C., and B.M. Haas, eds. 1997. Birds of note—October through December 1996. PB 10:220.
[7] Watts, B.D. 1995. Yellow-crowned Night-Heron *(Nyctanassa violacea)*. BNA (A. Poole and Gill 1992), no. 161.

Family Threskiornithidae: Ibises and Spoonbills

Birds in this family are primarily found in the tropics or subtropics. In North America, most species are found along the southern coasts. They closely resemble herons and have a long, narrow, decurved, or wide spatulate bill. Their legs and neck are long and extended in flight. Three species have been recorded in Pennsylvania. Of the three, Glossy Ibis is the most frequently reported and formerly bred in the state.

White Ibis *Eudocimus albus*

General status: These unmistakable white waders of saltwater marshes are resident along the Atlantic coast from North Carolina south to Florida and along the Gulf coast. They wander up the Delaware and Susquehanna valleys during postbreeding dispersal from populations south of Pennsylvania. They are rare regular visitors in the Coastal Plain and Piedmont. Nearly all birds reported are single immatures.

White Ibises are accidental elsewhere in the state. They have been recorded from only six sites north and west of the Piedmont. No records exist from either the Glaciated Northeast or the Lake Erie Shore.

Habitat: White Ibises are found along large rivers, ponds, and marshes.

Seasonal status and distribution

Spring: White Ibises have been recorded three times in spring: an adult was found at Minsi Lake after an early warm front in Northampton Co. on 30 Mar 1986,[1] and one was at Cook Forest State Park in Forest Co. on 26 May 1997;[2] one adult was seen along the Grove City–Leesburg Road in Mercer Co. on 5 May 1991.[3]

Summer: A bird flying over downtown Pittsburgh on 26 Jun 1972 was believed to have been carried there by Hurricane Agnes.[4] One was photographed at Wynnewood in Montgomery Co. on 30 Jun 1977.[5]

Fall: White Ibises begin wandering northward into southeastern Pennsylvania in late Jul or early Aug. Until the rookery at Washington Boro in Lancaster Co. was abandoned, this was the most reliable site in the state to find this species. During one invasion year, a total of 17 White Ibises was counted on 11 Aug 1977 on the Conejohela Flats in Lancaster Co. (Morrin et al. 1991). Except during invasion years, only one or two birds are reported annually. By early Sep they have usually retreated from the state. Late individuals have remained to the last week of Oct. Other records away from the Coastal Plain and Piedmont have come from the counties of Allegheny, Bedford, Berks, Clinton, Columbia, Luzerne, Northumberland, and Schuylkill.

History: White Ibises were listed as rare stragglers on the Delaware River near the middle of the nineteenth century Cassin (1862). There appear to have been no other records of this species in Pennsylvania until 1953, when an immature bird was found at the Springton Reservoir near Media in Delaware Co. on 2 Aug.[6]

[1] Boyle, W.J., Jr., R.O. Paxton, and D.A. Cutler. 1986. Hudson-Delaware region. AB 40:449.

[2] Haas, F.C., and B.M. Haas, eds. 1997. Birds of note—April through June 1997. PB 11:89.

[3] McKay, M. 1991. County reports—April through June 1991: Mercer County. PB 5:94.

[4] Hall, G.A. 1972. Appalachian region. AB 26:858.

[5] F.C. Haas, pers. comm.

White Ibis at Wynnewood, Montgomery County, June 1977. (Photo: Franklin C. Haas)

[6] Debes, V.A. 1953. White Ibis in Pennsylvania. Cassinia 1951–1952:24.

Glossy Ibis ***Plegadis falcinellus***

General status: Glossy Ibises breed locally from coastal Maine south along the coast to Florida and west along the Gulf coast to Louisiana. They winter in the southern part of their breeding range. Glossy Ibises are rare regular visitors in the Coastal Plain and Piedmont and casual in the Ridge and Valley and the Lake Erie Shore. A dark ibis, believed to be a Glossy Ibis, was found in the northwestern corner of the High Plateau in Warren Co. In the Glaciated Northwest, Glossy Ibises have been recorded about six times in the counties of Crawford and Erie. They are accidental elsewhere in the state. They nested in the state twice.

Habitat: Glossy Ibises prefer ponds, lakes, marshes, wet pastures or meadows, and rivers.

Seasonal status and distribution

Spring: Individuals have arrived as early as late Mar or early Apr east of the Allegheny Front and by mid- to late Apr west of the Allegheny Front. By late May, most have left the state, with stragglers occasionally remaining through the summer. Most of the birds seen west of the Allegheny Front occur in spring. The single record in the Southwest was of a bird identified as a Glossy Ibis on 1 Jun 1984 at Pittsburgh.[1]

Breeding: Six Glossy Ibis nests found on Rookery Island in Lancaster Co. in 1975 were the first breeding record in the state (Morrin et al. 1991). Only two nests were found at this site in 1976, and none have been reported there since then.[2] Because there has been no recent breeding in Pennsylvania, all birds reported during breeding season are either lingering spring migrants or early postbreeding arrivals.

Fall: Postbreeding dispersal begins in early to mid-Jul; most records come from east of the Allegheny Front in the southeastern half of the state. Far north of their usual range, a Glossy Ibis was found at Tunkhannock in Wyoming Co. on 23 Aug 1982.[3] Most sightings are in Aug, and by mid-Sep Glossy Ibises have receded south of the state. Lingering birds have been recorded to mid-Nov. Two exceptionally late birds were found on a CBC on 22 Dec 1985 in Bucks Co.[4]

Winter: The Glossy Ibis discovered on

Glossy Ibis at Green Pond, Northampton County, March 1998. (Photo: Rick Wiltraut)

22 Feb 1976 at Washington Boro in Lancaster Co.[5] was more likely an unusually early spring arrival than a wintering individual.

History: In the late nineteenth century Glossy Ibises were reported from the marshes of the Delaware River (Cassin 1862). The next reports were not until 1938 and 1939. It was another 11 years before they were reported again, and, from then on, sightings became more regular (Poole, unpbl. ms.). Most of the early records came from Tinicum marshes. Glossy Ibises were not recorded in western Pennsylvania until 1958, when one was found at PISP.[6]

Comments: With the recent occurrence of White-faced Ibises *(P. chihi)* in Ohio, New Jersey, and Delaware, and given the difficulty in separating the two species in the field, especially in fall, it is possible that White-faced Ibises have been identified as Glossy Ibises. There have been recent spring and fall records of *Plegadis* ibises in Butler, Erie, Indiana, Lawrence, and Warren cos.

[1] Hall, G.A. 1984. Appalachian region. AB 38:1019.

[2] Fingerhood, E.D. 1992. Glossy Ibis (*Plegadis falcinellus*). BBA (Brauning 1992):426–427.

[3] Paxton, R.O., W.J. Boyle Jr., and D.A. Cutler. 1983. Hudson-Delaware region. AB 37:161.

[4] Dyer, J. 1985. Lower Bucks Co. CBC. AB 39:525.

[5] Buckley, P.A., R.O. Paxton, and D.A. Cutler. 1976. Hudson-Delaware region. AB 30:698.

[6] Stull, J.H., and J.G. Stull. 1958. Field notes. Sandpiper 1:5.

Roseate Spoonbill *Ajaia ajaja*

General status: The combination of pink plumage and long spatulate-shaped bill make these waders unique in North America. Roseate Spoonbills are residents in central and southern Florida and along the Gulf coast of Texas and Louisiana, and they rarely wander north of their normal breeding range. They occurred at least once in Pennsylvania.

Seasonal status and distribution

Spring: On 24 May 1968, a moribund immature female Roseate Spoonbill was found in Albion, Erie Co. The bird was discovered in a weakened condition and was identified by a local resident. The bird was taken to the Glenwood Zoo in Erie, where it lived only a short time. At the time, no aviary was known to be missing a spoonbill.[1] The spoonbill was mounted and placed in the waterfowl museum at Pymatuning in Crawford Co., but it disappeared and could not be located for many years. The specimen was recently located by G. M. McWilliams with original data in a private collection in Erie.

History: Libhart (1869) stated that a specimen was shot on the Conestoga Creek in Lancaster Co. Beck (1924) mentioned a bird's having been shot at Elizabethtown in 1844; it may have been the same bird mentioned by Libhart. The specimen was still in existence in the North Museum in Lancaster Co. in 1960 (Poole, unpbl. ms.), and a Roseate Spoonbill with poor documentation still is housed in the North Museum.

[1] Stull, J.H., and J.G. Stull. 1968. Field notes. Sandpiper 10:50.

Family Ciconiidae: Storks

Crane- or heronlike in appearance, storks have a long bill that is very heavy at the base, and the Wood Stork's bill is decurved at the tip. Storks' wings are long and broad, and their wing beats are slow and labored. They fly with their neck and legs fully extended and often soar and circle like hawks.

Wood Stork *Mycteria americana*

General status: Wood Storks are listed as accidental in Pennsylvania. They are resident from coastal South Carolina south through Florida and along the

Wood Stork at Montrose, Susquehanna County, about 1960. (Photo: Edwin Johnson)

Gulf coast to Texas. Wood Storks occasionally wander north to Virginia. This species has been recorded about 12 times in Pennsylvania, with only four records since 1960.

Seasonal status and distribution

Summer: Only one record of Wood Stork falls within summer. H. McWilliams saw an immature bird flying over Little Spruce Lake in Wayne Co. on 17 Jun 1988.[1]

Fall: Two birds have been reported in fall. An immature Wood Stork was observed by E. Witmer on Hammer Creek at Pumping Station Road in Lancaster Co. on 22 Sep 1996 and was seen later that day upstream in Lebanon Co.[2] One seen by D. Pfoutz at Boalsburg in Centre Co. on 2–9 Sep 1985 was photographed.[3]

History: Reports were mainly from southeastern Pennsylvania in the late nineteenth century. Wood Storks were reported in Jun and Jul, with most records from Jul (Poole, unpbl. ms.). Pre-1900 records have been mainly from Lancaster Co., where one was shot out of a group of 10 birds on the Susquehanna River in Jul 1862 (Libhart 1869); three were killed near Elizabethtown in Jul 1883 (Beck 1924); and four were near Refton on 17 Jul 1896 (Poole, unpbl. ms.). They have also been recorded in Lycoming Co., where one was collected (Stone 1894), and a specimen was taken in Chester Co. in the late 1800s (Conway 1940). One was seen at Wildwood in Dauphin Co. in 1897 (Frey 1943). This species was not reported again until 1921, when one was collected on 11 Nov in Albany Township in Berks Co. Three were seen west of Carlisle for about a week prior to 29 Jun 1955 (Poole, unpbl. ms.). One was photographed at Montrose in Susquehanna Co. about 1960.[4]

[1] Paxton, R.O., W.J. Boyle Jr., and D.A. Cutler. 1988. Hudson-Delaware region. AB 42:1274.
[2] Miller, R. 1996. Local notes: Lebanon County. PB 10:165.
[3] Hall, G.A. 1986. Appalachian region. AB 40:112.
[4] W. Reid, pers. comm.

Family Cathartidae: American Vultures

American vultures, often mistakenly called buzzards, are large mostly black, long-winged birds with short legs. Because of their rather unattractive features and their preference for eating rotting flesh, they are disliked by many people. Unlike birds of prey, vultures have weak talons not suited for grasping. They are well adapted to feeding on carrion or refuse by having an unfeathered head, a short strongly hooked bill, and large open nostrils. They are frequently seen soaring high overhead, where they may spend hours in the air riding warm thermals without ever having to flap their wings. Vultures often roost together at night and may remain at roosts throughout the day during foul weather. Two species, Turkey and Black vultures, migrate through and breed in Pennsylvania. They do not build a nest but lay from one to three eggs on the ground, in caves, on cliffs, in stumps, or in hollow logs.

Black Vulture ***Coragyps atratus***

General status: Black Vultures are resident from southeastern Pennsylvania south to Florida and west to Texas and Arizona. They breed locally in the Piedmont and Ridge and Valley and are casual wanderers north and west of the Ridge and Valley in spring and fall. In winter Black Vultures remain local residents in the Piedmont and Ridge and Valley. They have been expanding their range north and west in Pennsylvania. Black Vultures frequently soar and roost with the more abundant Turkey Vultures.

Habitat: Black Vultures are usually seen soaring overhead or at roosts. They prefer open agricultural areas to wooded settings.

Seasonal status and distribution

Spring: During spring migration, a few Black Vultures join migrating Turkey Vultures and occasionally wander north of their breeding range. Black Vultures recorded west of the mountains may be birds wandering northeast from the Ohio River Valley rather than from populations in southeastern Pennsylvania. The closest breeding range to western Pennsylvania is Adams and Brown cos. in southwestern Ohio (Peterjohn 1989). Northernmost Black Vultures in the state are observed from the Lake Erie Shore with migrating Turkey Vultures.

Breeding: Black Vultures are regular, uncommon to fairly common summer residents throughout the Piedmont and rare residents north into the Ridge and Valley. Since the 1960s, the breeding range has expanded from Adams and Lancaster cos. north and west to include Berks, Huntingdon, and Northampton cos. Black Vultures are rare, nonbreeding vagrants in the northern Ridge and Valley and accidental in the High Plateau. They are less tolerant of urbanized areas than Turkey Vultures. For example, they were rarely observed in Philadelphia Co. until the mid-1990s.[1] Few nests have been found, but the clutch of two eggs is laid, with no nest structure, in stumps or under logs on the ground. Abandoned buildings[2] and caves[3] have also been used as nest sites.

Fall: A few southbound birds are detected annually flying along mountain ridges, especially along the Kittatinny Ridge, mostly during Oct and Nov. An unexpected fall record was of an individual found at Nysox in Bradford Co. on 28 Sep 1986.[4] This is the only known record for the northern-tier counties.

Winter: Birds usually withdraw southward from the northern limits of their summer range in winter, especially from the higher elevations. Black Vultures are fairly common during the winter in the Piedmont. In early winter some large concentrations have been recorded in southern Lancaster Co., such as 341 counted on the CBC on 19 Dec 1982 (Morrin et al. 1991). Roosts containing more than 100 Black Vultures have occurred at Gettysburg since the 1970s.[5] Birds are rare during winter, in the Ridge and Valley counties of Centre, Clinton, Dauphin, Huntingdon, Juniata, Mifflin, and Perry. Black Vultures wander less in winter than at any other time of year.

History: This southern species has experienced an expansion across southeastern Pennsylvania since 1900. The earliest reports of Black Vultures by Warren (1890) were disputed. These vultures expanded first into agricultural areas of Adams and Lancaster cos. and were first confirmed breeding in Adams Co. in 1952.[6] Black Vultures had been reported in most Piedmont counties by 1950 (Poole 1964).

[1] Fingerhood, E. 1996. Local notes: Philadelphia County. PB 10:103.
[2] F.C. Haas, pers. comm.
[3] Witmer, E. 1992. County reports—April through June 1992: Lancaster County. PB 6:83.
[4] Paxton, R.O., W.J. Boyle Jr., and D.A. Cutler. 1987. Hudson-Delaware region. AB 41:64.
[5] Wright, A.L. 1984. Winter habitat use and abundance of Black and Turkey vultures at Gettysburg. M.S. thesis. Penn State Univ.
[6] Grube, G.E. 1953. Black Vulture breeding in Pennsylvania. WB 65:119.

Turkey Vulture *Cathartes aura*

General status: Turkey Vultures breed across southern Canada and throughout the U.S. and winter primarily across the southern half of their breeding range. They are common to abundant regular migrants and probably breed throughout the state. Turkey Vultures are regular winter residents from the Coastal Plain to the Ridge and Valley. Elsewhere, Turkey Vultures are accidental in winter.

Habitat: Turkey Vultures are usually seen soaring overhead or perched at roosts, but they can also be found perched on the ground in open fields or farmlands.

Seasonal status and distribution

Spring: The earliest northbound birds are usually seen in the last week of Feb or the first week in Mar. Their typical migration period is from about the second week of Mar to the second week of May. Migration is less spectacular and less concentrated in the east than in the west, where numbers of birds peak sometime from the last few days of Mar to the third week of Apr, with the largest flights at the Lake Erie Shore. On 3 Apr 1974 more than 700 were counted along Lake Erie (Stull et al. 1985).

Breeding: Turkey Vultures are uncommon to fairly common summer residents statewide. They wander widely from nest sites, so May and Jun observations do not necessarily indicate local breeding. No nest is built, but eggs are generally laid in a scrape in a rock pile, cliff, or hollow log. Vultures are infrequently confirmed breeding because of the inaccessible location of nests and the species' low density. They probably breed in every county. Most clutches have been found from 18 Apr to 14 May.

Fall: Flights of southbound birds west of the mountains are very light; usually only singles or small groups are observed. The greatest number of migrants is seen from hawk watches in the Ridge and Valley region. The typical migration period is from late Aug to late Nov, with the peak usually occurring from the second week of Oct to the first week of Nov. West of the mountains the last birds are usually seen in mid-Oct, with stragglers to mid-Nov.

Winter: The greatest number of wintering Turkey Vultures is seen from the Piedmont north to the southern portion of the Ridge and Valley. At southern roosts they can be locally abundant. Early winter counts have yielded over 1000 vultures at Big Round Top near Gettysburg in Adams Co., where they prefer to roost in conifers.[1,2] On the Southern Lancaster Co. CBC, 635 were counted on 18 Dec 1983 (Morrin et al. 1991). In the Coastal Plain they are uncommon to fairly common. Turkey Vultures are rare to casual winter visitors north and west of the southern portion of the Ridge and Valley region. The only location in the Glaciated Northeast where they have been recorded in winter is near Athens, Bradford Co.[3] In the High Plateau, the only site where they have been recorded is at PNR in the Allegheny Mountains of Westmoreland Co., on 1 Feb 1989[4] and 1991 (no date).[5] In the Glaciated Northwest one was seen near Edinboro in Erie Co. on 3 Feb 1993.[6]

History: At the end of the 1800s Turkey Vultures were common in the Piedmont, scarce through the Ridge and Valley, and accidental in the northern portion of the High Plateaus (Warren 1890). Their range extended northward during the first half of this century (Todd 1940; Poole 1964). By the 1950s they were being observed statewide, with nests reported widely (Poole 1964). Expansion continued into the northeastern U.S., reaching Maine by 1982 (Erskines 1992). Todd (1940) did not cite any true winter records in western Pennsylvania but mentioned two individuals in Beaver Co. on the late date of 26 Dec 1926.

Comments: The expansion of Turkey Vultures has long been linked to the deer herd in Pennsylvania (Todd 1940). Although Black and Turkey vultures have not frequently been reported feeding, road kills of various species contribute to their food supply.

[1] Boyle, W.J., Jr., R.O. Paxton, and D.A. Cutler. 1981. Hudson-Delaware region. AB 35:283.
[2] Wright, A.L. 1984. Winter habitat use and abundance of Black and Turkey vultures at Gettysburg. M.S. thesis. Penn State Univ.
[3] T. Gerlach, pers. comm.
[4] Hall, G.A. 1989. Appalachian region. AB 43:313.
[5] Leberman, R., and R. Mulvihill 1991. County reports—January through March 1991: Westmoreland County. PB 5:51.
[6] G.M. McWilliams, pers. recs.

Order Phoenicopteriformes
Family Phoenicopteridae: Flamingos

Flamingos are well-known large wading birds normally found in the tropics. Their plumage is mostly pink. The downward bend of their large heavy bill gives it the shape of a boomerang. Their legs and neck are extremely long. Only one species occurs in North America, the Greater Flamingo. Most sightings of this species are in southern Florida; they very rarely wander north along the Atlantic and Gulf coasts. Because they are frequently kept in captivity in parks and zoos, most sightings are believed to be of escapees.

[Greater Flamingo] *Phoenicopterus ruber*

General status: Greater Flamingos are rare vagrants from the Caribbean islands north to southern Florida and very rare wanderers along the Atlantic coast to New Brunswick and Nova Scotia. Away from the coast, they have been recorded in Michigan, Ontario, and southern Quebec. They are listed as hypothetical in Pennsylvania, where all records are believed to have been escapees.

Comments: A specimen that was in the Old Lancaster Museum, shot prior to 1840 (Beck 1924), no longer exists. A flamingo was photographed in color on a PGC pond near Keller's Church in Bucks Co. on 27 Jul 1949. The color was described as quite brilliant and the bird did not look as if it had escaped from captivity. Another bird was present in the same area, and the two birds were observed flying off every evening to roost for a week before departing (Poole, unpbl. ms.).

Order Anseriformes
Family Anatidae: Geese, Swans, and Ducks

Waterfowl are perfectly structured for aquatic life, having short legs, webbed feet, a short tail (in most species), and a dense water-resistant outer layer of feathers with a heavy coat of down beneath. Thirty-nine species of waterfowl have been recorded in Pennsylvania, including five species of geese, two of swans, and thirty-two of ducks. All are highly migratory, and several species breed or winter in the state. Adults of many species of waterfowl undergo a complete molt in summer after the breeding season, when they may be flightless for a few weeks. Many species of ducks then acquire an adult nonbreeding plumage, which is called an eclipse plumage in males. By fall or early winter a complete body molt will begin returning their colorful breeding plumage. All waterfowl, except mergansers, have a lamellate and spatulate bill. Mergansers have a long, narrow, bony bill with toothlike serrations, but they share with all other waterfowl the bony nail at the tip. Geese and swans are characterized by a medium to very long neck, and they are larger and more robust than ducks. Their tarsi are somewhat longer than the tarsi of ducks, accounting for their less awkward gait when on land. Sexes are similar in plumage, and they usually mate for life. Most species of waterfowl are considered game birds and are hunted for sport.

Whistling-ducks are gooselike ducks with long legs. They feed primarily by foraging in fields along ponds, streams, and marshes. All other ducks are divided into two groups commonly referred to as puddle ducks and diving ducks. Puddle ducks prefer shallower water than diving ducks and usually feed by tipping up or picking from the surface. The legs of puddle ducks tend to be attached farther forward on the body, allowing more stable weight distribution for grazing in grain fields or along muddy shorelines. They usually take off from the water or from land by leaping into the air. Most have iridescent secondaries or specula, whereas diving ducks' secondaries are usually white or of dark somber colors. Like geese and swans, most puddle ducks feed on plant matter such as sprouts, grasses, and aquatic vegetation, roots, and seeds.

Diving ducks prefer deep water and are well equipped for diving, with shorter, narrower wings set slightly farther back on the body. Some species use their wings to propel themselves in the water. The legs are attached farther back on the body, the toes are longer, and the lobed hind toes assist them in maneuvering through the water. They usually take off from the water by running short distances rather than by leaping into the air as puddle ducks do. Their diet consists of more animal than plant matter; they feed on a variety of animal life such as fish, aquatic invertebrates, and mollusks. The three species of mergansers found in Pennsylvania feed mainly on fish.

Most waterfowl nest near water on or near the ground in depressions or on stumps, muskrat houses, or any other type of elevated structure. Fourteen species of ducks have been confirmed breeding in the state, and their status as nesting species in the state is often governed by human intervention in watersheds. After the Shenango River was dammed in Crawford Co., forming the Pymatuning Reservoir, several species nested there, at least temporarily, including Northern Pintail, Northern Shoveler, Gadwall, American Wigeon, Redhead, Ring-necked Duck, and Ruddy Duck. Most ducks' nests are constructed of sticks, grasses, cattails, reeds, leaves, or other plant parts and are lined with feathers and down. The females lay four to 18 white, cream, green, or buff-colored eggs. They may have one or two broods a year. Incubation for most waterfowl usually does not begin until egglaying is completed. Hatchlings immediately abandon the nest and accompany their parents to feeding sites.

Black-bellied Whistling-Duck ***Dendrocygna autumnalis***

General status: Widespread in the tropical lowlands of South America north through Mexico to Arizona, southern

Texas, and Louisiana, Black-bellied Whistling-Ducks rarely wander north of their breeding range. They have been recorded as far north as Ontario, Quebec, New York, and Pennsylvania. Two Pennsylvania records of this species have been accepted. They are accidental visitors in the state.

Seasonal status and distribution

Spring/summer: A flock of 11 Black-bellied Whistling-Ducks was discovered by M. A. Dunmire on 8 Jun 1993 on a farm pond north of Saltsburg, Indiana Co. On 9 Jun 1993 the birds were seen at Bush Recreation Area of the Loyalhanna Dam in Westmoreland Co. Three of the ducks remained until at least 22 Jun 1993 and were photographed.[1] Five were found by J. Book and other birders at the Howard Martin Farm near the intersection of Centerville and Charlestown roads in Lancaster Co. from 21 Jul to at least 12 Aug 1993 and were photographed.[2] All records were part of the northerly movement of Black-bellied Whistling-Ducks recorded during the summer of 1993 elsewhere in the U.S. and Canada. Reports also came from Maryland, New York, Virginia, Ontario, and Quebec.

Comments: A bird, potentially an escapee, was trapped by the PGC at Pymatuning Lake in Crawford Co. on 26 Aug 1979.[3]

[1] Dunmire, M.A. 1993. First documented record of Black-bellied Whistling-Ducks for Pennsylvania. PB 7:50.
[2] Haas, F.C., and B.M. Haas, eds. 1993. Rare and unusual bird reports. PB 7:103.
[3] Hall, G.A. 1980. Appalachian region. AB 34:161.

[Fulvous Whistling-Duck] ***Dendrocygna bicolor***

General status: In North America Fulvous Whistling-Ducks breed locally in central and southern Florida, southwestern Louisiana, and central and eastern Texas. In eastern North America records come from Michigan, Ontario, Quebec, Maine, New Brunswick, Prince Edward Island, and Nova Scotia. The only two records of this species in Pennsylvania are listed as hypothetical.

Comments: One was seen at the MCWMA in Lebanon Co. from 16 to 24 Mar 1975. It was seen by many and was photographed.[1] A flock of 13 was reported in York Co. on 29 Sep 1976.[2] The photograph of the bird from the MCWMA no longer exists, and there is apparently no documentation of the flock from York Co. It could not be concluded that these birds had not escaped from captivity.

Black-bellied Whistling-Ducks at Loyalhanna, Westmoreland County, June 1993. (Photo: Walter Shaffer)

[1] Scott, F.R., and D.A. Cutler. 1975. Middle Atlantic Coast region. AB 29:673.
[2] Paxton, R.O., P.A. Buckley, and D.A. Cutler. 1977. Hudson-Delaware region. AB 31:156.

Pink-footed Goose ***Anser brachyrhynchus***

General status: The Pink-footed Goose breeds in eastern Greenland, Iceland, and Spitsbergen and winters in northwestern Europe. Vagrants have occurred west to Newfoundland and Quebec. There is one accepted record in Pennsylvania.

Seasonal status and distribution

Spring: On 27 Mar 1997 a Pink-footed Goose appeared with a flock of 200 to 300 Snow Geese at Oley in Berks Co. The bird was observed and photographed by many people and left with the flock of Snow Geese on 12 Apr 1997.[1] Perhaps the same bird reappeared on 7 Mar 1998 at Lake Ontelaunee in Berks Co.[2]

[1] Keller, R. 1997. Local notes: Berks County. PB 11:26–27.
[2] F.C. Haas and B.M. Haas, pers. comm.

Greater White-fronted Goose ***Anser albifrons***

General status: Greater White-fronted Geese breed from Alaska across northern Canada and southern Greenland. They winter locally west of Texas and Montana. Greater White-fronted Geese are rare regular migrants and irregular winter visitors in the Piedmont. In the High Plateau they have been reported once from an impoundment area of Crooked Creek in Tioga Co.[1] West of the High Plateau they have been recorded only in the counties of Butler, Crawford, Erie, Indiana, and Westmoreland. Most birds discovered are single adults or small groups mixed with flocks of Canada Geese. The origin of some birds is questionable, especially those with domestic waterfowl, as lone birds, or those in late spring or in summer.

Habitat: They are found on lakes, ponds, grassy fields, or grain fields.

Seasonal status and distribution

Spring: White-fronted Geese usually arrive with flocks of migrating Canada Geese in early Mar. Most spring records are in Mar, with a few stragglers to the fourth week of Apr.

Fall: Arrival times of early migrants are less predictable at this season. Most birds believed to be of wild origin appear any time from late Oct through mid-Dec. As winter approaches, they become more predictable in the Piedmont in late Dec with wintering flocks of Canada Geese.

Winter: Greater White-fronted Geese are irregular visitors or residents in the Piedmont, where they have been recorded in the counties of Berks, Bucks, Chester, Delaware, Lancaster, Lehigh, and Montgomery. Birds found in Feb, especially after mid-Feb, may be early migrants. Outside the Piedmont, White-fronted Geese have been observed in winter only at Pymatuning, where they have been reported twice, from 7 Jan to 19 Feb 1984[2] and on 21 Jan 1989,[3] and at the Montour fly ash basin in Montour Co. in Jan 1997, where 18 birds were counted.[4]

History: Except for a record from the Delaware River in 1877 (Stone 1894), apparently all early records came from western Pennsylvania (Poole 1964). They have been reported from the Allegheny River near Tarentum in 1895 (Todd 1940) and at Conneaut Lake in 1908 and 1926 (Sutton 1928b).

[1] J. Stickler, pers. comm.
[2] Hall, G.A. 1984. Appalachian region. AB 38:316.
[3] Hall, G.A. 1989. Appalachian region. AB 43:313.
[4] Haas, F.C., and B.M. Haas, eds. 1997. Seasonal occurrence tables—January through March 1997. PB 11:44.

Snow Goose ***Chen caerulescens***

General status: Snow Geese breed from Alaska east along the Arctic coast and islands of Canada to Greenland south to James Bay. In the West they winter locally from Texas to California. In the east they winter primarily along the Atlantic coast from Chesapeake Bay to North Carolina. Snow Geese have become increasingly common in the past few decades in Pennsylvania. Huge concentrations, comparable to flocks observed on the Great Plains, are recorded annually in spring in areas of the central Piedmont, frequently numbering in the tens of thousands. Most of the Snow Geese that make up these concentrations are of the white morph. Both color forms, the blue morph and the white morph, once considered to be separate species, are recorded in the state. Snow Geese are uncommon to rare regular migrants elsewhere in Pennsylvania, but occasionally large flocks are observed flying along the mountain ridges of the Ridge and Valley. In the High Plateau, they are casual. Snow Geese frequently mix with Canada Geese, except in the Piedmont. Occasionally, nonbreeding individuals linger into summer. They are regular winter residents in the Coastal Plain and Piedmont and at Pymatuning in the Glaciated Northwest. Snow Geese are casual or accidental in winter elsewhere in the state.

Habitat: Snow Geese are found on lakes, ponds, and large slow-moving rivers, and they graze in cultivated fields, especially fields of winter wheat and corn.

Seasonal status and distribution

Spring: They are more frequently seen at this season than any other. Migration is early, often during the peak flights of Canada Geese, beginning some years as early as the first week of Feb. Their typical migration period is from about the third week of Feb to the first week of Apr. The migration peaks in early to mid-Mar. Snow Goose numbers in the past few years have increased during spring migration in the Piedmont. In Lancaster Co. it is not unusual to see flocks of tens of thousands of birds at Octoraro Lake, Muddy Run, Washington Boro, and MCWMA (Morrin et al. 1991). Single-day counts have reached over 125,000 Snow Geese at MCWMA during peak migration.[1] Areas away from the Piedmont have not seen the increase in Snow Geese except at the Pennsylvania Power and Light Company Fly Ash pond in Montour Co., where as many as 5000 were counted in Mar 1996.[2] Individual birds or small flocks are to be expected in the Glaciated Northwest, especially at Pymatuning, where they are seen mixed with flocks of Canada Geese and at the Lake Erie Shore. A few stragglers usually remain into summer.

Summer: Lingering birds will occasionally summer in areas where they are most numerous during migration. Records exist from the Glaciated Northeast and Ridge and Valley from Clinton, Lycoming, Susquehanna, and Wyoming cos., but the origin of the birds may be questionable. The only area in western Pennsylvania where birds believed to be of wild origin have been recorded in summer is in Crawford Co. at Pymatuning.

Fall: Snow Geese are less common in fall than in spring. Southbound flocks are usually seen from the hawk lookouts along the southeastern ridges of the mountains and at the same sites visited in spring. Hawk lookouts occasionally record large numbers, such as 1600 over the Tussey Mountain hawk-watch station on 12 Nov 1986.[3] Migration over the state may begin as early as mid-Sep, but usually not until mid-Oct. Peak migration occurs in Nov after the first major cold front has passed.

Winter: Most winter records are of late migrants or of visiting birds consisting of singles or small flocks. They are most common in the Piedmont, where flocks sometimes exceed 10,000 birds.[4] In the winter of 1995, an exceptional count of 70,000 Snow Geese was on the lower Susquehanna River in Lancaster Co.[5] Away from traditional wintering sites in the Piedmont, Snow Geese are usually found with wintering flocks of Canada Geese. Outside the Piedmont, winter records have come from the counties of Allegheny, Armstrong, Bradford, Clarion, Crawford, Indiana, Lycoming, Potter, Schuylkill, Somerset, Venango, Westmoreland, and Wyoming.

History: Snow Geese were listed as rare and irregular migrants statewide by most authors. However, flocks containing from 20 to over 200 birds occasionally were observed. Both blue- and white-morph geese were reported, but white morphs predominated. Snow geese were reported more frequently and in greater numbers in the southeastern and northwestern portions of the state than in any other area. Apparently there was no report of blue morphs in eastern Pennsylvania until after 1930. Snow Geese wintered occasionally at Lake Ontelaunee in Berks Co., but they were reported only as winter visitors at other scattered sites in eastern Pennsylvania (Poole, unpbl. ms.).

Comments: Two subspecies occur in Pennsylvania: the nominate 'Lesser Snow Goose' and *atlanticus*, the 'Greater Snow Goose.' The subspecies are not separable in the field, but most blue morphs are likely to be of the nominate subspecies, because blue morphs of *atlanticus* are very rare (Madge and Burn 1988). The proportion of blue morphs has increased in recent decades (Madge and Burn 1988) in Pennsylvania, as elsewhere.

[1] E. Gosnell, pers. comm.

[2] Brauning, D.W. 1996. Local notes: Montour County. PB 10:24.

[3] Hall, G.A. 1987. Appalachian region. AB 41:88.

[4] Boyle, W.J., Jr., R.O. Paxton, and D.A. Cutler. 1991. Hudson-Delaware region. AB 45:254.

[5] Paxton, R.O., W.J. Boyle Jr., and D.A. Cutler. 1995. Hudson-Delaware region. NASFN 49:131.

Ross's Goose *Chen rossii*

General status: Easily overlooked because of their similarities to Snow Geese, the smaller Ross's Geese breed in Mackenzie, Keewatin, Southampton Island, and south along the west coast of Hudson Bay to James Bay. They winter primarily in California and east locally to Missouri, Arkansas, and Louisiana. Most winter in the southwestern U.S., but a few birds are regularly seen with flocks of Snow Geese along the Atlantic coast. Ross's Geese had not been recorded in Pennsylvania before 1991, but they have occurred almost every year since. Though they have become rather regular in occurrence in the state, Ross's Geese are still very rare. Most sightings are of single birds, but two or more different birds in a single day have been reported. The MCWMA is the most reliable site to observe this species, where they are nearly annual during the peak flights of Snow Geese. Several Ross's Goose records have been documented with photographs that are in the files of PORC and VIREO.

Seasonal status and distribution

Fall through spring: The first accepted Pennsylvania record of Ross's Goose was found by Ed Pederson and Jo Pederson in Drumore Township, Lancaster Co. on 27 Feb 1991.[1] Several others have been observed annually since then. Most records of Ross's Geese have been from the Piedmont in Berks, Chester, Lancaster, Lebanon, Montgomery, and Northampton cos. The majority have been recorded during spring migration, when thousands of Snow Geese are passing through the state. Ross's Geese have been recorded as early as the fourth week of Jan through Feb, but most birds are observed during the first half of Mar. Birds have lin-

Ross's Goose at Montour Preserve, Montour County, October 1995. (Photo: Paul W. Schwalbe)

gered to at least 30 Mar. There were at least six reports of Ross's Geese between 2 Feb to 24 Mar 1996 in the Piedmont. At least that many reports came in 1997 just from MCWMA. The very rare blue-morph Ross's Goose was reported at Octoraro Lake in Lancaster Co. on 31 Jan 1995;[2] one was photographed on 24 Feb 1997 at MCWMA;[3] a single bird was at Fogelsville in Lehigh Co. on 13 Jan 1998[4] and another or possibly the same bird was at Nazareth in Northampton Co. on 21 Jan 1998.[5] Away from the Piedmont, one white morph with a flock of 30 to 40 Canada Geese at Montour Preserve in Montour Co. on 28 Oct remained to at least 2 Nov 1995.[6] Ross's Geese have been recorded only three times in the western half of the state. A bird was at Shawnee State Park in Bedford Co. on 9 Mar 1997;[7] one was with a flock of Canada Geese at Pymatuning in Crawford Co. on 14 Nov 1996;[8] and one was at Harrisville Road in Mercer Co. on 19 Mar 1997.[9]

Comments: As the number of Snow Geese in the southeastern part of the state increases, so do the reports of Ross's Geese.

[1] Haas, F.C. 1991. First Pennsylvania record of Ross' [*sic*] Goose, *Chen rossii*, Lancaster County. PB 5:19–20.

[2] Paxton, R.O., W.J. Boyle Jr., and D.A. Cutler. 1995. Hudson-Delaware region. NASFN 49:131.

[3] Haas, F.C., and B.M. Haas, eds. 1997. Photographic highlights. PB 11:25.

[4] J. Horn and R. Wiltraut, pers. comm.

[5] R. Wiltraut, pers. comm.

[6] Brauning, D.W. 1996. Local notes: Montour County. PB 9:216.

[7] Shaffer, J. 1997. Local notes: Bedford County. PB 11:26.

[8] Haas, F.C., and B.M. Haas, eds. 1997. Birds of note—October through December 1996. PB 10:220.

[9] McKay, M. 1997. Local notes: Mercer County. PB 11:32.

Canada Goose ***Branta canadensis***

General status: The Canada Goose is probably the best known waterfowl species in North America. The honking calls of migrating flocks of Canada Geese formerly announced the changing of the seasons. Since the introduction of the subspecies *B. c. maxima* as a resident breeding population, the thrill of seeing or hearing north- and southbound flocks of geese has lost much of its appeal. Canada Geese breed over most of North America, and they winter from Alaska south and east across southern Canada and across most of the U.S. Local populations have grown to the status of unwanted pests. In many areas of the state, especially around golf

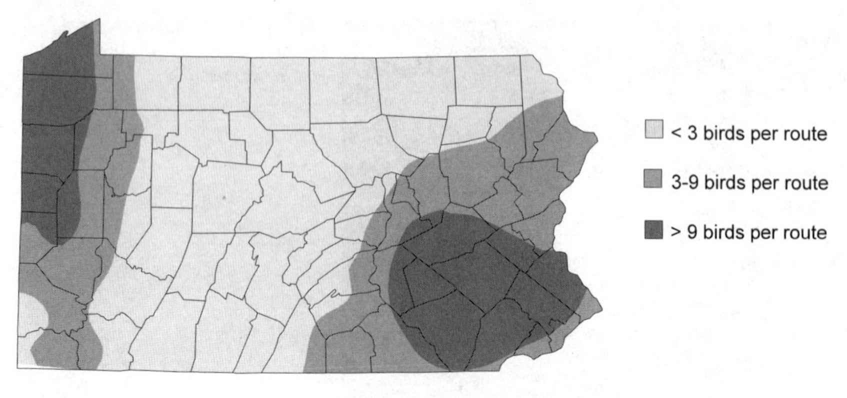

5. Canada Goose relative abundance, based on BBS routes, 1985–1994

courses and in city parks, they have become a health concern. They are now common to abundant permanent residents nearly statewide.

Habitat: Canada Geese are found almost anywhere there are water and sufficient areas to graze, particularly in farmland. They also have adapted well to city parks and urban and suburban areas.

Seasonal status and distribution

Spring: Because many Canada Geese are now resident birds, migration periods are less clear than in the past. The migration season is usually from late Feb to early Apr. Their numbers peak about mid-Mar.

Breeding: The breeding population experienced a dramatic increase since the early 1970s. By the mid-1980s, Canada Geese were nesting in all Pennsylvania counties. The population continued to expand in the early 1990s, occupying habitat increasingly removed from farming areas, where they are common. The relative abundance of summer residents reflects the historical source of birds; they are abundant in the Piedmont and Glaciated Northwest (Dunn and Jacobs 1995) and common in open habitats elsewhere (see Map 5). Geese are semidomestic in residential areas and parks. A peak population of 200,000 summer residents was estimated in 1995 (Dunn and Jacobs 1995). Geese are found in flocks most of the year, except during the breeding season, when pairs, which mate for life, rigorously defend small (a few hundred square feet) nest territories. White eggs, usually five or six, are placed in down-filled nests on uplands near water, on muskrat lodges, or at a diversity of other sites on the ground. They prefer small islands. Incubation occurs from early Apr to Jul. The gander does not incubate, but is very defensive of nest and young. An incubating goose extends her neck low over the ground or water to avoid detection.

Fall: Migration begins as early as late Sep, with most birds passing over the state during cold fronts in Oct and Nov. Migration is usually complete by late Dec or early Jan.

Winter: After early or mid-Jan, Canada Geese concentrate around open water. Except for Pymatuning, where thousands may winter, more geese are found in the Piedmont than in any other region in the state.

History: Canada Geese were historically common migrants and rare winter visitors, as Sutton (1928a) said, "merely passing by." The resident breeding population began with the release of 30 pinioned geese at Pymatuning, Crawford Co. in 1936. These birds nested in the sanctuary in 1937 (Trimble 1940) and have nested there each year since (Grimm 1952). Releases at MCWMA from Pymatuning stock in 1969 (Kosack 1995) contributed to the expansion of summer residents in the southeast. Since their introduction into the state, geese have spread and multiplied rapidly.[1]

Comments: The resident breeding population was derived largely from the subspecies *B. c. maxima*, the largest subspecies,[2] which formerly bred from Kentucky to North Dakota and at one time was believed to be extinct (AOU 1957). Breeding birds, remaining as year-round residents, have become pests, fouling lawns and ponds. Trap-and-transfer programs were begun in the 1970s to move problem flocks out of state,[3] but the programs were terminated in 1995 because they were expensive and ineffective.[4] Hunting seasons targeted at resident geese, begun in 1992, have slowed the expansion of the population.

Poole (1964) mentioned at least three birds that were described as being of the subspecies *hutchinsii*. In more recent years, small birds believed to be *hutchinsii* have been reported from several areas in the state. Most *hutchinsii* reports come from sites where the largest concentrations of geese are found, such as Pymatuning and MCWMA. There has been very little documentation to confirm individual sightings of the smaller subspecies of Canada Geese that are reported in Pennsylvania. Most Canada Goose subspecies are not reliably separable in the field.

[1] Hartman, F.E. 1992. Canada Goose *(Branta canadensis)*. BBA (Brauning 1992):66–67.

[2] Dunn, J.P. 1992. Pennsylvania's Canada Geese: Giant success or giant dilemma? Pennsylvania Game News 63:16–19.

[3] Gafney, B.J. 1993. Nuisance geese. Pennsylvania Game News 64:15.

[4] J.P. Dunn, pers. comm.

Brant *Branta bernicla*

General status: Brant have gone through severe declines and fluctuations throughout history, but their numbers have gradually recovered from historical lows. They breed from Melville, Ellesmere, and Baffin islands east to Greenland. Brant winter along the west coast from British Columbia south and along the Atlantic coast from New Brunswick south to Florida. Some of the largest flocks ever recorded in Pennsylvania have been in the last 20 years. Brant are regular fall migrants only on the Lake Erie Shore, where numbers vary from year to year depending on weather patterns. They are irregular to casual migrants with varying numbers elsewhere in the state, except in the northern-tier counties, where the species has not been recorded. Brant are accidental winter visitors in the Delaware River Valley, in the lower Susquehanna River, and on the Lake Erie Shore.

Habitat: They are usually found on large lakes, ponds, and large slow-moving rivers. They are occasionally observed in migration from hawk watches along the mountain ridges and along the Lake Erie Shore.

Seasonal status and distribution

Spring: The few spring records are widely scattered over the state. Apparently, most northbound Brant fly along the Atlantic coast rather than crossing overland (Bellrose 1980), so birds passing over Pennsylvania in spring are probably strays. Sightings begin around mid-Mar and continue to late May. One on 28 Feb 1996 at Peace Valley

Brant at Beltzville State Park, Carbon County, October 1989. (Photo: Rick Wiltraut)

Park in Bucks Co. was likely to have been an early spring migrant rather than an overwintering bird.[1] Usually no more than one or two birds are reported in spring; the 14 birds counted flying over Hamburg in Berks Co. on 25 May 1996[2] and the 9 birds on Harvey's Lake in Luzerne Co. on 13 May 1997[3] and later that day seen flying into Wyoming Co.[4] were unusual.

Fall: The fall migration corridor is a narrow band that crosses over New York State (Bellrose 1980), so a front from the north or northeast with strong wind from the same direction is usually required before a major movement of Brant is observed over Pennsylvania. Few are seen if the weather is mild with no major cold fronts. However, with strong cold fronts from the north, large flocks can be observed flying along the ridges at hawk watches and along Lake Erie. Their typical migration period is from the second week of Oct through Nov, with lingering birds to mid-Jan. The greatest number of migrants occurs from late Oct to mid-Nov. One notable flight was on the early date of 13 Oct in 1974, when 700 to 800 birds were seen flying over New Ringgold and HMS.[5] During severe storms with north winds, large flocks may stop on large lakes and rivers. Large flocks (containing 100 or more birds) were recorded in 1982, 1985, 1986, 1988, and 1989.

Winter: At least eight winter sightings of Brant have been documented. In Bucks Co., one bird was at Core Creek Park on 11 Feb 1990;[6] two birds were at Van Sciver Lake on 26 Jan 1992 and one was at Peace Valley Park on 1 and 8 Feb 1992;[7] and one was at Peace Valley Park on 17 Jan 1998.[8] A flock of 200 birds that appeared on the Conowingo Pond in Lancaster Co. in Feb 1976 and remained for two weeks (Morrin et al. 1991) was unusual. One bird was found at Oley in Berks Co. on 24 Jan 1998 and remained into spring 1998.[9] A bird was seen at Martin's Creek in Northampton Co. on 28 Jan 1982.[10] One was recorded in Erie Co. at PISP from 30 Jan to 1 Feb 1975.[11]

History: One of the major declines in Brant populations of the twentieth century occurred in the early 1930s. That decline was attributed to the reduction of eelgrass, Brant's principal food.[12] Populations increased steadily in later years, when Brant partially broadened their food habits to include sea lettuce and upland grazing (Bellrose 1980). Poole (1964) stated that most historical records were from east of the Appalachian Mountains.

[1] Kitson, K. 1996. Local notes: Bucks County. PB 10:17.
[2] Keller, R. 1996. Local notes: Berks County. PB 10:97.
[3] Koval, R. 1997. Local notes: Luzerne County. PB 11:97.
[4] Reid, W. 1997. Local notes: Wyoming County. PB 11:101.
[5] Scott, F.R., and D.A. Cutler. 1975. Middle Atlantic Coast region. AB 29:36.
[6] R. French, pers. comm.
[7] French, R. 1992. County reports—January through March 1992: Bucks County. PB 6:30.
[8] Kitson, K. 1998. Local notes: Bucks County. PB 12:16.
[9] Keller, R. 1998. Local notes: Berks County. PB 12:15.
[10] R. Wiltraut, pers. comm.
[11] G.M. McWilliams, pers. obs.
[12] Lincoln, F.C. 1950. The American Brant—Living bird or museum piece? Audubon Magazine 52:282–287.

[Barnacle Goose] ***Branta leucopsis***

General status: Barnacle Geese breed in eastern Greenland, Spitsbergen, and Novaya Zemlya, and they winter in the British Isles and the Netherlands. They are casual in Labrador, Baffin Island, Quebec, New Brunswick, Nova Scotia, and south along the Atlantic coast to South Carolina. Since 1982 numerous observations of this species have been reported in Pennsylvania. Sightings have come from Berks, Crawford, Chester, Lancaster, Lehigh, Montgomery, Northampton, and Washington cos. Dates range from the first week of Feb to the last week of Apr and in Nov. They are currently listed in Pennsylvania as hypothetical, because of their questionable origins.

Comments: Birds outside of their normal range are considered by most authorities to be escapees. They are frequently kept in captivity and are often released into the wild. It is possible that a few birds wander south of their normal range with flocks of eastern Greater White-fronted or Snow geese, therefore all sightings should be properly documented. Johnsgard (1978) stated that spring records are more likely than fall records to refer to wild birds and that this species is a long-distance migrant whose numbers have been increasing in Greenland.

Mute Swan ***Cygnus olor***

General status: Mute Swans were introduced in the U.S. from Europe and the British Isles. These common swans of parks and zoos are also kept in private waterfowl collections or to decorate backyard ponds. They are local residents along the central Atlantic coast and around the western Great Lakes. Free-flying birds are rare regular visitors, occasionally joining migrating flocks of Tundra Swans and becoming displaced over various parts of the state.[1] Their status and distribution as a migrant in the state are unclear. A few may become permanent residents and stay to breed and winter. Most of the records in the state are probably strays from the East Coast or western Great Lakes wild populations, though some may be escapees from captivity.

Habitat: Mute Swans are usually found on lakes, ponds, and large slow-moving rivers.

Seasonal status and distribution

Breeding: They have been confirmed breeding in the state only in recent years. Most breeding or suspected breeding records come from east of the Ridge and Valley, with local concentrations occurring across the Piedmont. Elsewhere in the state breeding has been confirmed in Mercer, Somerset, Westmoreland, and Wayne cos.[2] Wandering offspring from locally breeding pairs may account for some of the sightings.

Winter: Mute Swans are regular residents in the Coastal Plain and Piedmont, especially where they breed. They are casual winter visitors or residents at the Lake Erie Shore and accidental visitors or residents elsewhere in the state.

History: Mute Swans, native to Europe, were introduced into the U.S. in the late nineteenth century in New York City.[3] It is not clear when Mute Swans first appeared in Pennsylvania, but the appearance of one bird in 1932 at Lake Ontelaunee, Berks Co., and of three near Harrisburg in 1934 may have been the first wild scouts (Frey 1943). From that time, sightings have increased, with first confirmed breeding in 1939 (Wood 1958).

Comments: Mute Swans have not been a welcome addition to the avifauna of Pennsylvania because they may have an impact on native waterfowl species[4] (Long 1981). They are highly territorial, and they destroy large amounts of aquatic vegetation for nest building and while feeding. Unlike native ducks, geese, and swans, Mute Swans feed by ripping out the entire plant and its root system.[5]

[1] G.M. McWilliams, pers. obs.

[2] Master, T.L. 1992. Mute Swan *(Cygnus olor)*. BBA (Brauning 1992):64–65.

[3] Bump, G. 1941. The introduction and transplantation of game birds and mammals into the state of New York. Transactions of the North American Wildlife Conference 5:409–420.

[4] Reese, J.G. 1975. Productivity and management of feral Mute Swans in Chesapeake Bay. Journal of Wildlife Management 39:280–286.

[5] J.P. Dunn, pers. comm.

[Trumpeter Swan] *Cygnus buccinator*

General status: Trumpeter Swans breed locally in Alaska and portions of western Canada south to Montana. They winter along the Pacific coast from southern Alaska south to California and east to New Mexico. Trumpeter Swans have been reintroduced elsewhere in the east where they formerly bred. Because there is little documentation of wild Trumpeter Swans in the state, they are listed here as hypothetical.

Skeletal remains have been found in Pennsylvania in association with Indian remains in Huntingdon and Lancaster cos. Artifacts associated with the bones dated from 1600 to 1625.[1]

Comments: At this writing, it has yet to be proved that any sightings are of birds from the wild western migrant populations. Many birds marked with tags or bands from nearby Trumpeter Swan reintroduction programs have been sighted in the past few years. However, not all birds reported have been tagged or banded.

[1] Parmalee, P.W. 1961. A prehistoric record of the Trumpeter Swan from central Pennsylvania. WB 73:212–213.

Tundra Swan *Cygnus columbianus*

General status: These large, long-necked waterfowl breed from Alaska east across the Arctic coast to Baffin Island and south to Churchill, Manitoba. They winter in the U.S. along the Pacific coast and inland to Utah and in southern New Mexico and west Texas. They also winter in the Great Lakes region and along the Atlantic coast and in the Piedmont from southeastern Pennsylvania to North Carolina. Tundra Swans are common to abundant regular spring and fall migrants through the state. They are most often seen migrating during the day, flying overhead in large V-shaped flocks, or they are heard calling overhead at night. Spectacular concentrations can also be seen at several staging sites. The main migration over Pennsylvania is often a narrow band running diagonally from southeast to northwest, along which hundreds or thousands of birds can be recorded annually. North and south of this narrow band they usually are seen in fewer numbers. Weather patterns have been known to change or disrupt this migration path in some years. Tundra Swans are casual residents in summer and are regular, but local, residents in winter.

Habitat: Tundra Swans prefer large, slow-moving rivers, as well as lakes and large open fields, especially corn and winter wheat stubble near water.

Seasonal status and distribution

Spring: Like those of Canada Geese, the familiar calls of nocturnal migrating Tundra Swans are certainly one of the harbingers of spring. The largest flocks may be found grazing in flooded fields or resting on large rivers or lakes, primarily in the Piedmont and in the Glaciated Northwest. Up to 12,000 birds have been seen in Lancaster Co. at one time in Mar (Morrin et al. 1991). Their typical migration period is from late Feb to early Apr. Peak migration is during the first half of Mar. Stragglers remain to mid-May.

Summer: One or two birds have occasionally remained into summer at several locations in Pennsylvania, including the counties of Adams, Bedford, Centre, Clinton, Crawford, Cumberland, Dauphin, Elk, Erie, Indiana, Lancaster, Snyder, and York. Most unusual was a report of 31 Tundra Swans flying south on 15 Jun 1996 in Lebanon Co.[1]

Fall: Fewer Tundra Swans are observed in fall than in spring. They usually are not found in huge flocks grazing in fields at this season. However, over 5000 birds have been recorded from sites such as the Allegheny River in Pittsburgh and the Ligonier Valley in Westmoreland Co.,[2] and 30,000 were reported on the Allegheny River near Pittsburgh in Nov 1971.[3] Large flocks are also frequently seen from hawk watches along the eastern ridges of the Ridge and Valley. Some migrants arrive as early as mid-Oct. The greatest number of migrants passes through the state during the first half of Nov. At the regular wintering sites in Pennsylvania, numbers continue to build to a peak before freeze-up in mid- to late Dec.

Winter: Tundra Swans may be locally fairly common to abundant during winter. Most wintering Tundra Swans in the state are found along the lower Susquehanna River or at the MCWMA in the Piedmont, where up to several thousand may be found. In the Ridge and Valley, winter records have come from Blair, Centre, Clinton, Luzerne, Montour, and Snyder cos. Small numbers of swans have also been found in winter in the Glaciated Northwest. In the High Plateau they have been recorded only in Cambria, McKean, Somerset, Warren, and Westmoreland cos. West of the High Plateau most wintering swans away from Lake Erie are found on the Allegheny River from Allegheny Co. north to Warren Co. At the Lake Erie Shore up to 100 birds may be found wintering around PISP.

History: The distribution of Tundra Swans in Pennsylvania has changed little throughout history, but their status has changed from low numbers before the twentieth century, probably from the result of shooting, to high numbers in the last half of the twentieth century.

[1] Miller, R. 1996. Local notes: Lebanon County. PB 10:102.

[2] Hall, G.A. 1979. Appalachian region. AB 33:176.

[3] Freeland, D.B. 1972. Area bird summaries for November. 36:9–10.

Wood Duck ***Aix sponsa***

General status: One of the most attractive of all ducks, male Wood Ducks are popular among birders, hunters, and flytiers for their colorful plumage. They breed primarily in the eastern half of the U.S. and locally in the western half of the U.S. and southern Canada. They winter in the southern parts of their breeding range. In Pennsylvania Wood Ducks are fairly common to common regular migrants in spring and fall statewide. They nest throughout the state and are regular winter residents in southeastern Pennsylvania. Wood Ducks are irregular or casual winter residents north of the Piedmont, except in the High Plateau, where they are accidental.

Habitat: In migration Wood Ducks can be found on any body of water or wetland. During the breeding season they prefer wooded wetlands, especially beaver dams, which produce an abundance of dead trees for these cavity nesters. In winter Wood Ducks are found where there is available open water with some cover.

Seasonal status and distribution

Spring: Migration begins in early Mar and peaks in mid- to late Mar. Large concentrations of Wood Ducks can sometimes be found grazing in flooded fields, especially in corn stubble.

Breeding: Wood Ducks are fairly common statewide, nesting in diverse wetland habitats where nest cavities (trees or boxes) are available. Natural cavities and holes of the Pileated Woodpecker are used, often a distance from water.[1] Wood Ducks are relatively inconspicuous during the breeding season, but many boxes are occupied across the state. The PGC estimated 42,000 breeding pairs statewide (average 1989–1995), with the greatest densities in the Piedmont, Glaciated Northeast, and Glaciated Northwest (Dunn and Jacobs 1995). BBS data show a long-term increasing population trend (14% per year). Nests with eggs are found in Apr and May.

Fall: Their typical migration period is from mid-Sep until mid- to late Nov. Peak migration is in early to mid-Oct. Stragglers remain to early Jan or through the winter.

Winter: Wood Ducks winter almost anywhere they can find open water and a food source near the cover of brush or woods. Most sightings of wintering birds are from the Coastal Plain and Piedmont, where they are fairly common to uncommon. Winter records are rare in the remainder of the state, except in the Glaciated Northeast, where apparently Wood Ducks have not been reported. The winter distribution of Wood Ducks in the Ridge and Valley region is primarily along the Susquehanna River Valley and its major tributaries. In the High Plateau they have been found along the Allegheny River in Warren Co.[2] and in the Allegheny Mountains of Cambria, Somerset, and Westmoreland cos. In the Southwest and Glaciated Northwest in winter they have been recorded at only a few sites, in Allegheny, Butler, Crawford, Indiana, and Lawrence cos., and along the Allegheny River in Venango Co. At the Lake Erie Shore they occasionally winter around PISP.

History: Until the end of the 1800s Wood Ducks were widespread and fairly common to common across the state. Unrestricted hunting pressure along with logging and wetland losses seriously reduced numbers during the last decade of the nineteenth century. Harlow (1913) considered Wood Ducks "almost extinct as a summer resident." Federal prohibition of hunting Wood Ducks in 1914 as well as the return of the beaver helped restore the population (Kosack 1995). The resulting Wood Duck recovery was a major conservation success. By the 1920s Sutton (1928a) considered the species "fairly common as a summer resident throughout [Pennsylvania]," and by 1941 a hunting season was again permitted.

Comments: Wood Ducks were previously known as the "summer duck," because they were the dominant duck species to remain as a summer resident in the Mid-Atlantic states at the end of the nineteenth century. They benefit from nest boxes but may abandon nests when boxes are placed too close together.

[1] Fergus, C. undated. Wood Duck. Wildlife notes, no. 175–38. Pennsylvania Game Commission, Harrisburg.
[2] T. Grisez, pers. comm.

Gadwall ***Anas strepera***

General status: Gadwall breed primarily from coastal Alaska south to British Columbia and east across southern Canada and locally across the northern and western U.S. Wintering birds are found locally over most of the U.S. Gadwall are fairly common regular migrants at the Lake Erie Shore and are uncommon to rare regular migrants elsewhere in the state. They are often found mixed with other puddle ducks, especially American Wigeon, during migration. Though there is no recent breeding record, Gadwall formerly bred in Pennsylvania. They are regular winter residents on the Coastal Plain and the Piedmont and at the Lake Erie Shore. Elsewhere, they are irregular or casual in winter.

Habitat: They prefer large ponds, lakes, marshes, and slow-moving rivers.

Seasonal status and distribution

Spring: Their typical migration period is from early Mar to mid-Apr. Peak migration occurs from about the second to the fourth week of Mar. Stragglers remain to late May.

Breeding: The last known breeding in Pennsylvania was near Geneva in Crawford Co. in 1964.[1] They were not confirmed breeding during the BBA project (1983–1989).[2] Nonbreeding Gadwall have been found as rare summer visitors or residents in Armstrong, Berks, Bucks, Centre, Crawford, Erie, Lancaster, Luzerne, Somerset, and Wyoming cos.

Fall: Migration begins in late Aug or early Sep, with gradually increasing numbers building to a peak anytime after the passage of cold fronts from mid-Oct to mid-Nov. At the Lake Erie Shore numbers continue to increase with the approach of winter.

Winter: Gadwall are uncommon to rare winter residents in the Coastal Plain and Piedmont. The only site in the Glaciated Northeast where they have been found in winter is at Tunkhannock in Wyoming Co.[3] The only area in the High Plateau where they have been recorded is in the Allegheny Mountains of Somerset and Westmoreland cos. Most wintering Gadwall in Pennsylvania are found at PISP, where they are fairly common to common. They are casual winter visitors or residents in the remainder of the state.

History: Before 1960, most authorities considered Gadwall to be a rare transient and occasional winter visitor. Gadwall nested on several occasions in Crawford Co. after the Pymatuning Reservoir was created in 1934. The first nest was found in 1934, when R. L. Fricke and Todd (Todd 1940) collected unhatched eggs just west of Linesville. Nests and recently fledged broods were observed in 1934, 1938, 1939, and 1941 (Grimm 1952). Breeding was not noted again in Pennsylvania until 1957, when two nesting pairs were reported near Butler in Butler Co. Young were found in Jun and Jul of 1957 in this area[4] and again at a different location in 1960 and 1961.[5]

[1] Hall, G.A. 1964. Appalachian region. AFN 18:506.
[2] Fingerhood, E.D. 1992. Gadwall *(Anas strepera)*. BBA (Brauning 1992):427–428.
[3] W. Reid, pers. comm.

[4] Brooks, M. 1957. Appalachian region. AFN 11:405.
[5] Hall, G.A. 1961. Appalachian region. AFN 14:469.

Eurasian Wigeon ***Anas penelope***

General status: Eurasian Wigeon breed in Eurasia from Iceland east to central Russia. In North America they are regular in small numbers in winter and in migration. Recently in Pennsylvania, Eurasian Wigeon have become regular spring or fall visitors, but they are rare or accidental at all sites and in all seasons. The number of sightings fluctuates; several years pass without records, followed by several years of frequent sightings. In 1991, a total of at least 11 birds were reported from various parts of the state.[1] Most sightings are of single males associated with American Wigeon. Most records come from the Piedmont, where they are rare regular migrants and casual winter visitors. Eurasian Wigeon are casual migrants in the Coastal Plain, accidental during migration in the Ridge and Valley and High Plateau, and casual migrants in the Southwest and Glaciated Northwest. At the Lake Erie Shore they were casual before 1991 but have been recorded nearly every year since.[1]

Habitat: Eurasian Wigeon are found in marshes and open ponds, lakes, and flooded grain fields, especially cornfields.

Seasonal status and distribution

Spring: They usually arrive with flocks of American Wigeon beginning in late Feb or early Mar. Sometimes Eurasian Wigeon do not appear until late Mar or mid-Apr, long after the earliest American Wigeon. In the spring of 1991, five birds were recorded in Lancaster Co., more than are usually reported statewide in one calendar year.[2] They have been recorded during spring in the Ridge and Valley and High Plateau at least seven times. All spring records were from Lycoming, Montour, Somerset, and Tioga cos. Eurasian Wigeon occasionally linger to late Apr or early May.

Fall: Slightly fewer sightings are made in fall than in spring. Records begin the first week of Oct, with the majority of sightings occurring in late Oct through Nov. A few birds have been reported in Dec and into winter.

Winter: All winter records have come from the Coastal Plain and Piedmont. In the Coastal Plain they were recorded at Penn Manor in Bucks Co. from 10 to 17 Jan 1971.[3] In the Piedmont, they have been recorded near Hatboro in Montgomery Co. on 12 Feb 1974,[4] at Marion in Franklin Co. for a month during the winter of 1987,[5] at Lake Ontelaunee and a pond near Centerport in Berks Co. in the winter of 1990,[6] and in southern Lancaster Co. at Chestnut Level and Wakefield through the winters of 1990–1996.[7] Both males and females have been reported wintering in Lancaster Co.

History: The earliest mention of the occurrence of Eurasian Wigeon in Pennsylvania was from Turnbull (1869). Todd (1940) mentioned an influx of the species at Pymatuning during 1936–1937. In the southeastern part of the state they were reported from many sites from Berks, Bucks, Chester, Cumberland, Delaware, Lebanon, Philadelphia, and Snyder cos. In the western part of the state they were reported from Crawford, Elk, Erie, and McKean cos. Several specimens have been collected, including one placed at the State Museum in Harrisburg and one at the Pymatuning Museum in Crawford Co. (Poole, unpbl. ms.).

Comments: Though most birds are believed to be of wild origin, a few

Eurasian Wigeon are kept in captivity, increasing the possibility that some sightings are of escapees.

[1] G.M. McWilliams, pers. obs.
[2] Hall, G.A. 1991. Appalachian region. AB 45:444.
[3] Meritt, J.K. 1971–1972. Field notes. Cassinia 53:56.
[4] Scott, F.R., and D.A. Cutler. 1974. Middle Atlantic Coast region. AB 28:622.
[5] Hall, G.A. 1987. Appalachian region. AB 41:282.
[6] R. Keller, pers. comm.
[7] E. Witmer, pers. comm.

American Wigeon *Anas americana*

General status: American Wigeon breed from Alaska south across Canada and the western U.S. and locally east of Minnesota. In winter they are found from Alaska south to New Mexico and across the southern states and sporadically elsewhere in the U.S. American Wigeon are fairly common to common regular migrants statewide. They sometimes associate with American Coots and diving ducks and steal food brought to the surface. They have nested once in recent years in Crawford Co. American Wigeon are regular winter residents in the Coastal Plain and Piedmont and local and irregular elsewhere in winter.

Habitat: American Wigeon are found on ponds, lakes, rivers, and marshes and in flooded fields during migration. They may be found on deeper water than other puddle ducks, especially with feeding diving ducks.

Seasonal status and distribution

Spring: The earliest migrants have been recorded the second or third week of Feb, but the usual migration period is from the fourth week of Feb to mid-May. Migration usually peaks from mid- to late Mar. Lingering birds may be found to the fourth week of May.

Breeding: A family group of young wigeon was seen on a small pond in the Conneaut Marsh in Jul 1986.[1] This constitutes the only confirmed nesting since 1960. No nest has ever been found for this species in Pennsylvania. Individuals believed to be nonbreeders are found primarily in Erie and Crawford cos., but they have also been recorded in Berks, Cambria, Indiana, Lancaster, McKean, and Philadelphia cos.

Fall: They sometimes begin migrating as early as the last week of Aug. The usual migration season is from the fourth week of Sep to mid-Dec. The peak number of birds occurs from mid-Oct to early Nov.

Winter: American Wigeon are uncommon during winter in the Coastal Plain and Piedmont and generally avoid the central mountains. They are rare in the Ridge and Valley and the High Plateau, but sightings have come from the counties of Centre, Clinton, Huntingdon, Luzerne, Lycoming, Schuylkill, Snyder, Somerset, Union, Warren, Westmoreland, and Wyoming. West of the Allegheny Front, they are also rare and are usually found with other species of wintering ducks where there is open water.

History: The report by B. L. Oudette, who "told Mr. Todd of seeing adults with young" in Jul 1936 (Trimble 1940) was not reported by Todd (1940), so this, the first confirmed nesting report, is questionable. Still, Todd (1940) considered Oudette's observations of pairs sufficient evidence to consider them nesting.

[1] Hartman, F.E. 1992. American Wigeon *(Anas americana)*. BBA (Brauning 1992):80–81.

American Black Duck *Anas rubripes*

General status: The American Black Duck has decreased in recent years because of habitat destruction, competition, and interbreeding with Mallards. They breed in the northeastern quarter of North America from northern Quebec south to northern Illinois and east to coastal North Carolina and winter within their breeding range south to Texas. They are common to abundant regular spring and fall migrants in Pennsylvania and are usually associated with Mallards. American Black Ducks breed locally in the state and are regular winter residents.

Habitat: They can be found on any body of water or wetland in migration. In the breeding season they prefer wooded wetlands and slow-moving streams near woods or brush, especially those dammed by beavers.

Seasonal status and distribution

Spring: One of the earliest spring waterfowl migrants, American Black Ducks usually appear with the first spring thaw in mid- to late Feb, when hundreds frequently can be seen feeding in flooded corn stubble fields. Birds occasionally arrive before there is open water and can be seen huddled together on frozen lakes and rivers. Migration peaks in early to mid-Mar. The number of birds steadily declines by early Apr, when migration is nearly completed. A few migrants linger to mid-May.

Breeding: They are widely distributed and rare over most of the state but may be locally uncommon breeding birds. They have been observed in nearly every county, but some summering birds could be nonbreeders. The Pocono section contains the highest nesting densities in the state.[1] Hybridization with Mallards further confuses their status. Black ducks nest in a variety of wetlands, favoring shrubbier or more wooded settings than other puddle ducks. Nests rarely are documented in Pennsylvania; waterfowl surveys estimated an average of 1500 breeding pairs from 1989 to 1995 (Dunn and Jacobs 1995). American Black Ducks are on eggs from late Apr through May, with renesting into Jun. Todd (1940) stated that they nest earlier than Mallards.

Fall: Migration begins in early to mid-Sep, but most migrants accompany cold fronts in mid- to late Oct. Migration is usually completed with the arrival of heavy snow and ice in late Dec or early Jan. Most birds that remain at this time will probably winter.

Winter: Very hardy birds, American Black Ducks and Mallards are the two most common wintering puddle ducks in the state. They are widely distributed and have been found wherever food and open water are available. More winter in the Coastal Plain and Piedmont than in any other region in the state. They are common to abundant primarily on the Delaware and the lower Susquehanna rivers. Large wintering flocks can also be found at MCWMA, Pymatuning, and PISP.

History: American Black Ducks were rare and local as breeders before 1900. Harlow (1913) mentioned them among the "hypothetical" breeders but provided no supporting evidence. Poole (unpbl. ms.) stated that "prior to about 1930, when semi-domesticated Mallards started to usurp its place, it was the common breeding duck of eastern Pennsylvania." Black ducks responded to the impoundment of the Pymatuning Swamp with a dramatic increase in nesting activity (Trimble 1940).

[1] J.P. Dunn, pers. comm.

Mallard ***Anas platyrhynchos***

General status: Probably the most familiar duck to the casual observer, Mallards are as frequently encountered around city parks and farms as they are in the wild. Mallards, widespread in North America, breed from Alaska south through most of the U.S. except in the Gulf states and Florida, where they are local. They winter across the U.S. Mallards are common to abundant regular spring and fall migrants in the state and are the most abundant and widespread breeding ducks, as well as regular winter residents. They may hybridize with many species of ducks but most frequently with American Black Ducks.

Habitat: Any body of water or wetland will support Mallards in migration. In the breeding season, they are most common in farmlands. Mallards prefer ponds, lakes, or slow-moving streams where there is sufficient grass or brush along the shoreline for nesting. Nesting is less common in wooded wetlands. In winter Mallards will use any available open water.

Seasonal status and distribution

Spring: Migrants appear with the first spring thaw, usually around late Feb or early Mar. Peak migration occurs in mid-Mar, with decreasing numbers by early to mid-Apr. Mallards are frequently found with American Black Ducks feeding in flooded corn stubble fields.

Breeding: Mallards nest commonly across the state, regularly in every county, and often in association with humans. They are semidomestic in many cities, towns, parks, and farms. The summer population increased at a rate of 4% per year since 1966 on BBS routes and showed a strong increase on various waterfowl surveys (Dunn and Jacobs 1995). The average (1989–1995) estimated state population was 80,000 breeding pairs (Dunn and Jacobs 1995), with estimates above 100,000 in 1996–1997.[1] On BBS routes Mallards were most abundant in the Piedmont, and their lowest densities were associated with the High Plateau. Nests, usually placed on the ground a short distance from water, have been found with eggs between 1 Apr and mid-Jul.

Fall: Fall migration begins around mid-Sep. Most migrants appear in mid- to late Oct with the passage of cold fronts. Migration is usually completed with the arrival of heavy snow and ice in late Dec or early Jan.

Winter: Mallards are the most widely distributed wintering ducks in the state. They have been recorded in nearly all counties of Pennsylvania. Their wide distribution may be partly attributable to their association with humans; they often join domestic stock and readily accept handouts. The largest winter concentration of Mallards is in the Coastal Plain and Piedmont, where they are common to abundant. In the Glaciated Northeast, High Plateau, and Southwest they may be found where there is food and open water, but fewer birds are in these regions. Large wintering flocks also can be found at Pymatuning and at PISP.

History: Audubon (1840) mentioned Mallards as breeding in Pennsylvania at the beginning of the nineteenth century, but they became scarce during the later half of the 1800s (Warren 1890), the time of market hunting. Like American Black Ducks, they were mentioned in Harlow's (1913) thesis among the "hypothetical" breeders; Harlow stated that "in former years the Mallard was reported to breed." He listed Loyalsock Creek in Lycoming Co. and Presque Isle in Erie Co. as locations. Mallards, and Canada Geese, experienced some of the most dramatic expansions in breeding distribution of any species in the state, both abetted by semidomestic stock. Mallards were released privately in Chester Co. in 1911, from which they spread widely (Burns 1919). Sutton (1928a) considered them local and uncommon breeders across the state and common at Conneaut Marsh (Sutton 1928b). Since about the 1950s, they have become fairly common statewide. Captive-reared Mallards were released statewide by the PGC between 1951 and 1981, but these were not considered to have contributed to resident breeding populations (Dunn and Jacobs 1995).

Comments: Mallards are easily domesticated and are popular among waterfowl collectors, whose many strains have produced a variety of plumages. These Mallard strains occasionally hybridize with wild stock or with other species, producing unusually patterned offspring. These hybrid types are common in zoos and in city parks.

[1] Dunn, J.P., and K. Jacobs. 1997. 1996 duck and goose harvest summary and 1997 waterfowl populations for Pennsylvania and the Atlantic flyway. Pennsylvania Game Commission Report. Pennsylvania Game Commission, Harrisburg.

Blue-winged Teal ***Anas discors***

General status: Blue-winged Teal breed across the maritime provinces and south across the U.S. They winter mostly from coastal California across the southwestern and Gulf states and north along the Atlantic coast to North Carolina. Of the waterfowl, Blue-winged Teal are the latest spring and the earliest fall migrants to pass through Pennsylvania. They migrate through the state rather quickly during both seasons. Blue-winged Teal are fairly common spring and fall migrants and uncommon, regular breeders in the state. They are casual in winter.

Habitat: In migration, Blue-winged Teal inhabit the shorelines of shallow ponds, lakes, and marshes and flooded fields. During the breeding season, they are found in open marshes that have a diversity of aquatic vegetation and adjacent areas of grass for nesting.

Seasonal status and distribution

Spring: Extremely early spring arrivals have been recorded as early as 2 Mar.[1] Their usual migration period is from late Mar to late May. Peak migration is in early to mid-Apr.

Breeding: Blue-winged Teal are uncommon breeders in extensive emergent wetlands at Tinicum, locally in the Glaciated Northwest, and at PISP. They may occur casually in any region with emergent marshes and were confirmed breeding in each of the major physiographic regions during the BBA project.[2] Eggs have been found from mid-May to early Jun.

Fall: Migration begins in early Aug, peaks by mid-Sep, and usually ends by the second week of Oct. Some birds linger to early Nov or occasionally later if the weather is mild.

Winter: Most records after 5 Jan are from the Coastal Plain and Piedmont, although Wood (1983) stated, for Centre Co., "winters occasionally." Pennsylvania's harsh winters severely reduce suitable habitat for this species. A male spent the entire winter of 1969–1970 at a small pond near Sugar Lake, Crawford Co.,[3] and a bird was reported at Johnstown, Cambria Co., on 21 Jan 1957 (Poole, unpbl. ms.). Most Jan records are not wintering individuals but birds that have remained from fall migration until as late as mid-Jan because of unseasonably mild weather. Some birds linger because they were wounded by hunters. In the Coastal Plain, especially at Tinicum,[4] this species has survived during mild winters. Several records, all of a single bird, are known from the Piedmont: Allentown in Lehigh Co. on 13 Jan 1982;[5] Hammer Creek in Lancaster Co. to 5 Feb 1987 (Morrin et al. 1991); Green Lane Reservoir in Montgomery Co. on 16 Jan 1989;[6] and West Fairview in Cumberland Co. in the winter of 1993.[7]

History: Poole (unpbl. ms.) reported that Blue-winged Teal was "unknown as a breeding bird in Pennsylvania before 1932, when a nest was found at Presque

Isle." Previous summer records were presumed, but not confirmed, to be of breeding birds (Sutton 1928b). Blue-winged Teal were reported breeding regularly, but rarely, since the 1930s in Crawford Co. (Grimm 1952). They also nested in Berks, Philadelphia, Monroe, and Centre cos. and possibly were overlooked elsewhere.

[1] N. Pulcinella, pers. comm.
[2] Hartman, F.E. 1992. Blue-winged Teal *(Anas discors)*. BBA (Brauning 1992):76–77.
[3] Hall, G.A. 1970. Appalachian region. AB 24:503.
[4] J.C. Miller, pers. comm.
[5] Paxton, R.O., W.J. Boyle Jr., and D.A. Cutler. 1982. Hudson-Delaware region. AB 36:277.
[6] Paxton, R.O., W.J. Boyle Jr., and D.A. Cutler. 1989. Hudson-Delaware region. 43:290.
[7] Hoffman, D. 1993. Cumberland Co. report. PB 7:20.

Cinnamon Teal *Anas cyanoptera*

General status: Cinnamon Teal breed in southwestern Canada and the western U.S. and they winter from California to Texas. Cinnamon Teal are rare stragglers in the eastern U.S. There have been six or seven records of males identified in Pennsylvania. Some sightings may be escapees because some birds, especially males, are kept in captivity.

Seasonal status and distribution

Spring: Most records have been during spring. A bird seen by many observers was photographed at Mammoth Park, Westmoreland Co. on 27 Apr 1985, and the same or perhaps a different bird was found by R. C. Leberman about 10 miles away at Donegal Lake, from 8 to 31 May 1985.[1] On 18 Mar 1984 one was seen by D. Darney at MSP in Butler Co.,[2] and another was identified by J. Sweet at PISP in Erie Co. in Mar 1959.[3]

Fall: The only three fall records are of one bird identified by W. Reid and T. Reid at Tunkhannock in Wyoming Co. on 22 Aug 1974,[4] one identified by A. Zaid at Octoraro Reservoir in Chester Co. from 25 to 27 Nov 1994,[5] and another found by J. Fedak at MSP in Butler Co. on 5 Nov 1996.[6]

History: The first Pennsylvania record came from Quemahoning Reservoir in Somerset Co. on 16–17 Apr 1954 (Leberman 1988).

Comments: A hybrid male Blue-winged Teal × Cinnamon Teal was photographed at Lake Somerset in Somerset Co.[7]

[1] Hall, G.A. 1985. Appalachian region. AB 39:299.
[2] Hall, G.A. 1984. Appalachian region. AB 38:910.
[3] Stull, J.H., and J.G. Stull. 1959. Field notes. Sandpiper 2:9–10.

Cinnamon Teal at Donegal Lake, Westmoreland County, May 1985. (Photo: Dave Darney)

[4] Buckley, P.A., and R.P. Kane. 1974. Hudson St. Lawrence region. AB 28:616.
[5] Pulcinella, N. 1995. Rare bird reports. PB 8:218.
[6] Hess, P. 1997. Rare bird reports. PB 10:224.
[7] D. Darney, pers. comm.

Northern Shoveler ***Anas clypeata***

General status: The low swimming profile and the unique large spatulate bill easily identify this primarily western U.S. species. Shovelers breed primarily from Alaska south across Canada to northern Ontario and in the U.S. west of the Mississippi River. They winter from California and Utah south to Texas and east across the Gulf states to Florida and north along the coast to New Jersey. Shovelers are fairly common regular migrants in the northwestern and southeastern part of Pennsylvania. They can be locally common during migration at JHNWR, at Darby Creek, Delaware Co.,[1] and at Conneaut Marsh, Crawford Co. They are uncommon to rare regular migrants elsewhere in the state, most numerous statewide in spring. Shovelers are usually found mixed with other species of puddle ducks, especially Blue-winged Teal. They occasionally breed in Pennsylvania. Shovelers are regular winter residents in the Coastal Plain in small numbers and are casual in the Piedmont and at the Lake Erie Shore. They are accidental elsewhere in the state during winter.

Habitat: Like Blue-winged Teal, Northern Shovelers prefer the shorelines of shallow ponds, lakes, and marshes that have abundant aquatic vegetation and flooded fields.

Seasonal status and distribution

Spring: Their usual migration period is from mid-Mar to late Apr, with stragglers to late May. Peak migration occurs in late Mar or early Apr. Flocks are usually small, with concentrations of fewer than 50 birds.

Breeding: A nest with eggs was found on 15 May 1966, and fledged young were found the following year at Tinicum (J.C. Miller 1970). Nesting was confirmed again at the Tinicum wetlands during the BBA project in 1984. At MCWMA in Lancaster Co. a brood was located during the 1970s.[2] Shovelers are rare, regular summer residents in the Conneaut Marsh, with broods of young observed at Pymatuning and Conneaut Marsh in 1990.[3] Nonbreeding individuals occasionally have been found summering in wetland locations in Allegheny, Centre, Crawford, Dauphin, Elk, Lancaster, Lebanon, and Montgomery cos.

Fall: Fewer birds are seen in fall than in spring. Migration may begin as early as early Aug with the appearance of singles and pairs. Their usual migration period is from mid-Sep to late Nov. Flights are unpredictable, and small flocks may appear at any time in the fall with the passage of cold fronts. Stragglers remain through Dec and into early or mid-Jan.

Winter: If water remains unfrozen, a few shovelers will winter at Tinicum, where they are uncommon to rare.[1] They are casual winter residents in the Piedmont on unfrozen lakes. In the Ridge and Valley they were recorded from Luzerne Co. in 1968.[4] Wood (1983) listed them in Centre Co. as "winters occasionally." Apparently Northern Shovelers have not been recorded in the Glaciated Northeast in winter. The only winter records in the High Plateau come from Warren Co. at Kinzua Dam, where one remained to at least 11 Jan 1997,[5] and in the Allegheny Mountains in Somerset Co., where they have been recorded at Strongcreek River, at Quemahoning Dam, and from a spring-fed pond at Jennerstown.[6] In the Southwest there is a record of one on 17 Jan 1989 from

Monroeville in Allegheny Co.,[7] and one successfully wintered at the juncture of the Allegheny, Ohio, and Youghiogheny rivers in Pittsburgh.[8] In the Glaciated Northwest there is a record from MSP in Butler Co.[9] as well as in Crawford Co., where they wintered in 1995.[10] Shovelers are casual winter visitors or residents at PISP.[11]

History: R. L. Fricke found the state's first shoveler nest in 1935 at Pymatuning (Trimble 1940). The breeding population increased in subsequent years, but by the 1940s they were rare and since then have been found nesting in the region only irregularly (Grimm 1952).

[1] N. Pulcinella, pers. comm.
[2] Hartman, F.E. 1992. Northern Shoveler *(Anas clypeata)*. BBA (Brauning 1992):78–79.
[3] R.C. Leberman, pers. comm.
[4] M. Blauer, pers. comm.
[5] Grisez, T. 1997. Local notes: Warren County. PB 11:34.
[6] R. Sager and G. Sager, pers. comm.
[7] Grom, J. 1989. County reports—January through March 1989: Allegheny County. PB 3:23.
[8] Fialkovich, M. 1996. Allegheny Co. report. 10:16.
[9] P. Hess, pers. comm.
[10] Haas, F.C., and B.M. Haas, eds. 1995. Seasonal occurrence tables—January through March 1995. PB 9:41.
[11] G.M. McWilliams, pers. obs.

Northern Pintail *Anas acuta*

General status: Northern Pintails breed throughout North America except east and south of Oklahoma and they winter across most of the U.S. These puddle ducks with a long needle-like tail were once abundant migrants in the northwestern and southeastern portions of the state. Now, Northern Pintails are uncommon to fairly common regular migrants over most of the state, but rare to uncommon in the mountainous areas. In summer, Northern Pintails are irregular visitors or residents and accidental breeders. They are regular winter residents over most of the state, except in the High Plateau, where they are accidental.

Habitat: Northern Pintails frequent marshes, open ponds, lakes, and slow-moving streams. They also use flooded grain fields, especially cornfields, during migration.

Seasonal status and distribution

Spring: They are most common at this season. Northern Pintails are one of the earliest puddle ducks to appear in spring. They frequently accompany American Black Ducks in flooded fields, especially corn stubble. Birds appear as early as mid-Feb. Their typical migration period is from late Feb or early Mar until early to mid-Apr. Peak numbers of Northern Pintail occur from early to mid-Mar. Some birds linger to early May.

Breeding: Nonbreeding individuals are occasionally found in the Coastal Plain (at Tinicum), Piedmont, Glaciated Northeast, Glaciated Northwest (Conneaut Lake and Pymatuning), and Lake Erie Shore (PISP). Summer sightings in late Jul or early Aug may be very early migrants. Nesting was documented only twice since 1960: a female with six young on 7 Jun 1966[1] and one with nine young on 14 May 1969 were discovered at the Tinicum Refuge, Philadelphia Co. (J.C. Miller 1970). Banding operations during mid-Aug at Pymatuning and MCWMA have consistently captured both adult pintails and hatching-year young since 1989 (Dunn and Jacobs 1995). These perhaps are not local breeding birds but suggest that further investigation is needed.

Fall: Pintails are far less common in fall than in spring. Fall migration is difficult to discern, but individuals and small groups may enter the state as early as late Aug or early Sep. Small movements of birds accompany cold fronts throughout the fall. Most birds are gone by the arrival of severe winter weather in late Dec or early Jan.

Winter: They are usually found with wintering Canada Geese, American Black Ducks, or Mallards. Most wintering pintails in Pennsylvania are found east of the Ridge and Valley in the Coastal Plain, where they occasionally may be common at Tinicum. In the Piedmont they are fairly common. North of the Piedmont, most birds are found on the Susquehanna River, but they have been seen in winter as far north as Tioga Co.[2] In the High Plateau, wintering Northern Pintails have been recorded only in Bradford, Somerset, and Westmoreland cos. and along the Allegheny River at Warren in Warren Co.[3] Northern Pintails are rare and local in winter west of the High Plateau, with records in Allegheny, Armstrong, Crawford, Erie, Indiana, Venango, and Westmoreland cos.

History: Pintails were formerly more common. Thousands were once recorded at Tinicum; on the Glenolden CBC, 107,537 pintails were tallied on 22 Dec 1951 (Poole, unpbl. ms.). As many as 3000 were reported from Conneaut Marsh in Crawford Co. on 11 Mar 1961.[4] No evidence of this species' breeding in Pennsylvania was available until the Pymatuning Reservoir was filled in 1934. Along with several other waterfowl species, pintails' first documented breeding in Pennsylvania was at Pymatuning when R. L. Fricke found a nest in May 1934. Nesting was reported sporadically until 1952 (Grimm 1952) but has not been reported there since.[5]

[1] Miller, J. 1966. A nesting record of the pintail at Tinicum Wildlife Preserve. Cassinia 49:30.

[2] Haas, F.C., and B.M. Haas, eds. 1997. Seasonal occurrence tables—January through March 1997. PB 11:46.

[3] T. Grisez, pers. comm.

[4] Hall, G.A. 1961. Appalachian region. AFN 15: 329.

[5] Fingerhood, E.D. 1992. Northern Pintail *(Anas acuta)*. BBA (Brauning 1992):427.

Green-winged Teal *Anas crecca*

General status: The breeding range of Green-winged Teal extends from Alaska across most of Canada and the western U.S. and locally through the northern and Mid-Atlantic states to coastal Maryland. They winter over most of the U.S. The smallest waterfowl found in Pennsylvania, Green-winged Teal are fairly common to common regular spring and fall migrants. During migration they frequently associate with Blue-winged Teal. Green-wing Teal breed in the state. They are regular winter residents in the Coastal Plain and in the Piedmont. Green-winged Teal are casual elsewhere in the state in winter except in the High Plateau, where they are accidental.

Habitat: Green-winged Teal may be found on any body of water or wetland during migration. During the breeding season they prefer marshes, lakes, and ponds surrounded by grass or brush. In winter they are found where there is open, shallow water.

Seasonal status and distribution

Spring: Their usual migration period is from early to mid-Mar to late Apr or early May, with stragglers into summer. Peak migration is in late Mar or early Apr.

Breeding: These teal are rare and local breeders in northern-tier counties, but they also have nested in the Tinicum area. They are regularly observed during the breeding season at Splashdam Pond, Wyoming Co. Green-winged Teal may be overlooked at breeding areas; nonbreeding birds sometimes summer in larger wetland areas. Few nests have been found in the state.

Fall: Fall migration begins in mid-Aug and extends well into late Nov. Peak migration dates may vary from year to year but usually follow cold fronts from mid-Sep to mid-Nov. Usually two peaks occur in the fall, one in Sep and one in Nov. Early flocks may be mixed with Blue-winged Teal, but later flocks usually consist entirely of Green-winged Teal. Nonwintering birds sometimes remain into Dec or early Jan.

Winter: Most records of wintering birds are in the Coastal Plain at Tinicum, where they are fairly common to common,[1] and in the Piedmont, where they are uncommon. Away from the Coastal Plain and Piedmont they are rare. Elsewhere, Green-winged Teal have been recorded in winter in Allegheny, Butler, Carbon, Centre, Crawford, Dauphin, Erie, Indiana, Luzerne, Montour, Perry, Somerset, Venango, Warren, Westmoreland, and Wyoming cos.

History: Suggested to be a breeder in Lycoming Co. (Warren 1890), the first documented breeding of Green-winged Teal in Pennsylvania was a failed nest found by R. L. Fricke at Pymatuning in 1936 (Trimble 1940). They nested at Tinicum in the 1950s (J.C. Miller and Price 1959) and subsequently in 1966–1967 (Miller 1970). Green-winged Teal have been documented breeding in six counties (Crawford, Delaware, Erie, Philadelphia, Wayne, and Wyoming).

Comments: The Eurasian subspecies, *A. c. crecca*, was first recorded in the state on 3 Feb 1938 on a pond adjoining Lake Ontelaunee in Berks Co. (Poole, unpbl. ms.). They are casual spring and fall visitors to the Coastal Plain and Piedmont, usually with *A. c. carolinensis*. Most migrant records are from Tinicum. *Anas c. crecca* has been most recently recorded in Lancaster Co., at Long's Park (Morrin et al. 1991), Maiden Creek,[2] and the MCWMA.[3] They have also been recorded as a winter visitor or resident at Tinicum[4] and Peace Valley Park.[5] The only site where this subspecies has been reported away from the Coastal Plain and Piedmont in recent history is in Crawford Co., where a bird was seen on 8 Apr 1961 near Sugar Lake.[6] Possible intergrades between the two subspecies have been recorded at Lake Ontelaunee, Berks Co. (Poole 1947) and at PISP.[7]

[1] J.C. Miller, pers. comm.

[2] Scott, F.R., and D.A. Cutler. 1972. Middle Atlantic Coast region. AB 26:42.

[3] E. Witmer, pers. comm.

[4] Scott, F.R., and D.A. Cutler. 1966. Middle Atlantic Coast region. AFN 20:407.

[5] Boyle, W.J., Jr., R.O. Paxton, and D.A. Cutler. 1992. Hudson-Delaware region. AB 46:240.

[6] Leberman, R.C. 1961. Field notes. Sandpiper 3:67.

[7] McWilliams, G.M. 1995. Local notes: Erie County. PB 9:93.

Canvasback *Aythya valisineria*

General status: Canvasbacks suffered severe population losses during the twentieth century. The draining of prairie wetlands, droughts, and overshooting contributed to their decline. More recently they have slowly regained some of their former numbers. Canvasbacks breed from Alaska across Canada to Ontario and locally south to New Mexico in the West and New York in the East. They winter from British Columbia south to Colorado and along the Pacific coast, south through Mexico, east to Florida and north to the Great Lakes and New England. Canvasbacks are found across Pennsylvania on the larger bodies of water during migration but are rarely found elsewhere in the numbers that are found around PISP. They frequently associate with other diving ducks during migration and in winter. In summer, Canvasbacks are casual at the Lake Erie Shore and accidental elsewhere in the state.

Habitat: Canvasbacks prefer deep lakes and large, slow-moving rivers, but they may appear briefly on smaller bodies of water. They are occasionally found grazing in flooded fields during spring migration.

Seasonal status and distribution

Spring: The largest number of Canvasbacks is seen in spring. In late Feb, migrant flocks begin to join those that have wintered. Their typical migration period is from late Feb to early Apr. They peak during the second or third week of Mar. The greatest concentrations are usually found just offshore in Lake Erie and on Presque Isle Bay, where thousands are found annually. Fewer Canvasbacks are found in the remainder of the state, although as many as 555 have been recorded at MSP in Butler Co. in the spring of 1984.[1] On 5 Apr 1996, a very high count of 1700 birds was recorded at Lake Somerset in Somerset Co.[2] A few birds may linger in the state to late Apr.

Summer: At the Lake Erie Shore, Canvasbacks have been recorded several times from Presque Isle Bay, where singles or pairs have occasionally summered.[3] Two inland records come from Erie Co.: one at Edinboro Lake on 7 Jul 1974[4] and one bird from Siegel Marsh on 6 Aug 1980.[5] A Canvasback was at a fish farm in Chester Co. on 22 Jun 1991.[6]

Fall: Canvasbacks are among the last ducks to pass through the state in fall. Migration begins the last week of Oct or the first week of Nov, usually with the passage of the first cold front. Most flights accompany cold fronts, with numbers building at the Lake Erie Shore to a peak concentration in mid- to late Dec. Numbers during some years reach the thousands. On 1 Jan 1980, a total of 7000 Canvasbacks was estimated in Presque Isle Bay.[7] Migrants are observed mainly over Lake Erie.

Winter: The greatest number of Canvasbacks wintering within Pennsylvania is seen in the waters surrounding PISP, where several thousand may spend the winter if open water permits. Canvasbacks are uncommon to rare in the southeast, where most are encountered along the Delaware, Lehigh, and Susquehanna rivers. In the Ridge and Valley, Canvasbacks are found on the Susquehanna River in Clinton, Dauphin, and Perry cos. and on lakes in Carbon, Centre, Huntingdon, Montour, Lehigh, and Luzerne cos. In the High Plateau they have been found in winter at Prince Gallitzin State Park in Cambria Co.,[8] and along the Allegheny River at Warren in Warren Co.,[9] and at PNR.[10] They are rare and quite local in the western regions, most often found on the slower and deeper portions of the Allegheny River. Away from the Allegheny River winter sightings come from Allegheny, Beaver, Butler, Crawford, Indiana, Somerset, Washington, and Westmoreland cos.

History: During the beginning of the twentieth century, two important factors contributed to the decline of Canvasbacks: hunting pressure and the loss of an important food source, wild celery, within their staging and wintering sites. Since then, this species has experienced population fluctuations throughout North America. In the 1930s the population decreased after a series of drought years that seriously affected their breeding success. In the 1960s and 1970s the drainage of wetlands was an important factor in the reduction of the total population to 500,000 birds, a 50% decline from population estimates made in the 1940s and 1950s. Oil spills also have been a factor in the reduction of numbers wintering in recent years (Madge and Burn 1988). In Pennsylvania, during Alexander Wilson's day in the early nineteenth century, Canvasbacks were considered abundant on their wintering grounds on the lower Delaware and Susquehanna rivers

(Poole, unpbl. ms.). Todd (1940) stated that hunting pressure was responsible for the decline of this duck, especially at Presque Isle Bay.

[1] Hall, G.A. 1984. Appalachian region. AB 38:910.
[2] Bastian, S. 1996. Local notes: Somerset County. PB 10:104.
[3] G.M. McWilliams, pers. obs.
[4] Hall, G.A. 1974. Appalachian region. AB 28:900.
[5] Hall, G.A. 1981. Appalachian region. AB 35:182.
[6] Pasquarella, J. 1991. County reports—April through June 1991: Chester County. PB 5:82.
[7] Hall, G.A. 1980. Appalachian region. AB 34:272.
[8] G. Lamer, pers. comm.
[9] T. Grisez, pers. comm.
[10] R.C. Leberman, pers. comm.

Redhead ***Aythya americana***

General status: Redhead numbers declined during the twentieth century as a result of the draining of wetlands, overshooting, and droughts on their breeding grounds. The world population estimate during the 1970s was around 600,000 individuals and now seems to be stable (Madge and Burn 1988). Redheads breed locally in central Alaska and across southern Canada south to Texas in the West and to Illinois and New York in the East. They winter from Mexico and across the southern half of the U.S. north to the Great Lakes and New England. Like Canvasbacks, with which they frequently flock, Redheads are most numerous in Pennsylvania at the Lake Erie Shore. Redheads are common to abundant regular migrants in spring and late fall. Elsewhere in the state they are fairly common to rare regular migrants. They are casual summer visitors or residents at the Lake Erie Shore and are accidental elsewhere. Redheads nested historically. In winter, they are regular residents in the Coastal Plain, Piedmont, and at Lake Erie Shore and irregular to casual elsewhere.

Habitat: Redheads prefer deep lakes and large slow-moving rivers but may be found on smaller bodies of water.

Seasonal status and distribution

Spring: The greatest number of birds is seen during spring. The largest concentrations of Redheads are usually found just off shore in Lake Erie and on Presque Isle Bay. Migrant flocks of Redheads begin joining those that have wintered there in late Feb, with numbers building to a peak during the second or third week of Mar. The numbers rapidly diminish by late Mar or early Apr, and a few birds may linger to late Apr. Even fewer Redheads than Canvasbacks are found in the rest of the state at this season. However, exceptional concentrations at inland sites include 350 birds at MSP in Butler Co. in late Mar of 1980[1] and 900 at Pymatuning on 4 Apr 1972.[2]

Summer: A female Redhead was found on 24 Jun 1960[3] at Lake Ontelaunee in Berks Co., and a pair remained at Muddy Run Sewage Ponds in Lancaster Co. until 4 Jun 1978.[4] A late Redhead was at Pleasant Gap in Centre Co. to 6 Jun 1985,[5] and 6, which may have been early fall migrants, were at Kahle Lake in Clarion Co. on 29 Aug 1989.[6] At the Lake Erie Shore they have been recorded several times from Presque Isle Bay, where single individuals or pairs have occasionally summered.[7] Most summer records in the state are Jun birds that have lingered from spring migration.

Fall: Fall migration is not well defined. A few birds begin trickling through in early to mid-Oct. Most birds are seen beginning in late Nov with the passage of cold fronts; they gradually peak in mid- to late Dec.

Winter: Redheads may flock with Canvasbacks, but they tend to form tightly compacted flocks of their own kind. Most birds winter at the Lake Erie Shore in the waters surrounding PISP, where several hundred may remain if open water permits. They are uncommon to rare in the Coastal Plain and Piedmont, where most are encountered along the Delaware, Lehigh, and Susquehanna rivers. A flock of 34 that wintered in a lake near Allentown in 1977 was a very high inland count.[8] In the Ridge and Valley they have been recorded on the Susquehanna River and on lakes in Blair, Carbon, Centre, Dauphin, Huntingdon, Luzerne, and Perry cos. West of the Allegheny Front away from Lake Erie, they have been recorded during winter in Allegheny, Armstrong, Butler, Clarion, Crawford, Indiana, Mercer, Somerset, Westmoreland, and Warren cos.

History: So notable was the breeding of Redheads at Pymatuning after the flooding of the Reservoir in 1934 that Todd wrote a paper in the *Auk* describing the event.[9] He said elsewhere (Todd 1940) that the nesting came as "a great surprise." No nest was ever found, but as many as 20 pairs were present (Trimble 1940) during at least 1936 and 1937, and a few ducklings were collected to confirm the identity (Todd 1940). Unfortunately, Todd's (1940) confident assertion that the "Redhead seems likely to become established as a regular summer resident" failed to prove true (Grimm 1952). No further evidence of nesting by Redheads was obtained from Pymatuning or elsewhere in Pennsylvania. Todd (1940) stated that numbers had been depleted by excessive shooting and by drought.

[1] Hall, G.A. 1980. Appalachian region. AB 35:822.
[2] Robinson, R., and S. Robinson. 1974. Field notes. Sandpiper 14:64.
[3] R. Keller, pers. comm.
[4] E. Witmer, pers. comm.
[5] Hall, G.A. 1985. Appalachian region. AB 39: 911.
[6] Buckwalter, M. 1989. County reports—July through September 1989: Clarion County. PB 3:103.
[7] G.M. McWilliams, pers. obs.
[8] Buckley, P.A., R.O. Paxton, and D.A. Cutler. 1977. Hudson-Delaware region. AB 31:313.
[9] Todd, W.E.C. 1936. The Redhead and Ring-necked Duck breeding at Pymatuning Lake, Pennsylvania. Auk 53:440.

Ring-necked Duck *Aythya collaris*

General status: Most breeding Ring-necked Ducks range across Canada south from California east locally across the prairie states to the New England states. They winter primarily across the southern states and north along the Pacific and Atlantic coasts. Ring-necked Ducks are common to abundant regular spring migrants throughout Pennsylvania and are uncommon to fairly common to common regular fall migrants. Ring-necked Ducks are casual summer visitors or residents. They nested historically in Pennsylvania. In winter, Ring-necked Ducks are regular residents at the Coastal Plain, Piedmont, and Lake Erie Shore, and they are local elsewhere in the state. Ring-necked Ducks are often found with flocks of Lesser Scaup.

Habitat: They prefer lakes, ponds, swamps, marshes, and slow-moving streams. Ring-necked Ducks are also found in flooded fields in spring. Unlike other diving ducks, they prefer open areas of smaller bodies of water surrounded by high vegetation or woods rather than the centers of the largest lakes and reservoirs.

Seasonal status and distribution

Spring: Migration is first noted with the first spring thaw in late Feb or early Mar. The number of birds increases until peaking from mid- to late Mar. Concentrations of 1000 or more birds are not unusual during peak migration. Birds linger as late as early Jun.

Summer: Although nesting evidence has not been discovered recently in Pennsylvania, birds are occasionally found in suitable nesting habitat well into the summer. The only site in the Coastal

Plain where Ring-necked Ducks have been found in summer is Tinicum.[1] In the Piedmont they have been found in Berks, Chester, Cumberland, Dauphin, Lancaster, and Northumberland cos. In the Ridge and Valley they have been recorded in Blair, Huntingdon, Lycoming, and Montour cos. Two records are known from the High Plateau: one in the summer of 1991 from Lake Somerset in Somerset Co.[2] and one from 5 to 22 Jul 1996 at Prince Gallitzin State Park in Cambria Co.[3] In the Glaciated Northwest, summer records have come from Butler, Crawford, Mercer and Venango cos. The sole summer record in the Southwest appears to be at Fox Chapel, in Allegheny Co., on 21 Jun 1983.[4] At PISP they have been found in Jun and Jul.

Fall: Migration in fall is not well defined. Ring-necked Ducks are far less common in fall than in spring but females and males in molt or in nonbreeding plumage may be overlooked when mixed with flocks of scaup. Migration is first noted in early Oct with a gradual increase of birds on the water by late Oct, and a peak by mid-Nov, when numbers begin to decrease.

Winter: Ring-necked Ducks are local and uncommon where there is open water on lakes and ponds in the Coastal Plain and Piedmont. They are rare elsewhere in the state away from Lake Erie. In the Ridge and Valley, records are from Carbon, Centre, Clinton, Columbia, Huntingdon, Luzerne, Lycoming, Mifflin, Montour, and Wyoming cos. They have been recorded in winter from a single site in the Glaciated Northeast, at Lake Wallenpaupack in Wayne Co.[5] West of the Allegheny Front away from Lake Erie they have been recorded from Allegheny, Armstrong, Butler, Crawford, Greene, Indiana, McKean, Mercer, Somerset, Venango, Warren, and Westmoreland cos. At the Lake Erie Shore they are found only in the waters surrounding PISP, where they are uncommon to fairly common.

History: Nesting has been confirmed in Pennsylvania only at the Pymatuning Reservoir after its flooding, when a brood of 7 ducklings was found in early Jul 1936.[6] As many as 15 pairs were present from 1936 to 1939 (Trimble 1940), but no evidence of nesting was found in 1940 or subsequently (Grimm 1952).

[1] A. Guarente, pers. comm.

[2] Sager, R., and G. Sager 1991. County reports—July through September 1991: Somerset County. PB 5:139.

[3] Haas, F.C., and B.M. Haas, eds. 1996. Birds of note—July through September 1996. PB 10:150.

[4] Hess, P.D. 1983. Area bird summaries for June. ASWPB 48:6.

[5] J. Strasser, and V. Strasser, pers. comm.

[6] Todd, W.E.C. 1936. The Redhead and Ring-necked Duck breeding at Pymatuning Lake, Pennsylvania. Auk 53:440.

Tufted Duck *Aythya fuligula*

General status: The closest breeding and wintering range of the Tufted Duck to North America is Europe and Japan. They are regularly recorded as a migrant on the Aleutians, Pribilofs, St. Lawrence Island and east to southern Alaska. Tufted Ducks are rare vagrants elsewhere in North America. Because they are frequently found in waterfowl collections, the occurrence of this species in Pennsylvania as a wild vagrant is suspect. However, the inclusion of this species on the official list of birds of Pennsylvania[1] is based on two records of single males found with other wild ducks during the time of year when vagrancy is expected.

Seasonal status and distribution

Spring: The first and only spring record was of one found by W. O. Robinson and S. Robinson at MSP in Butler Co. on 16 Apr 1972.[2]

Fall: A male Tufted Duck discovered by D. Bird and colleagues at Lake Silk-worth in Luzerne Co. from 1 to 15 Dec 1985 is the only fall record.[3]

Comments: A male was found at Pymatuning in Crawford Co. and appeared on 19 Sep 1992 with semidomestic Mallards being fed bread.[4] Because the bird appeared tame and was with semidomestic ducks at a date when a vagrant Tufted Duck is not expected, it was listed as an escapee. The bird remained at that site until at least the end of that year.

[1] Kwater, E. 1990. Official list of the birds of Pennsylvania. PB 4:51–53.
[2] Hall, G.A. 1972. Appalachian region. AB 26:761.
[3] Boyle, W.J., Jr., R.O. Paxton, and D.A. Cutler. 1986. Hudson-Delaware region. AB 40:262.
[4] Haas, F.C., and B.M. Haas, eds. 1992. Rare and unusual bird reports. PB 6:128.

Greater Scaup ***Aythya marila***

General status: Greater Scaup breed from Alaska south across Yukon and southern Mackenzie to northern Quebec. In North America they winter along the Pacific Coast and the Great Lakes, Newfoundland, and south along the Atlantic coast to Florida and the Mississippi River valley south to and around the Gulf coast. Greater Scaup are abundant regular migrants at the Lake Erie Shore and are uncommon to rare regular migrants elsewhere in the state. In summer, they are visitors or residents at the Lake Erie Shore and are accidental elsewhere. Greater Scaup are regular winter residents in the Coastal Plain, Piedmont, and the Lake Erie Shore and irregular to casual elsewhere. They and Lesser Scaup frequently flock together.

Habitat: Greater Scaup prefer deep lakes and large slow-moving rivers. They settle on shallow or small bodies of water less frequently than the Lesser Scaup.

Seasonal status and distribution

Spring: The usual spring migration period is from late Feb to mid-Apr. The numbers peak by mid-Mar. Flocks usually reach several thousand birds at the Lake Erie Shore, but elsewhere in the state only singles or groups of up to a few hundred usually are found. A few stragglers remain to mid-May.

Summer: Except for one observation, all summer records away from Lake Erie are from the Piedmont. Three males, which may have lingered from spring migration, were seen on 6 Jun 1982 in Lancaster Co. Two were at Muddy Run along the Susquehanna River, and one was near Quarryville. The latter bird was injured and was last seen in mid-Jul (Morrin et al. 1991). A bird was observed from 28 Jul to 8 Aug 1997 at the Susquehanna River in Dauphin Co.[1] In Adams Co. 1 was reported at SGL 249 on 3 Jul 1996, and 3 were at Long Pine Dam on 19 Aug 1996.[2] In the Ridge and Valley a single bird was at Harvey's Lake in Luzerne Co. on 26 Jul 1992.[3] One or two occasionally reside or visit the waters surrounding PISP.[4]

Fall: Greater Scaup migrate later in the fall than do Lessers. A few birds may be mixed with Lesser Scaup flocks in Oct, but substantial flights of Greater Scaup normally begin in mid-Nov and continue to increase through the remainder of fall. Maximum numbers are usually seen at the Lake Erie Shore from late Dec into early Jan. Many remain to winter in Presque Isle Bay. Migrant Greater Scaup are even rarer in fall than in spring elsewhere in the state. In Delaware Co. they have decreased considerably since the 1970s. The average number of birds seen on the CBCs from 1973 to 1983 was 843 compared with the average number seen on CBCs from 1983 to 1993 of only two.[5]

Winter: The only site in Pennsylvania where large numbers of Greater Scaup winter is at the Lake Erie Shore around PISP, where at least a few hundred may be found every year. Most records away from the Lake Erie Shore are from the Coastal Plain and Piedmont. In the

Ridge and Valley records have come from Carbon, Centre, Clinton, Huntingdon, Juniata, and Luzerne Cos. In the Glaciated Northeast the only record is from Lake Wallenpaupack[6] in Wayne Co. In the High Plateau they have been reported in winter from two sites in the Allegheny Mountains: Donegal Lake and Trout Run Reservoir in Westmoreland Co.[7] In the Southwest they have been recorded in Allegheny, Armstrong, Butler, Indiana, Venango, and Washington cos. In the Glaciated Northwest they have been recorded in Crawford Co.

History: Early authors such as Todd (1940) believed Greater Scaup were more common than Lesser Scaup, especially in winter on Presque Isle Bay. However, most authorities seemed to agree on the difficulties in assigning most sight records to either species.

Comments: Many Greater Scaup may be misidentified as Lesser Scaup or reported as "scaup species" because of their similarity. At the Lake Erie Shore, where both species are more abundant than elsewhere in Pennsylvania, there has been ample opportunity to study the two species as well as to monitor their migration schedules. Lesser Scaup are less hardy than Greater Scaup. They arrive later in the spring and leave earlier in the fall than Greater Scaup. Overlap in migration occurs between the two species, but one species usually outnumbers the other. Wintering Greater Scaup always outnumber wintering Lesser Scaup at the Lake Erie Shore. In recent years the apparent decrease in numbers of Greater Scaup has raised concern about environmental contaminants, habitat change, and overhunting.

[1] Williams, R. 1997. Local notes: Dauphin County. PB 11:154.

[2] Haas, F.C., and B.M. Haas, eds. 1996. Birds of note—July through September 1996. PB 10:150.

[3] Haas, F.C., and B.M. Haas, eds. 1992. Rare and unusual bird reports. PB 6:128.

[4] G.M. McWilliams, pers. obs.

[5] N. Pulcinella, pers. comm.

[6] J. Strasser and V. Strasser, pers. comm.

[7] Leberman, R., and R. Mulvihill. 1991. County reports—January through March 1991: Westmoreland County. PB 5:51.

Lesser Scaup *Aythya affinis*

General status: The breeding range of the Lesser Scaup is from Alaska south across western and central Canada and the northwestern U.S. east to South Dakota. Their wintering range extends over most of the U.S. except the north-central and New England states south through the Appalachian Mountains. They are one of the most widely distributed diving ducks in Pennsylvania. They are abundant regular migrants at the Lake Erie Shore and common regular migrants elsewhere in the state. They are casual summer visitors or residents at the Lake Erie Shore and are accidental elsewhere. In winter Lesser Scaup are regular winter residents or visitors over most of the state. Most scaup identified at any season away from Lake Erie are of this species.

Habitat: Lesser Scaup are found on lakes, ponds, marshes, and slow-moving streams. Unlike Greater Scaup, they can frequently be found on farm ponds and other small bodies of water surrounded by emergent vegetation.

Seasonal status and distribution

Spring: A few may arrive with flocks of Greater Scaup in late Feb, but most do not arrive until about the second week of Mar. Peak migration range is from mid-Mar to early Apr. Flock sizes usually reach several thousand birds at the Lake Erie Shore; several hundred birds are common on lakes and rivers at other regions in Pennsylvania. Large numbers of birds occasionally remain until early May. A few Lesser Scaup linger to the first week of Jun.

Summer: One or two occasionally reside or visit the waters surrounding PISP during summer,[1] but they are rarely recorded in other regions of the state. Summer reports have come from at least 12 counties: Berks, Butler, Centre, Crawford, Delaware, Erie, Lancaster, Lycoming, Monroe, Northumberland, Venango, and Warren.

Fall: Migration may begin with the first cold front in early Oct but usually is not well under way until mid-Oct. Peak migration ranges from late Oct to mid-Nov. Thousands may be seen flying along the Lake Erie Shore during the passage of a cold front at this season. Fewer Lesser Scaup occur in fall than in spring away from the Lake Erie Shore. Migration usually ends by mid-Dec, with a few remaining to winter.

Winter: Most records are from the Coastal Plain, where only a few birds may be found along the Delaware River, especially at Tinicum[2] and from the Piedmont, along the lower Susquehanna River. In the Ridge and Valley records come from Centre, Huntingdon, Luzerne, Montour, and Perry cos. The two sites in the Glaciated Northeast where Lesser Scaup have been found are on the Susquehanna River at Athens in Bradford Co.[3] and at Lake Wallenpaupack in Wayne Co.[4] In the High Plateau they have been recorded only from the Youghiogheny River Lake in Somerset Co.[5] In the Southwest, records have come from Allegheny, Washington, and Westmoreland cos. In the Glaciated Northwest, they have been recorded in Butler and Crawford cos. Around PISP usually only a few birds winter, but several hundred can be found during mild winters.

History: Historical reports of Lesser Scaup during the breeding season have led authorities to believe that nesting was possible. Warren (1890) cited a report of a pair that remained through the summer in Northumberland Co. Todd (1940) thought that nesting could be expected in the Pymatuning region. However, there has never been any evidence of breeding Lesser Scaup in the state. An unusual summer concentration of 50 birds, including both males and females, was at Erie Bay on 21 Jun 1901 (Todd 1940).

[1] G.M. McWilliams, pers. obs.
[2] N. Pulcinella, pers. comm.
[3] T. Gerlach, pers. comm.
[4] J. Strasser and V. Strasser, pers. comm.
[5] R. Sager and G. Sager, pers. comm.

King Eider *Somateria spectabilis*

General status: Like Common Eiders, King Eiders are primarily birds of saltwater. They breed in North America along the Arctic coast and islands from Alaska east to Greenland. King Eiders winter primarily around the Aleutians in the West and from Greenland south along the Atlantic coast to as far south as coastal Maryland. They are more frequently found on freshwater lakes south of their normal breeding and winter grounds than are Common Eiders. In Pennsylvania they are casual visitors to PISP. Only four of the 19 birds recorded in this area have been males. Away from Lake Erie this species is accidental and has been recorded only three times in recent years.

Seasonal status and distribution:

Fall: A female was shot near Washington Boro in Lancaster Co. during the 1975 hunting season (Morrin et al. 1991). Two eiders were shot on the Susquehanna River below Selinsgrove in Snyder Co.: an immature female in 1959 and an immature male on 5 Nov 1960 (Poole, unpbl. ms.). At Lake Erie seven fall records range from 27 Oct to 19 Nov, with a single-day high count of 7 birds on 7 Nov 1981. Nearly all rec-

King Eider at Presque Isle State Park, Erie County, January 1993. (Photo: Rick Wiltraut)

ords have been of a lone female found after a cold front.[1]

Winter: Four winter records are all from Lake Erie. An adult male King Eider was discovered on 2 Feb 1977;[2] an immature male was present from 1 Jan to 16 Mar 1991;[3,4] a female was present on 2 Jan 1993; and up to two immature males were present from 14 to 20 Jan 1993.[5]

History: All King Eider records in the past have been in the fall and early winter. Apparently this species was first reported in Pennsylvania on 30 Nov 1889, when a flock of 18–20 birds appeared on Erie Bay after a storm (Warren 1890). They were recorded several more times at the same site until 1933 but were not reported again until 1976. Historical records have come from the Delaware River: one was shot near Tinicum, Delaware Co., on 4 Dec 1900;[6] 4 birds were taken from the Susquehanna River near Harrisburg (Poole, unpbl. ms.); and a female and an immature male were shot on the Conejohela Flats in Nov 1940 (Beck 1955).

[1] G.M. McWilliams, pers. obs.
[2] Hall, G.A. 1977. Appalachian region. AB 31:332.
[3] Hall, G.A. 1991. Appalachian region. AB 45:272.
[4] Hall, G.A. 1991. Appalachian region. AB 45:444.
[5] Haas, F.C., and B.M. Haas, eds. 1993. Rare and unusual bird reports. PB 7:17.
[6] No author. 1901. Abstract of the Proceedings of the Delaware Valley Ornithological Club. Cassinia 5:47.

[Common Eider] ***Somateria mollissima***

General status: This saltwater species breeds along the coastline and islands from Alaska, east across the Arctic, south around coastal Hudson and James bays, and along the coasts of Quebec, and south along the Atlantic coast to New Hampshire. They winter along coastal Alaska south to British Columbia and along the Atlantic coast south to New York. Common Eiders are listed as hypothetical in Pennsylvania.

Comments: In Pennsylvania, Common Eiders have been reported five times. A bird identified as a female Common Eider was shot on the Conejohela Flats in Lancaster Co., and the specimen was preserved (Beck 1924) but no longer exists. A bird, believed to be shot at the Pymatuning Reservoir in Crawford Co. in Apr 1945, is now in the Pymatuning Waterfowl Museum (Leberman 1988). Apparently this specimen has no collecting data.[1] An immature male was shot on the West Branch of the Susquehanna River at the mouth of Buffalo

Creek in Union Co. on 20 Oct 1952. The specimen was mounted and placed in a private collection (Poole, unpbl. ms.). It is uncertain whether this specimen still exists. A male was observed on the Conejohela Flats in Lancaster Co. in Nov 1981 (Morrin et al. 1991), and the most recent sighting (currently under review) was of one at Lake Ontelaunee on 12 Mar 1998.[2]

[1] R.C. Leberman, pers. comm.
[2] Haas, F.C., and B.M. Haas, eds. 1998. Birds of note—January through March 1998. PB 12:13.

Harlequin Duck ***Histrionicus histrionicus***

General status: There are two populations in North America: one in the West and one in the East. In the West it breeds in the northern Rockies and it winters along the coast. In eastern North America the Harlequin Duck breeds from Baffin Island, Quebec, Labrador, and Greenland, and it winters from Labrador south along the coast to New York and less commonly to the Great Lakes. Harlequin Ducks have been part of the avifauna of Pennsylvania only since 1967. They are casual visitors to the Lake Erie Shore, where all sightings have come from the waters surrounding PISP. They are accidental elsewhere in the state. Most records are of a single female. There have been about 15 reports of Harlequin Ducks in the state.

Status and distribution

Fall through spring: Harlequin Duck records range from 14 Nov to 5 Apr. The first report for the state was of 2 birds found by H. Johnson and colleagues on the Allegheny River near Warren in Warren Co. on 11–13 Feb 1967.[1] Other sightings away from the Lake Erie Shore have been from Bristol in Bucks Co. on 11–12 Jan 1971[2] and at New Hope in Bucks Co. on 21 Feb 1993.[3] Harlequin Ducks have been reported at Delaware Water Gap in Monroe Co. on 23 Dec 1978[4] and on the Ohio River near Pittsburgh through the spring of 1987[5] The most recent record was of a male and female photographed on the Susquehanna River at West Fairview in Dauphin Co. on 5 Apr 1997.[6]

Comments: Two Harlequins have been shot by hunters. An adult male in full breeding plumage was shot at Misery Bay at PISP on 10 Nov 1981 and was mounted and placed in a private collection. An immature male was shot at the same site on 14 Nov 1987. Identification of both birds was confirmed, and the adult male was photographed.[7]

[1] Hall, G.A. 1967. Appalachian region. AFN 21:419.
[2] Scott, F.R., and D.A. Cutler. 1971. Middle Atlantic Coast region. AB 25:559.
[3] Haas, F.C., and B.M. Haas, eds. 1993. Rare and unusual bird reports. PB 7:59.
[4] Smith, P.W., R.O. Paxton, and D.A. Cutler. 1979. Hudson Delaware region. AB 33:268.
[5] Grom, J. 1987. Allegheny County. PB 1:40.
[6] Williams, R. 1997. Local notes: Dauphin County. PB 11:94.
[7] G.M. McWilliams, pers. obs.

Surf Scoter ***Melanitta perspicillata***

General status: In North America, Surf Scoters breed from Alaska east across northern Canada to Labrador, and they winter along the Pacific and Atlantic coasts. Most sightings of these dark brown "sea ducks" are at the Lake Erie Shore in fall, where numbers vary from year to year. Surf Scoters are regular spring and fall migrants, irregular and rare in early winter at the Lake Erie Shore, and accidental elsewhere in the state in winter. The numbers observed in the state are less variable than are those of Black Scoters, with which they occasionally associate.

Habitat: Surf Scoters prefer deep lakes and large slow-moving rivers, but they have been found on smaller bodies of water that are in the open and not surrounded by high vegetation or trees.

Seasonal status and distribution

Spring: In spring, Surf Scoters are regular in the state only on the waters surrounding PISP. Most sightings at this season are of fewer than 6 birds. They may appear almost any time from early Mar to the third week of May, but most sightings are in Apr and May. They normally migrate along the Atlantic coastline in spring.

Fall: Surf Scoters are likely to be found in the greatest numbers during fall. At the Lake Erie Shore, flocks are regularly observed migrating offshore with the passage of cold fronts, but away from Lake Erie most flocks are seen only when storms force them down. Migration may begin in late Sep and continues through early Dec, with stragglers to early Jan. Most birds are seen from mid-Oct to mid-Nov. On 8 Oct 1987, 457 were counted flying past the tip of PISP,[1] an unusually early date for such a large number of birds. The majority of Surf Scoters recorded in the fall are females.

Winter: Away from Lake Erie they have been recorded at only five sites in winter: on Harvey's Lake in Luzerne Co. on 13–28 Jan 1997[2] and again on 10 Jan 1998;[3] at Beltzville Lake in Carbon Co. on 7 Feb 1998;[4] at the Quemahoning Dam in Somerset Co.;[5] on the Driftwood Branch in Cameron Co. on 20 Jan 1991;[6] and on the Allegheny River at Manorville in Armstrong Co. on 6 Feb 1992.[7] Most records of Surf Scoters during winter are from the waters surrounding PISP, where one or two birds normally may be found to late Jan, and during mild winters they may successfully winter. Off the outer beaches of PISP a flock containing an unprecedented 300 birds appeared in mid-Dec 1995. Their number dwindled through the winter, with only 30 remaining by 24 Feb.[8]

History: Surf Scoters have always been considered rare in the state. Todd (1940) stated that they occur rather commonly at Lake Erie but outside of Lake Erie had been only found at Conneaut Lake, in western Pennsylvania.

Comments: In recent years, Surf Scoters have become more common than White-winged Scoters at the Lake Erie Shore.[9] G. A. Hall also mentioned their increase for the entire Appalachian region in his winter report.[8]

[1] Hall, G.A. 1988. Appalachian region. AB 42:74.
[2] Haas, F.C., and B.M. Haas, eds. 1997. Birds of note—January through March 1997. PB 11:22.
[3] W. Reid, pers. comm.
[4] Haas, F.C., and B.M. Haas, eds. 1998. Birds of note—January through March 1998. PB 12:13.
[5] R. Sager and G. Sager, pers. comm.
[6] Hendrickson, B. 1991. County reports—January through March 1991: Cameron County. PB 5:34.
[7] M. Higbee and R. Higbee, pers. comm.
[8] Hall, G.A. 1995. Appalachian region. NASFN 49:147.
[9] G.M. McWilliams, pers. recs.

White-winged Scoter *Melanitta fusca*

General status: In North America White-winged Scoters, the largest of the three scoter species, breed from Alaska across Canada to Quebec, and they winter along the Pacific coast, on the Great Lakes, and along the Atlantic coast to South Carolina. In Pennsylvania they are uncommon to rare regular migrants. White-winged Scoters are accidental in summer. They are regular in early winter at the Lake Erie Shore and are casual to accidental elsewhere in the state in winter. These scoters do not often associate with the other two species.

Habitat: White-winged Scoters prefer deep lakes and large slow-moving rivers, but they have been found on open small ponds.

White-winged Scoter at Presque Isle State Park, Erie County, January 1982. (Photo: Gerald M. McWilliams)

Seasonal status and distribution

Spring: Migration may begin as early as late Feb. Their usual migration period is from late Mar to late Apr. During peak migration, Oldsquaw and White-winged Scoter may be found together. Large concentrations of White-winged Scoters have been recorded in spring; 70 birds were seen at Montour Preserve on 6 Apr 1979 (Schweinsburg 1988). The 47 at Bald Eagle State Park on 5 May 1990 was an unusual number for such a late date.[1] Single individuals or small groups may occasionally linger to late May. White-winged Scoters are more likely than Black and Surf scoters to be found in spring.

Summer: The only two summer records are of a bird summering at Tinicum in 1966[2] and one found at PISP on 14 Jun 1967.[3]

Fall: They are slightly more common in fall than in spring. White-winged Scoters may begin migrating as early as mid-Sep, but most birds migrate over the state from mid-Oct to early Dec, with stragglers to early Jan. They are occasionally found in sizable numbers at this season, such as counts of 55 at Beaver Run Reservoir on 5 Nov 1971[4] and of 75–100 at PISP on 11 Nov 1985.[5]

Winter: They are rare in winter at the Lake Erie Shore from the waters surrounding PISP. They may successfully winter at this site during mild winters but are usually not found past late Jan. Birds have also been reported in winter in Armstrong, Beaver, Butler, Centre, Clinton, Huntingdon, Lancaster, Lehigh, Luzerne, and Warren cos.

History: White-winged Scoters were always considered the most common of the three scoters in Pennsylvania. Singles or small groups were reported from various sites throughout the state. Poole (1964) considered them irregular transients, yet Sutton (1928b) reported them as fairly common transients at Conneaut Lake. An unusually high count of 150 was recorded flying over Hawk Mountain on 6 Oct 1935 (Poole, unpbl. ms.).

1 Hall, G.A. 1990. Appalachian region. AB 44:425.
2 Meritt, J.K. 1970. Field notes. Cassinia 52:22.
3 Stull, J.H., and J.G. Stull. 1967. Field notes. Sandpiper 10:13.
4 Hall, G.A. 1972. Appalachian region. AB 26:63.
5 G.M. McWilliams, pers. obs.

Black Scoter ***Melanitta nigra***

General status: The Black Scoter is generally the rarest of the three scoter species, but occasionally flocks in fall outnumber the others. Their breeding range in North America is in Alaska and in scattered localities in central and eastern Canada. In winter they range along the Pacific and Atlantic coasts. Black Scoters are regular fall migrants and irregular spring migrants over the state. Most sightings are at the Lake Erie Shore, where they may occasionally associate with Surf Scoters. The number of birds varies from year to year, from a few individuals to hundreds at one site. Black Scoters are irregular in early winter at the Lake Erie Shore and are accidental elsewhere in the state.

Habitat: Black Scoters prefer deep lakes and large, slow-moving rivers, but they have been found on small open ponds.

Seasonal status and distribution

Spring: Most sightings at this season are of fewer than 6 birds and may appear at almost any time from early Mar to the third week of May. A flock of 12 seen at Peace Valley Park in Montgomery Co. on 26 Mar 1992[1] was an unusual number for spring. The irregularity and scarcity of spring sightings may be because Black Scoters normally follow the Atlantic coastline in spring migration.

Fall: Black Scoters are more likely to be found during fall, and in larger numbers, than at any other time of year. Away from Lake Erie most flocks are seen only when storms force them down. Single birds may appear as early as the last week of Sep. Their typical migration period is from the second or third week of Oct to the first week of Dec. Flocks occasionally are downed from about mid-Oct to late Oct: for example, 400 from 18 to 20 Oct 1989 at Muddy Run area in Lancaster Co. (Morrin et al. 1991). Flocks of Black Scoters may be found migrating offshore along Lake Erie with the passage of cold fronts from late Oct to mid-Nov: for example, 180 on 6 Nov 1983 at PISP.[2] An unusually large flight, about 1000 scoters, containing mostly Black Scoters, passed Gull Point on the morning of 11 Nov 1985.[3] Stragglers may remain until early Jan. The majority of Black Scoters recorded in the fall are females.

Winter: Winter records away from Lake Erie have been from Springton Reservoir[4] and Marcus Hook in Delaware Co.,[5] Struble Lake in Chester Co. on 7 Feb 1995,[6] Harvey's Lake in Wyoming Co. to at least 29 Jan 1993[7] and again on 2–9 Jan 1998,[8] and at Quemahoning Dam in Somerset Co.[9] Black Scoters are found more frequently in the waters surrounding PISP during winter than at any other single site in Pennsylvania. Usually one or two birds may be found until late Jan in this area, and during mild winters they may successfully winter here. Off the outer beaches of PISP in 1995 a flock of 30 Black Scoters remained through the winter.[10] Unusual was a flock of 40 (early migrants?) seen flying past PISP on 17 Feb 1992.[11]

History: This species was considered rare and the least common of the three scoters by all authorities except Sutton (1928b). He found them to be fairly common migrants and believed that they may occasionally occur in winter at Conneaut Lake. Todd (1940) stated that all records in fall are from Nov and Dec at Erie, whereas Poole stated in his unpublished manuscript that all published observations elsewhere in the state are in Oct. Apparently there was a summer report at Tobyhanna in Monroe Co. on 4 Jul 1938.[12] No historical record exists of large flocks occurring in the state. The largest concentration reported before 1960 was of 13 birds on Bethlehem Steel Lake near Annville, Lebanon Co., on 18 May 1956 (Poole, unpbl. ms.).

[1] Boyle, W.J., Jr., R.O. Paxton, and D.A. Cutler. 1992. Hudson-Delaware region. AB 46:398.

[2] Hall, G.A. 1984. Appalachian region. AB 38:201.

[3] Hall, G.A. 1986. Appalachian region. AB 40:112.

[4] A. Guarente, pers. comm.

[5] Pulcinella, N. 1998. Local notes: Delaware County. PB 12:18.
[6] Haas, F.C., and B.M. Haas, eds. 1995. Birds of note—January through March 1995. PB 9:36.
[7] W. Reid, pers. comm.
[8] Haas, F.C., and B.M. Haas, eds. 1998. Birds of note—January through March 1998. PB 12:13.
[9] R. Sager and G. Sager, pers. comm.
[10] Hall, G.A. 1995. Appalachian region. NASFN 49:147.
[11] McWilliams, G.M. 1992. County reports—January through March 1992: Erie County. PB 6:36.
[12] Ross, C.C. 1938–1941. Field notes. Cassinia 31:44.

Oldsquaw *Clangula hyemalis*

General status: Only this handsome diver shares the long central rectrices with Northern Pintail. In North America, Oldsquaws breed from the Arctic coast of Alaska east across Arctic islands and northern Canada to Labrador, and they winter along the Pacific coast, on the Great Lakes, and along the Atlantic coast south to South Carolina. Oldsquaws are rare regular migrants over the state and frequently are grounded by storms. Most migrants arrive on inland bodies of water during the night. The number of birds varies from year to year from a few individuals to hundreds at any one site. Oldsquaws are accidental in summer at the Lake Erie Shore. They are irregular in winter in the Coastal Plain and are casual winter visitors or residents at the Lake Erie Shore. Oldsquaws are accidental in winter elsewhere in the state.

Habitat: Oldsquaws prefer deep lakes and large slow-moving rivers, but they are occasionally found on open small ponds.

Seasonal status and distribution

Spring: Most Oldsquaws in Pennsylvania are recorded during spring migration. Migration may begin as early as late Feb or early Mar. Their typical migration period is from late Mar to the third week of Apr, when they often suddenly appear in large numbers overnight. They are quite vocal in the spring; their "yodeling" calls are audible from a great distance. Compact rafts containing from 200 to 400 birds are not unusual on many of the largest lakes in the state. In some years, severe storms force down very large rafts of ducks. One incident occurred in 1979, when on 8–9 Apr, an ice storm brought down over 6000 birds on the Susquehanna River at Lock Haven in Clinton Co. and 2500 at State College in Centre Co.[1] Lingering birds have been recorded to the first week of Jun.

Summer: Two summer sightings have been reported from the waters surrounding PISP: one bird on 29 Aug 1969[2] and one slightly oiled bird on 12 Aug 1987.[3]

Fall: Migration has been noted as early as mid-Sep, but it usually does not begin until late Oct. Peak migration occurs in the second or third week of Nov. The large flocks seen in spring usually are not seen in the fall, except at the Lake Erie Shore, where migrating flocks of up to 400 or more birds have been recorded flying over Lake Erie during the passage of cold fronts.[4] However, 181 birds were at Muddy Run in Lancaster Co. on 15 Nov 1984 (Morrin et al. 1991). Small flocks can occasionally be seen flying past hawk watches: for example, 27 were seen flying over Hawk Mountain on 16 Nov 1980.[5] A few may remain through Dec into early Jan or through the winter.

Winter: Most Oldsquaws in winter are reported from the lower Delaware River in the Coastal Plain and the waters surrounding PISP at the Lake Erie Shore. Other areas with records east of the Allegheny Front are Berks, inland Bucks, Centre, Clinton, inland Delaware, Lackawanna, Lancaster, Luzerne, Montgomery, and York cos. Derry in Westmoreland Co.[6] and MSP in Butler Co.[7] are the only sites west of the Allegheny Front away from the Lake Erie Shore where they have been recorded in winter.

History: Poole (1964) listed this species as an uncommon transient, with most occurring after severe storms. Some of the large rafts that Poole mentioned in his manuscript are comparable to the large rafts that sometimes occur today. Todd (1940) felt that before 1904, Oldsquaws were among the commonest ducks at Presque Isle but considered them to be the rarest by the late 1930s.

[1] Hall, G.A. 1979. Appalachian region. AB 33:770.
[2] G.M. McWilliams, pers. recs.
[3] Hall, G.A. 1988. Appalachian region. AB 42:74.
[4] Hall, G.A. 1985. Appalachian region. AB 39:53.
[5] Paxton, R.O., W.J. Boyle Jr., and D.A. Cutler. 1981. Hudson-Delaware region. AB 35:163.
[6] R.C. Leberman, pers. comm.
[7] P. Hess, pers. comm.

Bufflehead ***Bucephala albeola***

General status: The breeding range of Buffleheads is across Canada and south locally in the western U.S. to Colorado. Their winter range is along the Pacific coast, south through Mexico and across the southern U.S., north to the Great Lakes, and along the Atlantic coast to Newfoundland. In Pennsylvania Buffleheads are fairly common to abundant regular migrants and are casual summer visitors or residents. They are regular winter visitors or residents across the state.

Habitat: Buffleheads are found on rivers, lakes, ponds, and marshes.

Seasonal status and distribution

Spring: Migration usually begins in mid- to late Feb and peaks from about the first to the third week of Mar. The number of birds may remain high as late as mid-May. Some birds linger well into Jun.

Summer: Most summer records are from sites in the eastern half of the state, including Berks, Bradford, Centre, Clinton, Cumberland, Lancaster, Lehigh, Lycoming, Montour, Pike, Susquehanna, and York cos. In western Pennsylvania summer records come from only two sites: a bird summered at Scandia in Warren Co. in 1985,[1] and one or two have been recorded in summer on three occasions on the waters surrounding PISP.[2] One at Nanticoke in Luzerne Co. on 1 Sep 1997 may have been a lingering summering bird rather than an early migrant.[3]

Fall: A few Buffleheads usually arrive just before Common Goldeneye in mid-Oct. The largest number of birds starts moving south in early Nov; they continue to build in numbers through the fall and early winter. A peak in their migration is hard to define, but noticeable movements of ducks accompany cold fronts. The majority of Buffleheads are usually driven south by heavy freezes in Jan.

Winter: Having the same winter range as Common Goldeneye, Buffleheads are among the hardiest wintering ducks in Pennsylvania. Away from Lake Erie they are uncommon to rare but may be locally fairly common, especially on larger rivers such as the Allegheny, Delaware, Juniata, Lehigh, Ohio, and Susquehanna. Buffleheads may be found on the deepest open water available in all regions of the state except the central and north-central High Plateau. The greatest numbers winter in Presque Isle Bay or on Lake Erie, where they are common, surviving in the harshest conditions on the open water between ice floes.

History: Their status has changed little throughout recorded history, but Buffleheads have declined since the 1800s when Baird (1845) and Turnbull (1869) reported them as abundant on the larger rivers and lakes.

[1] Hall, G.A. 1985. Appalachian region. AB 39:911.
[2] G.M. McWilliams, pers. recs.

[3] Haas, F.C., and B.M. Haas, eds. 1997. Birds of note—July through September 1997. PB 11:148.

Common Goldeneye *Bucephala clangula*

General status: Common Goldeneyes breed from Alaska south across Canada, the northern and the New England states. They winter primarily along the Pacific coast, on the Great Lakes, in the Mississippi River valley, and along the Atlantic and Gulf coasts. Common Goldeneyes are fairly common to abundant regular migrants. In summer they are casual at the Lake Erie Shore and accidental elsewhere. In winter they are regular residents or visitors throughout the state.

Habitat: Common Goldeneyes are found on rivers, lakes, and ponds and in marshes.

Seasonal status and distribution

Spring: They are one of the first ducks to appear in spring. The migration begins as early as mid-Feb or as soon as the ice begins to melt. However, it is difficult to separate wintering birds from early migrants. Peak migration ranges from early to mid-Mar, when thousands can be found on the larger lakes. The largest concentrations are found in Presque Isle Bay at the Lake Erie Shore, where more than 10,000 birds are not unusual.[1] Migration continues to mid-Apr, and a few may linger to late May.

Summer: One or two birds occasionally spend the summer in Misery Bay at PISP. They also have made brief appearances or have summered at least once in Berks, Centre, Fulton, Montour, Tioga, Warren, Venango, and York cos.

Fall: A few birds make early appearances beginning the first week of Oct, but most arrive after the first major cold fronts have passed in the first or second week of Nov. The number of birds gradually peaks by late Nov. High counts away from Lake Erie have come from Pymatuning Reservoir, where 18,000 birds were recorded on 30 Nov 1989.[2] At the Lake Erie Shore the number of birds continues to grow until late Dec or early Jan, when most of these birds will winter.

Winter: Common Goldeneyes are among the hardiest wintering ducks in Pennsylvania. Away from Lake Erie, they are uncommon to rare, but they may be locally common, especially on the larger rivers such as the Allegheny, Delaware, Juniata, Lehigh, Ohio, and Susquehanna. They have been found on deep, open water in all regions of the state except the central and north-central High Plateau. The 200 counted in an open patch of water on the Delaware River at Portland on 12 Jan 1981 was a high number for that area.[3] They are abundant during winter in Presque Isle Bay or on Lake Erie, where they survive in open water between ice floes.

[1] G.M. McWilliams, pers. obs.
[2] Hall, G.A. 1990. Appalachian region. AB 44:89.
[3] Boyle, W.J., Jr., R.O. Paxton, and D.A. Cutler. 1981. Hudson-Delaware region. AB 35:283.

Barrow's Goldeneye *Bucephala islandica*

General status: In North America, Barrow's Goldeneyes breed from Alaska south through British Columbia and Washington and locally in Colorado and in northern Quebec. They winter along the Pacific coast, in the Gulf of St. Lawrence, and along the Atlantic Coast south rarely to New York. They are rare away from the eastern coast of North America at any season. In Pennsylvania, Barrow's Goldeneyes are casual late-winter and spring visitors at the Lake Erie Shore in the waters surrounding PISP, and they are accidental elsewhere.

Seasonal status and distribution

Winter to spring: Reports range from 11 Feb to 21 May. Nine of the 12 reports have been from mid-Feb to late Mar at the time of greatest concentrations of

Barrow's Goldeneye at Washington Crossing State Park, Bucks County, February 1996. (Photo: Rick Wiltraut)

Common Goldeneye, with which they are nearly always associated. An adult male was photographed in Bucks Co. along the Delaware River, where it was first discovered on 11 Feb at New Hope and was later seen at Scudders Falls Bridge until 21 Feb 1996.[1] At Bricksville in Lancaster Co., a male was seen on 27 Mar 1978.[2] The Barrow's Goldeneye has also been reported from Highland Park in Pittsburgh on 26 Mar 1959 (Poole, unpbl. ms.) and on the Allegheny River between Franklin and Reno in Venango Co. on 4 Jan 1964 (Poole, unpbl. ms.). Despite the numerous reports, the occurrence of this species was not verified in Pennsylvania until 20 Feb 1992, when an adult male was photographed in Presque Isle Bay;[3] another adult male joined this bird for a day on 26 Feb.[3] A female was observed from 13 to 20 Feb 1993 at PISP and was photographed,[4] and an adult male was there on 23 Mar 1994.[5]

History: Barrow's Goldeneyes were first reported in Pennsylvania in 1924. There were at least six reports from 1924 to 1946. One was reported by Frey (1943) in Harrisburg on 18 Mar 1934. One shot near Williamsport in Nov 1956 proved to be an escapee (Poole, unpbl. ms.). This species was reported by P.B. Street (1954) at Lake Wallenpaupack on 8 Apr 1946. In Crawford Co., Sutton (1928b) reported a male shot at Conneaut Lake in Nov 1924, and Grimm (1952) reported single males at Hartstown Marsh on 1 Apr 1944 and again on 3 Apr 1949.

[1] Kitson, K. 1996. Local notes: Bucks County. PB 10:17.
[2] Paxton, R.O., P.A. Buckley, and D.A. Cutler. 1978. Hudson-Delaware region. AB 35:984.
[3] McWilliams, G.M. 1992. County reports—January through March 1992: Erie County. PB 6:36.
[4] Haas, F.C., and B.M. Haas, eds. 1993. Rare and unusual bird reports. PB 7:17.
[5] Haas, F.C., and B.M. Hass, eds. 1994. Rare and unusual bird reports. PB 8:41.

Hooded Merganser ***Lophodytes cucullatus***

General status: Hooded Mergansers breed from Alaska south to Oregon, from central Canada south along the Appalachian Mountains, and west to North Dakota and Arkansas. In winter they are found primarily along the Pacific, Atlantic, and Gulf coasts and locally inland. Hooded Mergansers are fairly common to abundant regular migrants in Pennsylvania. They breed primarily across the northern tier of counties and are regular summer visitors away from breeding sites. Hooded Mergansers are regular winter residents.

Habitat: They are found on lakes, ponds, swamps, marshes, and streams during migration and in winter. During the breeding season they inhabit clear food-rich small bodies of water, especially swamps and beaver dams with tree-lined shores.

Seasonal status and distribution

Spring: Migration usually begins in early Mar and extends to mid-Apr with stragglers to early May. Peak migration occurs from the second to the fourth week of Mar. The greatest concentrations of birds are usually found at Pymatuning in Crawford Co. and at PISP in Erie Co., where concentrations of 200–300 birds are not unusual. A total of 175 at MSP in Butler Co. in the spring of 1984 was a high count away from those sites.[1]

Breeding: Hooded Mergansers are uncommon to rare regular summer residents in wetlands of Crawford, Erie, and Pike cos. and are local in the rest of the state. Their breeding range has expanded statewide since 1960, including Philadelphia Co. in 1992.[2] Breeding was confirmed for the first time during 1996 in six widely scattered counties: Allegheny, Bradford, Columbia, Huntingdon, Indiana, and Venango.[3] They are cavity nesters, laying 10–14 glossy white eggs in a tree cavity near water or in a Wood Duck box (see Comments). Eggs are rounder and whiter than those of Wood Ducks. Nests are rarely reported in Pennsylvania. Most nesting is confirmed by observations of females with broods. The males leave the females after incubation has begun and are rarely seen until the young have fledged.

Fall: Southbound birds arrive away from their nesting sites in late Oct or early Nov. The number of birds gradually increases through the fall, with few noticeable peak migration periods. The largest concentrations reach the thousands at Pymatuning and at PISP. On 28 Nov 1991, 4000 birds were estimated at Pymatuning.[4] By late Dec or early Jan migration has ended, with a few remaining to winter.

Winter: Hooded Mergansers are uncommon to rare in all regions where there is open water. Most overwinter on the larger rivers, including the Allegheny, Delaware, Juniata, Lehigh, Ohio, and Susquehanna. A high count of 26 wintered in 1973 on a small lake near New Cumberland in Cumberland Co.[5] In the Glaciated Northeast, they have been recorded in Wyoming and Wayne cos. In the High Plateau, Hooded Mergansers have been recorded in winter only in Cambria, Somerset, Westmoreland, and Warren cos.

History: They were regarded as local, rare breeders since earliest ornithological records, particularly in southern counties such as Cumberland (Baird 1845) and Lancaster (Libhart 1869). Harlow (1913) stated that "no nests have ever been found in the state" but cited summer records in Chester, Greene, Perry, and Lycoming cos. That no nest has ever been found is not surprising because only when these cavity nesters use Wood Duck boxes is a nest likely to be encountered. Todd (1940) cited the first breeding evidence in Pennsylvania in Clearfield Co. E. Decker in 1957 (Poole 1964) reported the use of Wood Duck boxes by Hooded Mergansers near Cambridge Springs, Crawford Co.

Comments: Hooded Mergansers commonly use boxes placed for Wood Ducks. R. Criswell provided many confirmed breeding records in Wood Duck nest boxes in Crawford Co. during the BBA project.[6] Annual Wood Duck productivity surveys by the PGC document regular Hooded Merganser occupancy in Pike and Crawford cos. An average of 16% of boxes placed for

Wood Ducks were occupied by Hooded Mergansers in Pike Co. Game Lands, compared with just 7% of those in Crawford Co.[7] Hooded Mergansers are also known to share incubation with Wood Duck hens (H.H. Harrison 1975).

[1] Hall, G.A. 1984. Appalachian region. AB 38:910.
[2] Fingerhood, E. 1992. County reports—April through June 1992: Philadelphia County. PB 6:89.
[3] Floyd, T. 1996. A note on Hooded Merganser breeding in Pennsylvania and nearby states. PB 10:205.
[4] Leberman, R.F. 1992. County reports—October through December 1991: Crawford County. PB 5:170.
[5] Scott, F.R., and D.A. Cutler. 1973. Middle Atlantic Coast region. AB 27:597.
[6] Gross, D.A. 1992. Hooded Merganser (*Lophodytes cucullatus*). BBA (Brauning, 1992): 82–83.
[7] Dunn, J.P. 1995. Wood Duck population monitoring program. Annual Job Report no. 59201. Pennsylvania Game Commission, Harrisburg.

Common Merganser ***Mergus merganser***

General status: The largest of the mergansers, Common Mergansers' breeding range extends from Alaska south through Canada and the western U.S. and east across the northern states and south to the Mid-Atlantic states. They winter throughout the U.S. Common Mergansers are uncommon to fairly common regular migrants throughout Pennsylvania, except along the lower Susquehanna River and PISP, where they are abundant. Most breeding records are across the northern tier of counties. Common Mergansers are regular winter residents or visitors over most of the state.

Habitat: Common Mergansers are found on large lakes, ponds, and slow-moving streams during migration and in winter. During the breeding season they prefer clear, slow-moving wooded streams with mature trees nearby for nesting.

Seasonal status and distribution

Spring: An increase in the numbers of birds marks the beginning of spring migration. The typical migration period is from the first week of Feb or as soon as the ice begins to melt until mid-Apr. The peak number of Common Mergansers may vary from year to year depending upon availability of open water and food. During most years, hundreds or thousands may be seen in late Feb or early Mar. Larger concentrations are seen at the Lake Erie Shore in the water surrounding PISP than at any other site in the state. On 4 Mar 1990, about 20,000 were estimated at PISP.[1] A few nonbreeding birds may remain into summer away from their usual breeding range in the northern half of the state.

Breeding: During the 1980s and 1990s, the nesting range expanded dramatically southward, and populations increased. Common Mergansers now are regular breeders on the Delaware River drainage south to New Hope, Bucks Co.; on the Susquehanna River to Perry Co.; and on the Allegheny River to Armstrong Co. They breed irregularly south on these rivers to Beaver and Lancaster cos. Nests are placed on the ground in undercut banks, in cavities of large trees, and occasionally in a cabin chimney, but few nests have been found in the state. Mergansers are often seen in multifamily groups from Jun through the summer, when broods are maturing. Five fledgling Common Mergansers were discovered on the early date of 26 Apr 1997 on the Delaware River.[2] The males leave the females after incubation has begun and are rarely reported until the young have fledged. On a 4-mile stretch of the Allegheny River in Warren Co. in Aug 1985, 90 were counted.[3] Nonbreeding Common Mergansers are occasionally found away from breeding sites in summer.

Fall: Common Mergansers are the last of the waterfowl to move through the state in fall. A few migrants appear in early Oct, but migration does not get under way until mid- to late Nov, after which numbers gradually increase to maximum highs in the hundreds or thousands by late Dec. The largest concentrations at this season occur on the lower Susquehanna River; up to 12,000 birds were recorded on the Southern Lancaster Co. CBC in 1986 and 1988 (Morrin et al. 1991). As the weather turns harsh in Jan, numbers decline.

Winter: Common Mergansers are locally abundant in winter on the Lower Delaware and Susquehanna rivers and at the Lake Erie Shore, primarily in early to mid-winter. They are uncommon to rare in winter elsewhere. Most overwinter on streams or lakes where open water permits. When other sources of open water are not available, Common Mergansers may congregate in large numbers at warm-water discharges, such as an industrial plant at Penn Manor on the lower Delaware River and the Peach Bottom Nuclear Power Plant on the lower Susquehanna River. Thousands also winter in large rafts on Lake Erie on open water between ice floes.

History: The breeding history of Common Mergansers follows the pattern of other wetland and fish-eating birds. They were reported breeding during the mid-nineteenth century but became scarce at the beginning of the twentieth century. Breeding records were widely scattered in northern counties; Luzerne Co. (Warren 1890) and Clinton Co. (Harlow 1913) are two examples. At a low point, Harlow (1913) stated that "it would seem that the time is past when this bird nested" (in Pennsylvania). Common Mergansers went largely unreported as breeders during the first half of the twentieth century. They then began expanding their breeding range, exceeding the historically reported range and abundance by the mid-1990s.

Comments: This species visually pursues fish underwater, making it particularly sensitive to water quality and clarity.

[1] Hall, G.A. 1990. Appalachian region. AB 44:425.
[2] Wiltraut, R. 1997. Local notes: Northampton County. PB 11:98.
[3] Hall. G.A. 1986. Appalachian region. AB 40:112.

Red-breasted Merganser ***Mergus serrator***

General status: The breeding range of Red-breasted Mergansers is from Alaska across all but southwestern Canada. They winter primarily along the Pacific, Atlantic, and Gulf coasts. The least widely distributed of the mergansers in Pennsylvania, Red-breasted Mergansers are common to abundant regular migrants at the Glaciated Northwest and Lake Erie Shore. Their numbers vary from year to year elsewhere in the state, but they are usually not common. In summer they are regular at the Lake Erie Shore and are accidental elsewhere. They were reported to have nested in the state historically. Red-breasted Mergansers are regular but local in winter.

Habitat: They are found on large lakes, ponds, and slow-moving streams.

Seasonal status and distribution

Spring: Over most of the state they are more common in spring than in fall. Migration begins after the first spring thaw in late Feb or early Mar, and the peak ranges from mid-Mar to mid-Apr. Away from Lake Erie, migration has usually ended by late Apr. Some remarkable numbers have been recorded away from Lake Erie in spring, such as the 1000 that were at MSP in Butler Co. in the spring of 1982.[1] Huge flocks of Red-breasted Mergansers—10,000 birds is not unusual—may be seen in Presque Isle Bay each year. On 11 Apr 1991, over 20,000 were estimated in the bay.[2] During some years, hundreds may still be present in Presque Isle Bay well into the second week of May. Nearly all birds have departed from this site by the fourth week of May, but stragglers may remain to the first week of Jun or later.

Summer: In Warren Co., 23 birds were at Kinzua Dam on 15 Jul 1986.[3] Birds away from Lake Erie have been reported in summer from Allegheny, Berks, Bucks, Cambria, Carbon, Centre, Lancaster, Lehigh, Lycoming, and Warren cos. Nearly every year one or two birds remain after spring migration to summer on the waters surrounding PISP.

Fall: Single birds may appear as early as the second week of Sep, but an unusual number of 29 was seen on the early date of 18 Sep 1971 over Bake Oven Knob (Morris et al. 1984). Their typical migration period is from the third week of Oct to the third week of Dec. Peak migration occurs during the last three weeks of Nov. Heavy flights occur annually along the Lake Erie Shore, where thousands per day may be counted passing the eastern tip of PISP. On 29 Nov 1985 an estimated 15,000 migrating birds were recorded at this site.[4]

Winter: Most winter reports from the Coastal Plain and Piedmont are from the lower Delaware and Susquehanna rivers, although records exist away from these rivers in Berks, Bucks, Chester, Cumberland, Lebanon, and York cos. In the Ridge and Valley, the only reports are from Canoe Lake in Blair Co.,[5] Raystown Reservoir in Huntingdon Co.,[6] and in Centre Co., where they "winter occasionally" (Wood 1983). In the Glaciated Northeast, Red-breasted Mergansers have been found only in Wyoming Co.[7] Very few records exist from the High Plateau; however, they have been reported in Cambria, McKean, Somerset, Tioga, Warren, and Westmoreland cos. In the Southwest and Glaciated Northwest, winter records come from Allegheny, Armstrong, Beaver, Butler, Clarion, Crawford, Indiana, Lawrence, and Mercer cos. The number of wintering birds around PISP varies from year to year. Several hundred may be seen at this site during mild winters, but relatively few birds are seen during harsh winters.

History: Red-breasted Mergansers reportedly bred in Pennsylvania before the turn of the twentieth century (Libhart 1869; Turnbull 1869). S. E. Bacon reported young birds on Erie Bay in Jul and Aug and believed one he shot on 27 Jul 1893 was of this species (Todd 1940). Sutton (1928a) reported young birds on Erie Bay in Jul and Aug. Poole (1964) mentioned the lack of conclusive nesting evidence, at least in recent years.

Comments: It is likely that at least some summer records of nonbreeding-plumaged Red-breasted Mergansers are misidentified nonbreeding-plumaged Common Mergansers.

[1] Hall, G.A. 1982. Appalachian region. AB 36:851.
[2] G. M. McWilliams, pers. obs.
[3] Hall, G.A. 1986. Appalachian region. AB 40: 1203.
[4] Hall, G.A. 1986. Appalachian region. AB 40:112.
[5] S. Kotala, pers. comm.
[6] G. Grove, pers. comm.
[7] Haas, F.C., and B.M. Haas, eds. 1995. Seasonal occurrence tables—January through March 1995. PB 9:44.

Masked Duck *Nomonyx dominicus*

General status: Very little is known about these elusive and secretive ducks. Masked Ducks are local over tropical South America through the West Indies and over Central America north to Central Mexico. This species is a rare visitor to southern Texas and along the Gulf coast to Florida, with vagrants occurring as far north as Massachusetts, Wisconsin, Tennessee, North Carolina, Maryland, and Vermont. A single sighting of this species has been recorded in Pennsylvania.

Seasonal status and distribution

Spring: An unbanded adult male Masked Duck was seen near Lake Ontelaunee in Berks Co. on 12–14 Jun 1984. The sighting was documented with a photograph by R. Cook on 14 Jun. Since 1964 Masked Ducks had not been kept in captivity anywhere in North America.[1] This bird was believed to be of wild origin and was accepted as the first and only state record.

[1] Paxton, R.O., W.J. Boyle Jr., and D.A. Cutler. 1984. Hudson-Delaware region. AB 38:1002.

Ruddy Duck ***Oxyura jamaicensis***

General status: Ruddy Ducks breed primarily in southwestern Canada and the western U.S. In the East they breed sporadically at scattered sites from Ontario south to South Carolina. In winter they can be found along the Pacific coast through Mexico, east across the southern U.S., on the Great Lakes, and along the Atlantic coast north to New Jersey. These stiff-tailed ducks have declined in Pennsylvania since the early 1970s, especially the winter populations. However, they are still occasionally found in large flocks during migration. Ruddy Ducks are fairly common to common regular migrants statewide. Ruddy Ducks are casual visitors or residents in summer, but they have been known to breed in the state. They are regular visitors or residents in winter, but most are confined to the Coastal Plain. Several major oil spills during the 1960s and 1970s resulted in the direct loss of several thousand ducks and affected the tubificid worm populations, a major prey item.[1]

Habitat: Ruddy Ducks are found on lakes, ponds, marshes, and large slow-moving rivers.

Seasonal status and distribution

Spring: Migration begins in early Mar and continues to late Apr, with lingering birds to early Jun. Peak migration ranges from the last week of Mar to the third week of Apr, with fewer than 100 birds at most sites. In some years, numbers reach several hundred or more birds. A notable concentration was 4000 seen at Conneaut Lake on 29 Apr 1971.[2]

Breeding: Three breeding records have been documented in Pennsylvania since 1960. A nesting pair produced several young on the Upper Lake at Pymatuning in 1969,[3] and, two half-grown young birds provided evidence of nesting at Glen Morgan Lake near Morgantown in Berks Co. on 6–7 Sep 1997.[4] A pair nested successfully at Glen Morgan Lake in 1998.[5] Nesting was suspected at Dunmore in southern Lancaster Co. when 11 adults were seen during the summer of 1976.[6] Other summer sightings are from Armstrong, Bucks, Butler, Centre, Crawford, Delaware, Erie, Lehigh, Montour, Northampton, Philadelphia, Warren, Westmoreland, and York cos.

Fall: Southbound birds appear as early as the last week of Sep, but the majority of birds will move in late Oct or early Nov. Migration peaks by the last week of Nov. Most of those that do not remain in the state through the winter leave by the last week of Dec or the first week in Jan.

Winter: The state's greatest winter concentration of Ruddy Ducks is on the lower Delaware River south of Philadelphia, where they once were common to abundant. The number of ducks has been drastically reduced in this area since the 1970s, at least partially because of oil spills.[7] Tens of thousands formerly wintered on the lower Delaware River near Philadelphia.[8] On the Glenolden CBC in the early 1970s, counts of 40,000 birds were not uncommon.[9] In contrast, the average Glenolden CBC from 1983 to 1993 was only 472 birds.[7] Winter observations of Ruddy Ducks away from the lower Delaware River are rare. In the Piedmont, winter sightings have been reported from Berks, Bucks, Chester, Dauphin, Delaware, and York cos. In the Ridge and Valley they have been recorded on Harvey's Lake in Luzerne Co.,[10] the State College area of Centre Co. (Wood 1983), and in Wyoming Co.[11] Ruddy Ducks have not been reported from the Glaciated Northeast or the upper High Plateau region at this season. West of the Allegheny Front they have been recorded in winter in Armstrong, Butler, Cambria, Erie, Indiana, Mercer, Somerset, and Westmoreland cos.

History: Before 1900, Ruddy Ducks were considered abundant, but Beck (1924) and Todd (1940) believed that they had declined in numbers since then. Shortly after the Pymatuning Reservoir was created in Crawford Co. in 1934, R. L. Fricke (fide Todd 1940) reported having seen displaying males and pairs, leading to the suspicion of breeding. In 1935 at the Pymatuning Reservoir, J. K. Terres saw flightless young near Linesville in Jul (Trimble 1940). Finally in 1936, Fricke found the first (and only) nest at this site, containing four eggs, which were collected on 22 Jun (Todd 1940). Summer observations continued into the 1940s, but no further evidence of nesting was obtained (Grimm 1952). During migration, Ruddy Ducks were usually reported in small bands, but occasionally larger flocks were found, such as the 2000 at Pymatuning on 15 Apr 1946 (Grimm 1952).

[1] Stark, R.T. 1978. Food habits of the Ruddy Duck *(Oxyura jamaicensis)* at the Tinicum National Environmental Center. M.S. thesis. Penn State Univ., State College.

[2] Leberman, R.F. 1971. Field notes. Sandpiper 13:66.

[3] Leberman, R.C., and R.F. Leberman. 1969. Field notes. Sandpiper 7:14.

[4] Keller, R. 1997. First nesting of Ruddy Duck *(Oxyura jamaicensis)* in Berks County, Pennsylvania. PB 11:142–143.

[5] F.C. Haas and B.M. Haas, pers. comm.

[6] Buckley, P.A., R.O. Paxton, and D.A. Cutler. 1976. Hudson-Delaware region. AB 30:935.

[7] N. Pulcinella, pers. comm.

[8] Boyle, W.J., Jr., R O. Paxton, and D.A. Cutler. 1981. Hudson-Delaware region. AB 35:283.

[9] Guarente, A. 1989. County reports—October through December 1989: Delaware County. PB 3:143.

[10] M. Blauer, pers. comm.

[11] Haas, F.C., and B.M. Haas, eds. 1997. Seasonal occurrence tables—January through March 1997. PB 11:46.

Order Falconiformes
Family Accipitridae: Ospreys, Kites, Eagles, Harriers, and Hawks

Birds of prey have a hooked bill and strong feet equipped with long curved talons for grasping prey. Their predatory behaviors have given them a bad reputation among many people. During the first half of the twentieth century, birds of prey were shot for bounty in large numbers because people believed that they eliminated populations of game birds and threatened farmers' poultry. Migrating hawks in the genus *Accipiter* were especially persecuted and shot for sport by the hundreds, even thousands, along the ridges. Hawk Mountain was one of the most popular hawk-shooting sites along the Kittatinny Ridge. There was such an outcry from a small group concerned about the slaughter of hawks that the mountain was purchased, protected, and named Hawk Mountain Sanctuary. This turn of events exerted a strong influence on raptor conservation worldwide. Fish-eating and bird-eating birds of prey declined during the 1950s and 1960s from the widespread use of DDT as an insecticide. Since DDT was banned in the U.S. many species have begun to recover.

Osprey, buteos, and eagles have rather long, broad wings and a short tail. They typically soar and feed primarily on rodents, except for Osprey and Bald Eagles, which feed mainly on fish. Spectacular flights of buteos, especially Broad-winged Hawks, are seen along the mountain ridges of Pennsylvania in fall and along the Lake Erie Shore in spring. Two species of kites have been recorded in Pennsylvania; both are rare vagrants from the southern U.S. They have a slim body and long, pointed, narrow wings. Kites frequently hover and hawk (attack by swooping and striking) for insects. Harriers have long narrow wings, usually held at a dihedral when gliding, and a long narrow tail. Except during migration, harriers are usually found flying low over open fields in search of small rodents. Accipiters have short, rounded wings and a long narrow tail. They usually fly in a direct line, with a series of flaps and glides. Their short wings and long tail give them agility in their pursuit of birds, their primary food source, through woodlands and thickets. They often hunt small birds at bird feeding stations in winter. Most members in this family aggressively defend their nesting territory and will attack any intruder, including humans. Some are known to mate for life and often return to the same nest site year after year. Most build a nest on or near the top of a large tree or near the trunk at lower levels. Some species build a large nest. One such species, Bald Eagle, adds material to the same nest each year. Unlike other species in this family, harriers build their nest on or near the ground.

Osprey ***Pandion haliaetus***

General status: Ospreys, widely known as fish hawks, declined during the 1950s and 1960s as a result of the use of pesticides, particularly DDT. After these deadly chemicals were banned in the late 1960s, Ospreys quickly began to recover. In North America they breed from Alaska across Canada south through the western states, around the upper Great Lakes, and along the Atlantic and Gulf coasts. They winter along the Gulf coast, in southern Florida and from southern California south through Central and South America. In Pennsylvania Ospreys are uncommon to fairly common regular migrants. They breed locally in the state, and a few nonbreeding birds remain during summer. Three reintroduction programs were successful at expanding nesting populations in Pennsylvania. Ospreys are rarely encountered into early winter.

Habitat: Ospreys prefer lakes, ponds, rivers, and marshes bordered by trees used for perches. They are also observed along mountain ridges in migration.

Seasonal status and distribution

Spring: Ospreys are more conspicuous in spring than in fall. Migrants have been recorded as early as the fourth week of Mar. Their typical migration period is from the last week of Mar to the third week of May. Fewer are seen at hawk watches in spring than in fall, but large movements have been recorded. The peak number of birds passes from the first to the third week of Apr. At most sites, daily highs never reach 100 birds, but at Baer Rocks in Lehigh Co., 142 were counted on the late date of 30 Apr 1983,[1] and an unprecedented 171 were seen at the same location on 14 Apr 1990.[2] Stragglers may remain until early Jun.

Breeding: Ospreys are rare during summer across the state but are regularly and easily observed near nesting sites. The state breeding population grew as a result of reintroductions from a single nest in the Poconos in 1986 to over 30 pairs in 1997. From 1980 to 1996, 265 Ospreys had been released into the state (see the accompanying table). Nests have been established in each reintroduction area, including at least 13 nest sites in the Poconos, five at U.S. Army Corps of Engineers reservoirs in

Tioga Co., and one nest each at Moraine (Butler Co.) and M. K. Goddard (Mercer Co.) state parks. Since a pair hacked in West Virginia began nesting at Cranberry Glades Lake in Somerset Co. in 1990 the southwestern Pennsylvania population has grown to at least five pairs in 1997, including nests at Lake Somerset, Donegal Lake, and Loyalhanna Lake in Westmoreland Co. and on the Conemaugh River in Indiana Co.[3] Ospreys also expanded their breeding population into Pennsylvania away from hack sites, including up to five nesting pairs along the lower Susquehanna River in Lancaster Co. and a pair that nested successfully for the first time in 1997 in Bucks Co. Confirmed nesting has also been reported in Northampton Co. along the Delaware River at Lake Wallenpaupack in Pike Co. A large nest is built close to, or over water. Two or three white or pinkish white eggs spotted or blotched with reddish brown are usually laid in Apr; young may be found in the nest through Jul.

Fall: Most sightings are along rivers and from hawk watches along mountain ridges. The typical migration period is the second or third week of Aug to the fourth week of Oct. Peak passage occurs during the second and third weeks of Sep. Hundreds of Ospreys are recorded in fall at HMS annually, with the seasonal total reaching 873 birds during the fall of 1990.[4] Daily counts exceeding 100 are unusual, but on 11 Sep 1965, 102 were counted at Bake Oven[5] and 175 were recorded on 23 Sep 1989 passing by HMS.[6] Stragglers may remain into Jan.

Winter: Ospreys normally do not stay in Pennsylvania through the entire winter. The latest record appears to be 1 Feb 1972.[7] Ospreys in late Feb may be very early northbound migrants rather than birds that have wintered. All but one midwinter record have been in the southern half of the state from the counties of Bedford, Berks, Bucks, Butler, Carbon, Centre, Delaware, Franklin, Huntingdon, Mifflin, Northumberland, Philadelphia, Somerset, and Westmoreland. The only winter record in the northern half of the state is from the National Fish Laboratory in Tioga Co. on 29 Jan 1991.[8]

History: Warren (1890) listed correspondents from widely scattered counties that considered this species a breeder at the end of the 1800s and stated that they bred "more or less regularly . . . in the vicinity of large streams." Nests were rarely reported, however, and the absence of eggs from museum collections suggests their rarity. Todd (1940) stated that breeding records in western Pennsylvania were "surprisingly few," even though he considered Ospreys fairly common and regular around Pymatuning and along the upper Allegheny River in Warren Co. Some sources (e.g., Warren 1890) that listed the species as "breeding" may, in fact, have been reporting summering individuals that apparently were not breeding, such as has been observed in recent years. Histori-

Table 1. Summary of Osprey reintroductions in Pennsylvania

Release area	Years	Total released	First year of nesting	Number of nests in 1996
Moraine State Park	1993–1996	95	1996	1
Poconos	1980–1989	110	1986	13
Tioga Co. Reservoirs	1990–1994	60	1994	5

cally, nests were confirmed only from Beaver, Bucks, Clarion, Delaware, and, possibly, Wyoming cos. Most of these were in only one location and reported in only one year. The last definite historical nesting occurred in 1935 (Poole 1964).

Comments: The Osprey's listing as Extirpated in 1979 (Genoways and Brenner 1985) set the stage for the reintroduction program. The PGC upgraded their status from Endangered to Threatended in 1997.

[1] Boyle, W.J., Jr., R.O. Paxton, and D.A. Cutler. 1983. Hudson-Delaware region. AB 37:852.
[2] Morris, B. 1990. County reports—April through June 1990: Lehigh County. PB 4:74.
[3] Brauning, D.W. 1997. Osprey nesting surveys. Annual report. Pennsylvania Game Commission, Harrisburg.
[4] Goodrich, L. 1991. 1990 Hawk watch reports. PB 4:173.
[5] Heintzelman, D.S. 1968–1969. Autumn birds of Bake Oven Knob. Cassinia 51:17.
[6] Goodrich, L. 1989. Hawk watch report. PB 3:158.
[7] Scott, F.R., and D.A. Cutler. 1972. Middle Atlantic Coast region. AB 26:586.
[8] J. Stickler, pers. comm.

Swallow-tailed Kite *Elanoides forficatus*

General status: Swallow-tailed Kites breed along the eastern Gulf coast, through most of Florida, and along the southern Atlantic coast as far north as South Carolina, and they winter in South America. Strays have been seen as far north as Ontario and Nova Scotia. This species has been reported at least 19 times in Pennsylvania, but only seven times since 1960. All but one of those sightings was of a single bird, and most records were from the Piedmont.

Seasonal status and distribution

Spring and early summer: Six records of Swallow-tailed Kites fall within spring, from May 23 to Jun 22. The first modern record of Swallow-tailed Kite was at Birchrunville when R. Kuch observed one in Chester Co. from 7 to 13 Jun 1980[1] Others include single birds near Shippensburg on the Cumberland-Franklin line on 15 Jun 1981,[2] Media in Delaware Co. on 11 Jun 1982,[3] Jonas in Monroe Co. on 24 May 1991,[4] and the Latrobe Reservoir in Westmoreland Co. on 22 Jun 1991.[5] The most recent report involved up to three Swallow-tailed Kites intermittently seen between Bowmansville and Knauers on the Berks-Lancaster co. line from 23 May to 10 Jun 1995 (see also Mississippi Kite account).[6]

Fall: Only two sightings are from this season. Apparently one bird was photographed on 5 Sep 1976 at Cadogan Flats in Armstrong Co.[7] The most recent record was of one seen above Pinetown Road near Ski Roundtop in York Co. on 22 Sep 1996.[8]

History: Swallow-tailed Kites formerly bred over a much larger area of the eastern U.S., as far north as Minnesota, Wisconsin, and Ohio. In Pennsylvania the first record was from an account by Dr. Benjamin Barton (1799), which stated that the swallow-tailed falcon arrived (at Philadelphia) 4 Jul 1791 (Poole, unpbl. ms.). Turnbull (1869) reported that they had been seen once or twice in Pennsylvania; one was shot near Philadelphia in 1857. Libhart (1869) listed them as rare in the southern portion of Lancaster Co. Other records before 1900 occurred in 1857, 1888, and 1894 (Poole, unpbl. ms.). More recently, one was observed over the Wissahickon Creek at Chestnut Hill in Philadelphia Co. on 11 May 1952.[9]

Comments: Frequent recent sightings at Cape May Point in New Jersey and Point Pelee in Ontario, Canada, suggest a greater potential for strays to occur in Pennsylvania.

[1] Paxton, R.O., W.J Boyle Jr., D.A. Cutler, and K.C. Richards. 1980. Hudson-Delaware region. AB 34:759.
[2] Hall, G.A. 1981. Appalachian region. AB 35:939.
[3] Paxton, R.O., W.J. Boyle Jr., and D.A. Cutler. 1982. Hudson-Delaware region. AB 36:959.

[4] Boyle, W.J., R.O. Paxton, and D.A. Cutler. 1991. Hudson-Delaware region. AB 45:421.
[5] Hall, G.A. 1991. Appalachian region. AB 45: 1115.
[6] Heller, J. 1995. Kites and more. PB 9:87.
[7] M. Higbee and R. Higbee, pers. comm.
[8] Spiese, A. 1996. Local notes: York County. PB 10:168.
[9] Ross, C.C. 1953. General notes. Cassinia 1951–1952:25.

Mississippi Kite ***Ictinia mississippiensis***

General status: Mississippi Kites range primarily across the southern half of the U.S., from Arizona east to South Carolina, and they winter primarily in Central America. The first West Virginia nesting was reported in 1996.[1] In Pennsylvania they are listed as casual visitors. At least 15 sightings have been reported from the state since 1974; all but two were in spring and most were from the Piedmont. This species has been reported once in the summer and once in the fall. Nearly all sighting are of birds in flight.

Seasonal status and distribution

Spring: All but one report are of single birds observed for only a single day. Records east of the mountains may be strays from the southeastern population, while birds west of the mountains may be strays from the Mississippi Valley population. Dates range from 27 Mar to 10 Jun. The first modern record of a Missisissippi Kite was one found in poor conditon by J. J. Thouron on 31 May 1974 at Glenroy in the southeastern corner of the state and later deposited in the ANSP.[2] Two or three Mississippi Kites were present from 27 May to 10 Jun 1995 between the towns of Bowmansville in Lancaster Co. and Knauers in Berks Co.[3]

Summer: One was sighted near Baden in Beaver Co. on 26 Jun 1988.[4]

Fall: A Mississippi Kite was seen at Berwick in Luzerne Co. on 20 Oct 1989.[5]

History: A male Mississippi Kite, now at the American Museum of Natural History, was collected by W. T. Smith at Philadelphia on 2 Jul 1886 (Poole 1964). Poole (unpbl. ms.) mentions another bird collected in Cumberland Co. in Sep 1892, but that specimen was later lost.

Comments: A report of a Mississippi Kite seen on 19 Jun 1998 near Mt. Cobb in Lackawanna Co. is currently under review.[6]

[1] Quezan, A.J. 1997. Raven 68:85–88.
[2] Scott, F.R., and D.A. Cutler. 1974. Middle Atlantic Coast region. AB 28:785.
[3] Haas, F.C., and B.M. Haas, eds. 1995. Birds of note—April through June 1995. PB 9:87.
[4] Kwater, E. 1988. Mississippi Kite in Beaver County. PB 2:82.
[5] Paxton, R.O., W.J. Boyle Jr., and D.A. Cutler. 1990. Hudson-Delaware region. AB 44:62.
[6] Clauser, T. 1998. Local notes: Lackawanna County. PB 12:76.

Bald Eagle ***Haliaeetus leucocephalus***

General status: Bald Eagles, the most celebrated of birds of prey in North America, recently have made a dramatic return from decades of depressed numbers. They breed from Alaska south through Canada and at scattered sites in the West and around the Great Lakes, along the Atlantic and Gulf coasts, and in Florida. They winter generally thoughout their breeding range. In Pennsylvania, they are found year-round as uncommon to rare regular migrants and permanent residents around nesting areas. They are regular nonbreeding summer visitors away from nesting sites. In winter, eagles may become residents, sometimes forming roosts of dozens of birds, especially along the Delaware and lower Susquehanna rivers.

Habitat: Bald Eagles are found along large rivers and lakes with large trees used for perches. They are also seen along mountain ridges during migration.

Seasonal status and distribution

Spring: Fewer birds are seen in spring

than in fall. Away from wintering sites, the first northbound migrants are noted in late Feb or early Mar after the first major spring thaw. Migration continues to late Apr, with stragglers until early Jun. Greater numbers of birds are seen migrating along the Lake Erie Shore than anywhere else in spring in the state.

Breeding: Bald Eagles are rare but locally regular summer residents near nesting areas and rare and irregular away from breeding areas. Six or more nests are found in Crawford and Lancaster cos., where eagles are easily seen in all seasons. Nesting pairs also are found in Butler, Dauphin, Forest, Mercer, Pike, Tioga, Venango, Warren, and York cos. Pennsylvania's nesting population increased from two nests in Crawford Co. in 1980 to 28 widely distributed nests in 1998. Reintroduction efforts by the PGC from 1983 to 1989 released 88 young imported from Saskatchewan at hack sites in Dauphin and Pike cos. (Kosack 1995). The nesting season varies among pairs, with incubation beginning as early as late Feb or starting as late as Apr, but most pairs are on eggs by mid-Mar. Most nests are placed high in tall white pines or sycamores; some are large structures accumulated over many years of use. One to three dull white eggs are laid. Active nests have averaged 1.3 young per year since 1985.[1]

Fall: They are among the earliest raptors to begin migrating in fall. Most migrants are seen flying southwestward along the mountain ridges from late Aug through early Dec. Adult Bald Eagles usually pass through the state during a shorter period than the extended migration season of immatures. Peak adult migration occurs during the first two weeks of Sep, whereas immatures may peak any time from the second week of Sep to the second week of Nov. Daily counts of migrants at the hawk watches usually total fewer than 10 birds.

Winter: Most winter concentrations of Bald Eagles are located along the Delaware River and the lower Susquehanna River and around nesting areas in the Pymatuning-Conneaut areas in Crawford Co. In the Piedmont and the Ridge and Valley, most winter sightings are in the valleys along the Susquehanna River and its larger tributaries. On the Southern Lancaster Co. Christmas Bird Count, as many as 23 birds have been counted (Morrin et al. 1991). Most wintering Bald Eagles in the Glaciated Northeast are along the Delaware River, where they may be locally fairly common. A high tally of 44 Bald Eagles, 24 immatures and 20 adults, was counted within one hour along 8 miles of the Lackawaxen River on 13 Jan 1996.[2] West of the mountains and away from the Pymatuning-Conneaut area they are rare but regular in winter, mainly along the Allegheny River, French Creek, and Lake Erie.

History: Bald Eagles historically nested in low numbers along natural lakes and rivers statewide, particularly the upper Delaware in Pike Co., the lower Susquehanna River valley, and Lake Erie (Harlow 1913). Nesting areas not occupied in recent years include the borough of West Chester (Burns 1919), the North Valley hills near Valley Forge, and various points in Erie Co. (Stull et al. 1985). Despite recognition as the nation's symbol, eagle numbers declined from the use of pesticides, direct persecution, and loss of habitat. In 1940, the U.S. Bald Eagle Protection Act became law, but populations continued to decline. The federal Endangered Species Act and banning of DDT paved the way for the restoration of this and other species.

Comments: Bald Eagles were listed by the U.S. Fish and Wildlife Service as Endangered until 1995, when they were upgraded to Threatened. They are listed by the PGC as Endangered.

[1] Brauning, D.W. 1996. Bald Eagle nesting studies. Annual report. Pennsylvania Game Commission, Harrisburg.
[2] Brauning, D.W. 1996. Mid-winter Bald Eagle surveys. Annual report. Pennsylvania Game Commission, Harrisburg.

Northern Harrier ***Circus cyaneus***

General status: In North America, Northern Harriers breed from Alaska south across the U.S. as far south as southern California, northern Texas, and Kentucky. They winter throughout the U.S. These hawks of extensive open fields and marshes are usually seen flying low over fields or in migration along the mountain ridges of Pennsylvania. Northern Harriers are uncommon to fairly common regular migrants statewide and breed locally over most of the state, regularly on reclaimed strip mines. Nonbreeding birds are occasionally recorded in summer away from known breeding sites. Northern Harriers regularly winter in the state.

Habitat: Harriers prefer extensive open grassy fields, marshes, or scrub wetlands for feeding and nesting and are noted along mountain ridges in migration.

Seasonal status and distribution

Spring: The usual migration period is from late Feb to mid-May. Peak migration ranges from the last week of Mar to the third week of Apr. The greatest number of Northern Harriers is seen in spring migration at the Lake Erie Shore. Usually fewer than 10 per day are observed migrating near the shore of Lake Erie.

Breeding: Northern Harriers are rare to uncommon regular summer residents in northern and western counties and casual elsewhere statewide. A few pairs nested in the Philadelphia airport area until the 1980s. Northern Harriers are most frequently found on reclaimed strip mines (Yahner and Rohrbaugh 1996). Nesting is seldom confirmed, and some summer birds may be nonbreeders. For example, only 7% of BBA records were confirmed.[1] A steady decline in our summer population has been noted for decades (Poole 1964). Little is known of the breeding biology of Northern Harriers in the state except that nests are placed on the ground. Four to six pale white or bluish white unmarked eggs are laid. Few nests have been reported; five egg sets in collections were collected in mid- to late May.

Fall: Migration begins in early Sep and stretches well into Dec. The migration peak tends to vary from year to year and from hawk watch to hawk watch. However, Oct totals from most hawk watches average higher than the totals from other months.

Winter: Northern Harriers are uncommon to rare and can be found in open grassy fields during most winters in every region in the state. They are absent from the heavily forested mountains of the Ridge and Valley and High Plateau. Northern Harriers may be locally fairly common on reclaimed strip mines, particularly in the Southwest region. They are usually seen perched on fence posts or flying low over fields, especially at dusk.

History: Historical records indicate that Northern Harriers were rare and probably restricted to the northern swampy areas of Pennsylvania in summer. Nests were found in swampy areas of scattered northern counties, notably Crawford, where most early egg sets were taken. Sutton (1928b) described the harrier as ubiquitous and an abundant summer resident, "not only in the [Pymatuning] Swamp and about Conneaut Lake, but throughout the countryside, where it may nest in brookside weed-patches, or near farm-yard ponds." Besides this remarkable description, the Northern Harrier was considered generally rare but widespread across the state. As many as eight pairs once nested in marshes of the Tinicum area (J.C. Miller and Price 1959).

Comments: The Northern Harrier is listed as a Species of Special Concern, Candidate–At Risk, because of its low population and vulnerability to mowing and loss of wetlands.

[1] Goodrich, L. 1992. Northern Harrier (Circus cyaneus). BBA (Brauning 1992): 94–95.

Sharp-shinned Hawk ***Accipiter striatus***

General status: Sharp-shinned Hawks breed from Alaska south across Canada and in the East south through the Appalachian Mountains. They winter throughout the U.S. The smallest accipiter in our area, Sharp-shinned Hawks are the second most common migrant raptor in Pennsylvania. They are fairly common to abundant regular migrants, breed over most of the state, and are regular winter residents.

Habitat: Sharp-shinned Hawks are found in mixed forests during the breeding season. During migration the largest concentrations are observed primarily along mountain ridges and along Lake Erie. In winter, these hawks prefer mixed forests near bird feeding stations, where there is an abundance of prey.

Seasonal status and distribution

Spring: A few birds begin migrating as early as mid-Mar. The typical migration period is from the last week of Mar to about mid-May, with stragglers to late May or early Jun. Peak migration occurs from about the first to the third week of Apr. More migrants are observed along Lake Erie than at any other area in the state in spring.

Breeding: Sharp-shinned Hawks are uncommon regular breeders, widely distributed in wooded areas across Pennsylvania, except in the Piedmont and southwest of Pittsburgh, where they are rare and local. Most breeding records are in higher elevations of northern hardwood forests. Sharp-shinned Hawks are secretive and easily overlooked during summer, except for noisy mobs of fledged young begging for food. Nests are generally built in dense conifer stands in extensive forested areas, most often near a forest opening (Grimm and Yahner 1986). They are generally 20–40 feet up in a tree, often a conifer, next to the trunk. The nest is comparable in size to that of the larger Cooper's Hawk. Sharp-shinned Hawks nest later than many other raptors; most egg sets have been collected from 14 May to 6 Jun. Eggs are pale white or bluish white, variably splotched with rich brown markings.

Fall: More birds are seen along mountain ridges than elsewhere in the state during fall. The migration period is from early Sep to about the second week of Nov, with a peak usually during the first and second weeks of Oct. Hundreds may be seen in a single day at hawk watches. As many as 2616 have been recorded in one day at HMS in 1979.[1] High totals during the fall season have reached nearly 9000 birds at Waggoner's Gap in Perry Co. in 1982,[2] and nearly 10,000 birds have been recorded at HMS in 1989.[3] A few stragglers remain through the winter.

Winter: Sharp-shinned Hawks winter in every region in the state, but they are always uncommon to rare and local. As winter progresses and food becomes scarce in forested areas, they may move closer to human habitation to prey on small birds attracted to feeders.

History: Sharp-shinned Hawks were believed to nest in every Pennsylvania county until about 1900 (Warren 1890; Harlow 1913); egg sets were collected even south to Philadelphia and Greene cos. (Harlow 1913). Sharp-shinned Hawks were listed as common to abundant in summer ("resident") in southern counties such as Lancaster (Libhart 1869) and Cumberland (Baird 1845). They had declined significantly by the early 1900s as a result of persecution (Harlow 1913), a decline exacerbated by DDT in the 1950s. However, very large numbers continued to be reported during migration, and many were shot on the ridges before the birds were protected. Perhaps over 35,000 passed by Hawk Mountain in the fall of 1939.[4]

Comments: Hawks have been counted passing HMS since the sanctuary was established in 1934. A Sharp-shinned Hawk passing over the lookout on 8 Oct 1992 was identified as the one millionth raptor recorded there.[5]

[1] Paxton, R.O., K.C. Richards, and D.A. Cutler. 1980. Hudson-Delaware region. AB 34:145.
[2] Hoffman, D. 1987. Hawkwatching summaries. PB 1:144.
[3] Goodrich, L. 1989. 1989 Hawk watch reports. PB 3:158.
[4] Paxton, R.O., P.W. Smith, and D.A. Cutler. 1979. Hudson-Delaware region. AB 33:160.
[5] Goodrich, L. 1993. 1992 Hawk watch reports. PB 6:182.

Cooper's Hawk *Accipiter cooperii*

General status: Cooper's Hawks breed and winter across southern Canada and throughout the U.S. They are less common than Sharp-shinned Hawks and more common than goshawks in migration. Cooper's Hawks are fairly common to common regular migrants. They breed and winter throughout the state.

Habitat: Cooper's Hawks prefer mixed forests during the breeding season and are seen primarily along mountain ridges and along Lake Erie during migration. In winter they are usually found in mixed forests, near bird feeding stations in suburbs, and on farms where there is an abundance of prey.

Seasonal status and distribution

Spring: Northbound arrivals begin passing through the state somewhat later than the smaller Sharp-shinned Hawks. Migration usually begins in the last week of Mar and continues to the first week of May, with a peak within the first and third weeks of Apr. Migration is heaviest along the mountain ridges and Lake Erie, where fewer than 25 may be seen in a single day. Recently, counts of over 100 birds have been recorded in one day along Lake Erie (Stull et al. 1985).

Breeding: Cooper's Hawks are uncommon widespread breeders statewide, generally with a more southerly distribution than that of Sharp-shinned Hawks. They probably nest in every county, although a nest has not been reported in Philadelphia Co. in many years. Cooper's Hawks are scarce or overlooked in higher mountainous areas. They nest in forested areas with large trees, often near openings, and have been moving into suburban settings in recent years. Nests are substantial structures, placed in both deciduous and coniferous trees; frequently a white pine is selected (Grimm and Yahner 1986). BBS data show a significant increase of 8.5% per year (1966–1996), and many observers report an expansion of the nesting population. Cooper's Hawks continue to increase, in part as a response to a cleaner environment and better protection from indiscriminate shooting. Most egg sets have been found during the first three weeks of May. Eggs are pale white and sometimes scattered with light spots.

Fall: Cooper's Hawks are most frequently observed during fall migration. Their typical migration period is from the second week of Sep to the second week of Nov, with stragglers remaining through the winter. Peak migration is usually during the first and second weeks of Oct. They are more common along the mountain ridges than anywhere else in the state. Daily high counts at hawk watches usually do not exceed 40 or 50 birds. A seasonal high of 787 birds was counted at HMS in the fall of 1989.[1]

Winter: Cooper's Hawks have been recorded in all regions in winter. As winter progresses and food becomes scarce in forested areas, they may move into suburbs, parks, and farms to prey on small birds that are attracted to feeders.

History: Harlow (1913) stated that Cooper's Hawks were "undoubtedly the most common breeder, with the exception of the little Sparrow Hawk" at the beginning of the twentieth century. This situation rapidly changed, according to Todd's (1940) comment that, as a result of declines, Cooper's Hawks would "doubtless" become as rare as the Northern Goshawks, which were very rare. Direct persecution and contaminants reduced them from fairly common summer residents—for example in Berks Co. (Poole 1947) and Crawford Co. (Todd 1940)—to a rare and local bird by the 1960s.

[1] Goodrich, L. 1989. 1989 Hawk watch reports. PB 3:158.

Northern Goshawk ***Accipiter gentilis***

General status: They breed and winter in North America from Alaska south across Canada to western Mexico in the West and to Maryland in the East. The largest and the least common of the three accipiters, Northern Goshawks are fairly common to rare regular migrants. They are local breeders in the mountainous and extensively forested sections of the state. Northern Goshawks are rare but regular winter residents or visitors.

Habitat: In the breeding season, goshawks prefer mixed northern hardwood forests or conifer plantations. During migration they are found along the mountain ridges and Lake Erie. In winter goshawks may wander away from heavily forested areas into suburbs or around farms in search of food, especially where there are bird feeding stations.

Seasonal status and distribution

Spring: Goshawks are quite rare in spring; birds are most often seen in mi-

Northern Goshawk at Roderick Wildlife Reserve, Erie County, December 1993. (Photo: Gerald M. McWilliams)

gration along the Lake Erie Shore and at a few hawk watches along the mountain ridges. Usually fewer than 10 birds are seen through the spring season at any one site. Their typical migration period is from the last week of Feb to mid-Apr, with stragglers to early Jun. Most records fall from the third week of Mar to the second week of Apr.

Breeding: Most breeding records of goshawks are from the High Plateau. Goshawks are less frequent in the Ridge and Valley south to Dauphin Co. Nests have been documented in the Allegheny Mountains south to Somerset Co. Immature nonbreeders are casually reported outside northern counties during summer. Twenty-nine counties, including most across the northern third of the state, supported goshawk nests from 1974 to 1993. The state breeding population during the 1980s was estimated at 150–200 nest territories, although not all of these sites were active in any given year. Nesting densities were estimated at 1.2 pairs per 100 square kilometers of forested habitat in the northern-tier counties and about half that in the Ridge and Valley.[1] Secretive and found primarily in extensive woodlands, goshawks are probably overlooked more often than observed near nest sites. But they are extremely aggressive, particularly when young are in the nest in May and Jun. Breeding territories are associated with extensive older forests and higher elevations of plateaus and are usually away from roads. Nests may be in a conifer or deciduous tree, but southern sites are associated with groves of conifers. Two to four pale white or bluish white eggs are found from late Mar to early May.

Fall: Goshawks are most commonly reported in fall migration along the mountain ridges. Migration may begin as early as the fourth week of Sep but usually begins in mid-Oct. Peak migration may be at any time from the first week of Nov to the first week of Dec. At most hawk watches, daily highs rarely exceed 4 or 5 birds, but as many as 35 have been recorded (Morris et al. 1984). Sightings decline with the approach of winter.

Winter: They are widely distributed over the state, with winter records from every region. Goshawks are always rare and local, but their habitat preferences in winter are less specific than during the breeding season; birds have been observed perched in trees along highways or along open fields and even in suburban backyards.

History: Goshawks were more frequently observed as migrants during the first half than in the latter half of the twentieth century. During good flight years, as many as 60 once passed Hawk Mountain in a single day, as witnessed on 24 Nov 1935 (Poole, unpbl. ms.). Probably always rare, goshawks were thought to be regular breeders and were seen frequently in the northern tier of counties in the middle 1800s (Warren 1890). By the turn of the twentieth century they had become rare in summer. Harlow (1913) said "I have never been able to get a definite record in five years' search." Lumbering, human persecution, and loss of Passenger Pigeons (probably an important prey item) caused their decline (Todd 1940). As the state's forests recovered and matured, this species again became more widespread. Legislative protection of most raptors was established in 1937, but goshawks were particularly despised and bounties were paid in Pennsylvania until 1951 (Kosack 1995).

Comments: Immature Northern Goshawks are frequently confused with immature Cooper's Hawks during migration and in winter. Goshawks are listed by the OTC as Candidate–Rare.

[1] Kimmel, J.T., and R.H. Yahner. 1994. The Northern Goshawk in Pennsylvania: Habitat use, survey protocols, and status. Final report. School of Forest Resources, Penn State Univ., University Park.

Red-shouldered Hawk ***Buteo lineatus***

General status: Red-shouldered Hawks breed and winter along the coast of California in the West. In eastern North America they breed from Quebec west to Minnesota and south to Texas and Florida. They winter in the East primarily from New England south through their breeding range. Red-shouldered Hawks are more often observed in migration than at any other time of the year. They are rather secretive during the breeding season and in winter but are easily located by their far-carrying calls. They are fairly common to common regular migrants and regular breeders and local winter residents in lowlands over most of the state.

Habitat: They are found primarily in lowland deciduous or mixed forests interspersed with marshes or swamps but also in forested valleys of mountainous regions during the breeding season and in winter. Red-shouldered Hawks may be found in more open areas in winter than during the breeding season. During migration they are seen primarily along mountain ridges and Lake Erie.

Seasonal status and distribution

Spring: Migration begins the first or second week of Mar and usually extends to the last week of Apr. Peak migration ranges from the third week of Mar to the second week of Apr. More spring migrants concentrate along the Lake Erie Shore, where daily highs have exceeded 200 birds (Stull et al. 1985), than at any other site in the state.

Breeding: Breeding Red-shouldered Hawks are uncommon statewide. They were most frequently reported during the BBA project in northwestern counties and least frequently in the southeastern (Chester, Delaware, Philadelphia cos.) and southwestern (Washington and Greene cos.) portions of the state. They generally nest near a stream or wetland and occasionally are found in old suburban areas with large trees. Nests are solid structures, placed near the trunk high in a deciduous tree. Most eggs are found from 7 Apr to 15 May, but extreme dates are 28 Mar and 5 Jun. Two to four pale bluish eggs with pale brown blotches or streaks are laid. The breeding population is not well monitored but is thought by some observers to be declining.

Fall: Other than the Rough-legged Hawk, the Red-shouldered Hawk is the least common of the regularly occurring buteos along the mountain ridges during fall migration. A few early migrants begin to be seen in mid-Sep. Their usual migration period is from the fourth week of Sep through late Nov or early Dec. Peak migration is from the third and fourth weeks of Oct. Daily highs have reached the upper 60s at some hawk watches, but usually fewer are recorded. A few remain to winter.

Winter: Most winter sightings of Red-shouldered Hawks are from east of the Allegheny Front (in the river valleys of mountainous areas) and in the Southwest and Glaciated Northwest. Fewer winter records have been reported in the Glaciated Northeast and High Plateau than in the rest of the state.

History: Red-shouldered Hawks were believed to have declined with the clearing of the original forest. The nesting population disappeared from the populated southeastern and southwestern corners of the state by the turn of the nineteenth century. Todd (1940) suggested that they were more numerous in the Pymatuning Swamp than anywhere else in the state. Legal protection, reforestation, and banning of DDT contributed to the general recovery of this and other raptors.

Comments: The association with water is related to their diet. These handsome buteos feed more on amphibians and reptiles than do other diurnal raptors.

Broad-winged Hawk ***Buteo platypterus***

General status: Broad-winged Hawks breed across southern Canada and east of North Dakota and Texas. They winter in Florida and Central and South America. They are the most abundant buteos during migration in Pennsylvania. Spectacular flights containing hundreds or even thousands of birds are recorded annually during fall. Migrant Broad-winged Hawks are less confined to the mountain ridges than are other species of hawks, and, unlike other raptors, they migrate in groups. They are widely distributed as breeders in the state and are accidental in winter.

Habitat: In the breeding season they prefer heavily forested areas. In migration they are observed mainly along the mountain ridges and Lake Erie.

Seasonal status and distribution

Spring: Broad-winged Hawks migrate the farthest of our nesting raptors, not returning from South America until late Apr. They are more dependent than other raptors on favorable weather conditions for migration. Northbound birds concentrate along the Lake Erie Shore, where daily counts sometimes exceed 3000 birds (Stull et al. 1985). Birds have been recorded as early as the second week of Mar, but the main migration period is usually during a very brief period, the third and fourth weeks of Apr, typically between 20 and 25 Apr. Migration usually extends into the first week of May. A few stragglers, mainly subadults, pass through the state until late May or early Jun.

Breeding: Probably the most widespread forest raptors during summer, Broad-winged Hawks are uncommon statewide. Considered "area sensitive" and strongly associated with extensive wooded cover, they probably nest in every county. They are not as likely to occupy isolated woodlots as extensive forests (Grimm and Yahner 1986). Not conspicuous on territory, they are generally overlooked on BBS routes, reported at just 0.1 birds per route on 59 of 122 routes. No population trend is apparent. Nesting is commenced promptly, and two or three creamy white or bluish white eggs, blotched or spotted with brown, are laid. Eggs have been collected as early as 30 Apr, but most clutches are completed in the second and third weeks of May, with Harlow's (1918) average on 15 May. The nest may be in a deciduous tree or conifer, often within a mixed stand (Grimm and Yahner 1986).

Fall: As in spring migration, Broad-winged Hawks wait for favorable weather conditions before migrating. Along the mountain ridges and in southeastern Pennsylvania, they are common to abundant migrants. Broad-winged Hawks are uncommon to rare in northwestern Pennsylvania as fall migrants. Broad-winged Hawks are less restricted to the mountain ridges than are other raptors. Large flights are annually seen at various locations in Delaware and Philadelphia cos. far from the mountains. Migrants pass through rather quickly. They usually pass through the state from the third or fourth week of Aug to the second week of Oct. Peak migration is during the second and third weeks of Sep. Some exceptional daily highs have been recorded at hawk watches, such as the 21,447 that passed over HMS on 14 Sep 1978.[1] Stragglers have been recorded until at least the second week of Dec.

Winter: Although there have been numerous reports of this species in winter, just five winter records seem acceptable: one bird banded at Caledonia State Park in Franklin Co. on 13 Jan 1963,[2] one at Newburg on 25 Jan 1981,[3] one at the Erie National Wildlife Refuge in Crawford Co. on 27 Jan 1983,[4] one near Berwick in Columbia Co. on 20 Feb 1991,[5] and a reference to Broad-winged Hawks at Tinicum in Philadelphia Co. in winter.[6]

History: During the nineteenth century the Broad-winged Hawk was considered rare, both as a transient and as a breeder. Harlow (1913) suggested that they had "expanded" since about 1900. Descriptions by Sutton (1928a) are very similar to the present status.

Comments: It is likely that some immature Red-shouldered Hawks or even soaring accipiters are misidentified as Broad-winged Hawks during the winter.

[1] Paxton, R.O., P.W. Smith, and D.A. Cutler. 1979. Hudson-Delaware region. AB 33:161.
[2] K. Gabler, pers. comm.
[3] Hall, G.A. 1981. Appalachian region. AB 35:299.
[4] Hall, G.A. 1983. Appalachian region. AB 37:301.
[5] Gross, D.A. 1991. County reports—January through March 1991: Columbia County. PB 5:36.
[6] J.C. Miller, pers. comm.

Swainson's Hawk *Buteo swainsoni*

General status: Swainson's Hawks breed primarily in the western Great Plains and winter primarily in South America, with a few wintering in Florida. They wander to eastern North America and are recorded annually at Cape May, New Jersey. Hawk-watch reports suggest that they are regular in Pennsylvania, but there is a lack of documentation for this species. All but one record have been since 1966 and most are in fall at hawk watches along the Kittatinny Ridge. All sightings of Swainson's Hawks have been of single birds. Both light- and dark-morph plumages and both adult and immature birds have been reported.

Seasonal status and distribution

Spring: There are apparently only two spring reports of Swainson's Hawk: one on 21 Apr 1984 along Lake Erie[1] and one on 3 May 1989 at Rocky Ridge County Park in York Co.[2]

Fall: One or two birds are reported in most years from the first week of Sep to the second week of Nov. An immature was banded at Wind Gap in Northampton Co. on 19 Sep 1982.[3] Swainson's Hawks have been reported from Bedford, Berks, Bucks, Carbon, Cumberland, Huntingdon, Lancaster, Lehigh, Luzerne, Northampton, Philadelphia, Wyoming, and York cos.

History: The first and only record before 1966 was of a bird shot by J. A. Medsger in an open field at Jacobs Creek, Westmoreland Co., on 5 Sep 1901. It was preserved and placed in the CMNH (Todd 1940).

[1] Hall, G.A. 1984. Appalachian region. AB 38:910.
[2] Spiese, A. 1989. County reports—April through June 1989: York County. PB 3:83.
[3] Paxton, R.O., W.J. Boyle Jr., and D.A. Cutler. 1983. Hudson-Delaware region. AB 37:162.

Red-tailed Hawk *Buteo jamaicensis*

General status: Red-tailed Hawks are familiar hawks of open country. They breed across most of North America and winter across the U.S. Second only to the Broad-winged Hawk, they are the most common regular migrant buteo. Broun (1949) wrote that the spectacular flights of all these raptors at Hawk Mountain helped draw attention to the need for the conservation of raptors around the world. Spectacular flights can still be observed along the ridges of the Appalachian Mountains in fall and to a lesser extent along Lake Erie in spring. Red-tailed Hawks breed statewide and will nest almost anywhere there are trees, but they are most common outside heavily forested areas. They are winter residents throughout most of the state.

Habitat: Red-tailed Hawks are found in a wide variety of habitats. They prefer woodlots bordered by fields and are found in urban parkland but are uncommon in extensive forests. They frequently perch in trees or on utility poles along highways. In migration most are seen along mountain ridges and along Lake Erie.

Seasonal status and distribution

Spring: This is the first buteo to migrate in spring. Their typical migration period is from mid-Feb to about mid-Apr, with stragglers, usually immatures, into May. Peak migration is usually from the third week of Mar to the second week of Apr. Migrants occur across a broad front statewide but are concentrated most heavily along the Lake Erie Shore, where single-day totals have exceeded 250 birds (Stull et al. 1985).

Breeding: Red-tailed Hawks are fairly common breeders statewide, including the southeastern counties where Poole (unpbl. ms.) stated that they had "long since disappeared as a breeding species." In extensively forested regions they are usually observed around forest openings, along roads, or soaring high overhead. Their substantial nest is placed high in a prominent deciduous tree or conifer, generally more than 60 feet and sometimes over 100 feet from the ground (Grimm and Yahner 1986). The nest is lined with inner bark or grapevine strips and may be reused annually. Eggs are white, with variable brown splotches, most often found from 26 Mar to 6 May. Young remain in the nest four to six weeks. BBS data show a long-term increase of 5.2% per year.

Fall: Most migrant Red-tailed Hawks are observed along the mountain ridges. They are one of the hardiest of the raptors, and more migrate later in the fall than other raptors. However, the migration period is rather long and drawn out. Migration extends from mid- to late Aug through Dec, with stragglers remaining into winter. Peak migration ranges from the fourth week of Oct to the second week of Nov. Hundreds are seen passing by some hawk-watch stations on peak days. An exception, a high count of 914, passed Bake Oven Knob on 13 Nov 1982.[1]

Winter: Wide-ranging, Red-tailed Hawks tend to occupy open areas more frequently in winter than during the breeding season. They are uncommon to rare in large tracts of forests, especially at high elevations.

History: The earliest ornithological authors considered Red-tailed Hawks common (e.g., Libhart 1869; Warren 1890). The general decline of birds of prey during the first half of the 1900s extended even to this widespread species, primarily affecting the breeding population. Red-tailed Hawks were mercilessly persecuted as one of the "chicken hawks" through the early years of the twentieth century. Many authors (e.g., Harlow 1913; Sutton 1928a; Todd 1940) referred to declining numbers, particularly in the southeastern counties where contaminants and direct persecution took their toll. Todd (1940) also mentioned the absence of trees of sufficient size in many areas to support nests, attesting to the extent of deforestation during those years. Legal protection in Pennsylvania came for this and most other raptors in 1934 (Kosack 1995).

Comments: Conspicuous albinos or leucistic birds are reported annually, often in the same area in successive years. Most sightings probably pertain to the subspecies that breeds in the state and elsewhere in the eastern U.S., *B. j. borealis*. In recent years, two dark Red-tailed Hawks, perhaps *B. j. calanus*, have been reported. One found in Northampton Co. on 13 Jan 1991 was present to at least Nov 1997,[2] and one was in Centre Co. in four consecutive winters from 1992[3] to 1995.[4] Several subspecies have been reported, but studies have shown intergrades with adjacent subspecies,[5] so most subspecies are probably not safely identifiable in the field. Some collected specimens have proved to be subspecies intergrades. Carefully detailed written descriptions should be made and photographs should be taken of all Red-tailed Hawks that appear to be other than "normal" *B. j. borealis*. They may be indicators of the geographic origin of migrant and wintering Red-tailed Hawks.[6]

[1] Paxton, R.O., W.J. Boyle Jr., and D.A. Cutler. 1983. Hudson-Delaware region. AB 37:162.

[2] Wiltraut, R. 1991. County reports—October through December 1991: Northampton County. PB 5:46.

[3] Floyd, T. 1992. Dark-morph (western) Red-tailed Hawk. PB 6:6.

[4] Peplinski, J., and B. Peplinski. 1995. Local notes: Centre County. PB 9:32.

[5] Preston, C.R., and R.D. Beane. 1992. Red-tailed Hawk *(Buteo jamaicensis)*. BNA (A. Poole and Gill 1992), no. 52.

[6] Parkes, K.C. 1997. Subspecies and intergrade Red-tailed Hawks in western Pennsylvania. PB 10:203–205.

Rough-legged Hawk ***Buteo lagopus***

General status: Rough-legged Hawks breed in North America from Alaska, northern Yukon, and the Arctic islands east across northern Ontario, Quebec, and into Newfoundland. They winter across southern Canada and throughout the U.S. except in the Gulf states and Florida. Rough-legged Hawks are raptors of open country, most often seen hovering over large open fields in search of rodents. They move into Pennsylvania in small numbers and are uncommon to rare regular migrants and regular winter visitors or residents. Dark-, light-, and intermediate-morph birds are regularly seen in Pennsylvania, but light morphs usually are the most frequently seen.

Habitat: In winter Rough-legged Hawks are usually seen perched on utility poles or trees in large open fields or marshes. During migration they are observed along mountain ridges and Lake Erie.

Seasonal status and distribution

Spring: Rough-legged Hawks that have wandered into or south of Pennsylvania in the winter begin migrating north about the first week of Mar until the second week of Apr. Most birds seen migrating in spring are along the Lake Erie Shore, but a few are recorded every spring at some of the hawk watches along the mountain ridges. Fewer than 10 birds usually are recorded annually at the Lake Erie Shore. During some winters, Rough-legged Hawk numbers may continue to build well into Mar, becoming fairly common locally before moving north. Stragglers may remain to the third week of May.

Fall: Rough-legged Hawks are the rarest of the regularly occurring raptors recorded at the hawk watches. Birds may begin to appear from the far north as early as the last week of Sep but usually not before the second week of Oct. Only one or two birds are usually observed in a day along the mountain ridges during migration. The majority of migrants are recorded during the first three weeks of Nov. An unusual concentration of 35 to 40 birds was counted within a 5-mile radius of New Holland in Lancaster Co. in the fall of 1981.[1] Migration has usually ended by early Dec, with stragglers remaining through winter.

Winter: Rough-legged Hawks are usually rare in winter, but they may become locally fairly common. Up to 15 or 20 birds can sometimes be found in one area during some years. A high count of 43 was tallied on the Lewisburg Christmas Count on 2 Jan 1982 (Schweinsberg 1988). Rough-legged Hawks have been found in winter in every region.

History: The status and distribution of Rough-legged Hawks has changed little throughout history. However, Poole (unpbl. ms.) stated that Rough-legged Hawks had diminished in recent years and that they were rarer in western counties than in the eastern part of the state.

[1] Boyle, W.J., Jr., R.O. Paxton, and D.A. Cutler. 1982. Hudson-Delaware region. AB 36:159.

Golden Eagle ***Aquila chrysaetos***

General status: In North America, Golden Eagles breed primarily from Alaska across most of Canada and south through the western U.S., and they winter throughout most of their breeding range. In Pennsylvania, Golden Eagles have become the most frequently observed eagle in the Appalachian Mountains during migration, where they are uncommon to rare regular migrants. They may have bred historically in Pennsylvania, as noted by Beck (1924) and Todd (1940), but there is only one modern summer record. They occasionally winter in the state.

Habitat: In migration, Golden Eagles are most often found along the mountain ridges. When present in winter, they are usually found in heavily forested areas or along large rivers with Bald Eagles.

Seasonal status and distribution

Spring: Golden Eagles may appear any time in spring, often near water. Individuals may linger until the second week of Jun. Most birds are recorded in Mar and Apr at hawk watches along the mountain ridges, with spring totals of 30 or more birds at some sites. The 7 birds that passed Tuscarora Mountain on 22 Apr 1981 constituted an unusually high daily spring count.[1] Golden Eagles have not been reported in spring hawk flights along the Lake Erie Shore.

Summer: The only modern record during summer was of one that summered in 1968 near HMS.[2]

Fall: Most Golden Eagles are seen during fall in migration along the mountain ridges. Migration usually begins in mid-Oct and continues through Nov in Pennsylvania. Peak migration ranges from the fourth week of Oct to the third week of Nov. The daily high is usually fewer than 10 birds at hawk watches. One hawk watch has reported more than 20 birds on a single day; on 6 Nov 1990 a total of 22 were seen flying past the Bald Eagle Mountain Fire Tower in Centre Co.[3] Stragglers have been reported until the second week of Dec.

Winter: Most winter records of Golden Eagle are from the mountains and ridges. East of the mountains, winter observations have come from Delaware Co.[4] and from MCWMA in Lancaster Co.[5] and along the lower Susquehanna River south of Safe Harbor (Morrin et al. 1991). Golden Eagles have been reported in winter around HMS in Berks[6] and Schuylkill cos.[7] Along the Delaware River they have been observed in winter in northern Northampton Co.[8] In western Pennsylvania, birds have been recorded in winter along the Allegheny River in Warren Co.,[9] in the Laurel Hill area in Westmoreland Co.,[10] and at Pymatuning in Crawford Co.[11]

History: Poole (1964) listed Golden Eagles as rare transients and winter visitors. Beck (1924) wrote that before 1890 a pair nested on a cliffside of the Susquehanna River opposite the mouth of the Pequea River. Todd (1940) said that W. Van Fleet included this species in his list of breeding birds of Clearfield Co., but documentation of nesting in Pennsylvania is sketchy.

Comments: The apparent increase in Golden Eagle populations is probably attributable to the fact that there are more observers at hawk watches than there were in the past. In addition, many sightings may be misidentified immature Bald Eagles, especially away from established hawk watches.

[1] Hall, G.A. 1981. Appalachian region. AB 35: 822–825.

[2] Scott, F.R., and D.A. Cutler. 1968. Middle Atlantic Coast region. AFN 22:595.

[3] Peplinski, J., and B. Peplinski. 1991. 1990 Hawk watch reports. PB 4:173.
[4] N. Pulcinella, pers. comm.
[5] E. Witmer, pers. comm.
[6] R. Keller, pers. comm.
[7] Scott, F.R., and D.A. Cutler. 1968. Middle Atlantic Coast region. AFN 22:426.
[8] Boyle, W.J., Jr., R.O. Paxton, and D.A. Cutler. 1983. Hudson-Delaware region. AB 37:285.
[9] T. Grisez, pers. comm.
[10] Leberman, R., and R. Mulvihill. 1990. County reports—January through March 1990: Westmoreland County. PB 4:39.
[11] Hall, G.A. '1978. Appalachian region. AB 32: 350.

Family Falconidae: Falcons

Falcons are among the fastest birds in the world. They fly with quick powerful wing beats. Ranging in size from the small American Kestrel to the large and robust Gyrfalcon in North America, they have long, narrow, pointed wings, usually bent sharply back at the wrist when in flight. Most feed on birds that they pursue and catch in flight. Because of their swift and powerful flight, falcons, especially the Peregrine Falcon and Gyrfalcon, are among the most sought by falconers. They also have suffered the most from contaminants. Four species have been recorded in Pennsylvania. American Kestrels and Peregrine Falcons breed in the state; Merlins are regular migrants; and Gyrfalcons are irregular visitors.

American Kestrel *Falco sparverius*

General status: These small falcons of meadows and fields breed from Alaska south across Canada and throughout the U.S. They winter throughout their breeding range. American Kestrels are common to abundant regular migrants in Pennsylvania. They are regular breeders and winter residents except in heavily forested areas throughout the state.

Habitat: Kestrels are frequently seen perching on utility wires or poles adjacent to fields. During the breeding season they are found in open areas such as agricultural fields (particularly abandoned fields) where perches and nest cavities are available. Kestrels can also be found around suburban and urban areas, especially around vacant lots and parks. In migration they are mainly observed along mountain ridges and along Lake Erie.

Seasonal status and distribution

Spring: Their typical migration period is from mid-Mar to about the third week of Apr. Peak migration ranges from the third week of Mar to the second week of Apr. In spring more migrants are regularly observed along the Lake Erie Shore than at any other single site in the state, where single-day counts average between 40 and 50 birds. Most hawk watches average between 20 and 30 birds per day during the peak season. High daily counts along Lake Erie have exceeded 100 birds (Stull et al. 1985). Some hawk watches along the Kittatinny Ridge have had large spring flights. On 2 Apr 1974, 154 were counted passing Baer Rocks in Lehigh Co. (Morris et al. 1984).

Breeding: American Kestrels are fairly common breeding residents statewide. They are the most frequently seen birds of prey in open country and the second most widespread species, after Red-tailed Hawks. BBS and BBA data demonstrate that kestrels are absent or rare only in extensively forested areas. Adaptable, kestrels nest in towns and even in the center of large towns and cities. Our only cavity-nesting hawk, American Kestrels readily occupy nest boxes with an entrance hole of 3 inches and frequently use holes and crevasses in buildings.[1] They also occupy vacant woodpecker holes and natural cavities high in trees. European Starlings compete aggressively with this species for nest sites. Very little, if any, nesting material is added to the cavity. Most egg sets are found from 25 Apr to 25 May. Kestrels lay four or five white to pale cinnamon eggs that are evenly speckled or spotted with brown.

Fall: The greatest number of American Kestrels is observed during fall migration along the mountain ridges. Their typical migration period is from mid- to late Aug to mid-Oct, with stragglers to the fourth week of Oct. Usually, daily counts are fewer than 20 or 30 birds at hawk watches. Peak migration ranges from the second week of Sep to the first week of Oct. A notable count of 50 birds was made at the Little Gap Bird Observatory in Northampton Co. on 6 Oct 1990.[2]

Winter: Fewer birds are seen in winter in the northern counties than elsewhere in the state, but in most areas they are still uncommon to fairly common. The majority of birds observed in winter are in the Piedmont and in the valleys of the Ridge and Valley. They are rare in winter in the higher elevations of the High Plateau and in heavily forested areas, where they are local in clearings or around farms.

History: One can only assume that kestrels were much less common when extensive forests covered most of Pennsylvania. However, since the earliest ornithological records, they have been regarded as common year-round, at least in the settled areas.

[1] Brauning, D.W. 1983. Cavity availability and nest site selection of the American Kestrel *(Falco sparverius)*. Raptor Research 17:122.

[2] Schall, M., Little Gap Bird Observatory staff 1991. 1990 Hawk watch reports. PB 4:174.

Merlin *Falco columbarius*

General status: In North America, Merlins breed primarily across boreal Canada, and they winter mostly from the Pacific coast across the southern U.S. They are uncommon to rare regular migrants and are accidental in summer. In winter they are irregular visitors or residents. Merlins are more commonly detected now than in the past.

Habitat: In migration Merlins are found primarily along the mountain ridges and along Lake Erie. In winter they prefer rather open settings, especially near water and urban areas.

Seasonal status and distribution

Spring: More migrants are observed in spring along the Lake Erie Shore, especially at PISP, than at any other single site in the state. Migration may begin as early as the second week of Mar. Their typical migration period is from the last week of Mar until mid-May. Peak migration ranges from the first to the third week of Apr. Stragglers have been reported to mid-Jun.

Summer: Apparently the only modern records of Merlins in summer are from Crawford, and Erie cos. In Crawford Co. one was reported on 30 Jun 1983,[1] and in Erie Co. on 22 Jul 1992.[2]

Fall: Most fall sightings are of migrants along the mountain ridges. Migration may begin as early as the last week of Aug. Merlins continue passing through the state into early Dec, with stragglers remaining to winter. Peak migration occurs during the first half of Oct. Daily highs at the larger hawk watches are usually fewer than 15 birds, but on 22 Oct 1989, 34 passed HMS and 44 passed Baer Rocks.[3]

Winter: Records are scattered throughout the western and southeastern portions of the state. Most winter sightings are from the Coastal Plain and Piedmont. East of the Allegheny Front, Merlins have been reported in winter from Bucks, Centre, Chester, Dauphin, Delaware, Franklin, Lancaster, Luzerne, Montgomery, Northampton, Northumberland, and Philadelphia cos. West of the Allegheny Front, winter sightings have come from Allegheny, Clarion, Erie, Indiana, Mercer, and Westmoreland cos. In winter, Merlins tend to frequent urban sites, where they are often seen perched on buildings. They have been observed in small communal roosts in winter. One recent roost of up to 5 birds was found in Pittsburgh during the winter of 1998.[4]

History: L. E. Hicks believed that they bred in Erie Co. because he saw one on 1 Jul 1928 (Todd 1940), but Poole (unpbl. ms.) believed that most sightings were of misidentified Sharp-shinned Hawks or American Kestrels.

[1] Hall, G.A. 1983. Appalachian region. AB 37:988.
[2] Hall, G.A. 1992. Appalachian region. AB 46: 1135.
[3] Paxton, R.O., W.J. Boyle Jr., and D.A. Cutler. 1990. Hudson-Delaware region. AB 44:63.
[4] M. Fialkovich, pers. comm.

Gyrfalcon ***Falco rusticolus***

General status: These are the largest falcons in the world. In North America, Gyrfalcons spend most of their lives in the Arctic tundra. They regularly wander south into the northern U.S., where they are rare. In Pennsylvania they are casual fall migrants and winter visitors or residents. Most records are of single birds from the Ridge and Valley and Lake Erie Shore. They probably occur more frequently than records indicate, but documentation is lacking for many sightings.

Habitat: Gyrfalcons are reported in migration primarily along the mountain ridges. In winter they have been found around large open areas such as extensive agricultural areas and ice floes on PISP, where there are concentrations of large birds such as ducks, gulls, or Rock Doves, on which they prey.

Seasonal status and distribution

Fall through spring: Most Gyrfalcon reports are in fall and early winter along the Kittatinny Ridge at hawk watches from southwestern Schuylkill Co. to southern Carbon Co. or on the ice surrounding PISP in Erie Co. Sightings have come from Bedford, Berks, Carbon, Centre, Crawford, Cumberland, Erie, Jefferson, Lancaster, Lehigh, Northampton, Schuylkill, Tioga, and Wayne cos. Dates range from 10 Oct to about 9 Apr. Most spring sightings are of birds that wintered. From the fall of 1981 to the spring of 1982 and again in the fall of 1982 perhaps as many as 8 different Gyrfalcons were reported: up to 5 in Lehigh Co.[1] and 3 in Lancaster Co. (Morrin et al. 1991). One of the Lancaster birds died and is on display at the MCWMA Museum. In 1983, a gray-morph bird was found on 28 Feb and remained through 25 Mar, feeding on Ring-necked Pheasants at a state game farm near Cambridge Springs in Crawford Co.[2] Despite the numerous

Gyrfalcon at New Holland, Lancaster County, November 1981. (Photo: Alan Brady)

reports, the only accepted records since 1983 have come from PISP in 1991, 1994 and 1996.[3]

History: Records date back to 1913, when a female Gyrfalcon captured alive was kept in captivity for several months and later died. The specimen was evidently destroyed (Poole 1964). Poole (unpbl. ms.) mentioned three other birds that were collected, mounted, and placed in the North Museum at Lancaster, the Reading Public Museum, and the PGC collection in Harrisburg. All Gyrfalcon sightings before 1960 were from the Piedmont or Kittatinny Ridge, except a bird collected in Sullivan Co. and in the possession of the PGC (Poole unpbl. ms.).

Comments: Many plumage variations have been reported in the state, from white birds to various shades of gray and brown to very dark brown. Most Gyrfalcons reported are of the gray morph. It is likely that a few Peregrine Falcons are misidentified as Gyrfalcons.

[1] Paxton, R.O., W.J. Boyle Jr., and D.A. Cutler 1983. Hudson-Delaware region. AB 37:162.
[2] Hall, G.A. 1983. Appalachian region. AB 37:301.
[3] PORC files.

Peregrine Falcon *Falco peregrinus*

General status: The Peregrine Falcon is a well-known bird of prey throughout the world. Its recovery from the days of pesticide contamination has been a gradual but successful one. Peregrines are more frequently observed today than in the past 40 years. They breed at scattered locations mostly from Alaska across northern Canada and in the western U.S. They winter locally in the U.S. rockies in the West and primarily along the Atlantic and the Gulf coasts in the East. At least two populations occur in Pennsylvania: migrants and reintroduced residents. Peregrine Falcons are uncommon to rare regular migrants over most of the state, but very few records have come from the High Plateau. Peregrines formerly nested on cliffs in central and eastern Pennsylvania; they are now year-round residents near urban nest sites. They are casual winter visitors in other urban areas in Pennsylvania where they do not nest.

Habitat: In migration they are observed mainly along ridge-top hawk watches and along Lake Erie. In the breeding season they are found around urban areas near water, where they nest on bridges and buildings. In winter they are in open country near water and in urban areas.

Seasonal status and distribution

Spring: Peregrines are far rarer in spring than in fall. The usual migration period for this species is from the last week of Mar to the second week of May. Late Feb sightings may be very early migrants rather than wintering individuals. There does not appear to be a peak migration period, but most Peregrine Falcons are recorded in Apr or early May. Stragglers remain to the last week of May.

Breeding: Peregrine Falcons did not nest in Pennsylvania from about 1959 until 1987. The species was reestablished in the eastern U.S. by the Peregrine Fund in a nationwide restoration program. The program released Peregrine Falcons in 1976 and 1977 from cliffs near Dauphin, Dauphin Co., and near Towanda, Bradford Co., and in 1981 from a building in Philadelphia. The first nest of the reintroduced population was discovered in 1987 on the Walt Whitman Bridge in Philadelphia.[1] By 1990 the nesting population had grown to include five major bridges over the Schuylkill and Delaware rivers in Bucks, Delaware, and Philadelphia cos. and downtown buildings in Philadelphia and Pittsburgh. During the 1990s,

Peregrine Falcons were released in four urban areas: Allentown from 1995 to 1997; Harrisburg from 1992 to 1995; Reading from 1993 to 1995; and Williamsport from 1993 to 1997. Subadult pairs established territories in Harrisburg, Dauphin Co., in 1997 and in Wilkes-Barre, Luzerne Co., and Williamsport, Lycoming Co., in 1998. The population is increasing throughout the region, but reproductive success at bridge-nesting sites has been poor.[2] No nest is constructed; eggs are laid, beginning in late Mar, on existing gravel. Second (replacement) clutches are laid in mid-May. The usual clutch is four creamy white eggs, heavily blotched and spotted with various shades of brown. Nonbreeding birds occasionally appear in the state. During the BBA project, birds were seen in Bucks, Huntingdon, Northumberland, and Wayne cos.[3] In Erie Co. there have been at least three summer records since 1971.[4] A single record exists from Lake Arthur in Butler Co. in 1983.[5]

Fall: Most fall sightings of Peregrine Falcons are along the mountain ridges of birds en route from Arctic breeding areas to South American wintering grounds. They arrive as early as the last week of Aug and continue to about the third week of Oct. Peak migration occurs during the first and second weeks of Oct. Daily highs normally are fewer than 10 birds at hawk watches. Stragglers may remain in the state until early Dec.

Winter: The majority of winter birds are found near lakes or along the larger rivers or near nest sites. Sightings are scattered throughout the state in open areas. East of the Allegheny Front, they are regular in winter in the Coastal Plain along the Delaware River and in urban Philadelphia.[6] Peregrines have been reported in winter in the Piedmont from Lancaster and York cos. In the Ridge and Valley, winter sightings have come from Carbon, Centre, Luzerne, Lycoming, and Union cos. The only record of a Peregrine Falcon in the Glaciated Northeast was one found at Dimock in Susquehanna Co. on 10 Feb 1976.[7] No winter record exists from the High Plateau. In western Pennsylvania, winter records of Peregrine Falcons have come from Allegheny, Crawford, and Erie cos.

History: As many as 44 Peregrine nesting sites in more than 21 Pennsylvania counties were known during the early decades of the twentieth century. Nesting was confirmed even on Philadelphia's City Hall for several years.[8] The decline of the species here and worldwide was largely attributed to DDT contamination. The last successful nesting of native birds in Pennsylvania apparently occurred in 1957, with pairs present until 1959 or possibly later at prominent cliffs along the North Branch Susquehanna River.[9]

Comments: Pennsylvania was at the heart of the range of the *anatum* subspecies, formerly found across eastern North America. The eastern U.S. population totaled an estimated 350 nesting pairs in the early decades of the twentieth century.[10] They are listed as Endangered by the state and federal governments.

[1] Cade, T.J., and P.R. Dague. 1987. The Peregrine Fund Newsletter, no. 15.

[2] Brauning, D.W. 1997. Peregrine Falcon research/management. Annual report. Pennsylvania Game Commission, Harrisburg.

[3] Brauning, D.W. 1992. Peregrine Falcon *(Falco peregrinus)*. BBA (Brauning 1992): 110–111.

[4] G.M. McWilliams, pers. recs.

[5] Hall, G.A. 1983. Appalachian region. AB 37:988.

[6] J.C. Miller and E. Fingerhood, pers. comm.

[7] W. Reid, pers. comm.

[8] Goskin, H. 1952. Observations of Duck Hawks nesting on man-made structures. Auk 69:246–253.

[9] Rice, J.N. 1969. The decline of the Peregrine population in Pennsylvania. Pages 155–163 *in*

Peregrine Falcon population: Their biology and decline, ed. J.J. Hickey. Univ. Wisconsin Press, Madison.

[10] Hickey, J.J. 1942. Eastern population of the Duck Hawk. Auk 59:176–204.

Order Galliformes
Family Phasianidae: Pheasants, Grouse, and Turkeys

The family Phasianidae includes ground-dwelling fowl-like birds. Most species are hunted and prized for their tasty flesh. They have short legs, and short rounded wings, and some species have spurs. Most have a short to moderate rounded, fanlike tail. Most feed on the ground by scratching or by picking for seeds, fruits, or insects from low plants reached from the ground. When escaping danger, they usually run before flying. Their flight is short, rapid, and direct, and after landing they often resume running. About 15 species are found in North America, three of which are Pennsylvania residents. Unsuccessful attempts have been made to introduce several other species. Buoyed by the success with Ring-necked Pheasants, state and private groups released Sharp-tailed Grouse *(Tympanuchus phasianellus)*, but those attempts were not successful. Nests of galliformes are placed on the ground. Chicks leave the nest shortly after hatching and begin to feed on their own while accompanying their parents.

Ring-necked Pheasant *Phasianus colchicus*

General status: Ring-necked Pheasants are widely introduced and established residents in North America. They were introduced into Pennsylvania for the purpose of hunting. Ring-necked Pheasants are permanent residents statewide. Populations have declined in recent years at least in part because of changing farming practices. In areas of the state where there is heavy snow cover, many do not survive the winter; local populations are replenished annually with additional stocking. They are usually seen as singles or in pairs but are occasionally seen in flocks, especially after recent stockings.

Habitat: Pheasants inhabit grasslands and farmlands planted with various grains, especially corn, and are found along uncut fence rows and brushy areas surrounding fields. They also may be found in brushy abandoned fields in suburban and urban areas.

Seasonal status and distribution

Breeding: Ring-necked Pheasants were formerly widespread and common in agricultural areas statewide but were noticeably absent from extensive wooded regions of the central and northern counties. Annual releases of more than 200,000 birds by the PGC, including some in spring and fall, obscure the status of the resident breeding population. Breeding populations, as documented by the BBS, have declined precipitously over the past 30 years. Now Ring-necked Pheasants are uncommon over most of the state. The highest densities remain in the southeastern counties, with an average of more than 8 birds per route reported in the Piedmont, but few areas supported viable resident populations by the early 1990s. Loss of brushy edges, weedy fields, and winter cover all contributed to the decline of Ring-necked Pheasants. Nests are placed on the ground in well-concealed, grass-lined hollows, but mowing frequently kills eggs and females. As many as 15 brownish, unmarked olive eggs are laid from Apr and (presumably second clutches) to Jul.

History: The PGC first tried to introduce this Asiatic exotic in 1915 (Sutton 1928a, Kosack 1995). Prior releases by private groups were not thought to successfully establish the species. Other continental U.S. releases began in the late nineteenth century. The species apparently acclimated and became established in agricultural areas by the mid-1920s and became very common by the 1950s. By the late 1970s the number of birds had began to decline (Palmer 1988).

Comments: Various subspecies and strains have been stocked in Pennsylvania with varying degrees of success. During the 1990s, a form lacking the white neck collar, for which the species is named, was released in western and central counties.

Ruffed Grouse ***Bonasa umbellus***

General status: Ruffed Grouse are easily recognized by their territorial drumming and the roar of their wings when they take to flight. They are residents from Alaska across Canada to the northwestern and northeastern states south through the Appalachian Mountains. Ruffed Grouse are uncommon to fairly common year-round residents, primarily in heavily forested portions of the state. Grouse are least common in the lowland open valleys of the Piedmont, and they are absent from areas of urban development. They are subject to periodic fluctuations in numbers[1] (Gullion 1970). Ruffed Grouse are hunted as game in Pennsylvania.

Habitat: They are found in various extensive forested settings as well as in overgrown pastures in forested areas. They usually prefer forests or wooded ravines with grapevine tangles, laurel thickets, or patches of small conifers, especially hemlock, for winter cover.

Seasonal status and distribution

Breeding: Ruffed Grouse are fairly common permanent residents in forested settings over most of the state. In the Piedmont they are rare, except in the more extensively wooded hills such as those found at French Creek State Park, Berks Co. Numbers fluctuate, rising and falling in response to food, cover, or unknown causes at 5- to 10-year intervals. Logging enhances grouse habitat by providing open areas with greater insect populations. Aspen stands are promoted in some areas for grouse. The male's territorial drumming announces the Mar to May breeding season. A dozen buffy eggs, lightly spotted with brown, constitute a typical clutch. Sets are found from late Apr through May, with some in Jun. The nest of leaves is placed on the ground in a well-concealed location.

History: Ruffed Grouse, variously known as the "pheasant" and "partridge" in early ornithological writings, were the targets of market hunters but survived unregulated hunting better than other species because they do not congregate into groups. The species was generally considered widespread and common across Pennsylvania.

Comments: Most individuals observed in Pennsylvania are of the red morph. Little seems to be known of the status and distribution of the gray morph in the state, but they have been reported from Bedford, Butler, Clarion, Crawford, Jefferson, and Lancaster cos. Ruffed Grouse were named Pennsylvania's state bird in 1931.[2]

[1] Stoll, R.J., Jr., and M.W. McClain. 1986. Distribution and relative abundance of Ruffed Grouse in Ohio. Ohio Journal of Science 86:182–185.

[2] Fergus, C. undated. Ruffed Grouse. Wildlife notes, no. 175–1. Pennsylvania Game Commission, Harrisburg.

Greater Prairie-Chicken ***Tympanuchus cupido***

General status: Extirpated. Greater Prairie-Chickens, now birds primarily of the Great Plains, apparently disappeared from Pennsylvania about 125 years ago. Very little was documented about their status and distribution, but they were believed to be widespread and locally numerous. Most accounts came from the pine and scrub oak barrens of the Poconos and from southeastern counties.

History: The earliest references to Greater Prairie-Chickens date back as far as the mid-1600s. Authorities such as Bartram,[1] Barton (1799), and Wilson (1808–1814) noted the bird's presence in Pennsylvania. In 1829, Audubon (1831–1839) found them to be abundant in Carbon Co. Three years later, Audubon wrote that they had become very rare in the markets of Philadelphia. The last reliable report in Pennsylvania came from Turnbull (1869), who wrote that they were very rare and that he had shot the species in Monroe and Northampton cos. Apparently the bird's range was throughout the Pocono Plateau and Broad Mountain in northeastern Pennsylvania, a range that included the counties of Carbon, Lehigh, Monroe, Northampton, Schuylkill, and Wayne. There were also records from the southeast in the counties of Chester, Lancaster (specimen in the North Museum; Morrin et al. 1991), Montgomery, and York.[2] Apparently they had been found in Clinton and Union cos. as well (Poole, unpbl. ms.). Most authorities presume that of the three recognized subspecies of Greater Prairie-Chicken, the subspecies found in Pennsylvania was *T. c. cupido*, the Heath Hen. Poole (1964) found that the five specimens he examined were closer to the Western subspecies *T. c. pinnatus*.

[1] Bartram, W. 1791. Travels through North & South Carolina, Georgia, East & West Florida. Privately published, Philadelphia.

[2] Fingerhood, E.D. 1992. Greater Prairie-Chicken *(Tympanuchus cupido)*. BBA (Brauning 1992): 431–432.

Wild Turkey ***Meleagris gallopavo***

General status: The Wild Turkey is by far the largest avian resident native to North America. Wild Turkeys are widely scattered residents throughout the U.S. Populations have fluctuated throughout their history in Pennsylvania; they were probably widespread and frequently encountered before forests were cleared for farming but subsequently disappeared from many parts of the state. Regrowth of forest, protection from overhunting, and restocking contributed to population recovery and range expansion since the 1950s. Wild Turkeys are uncommon to fairly common residents across the state, except in the lower Coastal Plain and in the Piedmont. They are usually seen in flocks consisting of a few birds to 50 or more. Turkeys are hunted as game throughout the state.

Habitat: Wild Turkeys are found in forested areas, especially those adjacent to fields. Small populations may survive in isolated woodlots surrounded by fields or in city parks.

Seasonal status and distribution

Breeding: They are widespread and fairly common permanent residents over most of the state. Wild Turkeys became more abundant and widespread in the mid-1990s than at any other time during the century, expanding into Bucks Co. in the early 1990s. They have been observed even in Philadelphia Co. but are least common in the Piedmont and around heavily urbanized areas. Turkey flocks disperse during the nesting season, but broods are frequently seen in open areas foraging for grasshoppers throughout the summer. A turkey flock may range over more than 1000 acres through a year.[1] The nest is a leaf-lined depression in the ground. It typically contains more than a dozen eggs that are pale-buff and evenly spotted with reddish brown. Incubation lasts 28 days. The trap, transfer, and release of more than 2000 Wild Turkeys within Pennsylvania between 1958 and 1984 greatly promoted the species' range expansion.[1] That program and careful hunting regulations enabled the turkey population to dramatically recover and expand statewide.

History: The Wild Turkey's range retracted rapidly in response to expanding European settlement. Probably once distributed statewide, they were gone from counties east of Lancaster by the beginning of the eighteenth century (Wilson 1808–1814). Unregulated hunting and intensive lumbering pushed the species back to remnant pockets in forested patches, predominantly in the southern Ridge and Valley, by the start of the twentieth century. The remaining population was estimated at 5000 birds in 1900.[2] The PGC released game-farm Wild Turkeys for more than 30 years after establishing the first farm in 1930, but the stocking is not credited with expanding the species' distribution. The recovery of Wild Turkeys is one of wildlife management's great success stories.

[1] Drake, B. 1997. Wild Turkey trap and transfer. Annual report. Pennsylvania Game Commission, Harrisburg.

[2] Fergus, C. undated. Wild Turkey. Wildlife notes, no. 175–17. Pennsylvania Game Commission, Harrisburg.

Family Odontophoridae: New World Quail

Quail are primarily birds of western scrubland, deserts, and mountains. Most species are hunted for game. The Northern Bobwhite is the only species native to Pennsylvania. State and private groups attempted to introduce Gray Partridge *(Perdix perdix)* and Gambel's Quail *(Callipepla gambelii)* without success. Other species, such as Chukars, are also released for dog training. Quail have short legs, short rounded wings, and a very short, rounded tail. Some species have a crest; others are adorned with head plumes. They feed on the ground by scratching or by picking for seeds, fruits, or insects at low plants reached from the ground. When escaping danger, they usually run before flying. Their flight is short, rapid, and direct, and after landing they often resume running.

Northern Bobwhite *Colinus virginianus*

General status: Northern Bobwhites are resident in the U.S. west of the Rocky Mountains. Frequently hunted as game in Pennsylvania, they have declined to a point where self-supporting populations have almost been eliminated. They are uncommon residents in the southern tier of the Piedmont west to Franklin Co., but they were formerly widespread. Northern Bobwhites have been released from private or commercially raised stock for many years (Sutton 1928b; Grimm 1952; Stull et al. 1985) and continue to be released to train dogs, making it difficult to determine the presence of resident populations. Bobwhites are rarely reported across the northern portion of the state and in the mountains; most birds reported from those area are released birds.

Habitat: Northern Bobwhites prefer farmland with fence rows or borders. They need a mixture of grass, croplands, abandoned fields, and brushy woodland margins for cover.

Seasonal status and distribution

Breeding: A remnant resident population of bobwhites is probably restricted to scattered coveys in the south-central counties, from Chester west to Franklin. Populations are limited by weather and habitat and have been declining as the result of changing agricultural practices. BBS data show a severe decline (8% per year), but even this estimate may be confused by birds released for hunting and to train dogs. Bobwhites are also susceptible to harsh winters. Although social through the rest of the year, the monogamous pairs nest solitarily in tall grass, often near a hedgerow. The female lays a large clutch, typically 15 to 20 white eggs, in a well-concealed grass nest on the ground. Clutches have been reported from late-May through Aug.

History: The earliest ornithological records in Pennsylvania listed the partridge (probably quail), sometimes as abundant. Quail have long been associated with open agricultural areas in the East. Birds were stocked in Pennsylvania as early as 1859 (Poole, unpbl. ms.). At the peak of agricultural production in Pennsylvania, quail were thought to have been nearly statewide (Warren 1890; Harlow 1913) but have always been most common in southern counties. Historical accounts reflect pronounced population fluctuations. The comment that birds in Delaware Co. (Cassin 1862) were "less common than formerly," is echoed in other publications. Bobwhites were a target of market hunting, which may have affected resident populations before game laws were established at the end of the nineteenth century (Kosack 1995). Quail populations dramatically declined and their range retracted throughout the twentieth century. The PGC's quail introductions began as early as 1906, with most of the birds obtained from far south (e.g., Alabama and Mexico). The PGC discontinued, in the 1970s, its 42-year program of annually releasing 5000 to 10,000 captive-raised birds after it became clear that this effort had failed to sustain a resident population (Kosack 1995).

Comments: Captive-bred birds never contributed to the establishment of a viable quail population, and importation of birds from the South may have weakened native populations. The hunting season has been closed in the southern counties of the state that may support a resident population. Bobwhites are listed by the OTC as Candidate–At Risk to reflect their precarious status.

Order Gruiformes
Family Rallidae: Rails, Gallinules, and Coots

Rails and moorhens are shy skulkers of marshes, staying within dense vegetation; where they move about with ease aided by their laterally compressed bodies. They have strong legs and long toes that carry them swiftly through vegetation; they prefer to run rather than fly to escape danger. Their tail is short, and their wings are rounded and weak, making flight labored. The birds normally stay out of sight or fly short distances, quickly dropping in the cover of marshes. Most species are quite vocal during the breeding season and are more often heard than seen. Coots congregate in large flocks during migration and in winter, often in open water far from marshes, whereas rails and moorhens are usually found as singles or in pairs in dense vegetation. Coots share the short tail of rails and moorhens, but they have long toes that are lobed. Ten members in this family have been recorded in Pennsylvania, and four regularly breed in the state: the Virginia Rail, Sora, Common Moorhen, and American Coot. Their nests are usually concealed in vegetation over water. Eight to 12 nearly white to rich buff or brown-spotted eggs form a normal clutch.

Yellow Rail *Coturnicops noveboracensis*

General status: Yellow Rails breed across Canada from Alberta east to Nova Scotia and south to Oregon, North Dakota, Minnesota, Wisonsin, and Michigan. They winter along the Atlantic coast from North Carolina south to Florida and west along the Gulf coast to Texas. Most authorities considered them to be rare spring and fall transients in Pennsylvania, with most records in Apr,

May, Sep, and Oct. These very secretive rails have been reported once in the state since 1985; their status and distribution remain unknown. Most records were from before 1910. There have been about five records since 1959.

Seasonal status and distribution

Spring: There are two reports of Yellow Rails in spring. One was seen by Russell at Tinicum on 26 Mar 1963,[1] and one was identified by R. Capana at Conneaut Marsh on 11 May 1985.[2]

Fall: On 4 Oct 1977 B. Haas and J. C. Miller heard a Yellow Rail calling at Tinicum.[3] In 1966, W. S. Clarke found one at State College in early Oct.[4] A Yellow Rail was found dead in the latticework of a steel tower on Laurel Hill in Westmoreland Co. in early Nov 1959 and is now in the CMNH (Leberman 1976).

History: Poole (1964) believed that they could have nested in the state, on the basis of specimens taken in the summer. Two were captured in Jul 1882 in Delaware Co. near Chester (Warren 1890). Historical records in the eastern half of the state have come from Bucks, Cumberland, Delaware, Lancaster, Luzerne, and Philadelphia counties (Poole, unpbl. ms.). In western Pennsylvania all historical records of Yellow Rails have come from Crawford and Erie counties, except one in Centre Co. from Millbrook Swamp on 16 May 1948 (Wood 1983). L. E. Hicks saw a half-grown bird on 9 Aug 1932 in the Ohio portion of the Pymatuning Swamp near the state line (Todd 1940). Eight specimens were collected in 1900–1901 along Lake Erie at the mouth of Mill Creek in Erie (Todd 1904).

Comments: A recent record of a Yellow Rail picked up on 18 May 1998 by a farmer who was cutting a hay field in Juniata Co. is currently under review.[5]

[1] Scott, F.R., and D.A. Cutler. 1963. Middle Atlantic Coast region. AFN 17:315.

[2] Hall, G.A. 1985. Appalachian region. AB 39:300.

[3] Miller, J.C. 1986–1987. Birds of the Tinicum Wildlife Refuge and adjacent areas, Philadelphia, Pennsylvania. Cassinia 62:45.

[4] Hall, G.A. 1967. Appalachian region. AFN 21:34.

[5] Troyer, A. 1998. Local notes: Juniata County. PB 12:76.

Black Rail *Laterallus jamaicensis*

General status: In the West, Black Rails are resident in California, and they breed locally in Arizona and Kansas. In the eastern U.S. their breeding range extends primarily along the East coast from New York south to Florida and along the Gulf coast to Texas, with some very irregular local breeding in Illinois and Ohio. They winter along the Atlantic coast from North Carolina south through Florida and along the Gulf coast. Black Rails, considered accidental in the state, are usually discovered by their distinctive "ki-kee-doo" call. There have been six records of Black Rails in Pennsylvania in recent years.

Seasonal status and distribution

Spring through summer: One was seen and heard in Iron Spring Swamp at PNR from 23 May through 21 Jun 1967. The bird's call was taped, and the recording was deposited with the Library of Natural Sounds at Cornell University (Leberman 1976). A Black Rail was found by P. DeAoun at Tinicum on 14 Apr 1970.[1] In 1985 one was present in a small marsh near Quarryville in Lancaster Co. from 16 May to at least 25 Jun (Morrin et al. 1991). Two birds were found by J. Peplinski in a marsh in Centre Co. on 25 May 1986, and at least one remained though the summer.[2,3] The most recent record was a vocalizing male, which was tape recorded, discovered by S. Rannels at MCWMA in Lebanon Co. from 30 May to 12 Jun 1994.[4]

Fall: The only fall record of the Black Rail is of a bird found dead under wires by

T. Dougherty in Philadelphia on 26 Sep 1968.[5]

History: According to Warren (1890) Black Rails were found breeding many years ago near Philadelphia. He saw two that were shot in Sep along the Delaware River near Chester and mentioned that a Black Rail had been captured near Bethlehem and two were collected in Lebanon Co. in Aug 1879 and 1880. In Delaware Co. Cassin (1862) listed them as rare, but occasionally occurring. According to Beck (1924) a bird was caught by hand as it was running through the grass in Lititz in 1900, but the specimen was accidentally destroyed. A bird was seen on 27 Aug 1950 at Tinicum (Poole, unpbl. ms.). In western Pennsylvania, Sutton (1928b) recorded this species from Crystal Lake near Hartstown, Crawford Co., on 7 Sep 1925.

[1] Meritt J.K. 1970. Field notes. Cassinia 52:53.

[2] Hall, G.A. 1986. Appalachian region. AB 40:470.

[3] Hall, G.A. 1986. Appalachian region. AB 40: 1203.

[4] Rannels, S. 1994. Black Rail, Lebanon County. PB 8:89.

[5] Scott, F.R., and D.A. Cutler. 1969. Middle Atlantic Coast region. AFN 23:30.

Clapper Rail *Rallus longirostris*

General status: They are resident along coastal california and Mexico in the West. In the eastern U.S., Clapper Rails are resident in salt marshes along the Atlantic coast from Connecticut south to Florida and west along the Gulf coast to Texas. They occasionally stray inland from the coast. All but the most recent record are from the southeastern corner of Pennsylvania.

Seasonal status and distribution

Fall: The only Clapper Rail reported in the state since 1960 was found dead by M. Guter in Latrobe, Westmoreland Co., on 13 Sep 1988.[1]

History: Libhart (1869) reported that G. W. Hensel obtained a Clapper Rail from the Conestoga River in Lancaster Co. One was shot in Chester, Delaware Co., by G. R. Hoopes in Sep 1880 (Poole, unpbl. ms.). Burns (1919) mentioned a Clapper Rail taken by C. J. Pennock at Kennett Square in Chester Co. on 28 Apr 1908, which is in the ANSP. An individual was captured on the Franklin and Marshall College campus in Lancaster on 25 Apr 1946 by K. Corbett and is now a specimen in the North Museum. A Clapper Rail found injured in Kensington, Philadelphia, on 17 Apr 1956 and taken to the Philadelphia Zoo died two days later. Three birds were picked up in 1958: one was found on 23 Aug in Philadelphia; the second was captured in a garage at Strafford, Chester Co., on 31 Aug; and the third was also captured in a garage in Philadelphia (Poole, unpbl. ms.).

[1] Mulvihill, R.S., and R.C. Leberman. 1989. Another unique rail specimen from southwestern Pennsylvania: A Clapper Rail. PB 3:14–15.

King Rail *Rallus elegans*

General status: King Rails breed locally in the U.S. from South Dakota and Texas east, except in the Appalachian Mountains, to the Atlantic coast. They winter primarily in Georgia, Florida, and the southern portions of the Gulf states to south Texas. In Pennsylvania, King Rails are rarely reported and very little is known of their habits. Some may be overlooked or mistaken for the more common Virginia Rail, which has a similar call. They are casual in scattered wetlands and irregular in summer only in Crawford Co. King Rails have suffered from loss of habitat in Pennsylvania as well as in other areas within their breeding range in the eastern U.S.

Habitat: King Rails are nearly always found in emergent wetlands, but they have been reported in a variety of habitats in other areas.

King Rail at Alcoa Marsh, Lancaster County, June 1982. (Photo: Franklin C. Haas)

Seasonal status and distribution

Spring: There are at least 10 spring records since 1959 away from confirmed breeding locations, with dates ranging from 30 Mar to 31 May. They have been reported from Berks, Cambria, Centre, Clinton, Erie, Lancaster, Lycoming, Westmoreland, and York cos.

Breeding: This species is rare at the northern periphery of its range in Pennsylvania; breeding-season records since 1960 are known only from Butler, Cambria, Crawford, Delaware, Lancaster, Mercer, Philadelphia, Somerset, and Tioga cos. Confirmed breeding was restricted to two locations in the state since the 1980s, neither of which were known historically. These were a marshy field along Route 15 south of Lawrenceville in Tioga Co. and SGL 284 in Mercer Co. Until the early 1990s, one or two pairs nested in the marshes around Tinicum, Philadelphia Co.[1] Reports have come irregularly from Conneaut Marsh in Crawford Co. Further surveys are needed to determine the status of populations at known sites and in other areas of appropriate habitat.

Fall: Apparently only five King Rail records are known since 1960 for this season away from known breeding sites: one bird at Octoraro Reservoir in Lancaster Co. on 2 Nov 1990,[2] a very late bird on 29 Dec 1973 at Lock Haven in Clinton Co.,[3] one at MSP in Butler Co. on 1 Sep 1971,[4] and another on 22 Oct 1967 at PISP in Erie Co. (Stull et al. 1985).

History: King Rails formerly were reported in summer in Berks, Bucks, Chester, Crawford, Delaware, Northampton, Philadelphia, and Union cos. (Poole, unpbl. ms.). Two observations of young confirmed at least occasional nesting in the Pymatuning region of Crawford Co. before the creation of the Pymatuning Reservoir (Grimm 1952), but not since. The only locations historically in which King Rails were regular in summer were the tidal marshes of Delaware and Philadelphia cos., of which Tinicum is a small remaining fragment. J. C. Miller and Price (1959) said that King Rails were seen every year, and three nests were found in 1953. Little more is known of the species.

Comments: The loss of emergent wetlands is described as the single most critical threat to this species, which is listed by the PGC as Endangered.

[1] J.C. Miller, pers. comm.

[2] Paxton, R.O., W.J. Boyle Jr., and D.A. Cutler. 1991. Hudson-Delaware region. AB 45:81.

[3] Hall, G.A. 1974. Appalachian region. AB 28:639.
[4] Hall, G.A. 1972. Appalachian region. AB 26:64.

Virginia Rail ***Rallus limicola***

General status: In North America, Virginia Rails breed locally across southern Canada south through the western U.S. and east across the central states to the Atlantic coast. They winter along the Pacific coast states south to coastal Mexico and along the Atlantic coast from New England and south to Florida and the Gulf states. Virginia Rails are smaller versions of King Rails but are far more common and widely distributed in Pennsylvania. They are uncommon to fairly common regular migrants and local breeders at scattered sites throughout the state. Virginia Rails are secretive but quite vocal during the breeding season. They are accidental in winter.

Habitat: Emergent wetlands, a few acres or larger, support this species, but birds may be irregular in smaller patches. They require less extensive habitat than the Sora.

Seasonal status and distribution

Spring: Migrants may arrive the third week of Mar, but most are found beginning in mid-Apr. By late Apr vocal activity is well under way. It is uncertain how many of the birds found in spring remain to breed or whether most migrate farther north to more suitable nesting sites.

Breeding: Virginia Rails are uncommon to rare in the summer statewide. Breeding records come from across the state, except a few central counties that lack habitat. Rails usually are overlooked unless a specific effort is made to find them. The state's largest population of Virginia Rails is in the extensive wetlands of the Glaciated Northwest, notably in Crawford Co. The largest wetland, Conneaut Marsh, supports several hundred pairs.[1] The Glaciated Northeast also supports nesting pairs in widely scattered marshes. The well-concealed nests are suspended over water or are placed in the drier areas of the marsh. Virginia Rails produce large clutches; typically 8 to 10 eggs. Most clutches are completed by 15 May to 11 Jun, but small downy young have been observed as late as 14 Aug (Grimm 1952).

Fall: The timing of fall migration is unclear. Virginia Rails probably remain until the ground and water freeze. Evidence was witnessed in a small marsh east of Waterford in Erie Co. during the first or second week of Sep 1970. Eight to 10 birds were seen on the day before a hard freeze, but not after.[2] Virginia Rails are usually not found after early to mid-Oct, but occasionally stragglers remain as late as early Jan.

Winter: Virginia Rails are not known to survive through the winter in Pennsylvania, but they have been recorded near the beginning of the winter at least nine times on the CBC near Tinicum in Delaware Co.[3] One remained until 6 Jan 1969 at Tinicum in Philadelphia Co.[4] On 1 Jan 1988, one was at the Alcoa Marsh in Lancaster Co. (Morrin et al. 1991). One was seen at Canoe Lake in Blair Co. on 19 Jan 1992.[5] and fresh remains of a bird were found at Fairview in Erie Co. on 2 Jan 1992.[6]

History: Little historical information for this secretive species exists. Poole listed 21 widely scattered counties with confirmed or probable nesting, similar to BBA results. Records in the northeastern counties were scarce in early accounts, undoubtedly because of lack of effort. Virginia Rails have suffered from the loss of 50% of Pennsylvania's wetlands during the twentieth century (Tiner 1987). This species probably occupied some of the many smaller wetlands that were drained for farming or were flooded.

[1] Brauning, D.W. 1997. Wetland nesting bird population survey. Annual report. Pennsylvania Game Commission, Harrisburg.
[2] G.M. McWilliams, pers. obs.
[3] N. Pulcinella, pers. comm.
[4] E. Fingerhood, pers. comm.
[5] S. Kotala, pers. comm.
[6] Hall, G.A. 1992. Appalachian region. AB 46: 264.

Sora ***Porzana carolina***

General status: Soras breed across Canada south through the western U.S. and east across the central states to the Atlantic coast. They winter primarily across the southern U.S. The status and distribution of Soras in Pennsylvania are very similar to the status and distribution of Virginia Rails. Soras are uncommon to fairly common regular migrants and local breeders in scattered sites throughout the state. They are skulky but quite vocal in the breeding season. Soras routinely wander into the open to feed, especially along muddy edges of marshes.

Habitat: They are found in marshes or along wetlands with sufficient sedges or cattails and occasionally in wet grassy fields or along streamsides. They are frequently found in the same habitats as Virginia Rails, but Soras usually prefer more extensive marshes with slightly deeper water.

Seasonal status and distribution

Spring: Birds have been recorded as early as mid-Mar, but most arrive in mid- to late Apr. By late Apr vocal activity is well under way. It is uncertain as to how many of the birds found in spring remain to breed or migrate farther north.

Breeding: Like Virginia Rails, Soras are uncommon in suitable habitat statewide but are easily overlooked. Soras and Virginia Rails are much alike in many ways. The nesting season is nearly identical; Harlow's (1918) dates from central Pennsylvania's nests were 10 May to 11 Jun, with an average date of 17 May. Soras are reported about as frequently as are Virginia Rails in the Glaciated Northwest, but they are slightly less common statewide. Clutches of 10–13 buff-colored eggs, spotted with brown, are laid in a nest above the water in cattails or bulrushes.

Fall: The timing of fall migration is unclear. Notable numbers of rails of any species are rarely reported, so the estimated 200 Soras at Tinicum in Philadelphia Co. on 6 Oct 1967 was remarkable.[1] Like Virginia Rails, they may remain until the ground and water freeze, but Soras appear to be less hardy than Virginia Rails, leaving the marshes and wetlands earlier, usually by late Sep or early Oct, with stragglers remaining as late as early Dec. One remained to 1 Jan 1970 at Harrisburg,[2] and another was found on 1 Jan 1988 at MCWMA (Morrin et al. 1991).

History: Like most wetland birds, Soras have lost much habitat to the draining or flooding of wetlands. They were formerly considered abundant in marshes of the lower Delaware River (Poole, unpbl. ms.), where only a fraction of the historical wetlands remains and Soras are now rare.

Comments: The bills identify the greatest ecological difference between the two species: the Virginia's is equipped to probe for insects, the Sora's to crush seeds.

[1] Scott, F.R., and D.A. Cutler. 1968. Middle Atlantic Coast region. AFN 22:19.
[2] Scott, F.R., and D.A. Cutler. 1970. Middle Atlantic Coast region. AFN 24:490.

Spotted Rail ***Pardirallus maculatus***

General status: The Spotted Rail is a tropical species; little is known of its status and distribution. It is found locally from central Mexico and Hispaniola to northern Argentina. Remarkably, there is one record of this tropical species in Pennsylvania.

Seasonal status and distribution

Fall: On the morning of 12 Nov 1976, D. P. Kibbe found a dead large, dark rail while making his regular weekly survey for avian migrant mortalities at a nuclear power plant in Beaver Co. The bird was discovered on a walkway about 60 feet above the base of a 500-foot cooling tower along the south bank of the Ohio River at Beaver Valley. Dr. Kenneth Parkes identified the bird as an adult male Spotted Rail. This was the first record for the U.S., and the only Pennsylvania record. The skin and skeleton were preserved by Parkes and placed in the CMNH.[1]

[1] Parkes, K.C., D.P. Kibbe, and E.L. Roth. 1978. First records of the Spotted Rail for the U.S., Chile, Bolivia, and western Mexico. AB 32:295–299.

Purple Gallinule ***Porphyrula martinica***

General status: Purple Gallinules are distributed along the Atlantic coast from North Carolina south into Florida, west along the Gulf coast to Texas, and north along the Mississippi Valley to Tennessee. They winter in Florida and southern Texas. They occasionally wander and breed well north of their normal breeding range. They sometimes appear in unusual places, such as backyards on city streets. One flew through the open window of a powerhouse at Holtwood Dam in Lancaster Co. (Morrin et al. 1991). Purple Gallinules are accidental in Pennsylvania.

Seasonal status and distribution

Spring through summer: Purple Gallinules have been recorded in the state at least 11 times since 1960, in the Coastal Plain and Piedmont counties of Berks, Bucks, Chester, Delaware, Lancaster, and Philadelphia. Most sightings of Purple Gallinules have been from 15 Apr through 27 Jun. A later record, one bird on 7 Jul 1986, was seen by many at a Morrisville, Bucks Co., nursery.[1] Only three records are from outside southeastern Pennsylvania: one seen from 24 Apr to 8 May 1977 at Bells Springs near Mackeyville, Clinton Co. (Wood 1983), a bird found moribund at Johnstown in Cambria Co. in early May 1984,[2] and a dead bird picked up at YCSP in Indiana Co. on 23 April 1983.[3]

Fall: The only fall record is from a Philadelphia backyard on 22 Oct 1983.[4]

History: There were about 10 records prior to 1960. Historically, birds were found in unusual places as well. All but one record was from the Coastal Plain or Piedmont: a bird killed with a rock in the athletic park at Washington in Washington Co. on 23 Apr 1896. The specimen was placed in the CMNH (Todd 1940). Other odd settings include a Purple Gallinule found walking along a street in Harrisburg on 14 Apr 1932[5] and one struck by a car in Tinicum Township in Delaware Co. on 15 Jun 1934.[6]

[1] Paxton, R.O., W.J. Boyle Jr., and D.A. Cutler. 1986. Hudson-Delaware region. AB 40:1183.
[2] Hall, G.A. 1984. Appalachian region. AB 38:910.
[3] M. Higbee and R. Higbee, pers. comm.
[4] Paxton, R.O., W.J. Boyle Jr., and D.A. Cutler. 1984. Hudson-Delaware region. AB 38:182.
[5] Cramer, W.S. 1932. Purple Gallinule at Harrisburg, Pennsylvania. Auk 49:348–349.
[6] Worth, C.B. 1934. Purple Gallinule *(Ionornis martinica)* in Pennsylvania. Auk 51:519.

Common Moorhen ***Gallinula chloropus***

General status: Less secretive than other marsh birds, Common Moorhens can frequently be observed swimming on open marsh pools or along the muddy edges of marshes at dawn or dusk. In western North America, moorhens are residents from California to New Mexico south through Mexico. In eastern North America they breed from Minnesota and Texas east to the Atlantic coast and winter mostly from South

Carolina south along the Atlantic and Gulf coasts. In Pennsylvania they are uncommon to fairly common regular migrants and are local breeders, primarily in the Coastal Plain and Glaciated Northwest. Common Moorhens are accidental in winter in the Coastal Plain. They have benefited from creation of manmade wetlands.

Habitat: Common Moorhens are usually found in marshes but also on the edges of ponds, lakes, and streams where there is sufficient emergent vegetation.

Seasonal status and distribution

Spring: Birds arrive by mid-Apr and are most easily observed in spring. As spring progresses, growing vegetation makes moorhens increasingly difficult to see.

Breeding: Moorhens are fairly common in the marshes of eastern Butler and western Crawford cos. Formerly moorhens were common in wetlands along the lower Delaware River, but they declined during the 1990s to just a few pairs in marshes at Tinicum and around the Philadelphia sewage treatment plant. Opportunistic and sporadic elsewhere, they may be found occasionally in large wetlands across the state. Breeding has been confirmed in Berks, Bucks, Butler, Centre, Crawford, Delaware, Erie, Mercer, Lawrence, Snyder, Somerset, Sullivan, Wyoming, and York cos. Common Moorhens are more conspicuous and vocal while foraging in beds of aquatic vegetation than are other rails during the day. The large (15-inch diameter) nest of cattails is built over the water. Complete clutches of 8 to 10 eggs have been reported from the second half of May through the first half of Jun, but pairs are on territory in late Apr.

Fall: Fewer moorhens are observed in fall than in spring. The timing of fall migration is unclear, but most birds have left the state by the last week of Oct. Stragglers may remain to the first week of Jan.

Winter: The only known wintering site in Pennsylvania is in the Coastal Plain at Tinicum, where there are at least four records since 1960: one from 1 to 3 Feb 1965,[1] one on 23 Feb 1991,[2] two on 11 Jan 1992,[3] and one on 17 Jan 1993.[4]

History: Moorhens (then known as Florida Gallinules) were unknown as a nesting species in Pennsylvania by earlier ornithologists. The state's first nest was found in 1904, in Philadelphia (R. F. Miller 1946). Todd (1904), citing several records in May and early fall, expressed surprise that moorhens did not breed in the state. The species expanded rapidly and became a locally common breeder in the Philadelphia and Delaware co. marshes. Similarly, moorhens were first reported nesting in Crawford Co. in 1925 by Sutton (1928b). They may occur in substantial densities: e.g., 16 nests were found in 4 acres of Tinicum marsh, and 50 pairs were in the area of Tinicum in 1957 (Poole, unpbl. ms.). One remained through the winter of 1954–1955 at Centre Furnace Pond in Centre Co. (Wood 1983).

1 Meritt, J.K. 1965. Field notes. Cassinia 49:37.
2 Guarente, A. 1991. County reports—January through March 1991: Delaware County. PB 5:37.
3 Boyle, W.J., Jr., R.O. Paxton, and D.A. Cutler. 1992. Hudson-Delaware region. AB 46:241.
4 Haas, F.C., and B.M. Haas, eds. 1993. Rare and unusual bird reports. PB 7:17.

American Coot *Fulica americana*

General status: In North America, American Coots breed from Yukon south across Canada and the U.S. They winter across the U.S. except in the north-central states. Coots are fairly common to abundant regular migrants in Pennsylvania. Unlike other marsh birds, which are quite secretive, they prefer to feed in the open. During migration they are frequently seen in huge rafts on lakes. However, migrant numbers reportedly declined since the early 1980s in most areas of the state. Coots are local breeders primarily restricted to the Glaciated Northwest. They are regular winter residents locally where open water remains.

Habitat: During the breeding season, American Coots usually nest in emergent vegetation in large marshes. During migration and in winter, coots usually prefer lakes and slow-moving rivers with some emergent vegetation along the shoreline. Coots are as much at home on open water, where they frequently dive for food, as they are on muddy edges of marshes.

Seasonal status and distribution

Spring: Their typical migration period is from mid-Mar to the second week of May. Peak migration occurs from late Mar to early Apr. The largest concentrations of birds may be found at PISP, where flocks containing 3000 or more birds are not unusual. The maximum flock size averages several hundred birds in other areas of the state at this season. Stragglers linger to the third week of Jun.

Breeding: Coots are rare breeding residents in the wetlands of Mercer and Butler cos. and uncommon but regular residents in large wetlands in Crawford Co., notably at Conneaut Marsh. They are accidental during summer in the rest of the state, except at Glen Morgan Lake, Berks Co., where breeding has been documented since 1995. Since 1960, confirmed breeding records also have come from Centre, Erie, Lawrence, and Snyder cos. They were reported in only 25 BBA blocks and confirmed in 11 of those.[1] Opportunistic, coots nest casually where suitable habitat exists, potentially statewide. The nest is a large structure, usually floating in the water or on a mat of vegetation; it contains 8–12 eggs. Most nests with eggs have been found in the second and third weeks of May.

Fall: Migration begins the last week of Aug or the first week of Sep. Most birds arrive the third or fourth week of Sep. The number continues to increase through the fall, but by the third week of Dec it declines rapidly. Flocks may reach a thousand birds for some areas in the state at this season. The greatest concentration of coots is found at PISP in Erie Co., where over 5000 birds have been estimated.[2]

Winter: The state's largest wintering concentrations, usually fewer than 50 coots, regularly occur at Tinicum, at PISP, and at YCSP. Presque Isle Bay hosted over 500 coots during the winter of 1998, when the bay did not freeze.[2] Single birds or small groups have been recorded in winter on the larger streams and rivers and on unfrozen lakes in the Piedmont, Ridge and Valley, Southwest, and Glaciated Northwest. In the Glaciated Northeast they have been recorded in winter only at Lake Carey in Wyoming Co., where aerators keep water around docks from freezing.[3] In the High Plateau coots have been recorded in winter from only one site in the northern-tier counties, with at least two records from the Allegheny River at Warren in Warren Co.[4] In the Allegheny Mountains they have been recorded in winter in Cambria, Somerset, and Westmoreland cos.

History: Warren (1890) refers to two "correspondents" that regarded coots as breeders. Notably, one of these was from Mercer Co., where the BBA project documented nesting. Harlow (1913) also cited R. F. Miller, who suggested "probable" breeding in Philadelphia Co. Pennsylvania's first documented nesting by American Coots was in a marsh north of Hartstown in 1923 (Sutton 1928b). Surveys by the Carnegie Museum found more than a dozen nests in the two years (1934, 1935) following the creation of Pymatuning Lake (Grimm 1952), but numbers quickly declined after the flush of vegetation died back. Thereafter they were considered common spring and fall migrants, rare winter residents, and rare, local breeders. Poole (1964) listed the above sites, plus Chester, Northampton, Pike, and Berks cos. as nesting-season locations before 1960.

Comments: The limited nesting range of American Coots prompted the OTC to list them as Candidate–Rare in Pennsylvania.

[1] Leberman, R.C. 1992. American Coot *(Fulica americana)*. BBA (Brauning 1992): 130–131.
[2] G.M. McWilliams, pers. obs.
[3] W. Reid, pers. comm.
[4] T. Grisez, pers. comm.

Family Gruidae: Cranes

Cranes are large, long-legged, long-necked, long-bodied, heavy-billed waders. They differ from herons by having more tightly packed tertials (long feathers that curve over their rump forming a bustle when standing) and a heavier body. Unlike herons, which fly with their neck pulled back in an S-shape, cranes fly with the neck straight and outstretched. Herons usually fly with slow, steady, deep wing beats, with the outer primaries held together; cranes fly with shallow quick wing beats, flicking their wing tips on the upstroke, with outer primaries held apart. Cranes frequently soar on thermals and fly in vocal flocks. During the breeding season they perform excited courtship dances. Only the Sandhill Crane is found in Pennsylvania, usually as a rare vagrant, but two pair recently have nested in the western part of the state.

Sandhill Crane ***Grus canadensis***

General status: Sandhill Cranes breed widely across the northern portion of the central and western U.S. north through Canada and Alaska. They winter primarily across the southwestern and Gulf states. They are rare but regular migrants through Pennsylvania. Most sightings are from the northwestern and southeastern portions of the state. They apparently have nested recently in Crawford and Lawrence or Mercer cos. Sandhill Cranes are accidental in winter.

Habitat: Sandhill Cranes prefer harvested cornfields or fields recently plowed and the shoreline of lakes and large rivers. They are often observed in flight during migration, especially along mountain ridges and Lake Erie. The suspected nesting habitat is extensive emergent marshes.

Seasonal status and distribution

Spring: Earliest migrants have arrived as early as the first week of Mar, but cranes usually are observed from the last week of Mar or the first week of Apr to the third week of May. Most sightings are of single birds or groups of fewer than six. Pairs have been seen in breeding areas performing courtship dances. Stragglers may remain through the summer.

Breeding: There is overwhelming evidence that two pairs bred regularly. The state's first evidence of local breeding came in Lawrence Co. in 1993. Two adults and one immature were found by L. Cooper and N. Rodgers near Plain Grove, close to the juncture of Butler, Lawrence, and Mercer cos., on 4 Aug 1993. A nest site has not been found but was believed to have been in SGL 151 in Mercer Co. A pair of Sandhill Cranes had been at this same site in 1991 and 1992.[1] A juvenile again was seen with two adults on 15 Aug 1994 in a wheat field with another pair of Sandhill

Sandhill Crane at Hartzell's Ferry, Northampton County, October 1994. (Photo: Rick Wiltraut)

Cranes.[2] In 1996 a pair with two young was found in the same area[3] and in 1997 a pair with a single young was near Plain Grove in Lawrence Co.[4] Similar nesting evidence was reported in fields adjacent to the Erie National Wildlife Refuge in Crawford Co. in 1995, when PGC officer J. McKollop reported seeing a single young in Aug 1997.[5] A displaying pair of cranes was found near Edinboro, Erie Co., on 26 Jun 1988, but no evidence of nesting could be found there. A pair was seen during the same year at West Springfield, Erie Co., from late Jul to 10 Sep but again without evidence of nesting.[6] Single cranes were reported at Tinicum on 4 Jul 1969[7] and at SGL 63 in Clarion Co. on 23 Jun 1995.[8]

Fall: Away from the breeding areas, most sightings east of the mountains are in this season and in the Piedmont. Nearly all sightings in fall are of individual birds. Migrants have arrived as early as the first week of Sep. Sandhill Cranes usually are reported from the third week of Sep to late Nov, with stragglers to late Dec. Most records are in Oct.

Winter: There have been six winter records from scattered locations in the state. These were at Penn Manor in Bucks Co. on 1 Feb 1992,[9] at the Octoraro Reservoir in Lancaster Co. from 3 to 29 Feb 1980,[10] at MCWMA from 2 Feb to 13 Mar 1997,[11] at Colyer Lake in Centre Co. on 2 Feb 1991,[12] near Raystown Dam in Huntingdon Co. from 10 to 15 Jan 1992,[13] and one with an injured leg at PISP in Erie Co. on 24 Jan 1989.[6]

History: Sandhill Crane records date back to 1600–1625, on the basis of skeletal remains found in an archaeological site at Washington Boro, Lancaster Co.[14] Barton (1799) wrote of immense flocks that were seen and heard flying over Pennsylvania, New Jersey, and New York, saying "the flocks which are immense, are seen and heard very high in the air." However, from 1840 to the present there have never been more than five or six birds reported at a time.

Poole (unpbl. ms.) cited fewer than 10 valid pre-1960 records.

[1] Wilhelm, G. 1993. First breeding record of Sandhill Crane for Pennsylvania. PB 7:91–92.
[2] Wilhelm, G. 1994. Second breeding record of Sandhill Crane for Pennsylvania. PB 8:136–137.
[3] Butcher, S. 1996. Local notes: Lawrence County. PB 10:165.
[4] Dean, B. 1997. Local notes: Lawrence County. PB 11:97.
[5] J. McKellop, pers. comm.
[6] G.M. McWilliams, pers. obs.
[7] Scott, F.R., and D.A. Cutler. 1969. Middle Atlantic Coast region. AFN 24:28.
[8] Haas, F.C., and B.M. Haas, eds. 1995. Birds of note—April through June 1995. PB 9:87.
[9] French, R. 1992. County reports—January through March 1992: Bucks County. PB 6:30.
[10] Richards, K.C., R.O. Paxton, and D.A. Cutler. 1980. Hudson-Delaware region. AB 34:258.
[11] Haas, F.C., and B.M. Haas, eds. 1997. Birds of note—January through March 1997. PB 11:22.
[12] Hall, G.A. 1991. Appalachian region. AB 45: 273.
[13] G. Grove, pers. comm.
[14] Guilday, J.E. 1962. Bird remains from Pennsylvania archaeological sites. Sandpiper 4:78.

[Whooping Crane] ***Grus americana***

General status: Whooping Cranes breed in south-central Mackenzie and adjacent northern Alberta and winter near the coast of southeastern Texas. Whooping Cranes are listed as hypothetical in Pennsylvania on the basis of historical records.

Comments: This species is believed to have occurred in Delaware Co. (Cassin 1862). Cassin listed these cranes as very rare visitors and stated that they have "occurred in a few instances." There were reports of Whooping Cranes in New Jersey: Poole (unpbl. ms.) stated that Wilson saw them in the marshes of Cape May in Dec and Turnbull saw three at Beasley's Point in 1857. Poole also stated that there were early references from the coasts of New York as well. Captain Philip Amadas and his fellow adventurers, who visited and explored the Atlantic coast in 1584, wrote in *Hakluyt's Voyages*, ed. 1589, folio 729, "having discharged their harquebus-shot, such a flocke of Cranes (the most part white) arose, with such a crye; redoubled by many ecchoes, as if an armie of men had showted altogether" (Turnbull 1869). The AOU (1957), listed Chester and Philadelphia cos. as sites where Whooping Cranes formerly occurred, but Poole (1964) questioned the validity of the Chester Co. record.

Order Charadriiformes
Family Charadriidae: Plovers

Plovers are small to medium-sized, plump-bodied shorebirds. The bill is short, straight, and rigid with a slightly bulbous outer half and horny tip. The behaviors of plovers help to separate them from other shorebirds; they walk or scurry a short distance, stop a moment, and then resume walking. This behavior is repeated especially when feeding. Most species flock with their own kind but may mix with other species when feeding or resting. Sixteen species of plovers are known to occur in North America, and seven of them have been found in Pennsylvania. Only one species, the Killdeer, regularly breeds in the state. The endangered Piping Plover formerly bred in Pennsylvania but now is a rare migrant.

Black-bellied Plover ***Pluvialis squatarola***

General status: Black-bellied Plovers breed in North America from northern Alaska east across the Arctic coast and islands. They winter along the Pacific, Atlantic, and Gulf coasts. Black-bellied Plovers are uncommon to fairly common regular spring and fall migrants statewide, except in the mountains of the Ridge and Valley and High Plateau, where they are rare or absent. They are most numerous in the Coastal Plain, Piedmont, and Lake Erie Shore.

Habitat: They inhabit sandy beaches, mudflats, muddy shorelines of lakes, ponds, rivers, and marshes, and occasionally flooded or freshly plowed fields. They have also been observed flying past hawk watches along the mountain ridges.

Seasonal status and distribution

Spring: Individuals have appeared as early as the second week of Apr, but Wood (1983) cited an extremely early record on 29 Mar. Their typical migration period is from the second week of May to the first week of Jun. Peak migration occurs during the third or fourth week of May. Daily highs are usually fewer than 20 birds, but over 100 birds have been reported. On 26 May 1977, 100 were counted in the Tinicum area in Delaware Co.[1] On 28 May 1973 at Longwood in Chester Co., 300 were seen in flight,[2] and on 26 May 1992 at the Holtwood Fly-ash Basin in Lancaster Co., 125 were counted.[3] Stragglers have been recorded to the fourth week of Jun.

Fall: Adult Black-bellied Plovers may appear the first or second week of Jul but usually do not appear until the last week of Jul or the first week of Aug. The earliest juveniles appear during the first week of Sep. Fall migration continues until about the third week of Oct, with stragglers until the first week of Dec. There have been several high fall counts: 54 were seen at the Montour Fly Ash Basin on 27 Sep 1993[4] and 75 were seen with over 30 American Golden-Plovers and hundreds of Killdeer west of Linesville in Crawford Co. on 13 Sep 1967.[5] One late Dec record since 1960 was of 3 birds on a CBC in Delaware Co. on 26 Dec 1970.[1]

[1] N. Pulcinella, pers. comm.

[2] Scott, F.R., and D.A. Cutler. 1973. Middle Atlantic Coast region. AB 27:755.

[3] Witmer, E. 1992. County reports—April through June 1992: Lancaster County. PB 6:83.

[4] D.W. Brauning, pers. comm.

[5] Leberman, R.C., and R.F. Leberman. 1967. Field notes. Sandpiper 10:18.

American Golden-Plover ***Pluvialis dominica***

General status: American Golden-Plovers are celebrated for the long-distance flights from their breeding grounds in the Arctic tundra to their wintering grounds in Argentina. Formerly called Lesser Golden-Plovers, they were recently split into two species named Pacific Golden-Plover (in North America principally migrating along the West coast) and American Golden-Plover (on the East coast). American Golden-Plovers are usually uncommon to rare regular fall migrants in Pennsylvania, but in some years they may be locally fairly common to common and occasionally are seen in large flocks. They are most numerous at the Coastal Plain, Piedmont, Glaciated Northwest, and Lake Erie Shore. American Golden-Plovers are casual spring migrants in the state but are not as common as in the past.

Habitat: They are usually found on mudflats—muddy shorelines of lakes, ponds, rivers, and marshes—sandy beaches, short grass (airports), or plowed fields. They have been reported flying past hawk watches along the mountain ridges.

Seasonal status and distribution

Spring: American Golden-Plovers are very rare in spring. Most sightings are of one or two birds, with dates ranging from 17 Mar to 11 Jun. A very high count for this season was a flock of 55 reported in Butler Co. on 11 May 1981.[1]

Fall: Adults arrive as early as the last week of Jul. A very early fall migrant was seen on 2–5 Jul 1990 at a turf farm in Wycombe, Bucks Co.[2] Most appear the last week of Aug or the first week of Sep. The majority of adults migrate through Pennsylvania during the first or second week of Sep. Juveniles have been recorded as early as the first week of Sep, but most arrive the third or fourth week of Sep. The peak number of juveniles occurs any time from the last week of Sep to the last week of Oct. Flocks containing 100 or more have been recorded at various sites on several occasions. Three hundred were seen at Leola in Lancaster Co. on 20 Oct 1981,[3] and, the same year, 350 were counted in a pasture near Mascot in Lancaster Co. on 28 Oct (Morrin et al. 1991). Most have left the state by the first week of Nov. Stragglers remain to the last week of Nov.

History: American Golden-Plovers were reported to have been abundant during some years before 1890. In 1860 flocks of hundreds covered the fields in the "Great Valley" and around West Chester (Warren 1890). Warren (1890) told of seeing flocks of 50–100 in plowed fields and grass fields around West Chester during the fall of 1880. Todd (1940) considered these plovers to be quite rare in western Pennsylvania but mentioned S. E. Bacon's report of flocks containing of 25–50 seen every season in the plowed fields west of the city of Erie.

[1] Hall, G.A. 1981. Appalachian region. AB 35:823. Bucks County. PB 4:107.
[2] French, R. 1990. County reports—July through September 1990: Bucks County. PB4:107.
[3] Boyle, W.J., Jr., R.O. Paxton, and D.A. Cutler. 1982. Hudson-Delaware region. AB 36:160.

Snowy Plover *Charadrius alexandrinus*

General status: Snowy Plovers are resident along the Pacific coast from Washington to Baja California and along the Gulf coast from Florida to Texas. They breed locally in the interior from Oregon and Arizona to Oklahoma. They have been recorded inland and north in the East to the Great Lakes. Only two records are known in Pennsylvania, one modern and one historical.

Seasonal status and distribution

Spring: One discovered by R. Koury, D. Hoffman, and G. M. McWilliams at PISP on 17 May 1986 was photographed and seen by many observers.[1] This adult Snowy Plover was observed actively feeding around the sandy pools and outer beach of Gull Point until 23 May.[2] Photographs are deposited in the files of PORC and VIREO.

History: A specimen was taken near The Pinnacle in Berks Co. on 29 Jun 1886 and was placed in the Reading Public Museum. It was originally mislabeled as a Piping Plover (Poole 1964).

[1] Hall, G.A. 1986. Appalachian region. AB 40: 470–471
[2] G.M. McWilliams, pers. obs.

Wilson's Plover *Charadrius wilsonia*

General status: This stout-billed plover breeds along the Atlantic coast from New Jersey south to Florida and along the Gulf coast to Texas. Wilson's Plovers winter along the west coast of Mexico and the coast of Florida. They have been recorded as far north as Minnesota, southern Ontario and Nova Scotia. There are four modern records of this coastal plover in Pennsylvania. All but one record is from Gull Point at PISP.

Seasonal status and distribution

Spring: A Wilson's Plover was photographed by S. Stull at PISP on 4 May 1968 (Stull et al. 1985). The photo was deposited in the files of PORC. Another bird was found by R. C. Leberman and S. Robinson at PISP on 29 May 1971 (Stull et al. 1985) and remained until at least 2 Jun.[1]

Fall: One was discovered by W. Reid and W. Evans at Exeter in Luzerne Co. on 21 Sep 1970,[2] and one was identified by D. Snyder on 10 Sep 1976 at PISP (Stull et al. 1985).

History: Poole (unpbl. ms.) listed this species as rare or casual, but it evidently occurred more frequently before 1890. Libhart (1869) considered Wilson's Plover to be frequent on the Susquehanna River in autumn. Warren (1890) reported birds captured on the

Susquehanna River in Cumberland, Dauphin, Lancaster, Northumberland, and York cos. Apparently three specimens were collected near Reading in Berks Co.: two at Pricetown Hills on 26 Sep 1886 and one at Fritz's Island on 2 Aug 1888 (Poole 1947). Two were seen by H. Arnett and J. Cadbury near Media in Delaware Co. on 10 May 1952 (Poole, unpbl. ms.).

Comments: Some birds reported as Wilson's Plovers have proved to be misidentified Semipalmated Plovers.

[1] J.H. Stull, pers. comm.
[2] Bollinger, R.C. 1972. Ontario-Western New York region. AB 26:754.

Semipalmated Plover ***Charadrius semipalmatus***

General status: Semipalmated Plovers breed from Alaska across northern Canada to Nova Scotia. They winter along the Pacific coast of California and Mexico, along the Atlantic coast from South Carolina south, and along the Gulf coast. These smallest of the regularly occurring plovers found in Pennsylvania are fairly common migrants over most of the state in suitable habitat. Semipalmated Plovers are local and rare in the mountains of the Ridge and Valley and the High Plateau.

Habitat: They inhabit sandy beaches, mudflats, and muddy shorelines of lakes, ponds, rivers, and marshes. These plovers are occasionally found on open level wet fields that have sparse vegetation or on freshly plowed fields.

Seasonal status and distribution

Spring: Their migration period extends from the last week of Apr to the first week of Jun. Peak migration occurs during the third and fourth weeks of May. Concentrations of over 30 birds in a day in spring are rare, but on 18 May 1984 a total of 45 birds was at Gull Point on PISP.[1] Stragglers have remained until late Jun.

Fall: Adult Semipalmated Plovers arrive as single individuals or in small groups the third week of Jul, sometimes as early as the first week of Jul. Adults are joined by juveniles beginning in the first week of Aug. By the third week of Aug the majority of birds in the state are juveniles. Most migrants pass over the state from the first week of Aug to the last week of Sep. Daily highs are usually fewer than 15 or 20 birds. A high of 50 was counted on 5 Sep 1988 at Gull Point on PISP.[1] Migration has usually ended by the second week of Oct, with stragglers to the first week of Nov. A very late bird was seen at Bloomsburg in Columbia Co. on 18–20 Dec 1988.[2]

History: The historical status and distribution of Semipalmated Plovers were much the same as today. However, some unprecedented historical concentrations of birds were recorded in the state. A very high count of 800 was made on 12 Aug 1955 along Darby Creek in Delaware Co.,[3] and 113 were on the Conejohela Flats in Lancaster Co. on 9 Sep 1951 and again on 4 Sep 1954 (Morrin et al. 1991).

[1] McWilliams, G.M. 1984. International Shorebird Survey (Erie, Pa. site). Files of Manomet Bird Observatory, Mass.
[2] Gross, D.A. 1988. County reports—October through December 1988: Columbia County. PB 2:142–143.
[3] N. Pulcinella, pers. comm.

Piping Plover ***Charadrius melodus***

General status: The inland population of Piping Plovers breed locally in Alberta and Manitoba, from Montana east to the Great Lakes, and in Iowa and Oklahoma. The coastal population breed locally from Newfoundland south along the Atlantic and Gulf coasts. They winter primarily along the Atlantic coast from North Carolina south and along the Gulf coast. In Pennsylvania, Piping Plovers are almost never reported away from the sandy beaches of PISP in Erie Co., where these small sand-colored plovers nested until the mid-1950s. Currently they are very rare but regular fall and casual spring migrants at PISP. Usually no more than one or two birds are sighted each year. Since 1961 there has been only one report away from the park.

Habitat: Piping Plovers are found almost exclusively on the outer sandy beaches and the sandy tip of Gull Point on PISP.

Seasonal status and distribution

Spring: Dates range from the third week of Apr to the second week of May, with stragglers to the first week of Jul. Nearly all sightings are of a single adult that rarely remains for more than a day or two. In 1992, however, an adult male arrived on 13 May and remained until 5 Jun. During this time it established a territory at Gull Point, where G. M. McWilliams observed the plover displaying and digging a nest scrape. A mate never appeared, and the male was not seen after 5 Jun.[1]

Fall: Single adults may appear in the second week of Jul but usually not until the third or fourth week of Jul. Earliest migrant juveniles are seen the second week of Aug. Most recent sightings of Piping Plovers have been in Jul or Aug, but birds have been seen until the third week of Oct. At the Conejohela Flats along the Susquehanna River, single birds were recorded on 21 Aug, 10 Sep, and 25 Sep 1960. At the same site, another bird was seen on 10 Sep 1961 (Morrin et al. 1991). The most recent fall record was at Woodcock Dam in Crawford Co., where an adult was observed on 22 Sep 1996.[2]

History: Warren (1890) said Piping Plovers were about as plentiful as Semipalmated Plovers as migrants in the Lake Erie Region. Todd (1940) believed these birds to be summer residents and in 1900 shot a female containing an egg almost ready to be laid. The state's first nest with eggs was confirmed on 31 May 1911. About 15 pairs nested annually on the outer shores of Presque Isle in May and Jun (Todd 1940). They nested annually until the mid- to late 1950s and became increasingly rare from then on (Stull et al. 1985). Elsewhere in the state, Libhart (1869) listed Piping Plovers in autumn in Lancaster Co. along the Susquehanna River. Warren (1890) wrote that stragglers had been taken in the fall along the Delaware and Lehigh rivers after severe storms from the Atlantic coast. Sutton (1928b) mentioned a specimen taken

Piping Plover at Presque Isle State Park, Erie County, September 1993. (Photo: Rick Wiltraut)

along French Creek near Meadville on 7 Sep 1908. Four were seen at Tinicum on 1 Aug 1950 (Poole, unpbl. ms.). Piping Plovers were reported in the Scranton area on 15 Sep 1951 (Poole, unpbl. ms.), and on the Conejohela Flats a bird was seen on 10 Sep 1955 (Morrin et al. 1991).

Comments: There are two described subspecies of Piping Plovers. The breast band of the nominate form *(C. m. melodus)*, found along the Atlantic coast, is usually broken; that of the subspecies found along the Great Lakes and on the prairies *(C. m. circumcinctus)* is usually complete (Hayman et al. 1986). The Great Lakes nesting population of Piping Plovers has dwindled drastically. In the 1930s it was estimated to be around 500 pairs,[3] and by 1982 there were only 17–19 pairs.[4] In 1988 only one nest was found on Lake Ontario, and none was found on Lake Erie in New York (Andrle and Carroll 1988). Their declining numbers have been attributed to the loss of habitat and increased human activity on beaches. The Great Lakes population is federally designated as Endangered, and throughout the rest of its range it is listed as Threatened.[5] Piping Plovers are listed as Endangered by the PGC as a breeding species in Pennsylvania.

[1] McWilliams, G.M. 1992. County reports—April through June 1992: Erie County. PB 6:79.
[2] Leberman, R.F. 1996. Local notes: Crawford County. PB 10:163.
[3] Wilcox, L. 1939. Notes on the life history of the Piping Plover. Birds of Long Island 1:3–13.
[4] Russell, R.P., Jr. 1983. The Piping Plover in the Great Lakes region. AB 37: 951–955.
[5] U.S. Fish and Wildlife Service. 1988. Great Lakes and northern Great Plains Piping Plover recovery plan. Twin Cities, Minn.

Killdeer *Charadrius vociferus*

General status: Killdeer breed over most of Canada and throughout the U.S., and they winter primarily across the southern part of their breeding range. They are the most common and widely distributed plovers in Pennsylvania. They are fairly common to common regular migrants but may occasionally be locally abundant. Killdeer breed throughout the state and are regular winter residents.

Habitat: In migration they use open areas of short grass or plowed fields; shorelines of lakes, ponds, and rivers; mudflats; and sandy beaches. In the breeding season Killdeer may be found in almost any open area such as in fields, pastures, golf courses, parking lots, and even rooftops. In winter they inhabit spring-fed wet meadows or fields.

Seasonal status and distribution

Spring: Nonwintering Killdeer often arrive after the first major spring thaw, usually in the last week of Feb or the first week of Mar. After severe winters with lingering heavy snowfall and freezing temperatures, they may not arrive until the second or third week of Mar, especially across the northern half of the state. Migration usually peaks within the last three weeks of Mar. Daily high counts may reach several hundred birds at some staging sites.

Breeding: Killdeer are fairly common statewide, except in extensively forested regions, where they are rare (e.g., northern Clinton Co.). More than most birds, Killdeer have acclimated to human activity. Killdeer are most common on BBS routes in the agricultural areas of western counties along the Ohio border (see Map 6) but may be found in open patches anywhere in the state. Nesting has been documented in every Pennsylvania county. Nests are made by scraping a shallow depression in the ground and may be lined with pebbles, pieces of grass, or wood chips, or they may be left unlined. Eggs are cream-colored, heavily splotched or scrawled with black, and narrower at

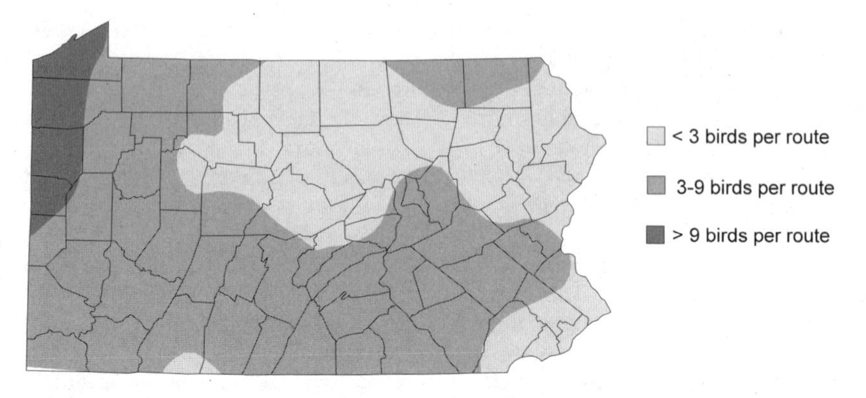

6. Killdeer relative abundance, based on BBS routes, 1985–1994

one end. The three to five eggs are most often found from 11 Apr to 16 Jun. Jun nests probably reflect second clutches. As do the young of other species in this family, Kildeer young leave the nest immediately after hatching and begin feeding on their own. BBS data do not show a change in the breeding population in Pennsylvania since 1966.

Fall: The number of birds begins to build by late Aug. Sep and Oct are the months of peak migration. As soon as the ground freezes, birds begin to migrate south, and by mid- to late Nov, only a few individuals or small flocks will remain to winter. In the northern half of the state, and in the higher elevations elsewhere, nearly all birds are gone by the first or second week of Jan.

Winter: Most Killdeer in Pennsylvania winter in the Piedmont, where they are uncommon. However, singles or small groups may also remain during mild winters at low elevations in the Ridge and Valley and in the Southwest. Across the northern-tier counties and through the central mountains of the High Plateau there are only two known winter sites near Athens in Bradford Co.[1] and at Starbrick in Warren Co.[2] In the Glaciated Northwest they are quite rare and are absent most winters.

History: Killdeer have been common birds in settled areas. Hunting was thought to reduce numbers in some areas (R. F. Miller 1949), but Warren (1890) made no reference to the impact of hunting on this species as he did regarding others. Harlow (1913) considered Killdeer rare and locally distributed in the northern counties (e.g., Warren) and had no breeding records from Pike, Wayne, or Monroe cos.

[1] T. Gerlach, pers. comm.
[2] T. Grisez, pers. comm.

Family Haematopodidae: Oystercatchers

Oystercatchers are large shorebirds, normally confined to coastal beaches, and rarely found far from saltwater. They have a robust body and a long, heavy, laterally flattened bill that is well suited for extracting mollusks from their shells. Eleven species are found worldwide and only two in North America. The American Oystercatcher is found on the East coast, and the Black Oystercatcher on the West coast. Only the former species has been reported in Pennsylvania.

American Oystercatcher ***Haematopus palliatus***

General status: American Oystercatchers breed along the coast from Massachusetts south to Florida and the Gulf coast, and they winter along the west coast of Mexico and from Maryland south along the Atlantic and Gulf coasts. Only one modern record exists for Pennsylvania.

Seasonal status and distribution

Spring: After a day of strong northeasterly winds on 2 Apr 1993, an American Oystercatcher was seen and photographed by R. Wiltraut along the shoreline of Beltzville Lake in Beltzville State Park, Carbon Co.[1] A photograph has been deposited in the files of PORC.

History: Libhart (1869) reported oystercatchers as occasional stragglers in autumn in Lancaster Co. One was shot on Chester Island in the Delaware River on 14 May 1891 (Stone 1894).

[1] Wiltraut, R. 1994. First documented record of American Oystercatcher in Pennsylvania. PB 7:51.

Family Recurvirostridae: Stilts and Avocets

Stilts and avocets are slender-bodied shorebirds with striking contrasting plumage; a sleek, long neck; long, thin, straight or recurved bill; and very long spindly legs. Birds in this family are very active while feeding, yet graceful in their behavior. They feed on mudflats or in shallow water by picking at the surface or by sweeping their bill from side to side. Two of the eight species known worldwide inhabit North America—the American Avocet and the Black-necked Stilt—and both species have been recorded in Pennsylvania. Black-necked Stilts recently nested in the southeastern corner of the state.

Black-necked Stilt ***Himantopus mexicanus***

General status: Black-necked Stilts breed locally throughout the west and in the east primarily from Delaware south along the Atlantic and Gulf coasts. They winter along the Pacific coast of Southern California and Mexico and along the Gulf coast east to Florida. The Black-necked Stilt has entered the state only in the Coastal Plain and Piedmont. Black-necked Stilts are listed as casual in Pennsylvania. There have been at least 12 records in Pennsylvania since 1960. They have nested in Philadelphia Co.

Habitat: Stilts prefer mudflats and shorelines of ponds with grassy or muddy

American Oystercatcher at Beltzville State Park, Carbon County, April 1993. (Photo: Rick Wiltraut)

edges and are often found at sewage treatment ponds.

Seasonal status and distribution

Spring: All but one sighting has been in the spring or early summer. Observations range from the last week of Apr to the fourth week of May. In the Coastal Plain they have been recorded in Philadelphia and Delaware cos. In the Piedmont they have been recorded from Nazareth in Northampton Co.[1] and at MCWMA,[2] Holtwood,[2] and the Conejohela Flats in Lancaster Co.[3] In York Co., two birds were seen at Spring Grove.[4] Singles or pairs usually are found, but as many as 10 individuals were photographed at the Holtwood Flash Pond in Lancaster Co. on 20 May 1990 (Morrin et al. 1991), and a high count of nine stilts was observed on 10 May 1990 at the Philadelphia sewage treatment plant; four birds remained to the end of the month.[5]

Breeding: One apparently nonbreeding Black-necked Stilt was found on 13 Jul in 1961 at Tinicum.[6] Black-necked Stilts nested successfully in Pennsylvania only at the Philadelphia sewage treatment plant in 1989.[7] A nesting pair was found by S. Santner and colleagues on 12 Jun 1989, and a fledgling was observed on 24 Jun. A second nest was found on 29 Jun, but heavy rains flooded the area in Jul and the chick was not seen again. Black-necked Stilts had been present in the area in 1988 and constructed, but abandoned, a nest in 1990. Nesting was attempted at least twice since then[3] at the sewage plant (not at JNHWR), but they have not been observed in this area since 1992. Black-necked Stilts expanded their range northward during recent decades; nesting was first confirmed in Delaware in 1969[8] and in Maryland in 1987.[9]

Fall: The only fall (nonbreeding) records in recent history were in York Co. at Brunner's Island on 23 Aug 1961[10] and at Lake Redman on 11 Jul 1996.[11]

History: A Black-necked Stilt was observed at Glenolden in Delaware Co. on 28 Sep 1938 (Poole 1964). Libhart (1869) reported stilts as summer stragglers in Lancaster Co.

[1] Paxton, R.O., W.J. Boyle Jr., and D.A. Cutler. 1981. Hudson-Delaware region. AB 35:807.

[2] Witmer, E. 1990. County reports—April through June 1990: Lancaster County. PB 4:72.

[3] Boyle, W.J., Jr., R.O. Paxton, and D.A. Cutler. 1992. Hudson-Delaware region. AB 46:399.

[4] Spiese, A. 1991. County reports—April through June 1991: York County. PB 5:104.

[5] Fingerhood, E. 1990. County reports—April through June 1990: Philadelphia County. PB 4:81.

[6] Scott, F.R., and D.A. Cutler. 1961. Middle Atlantic Coast region. AFN 15:457.

[7] Santner, S. 1992. Black-necked Stilt *(Himantopus mexicanus)*. BBA (Brauning 1992): 134–135.

[8] Scott, F.R., and D.A. Cutler. 1969. Middle Atlantic Coast region. AFN 23:646–649.

[9] Armistead, H.T. 1987. Middle Atlantic Coast region. AB 41:1420.

[10] A. Spiese, pers. comm.

[11] Spiese, A. 1996. Local notes: York County. PB 10:168.

American Avocet *Recurvirostra americana*

General status: American Avocets breed in the West and formerly bred in the East in Virginia and North Carolina. They winter along the Pacific coast of California and Mexico and along the Gulf coast from Louisiana to Florida. American Avocets are rare but regular visitors only on Gull Point at PISP. Recently they have been observed irregularly at sites away from PISP. They have been recorded in all regions of the state except the Glaciated Northeast.

Habitat: They are usually observed on sandy beaches, but can also be seen on muddy or grassy shorelines of ponds and lakes and on mudflats.

Seasonal status and distribution

Spring: There are at least seven spring records of American Avocet in Pennsyl-

American Avocet at Allentown, Lehigh County, November 1991. (Photo: Rick Wiltraut)

vania. Three birds were at Tinicum in Delaware Co. on 30 Apr 1969,[1] and 5 were found at Lake Arthur in Butler Co. on 25 Apr 1982.[2] At YCSP in Indiana Co., one was seen on 3 May 1996.[3] Four spring records came from PISP in Erie Co.: a flock of 9 was seen on 29 Apr 1986,[4] one flew in during a snow storm on 7 May 1989,[5] one was seen on 22 Apr 1992,[6] and a single flock containing 18 birds was reported on 30 Apr 1996.[7]

Fall: Fall sightings range from the third week of Jul to the second week of Nov. One to six birds are usually recorded in each sighting, and most do not stay for more than a day. East of the Allegheny Front they have been recorded in fall in Berks, Cumberland, Delaware, Fulton, Lancaster, Lehigh, Lebanon, Montgomery, Montour, Northampton, Philadelphia, and York cos. There are two sightings from the northern-tier counties: a bird at Wellsboro in Tioga Co. on 5 Nov 1979[8] and one at Dimock in Susquehanna Co. on 15 Sep 1996.[9] West of the Allegheny Front American Avocets have been recorded in fall in Allegheny, Butler, Cambria, Erie, Fayette, Mercer, Somerset, Washington, and Westmoreland cos.

History: Libhart (1869) reported the occurrence of this species along the Susquehanna River in Lancaster Co. but did not cite any specific records. The first specimen from Pennsylvania was a bird shot on 6 May 1905 at Conneaut Lake in Crawford Co. An American Avocet was found at Presque Isle in Erie Co. on 21 Sep 1936 (Todd 1940). There were three records from Berks Co., all at Lake Ontelaunee: one on 20 Sep 1947, one on 31 Oct 1953, and another on 25 Sep 1957 (Poole, unpbl. ms.).

[1] Merrit. J.K. 1970. Field notes. Cassinia 52:24.
[2] Hall, G.A. 1982. Appalachian region. AB 36:852.
[3] Higbee, M., and R. Higbee 1996. Local notes: Indiana County. PB 10:101.
[4] Hall, G.A. 1986. Appalachian region. AB 40:471.
[5] Hall, G.A. 1989. Appalachian region. AB 43:1315.
[6] Hall, G.A. 1992. Appalachian region. AB 46:422.
[7] Haas, F.C., and B.M. Haas, eds. 1996. Birds of note—April through June 1996. PB 10:90.
[8] Hall, G.A. 1980. Appalachian region. AB 34:161.
[9] W. Reid, pers. comm.

Family Scolopacidae: Sandpipers and Phalaropes

Scolopacidae, the largest family of shorebirds, consists of more than 60 species in North America that vary greatly in size, color, and shape. Thirty species have occurred in Pennsylvania, but only Spotted Sandpiper, Upland Sandpiper, Common Snipe, and American Woodcock breed in the state. The bill of birds in this family differs from the plovers' bill by being thinner and softer and flexible throughout its length and by lack of the bulbous outer half and the horny tip. Body size and color vary among species, as do bill length and curvature, from short and straight, to long and decurved (Whimbrel, curlews), or upturned (godwits). Most are gregarious and are found in flocks along exposed edges of ponds, lakes, and oceans, or wherever there are mud- or sand-flats. Some species may be found in the hundreds at some staging sites. Most species have three distinct plumages: winter, breeding, and juvenile (first complete molt after fledging). They feed on a variety of small invertebrates by picking or probing in water or mud. Many of the smaller sandpipers, collectively known as peeps, are a challenge to identify in the field.

The subfamily Phalaropodinae (phalaropes) differs from other members in this family by having lobed toes with webs at the bases. This adaptation is well suited for swimming, which they do the majority of time. Phalaropes are unique in that the males are smaller and duller than the females. When feeding, they rapidly swim in circles to bring food to the surface. Only three species are known worldwide, and all regularly occur in Pennsylvania.

Greater Yellowlegs ***Tringa melanoleuca***

General status: Greater Yellowlegs breed from southern Alaska south along the Pacific coast and east across central Canada to Newfoundland. They winter in the West along the Pacific coast of the U.S. and Mexico and in the East from southern New England south along the Atlantic and Gulf coasts. The larger and less common of the two species of yellowlegs, Greater Yellowlegs, are uncommon to fairly common regular migrants in Pennsylvania. They are widely distributed throughout the state and often are found in company with their smaller relative, Lesser Yellowlegs.

Habitat: In early spring they are frequently found in low-lying open pastures or fields with shallow pools of water formed by melting snow, heavy rain, or flooding. In late spring and fall they inhabit grassy edges of ponds, lakes, and rivers and mudflats.

Seasonal status and distribution

Spring: Greater Yellowlegs are among the earliest shorebirds to arrive, occasionally arriving as early as the first week of Mar. Their typical migration period is from the fourth week of Mar to the third week of May. Peak migration usually occurs from the first week in Apr to the second week of May. Stragglers have been recorded to late Jun.

Fall: This is one of the first shorebirds to return south from their breeding grounds. Adult Greater Yellowlegs may be seen as early as the first week of Jul, but most arrive the second or third week of Jul. The earliest juveniles are seen the second week of Aug, usually replacing the adults by late Aug or early Sep. Peak migration is difficult to discern, but the largest concentrations of birds are usually found in Aug or Sep. They are most common during years of drought, when water levels are low. At some sites high counts have reached well over a hundred birds, with as many as 185 recorded in Delaware Co. on 13 Aug 1977.[1] Migration is usually over by the second or third week of Nov, with stragglers to the last week of Dec.

History: Their status and distribution have remained relatively unchanged. At Tinicum, where large flocks have been known to occur, about 600 were present on 10 May 1953 (Poole, unpbl. ms.).

[1] N. Pulcinella, pers. comm.

Lesser Yellowlegs ***Tringa flavipes***

General status: The breeding range of the Lesser Yellowlegs is from Alaska south across central Canada to Quebec. They winter along the Pacific coast of Southern California and Mexico and from South Carolina south along the Atlantic and Gulf coasts. In Pennsylvania they are fairly common to common regular migrants and are accidental in winter. Lesser Yellowlegs are widely distributed throughout the state, often found in company with their larger relative, the Greater Yellowlegs. At most sites, Lesser are more common than Greater yellowlegs.

Habitat: In early spring they are frequently found in low-lying open pastures or fields with shallow pools of water formed by melting snow, heavy rain, or flooding. In late spring and fall they inhabit grassy edges of ponds, lakes, and rivers and mudflats.

Seasonal status and distribution

Spring: Lesser Yellowlegs arrive somewhat later than Greater Yellowlegs, with the first northbound birds appearing the third week of Mar. Migration continues to the fourth week of May. Concentrations containing 100 or more birds are not uncommon, especially at Tinicum in Delaware Co.[1] A high count of about 200 birds was observed on 5 May 1967 at Conneaut Marsh, Crawford Co.[2] Stragglers have remained to the last week of Jun.

Fall: The first southbound arrivals may be seen as early as the first week of Jul. Most begin migrating through the state the third or fourth week of Jul. First juveniles have been recorded as early as the first of Aug.[3] The majority arrive the second week of Aug and usually replace adults by late Aug or early Sep. Peak migration is difficult to discern, but largest concentrations of birds are usually found in Aug or Sep. During years of drought, when lake levels are low, they are more common. On 27 Aug 1961 at the Conejohela Flats in Lancaster Co., 250 were counted (Morrin et al. 1991). Migration continues to about the third week of Oct. Some birds have lingered until as late as the first week of Jan.

Winter: Entering the winter period were seven birds observed at Tinicum on 10 Jan 1966.[4] Two birds spent the winter of 1985 along Mill Creek near New Holland in Lancaster Co. (Morrin et al. 1991). One was present on 7 Feb 1998 in the "Amish farm area" in Lancaster Co.[5]

History: Their status and distribution have remained relatively unchanged. According to E. Manners,[6] in spring they were historically recorded as early as Feb. Poole (unpbl. ms.) gives a date of 22 Feb 1951 at Tinicum for this species.

[1] N. Pulcinella, pers. comm.
[2] Leberman, R.C., and R.F. Leberman. 1967. Field notes. Sandpiper 9:55.
[3] G.M. McWilliams, pers. obs.
[4] Scott, F.R., and D.A. Cutler. 1966. Middle Atlantic Coast region. AFN 20:407.
[5] Haas, F.C., and B.M. Haas, eds. 1998. Birds of note—January through March 1998. PB 12:14.
[6] Manners, E.R. 1945. Shorebirds along the Delaware. Cassinia 35:23–34.

Solitary Sandpiper ***Tringa solitaria***

General status: As the name implies, Solitary Sandpipers usually do not often associate with other species of shorebirds. They breed from Alaska south across most of Canada and east to Labrador. They winter from northern Mexico south to Argentina. In Pennsylvania, Solitary Sandpipers are occasionally found in pairs or in loose groups, rarely numbering more than 5 to 10 birds.

Solitary Sandpipers are uncommon regular migrants.

Habitat: Unlike most species of shorebirds, they can be found wherever there is a puddle of water, including parking lots, ditches, and lawns. They prefer more secluded sites than their relatives the yellowlegs and are not as frequently found in open mudflats. Grassy or muddy shorelines of marshes, swamps, ponds, lakes, woodland streams, and rivers, and flooded fields and pastures are used.

Seasonal status and distribution

Spring: Solitary Sandpipers may arrive as early as the first week of Apr. Their usual migration period is from the third week of Apr to the third week of May. Peak migration ranges from the last week of Apr to the second week of May. A high count of 55 birds was recorded at the Philadelphia sewage ponds on 16 May 1980.[1] Stragglers may linger to the first week of Jun.

Fall: The earliest southbound migrants appear the first week of Jul, and juveniles first appear the second week of Aug. Peak migration occurs in Aug, and migration ends by the third week of Oct. Lingering Solitary Sandpipers have been recorded to the first week of Dec.

History: On the basis of observations in Jun, several authorities believed, probably erroneously, that Solitary Sandpipers nested in the northern-tier counties. Harlow claimed to have seen three Solitary Sandpipers that were still too young to fly in western Pike Co. in the summer of 1905 (Harlow 1906). Todd (1940) saw a pair on a small pond adjacent to Two Mile Run in Beaver Co. in Jul and stated that "they appeared accompanied by what were presumably their young," but he went on to say that his observations did not conclusively indicate local breeding. Sutton (1928b) stated that he had seen what he believed was a female at Lake Pymatuning in 1922 entering a deserted nest of a Hairy Woodpecker, but neither eggs nor young were found. These and other observations suggest that Solitary Sandpipers irregularly summered in the state and may, in fact, have bred at one time. But definite evidence is lacking. The nearest known breeding range to Pennsylvania is in central Quebec and Ontario (Hayman et al. 1986).

[1] Paxton, R.O., W.J. Boyle Jr., and D.A. Cutler, and K.C. Richards. 1980. Hudson-Delaware region. AB 34:759.

Willet *Catoptrophorus semipalmatus*

General status: The western population of Willets breed mostly in south-central Canada and the northwestern states and they winter along the Pacific coast. The eastern population breeds locally along the Atlantic coast from Nova Scotia south along the Atlantic and Gulf coasts and winters from Virginia south along the coast. Willets are rare but regular vagrants in Pennsylvania. Most sightings are in the fall at Gull Point on PISP, but observations elsewhere have become more frequent in recent years. Most sightings are of single birds, but as many as 12 have been recorded at one time: YCSP on 30 Apr 1991.[1] There are usually no more than three to five reports of Willets annually in the state.

Habitat: Sandy beaches, mudflats, and grassy or muddy shorelines of lakes are preferred habitats.

Seasonal status and distribution

Spring: Recorded dates range from the first week of Apr to the first week of Jun. Most recent records are from the first to the third week of May. It is uncertain whether records from the latter part of the third week of Jun to the end of Jun are lingering spring vagrants or early fall migrants. No record exists for the second week of Jun. In spring Willets have been reported from Centre, Chester, Crawford, Erie, Franklin, Indiana, Lancaster, Northampton, Philadelphia, Venango, and Westmoreland cos.

Fall: Observations range from the first week of Jul to the last week of Oct, with most records in Jul and Aug. The following counties have recorded Willets only in the fall: Beaver, Bedford, Berks, Butler, Delaware, Dauphin, Lebanon, Lehigh, McKean, Mercer, Montgomery, Montour, and Wyoming.

History: Willets were considered a rare transient by most authorities. Libhart (1869) reported that they occasionally had been seen in large flocks in Aug.

[1] Higbee, M., and R. Higbee. 1991. County reports—April through June 1991: Indiana County. PB 5:88.

Spotted Sandpiper ***Actitis macularia***

General status: Their presence on gravel or rocky shorelines and constant bobbing motion make Spotted Sandpipers fairly easy to separate from other species of shorebirds. They breed widely across North America except in the southern portions of the Gulf states and from North Carolina south. They winter across the southern states south of their breeding range. In Pennsylvania, Spotted Sandpipers are uncommon to fairly common regular migrants and breeding residents.

Habitat: They are found near the edge of water, especially along the open shoreline of ponds, lakes, and streams in migration and during the breeding season. Unlike most other shorebird species, they are frequently found on gravelly or rocky edges of shorelines.

Seasonal status and distribution

Spring: Spotted Sandpipers have arrived as early as the second week of Apr, but their typical migration period is from the third week of Apr to the end of May, with lingering birds to early Jun. Peak migration may occur any time in May. The largest concentrations during migration may contain over 30 birds.

Breeding: They are widespread but uncommon breeders in Pennsylvania. Spotted Sandpipers are poorly represented on BBS routes, reported at a rate of 0.09 birds per route statewide, on just 16 routes, but they may be found in every Pennsylvania county. No population trend is evident, but they apparently are not as common anywhere in Pennsylvania as they were described to be during the nineteenth century. The nest is a grass-lined depression on the ground, sometimes away from water. A female may lay multiple clutches of heavily brown-splotched eggs. Most eggs were found in the second half of May to early Jun. Young are precocial, quickly on their feet and moving about after they hatch.

Fall: The overlap of resident families and migrant birds makes it difficult to discern the exact arrival dates of southbound migrants, but the number of birds increases during the last week of Jul. The number of birds remains high through Aug and to the first or second week of Sep. By the second week of Oct most have left the state. Some birds linger through Nov. During mild falls they have been recorded as late as the last week of Dec. An exceptionally late bird remained along the Schuylkill River to 7 Jan 1989.[1]

Winter: A hypothetical record exists of a bird, believed to be a Spotted Sandpiper, that was discovered on 7 Feb 1984 in Kutztown, Berks Co.[2]

History: Historically Spotted Sandpipers were considered very common (Baird 1845). Todd (1940) said, "There is no part of our region [western Pennsylvania] where it is not found." He further stated that it was more numerous (in the early 1900s) than ever because of its ability to adapt to changing conditions. For example, nests were found in grain fields. Grimm (1952) thought that the population was much greater around the flooded Pymatuning Reservoir than it had been in the 1930s or before. However, the lack of quantitative data and the loose use of terminology, such as the term *common* by many early authors make interpretation difficult. Some authors of local bird accounts, such as Burleigh (1931) for Centre Co., during this time listed the species as rare. Possibly the novelty of encountering a shorebird in central Pennsylvania biased the observer's impression of abundance. Spotted Sandpipers apparently responded favorably to European settlement, benefiting directly from the creation of lakes and ponds, damming of streams, and flooding of wetlands. In contrast, Frey (1943) suggested that they were less common in Cumberland Co. than indicated by Baird (1845). Poole (unpbl. ms.) cited one winter record: one seen at Lake Ontelaunee on 19 Jan 1947.

[1] Boyle, W.J., Jr., R.O. Paxton, and D.A. Cutler. 1989. Hudson-Delaware region. AB 43:291.
[2] Boyle, W.J., Jr., R.O. Paxton, and D.A. Cutler. 1984. Hudson-Delaware region. AB 38:300.

Upland Sandpiper ***Bartramia longicauda***

General status: Upland Sandpipers breed locally from Alaska through central Canada and the Great Plains east to Maryland and north to New Brunswick. They winter in eastern South America to Argentina. Upland Sandpipers have experienced dramatic population changes in Pennsylvania, increasing with the clearing of forests in the nineteenth century and then decreasing with the advent of the use of pesticides and changes in farming practices. They are now uncommon to rare regular migrants and local breeders, primarily in the west-central portion of the state.

Habitat: Upland Sandpipers prefer open pastures and grassy fields, especially reclaimed strip mines, airports, golf courses, and freshly plowed fields.

Seasonal status and distribution

Spring: It is difficult to determine how many birds actually migrate through Pennsylvania to breeding areas farther north. Birds usually appear in the state during the first or second week of Apr but have been seen as early as the third week of Mar. The majority of migrants are observed as individuals or groups of 5 to 10 birds. On 25 Apr 1989 three large flocks of Upland Sandpipers, one of 50 or 60 birds, were seen landing in a large field along White Chapel Road in Mercer Co.[1]

Breeding: Upland Sandpipers are rare breeding birds, with scattered nesting sites almost statewide. They are more likely to be found in fields 150 acres or larger (Jones and Vickery, undated) than in smaller fields. Upland Sandpipers are regular at traditional sites in Adams, Clarion, Erie, Lawrence, Somerset, and Westmoreland cos. Most regularly occupied areas now are on reclaimed strip mines. Rarely are more than one or two pairs found in a field until migration, when family groups gather into flocks or are joined by migrants. The nest is in a depression on the ground, well hidden in a clump of grass. Eggs are marked similarly to, but are nearly twice the size of, meadowlark eggs. Most nests are active by the third week of May, with one brood per year. Young are precocial.

Fall: More fall migrants are reported from the southeastern portion of the state than from any other area. The usual migration period is from the third or fourth week of Jul to the second week of Sep. The greatest number of birds is recorded during the first half of Aug. Flocks rarely exceed 10 to 15 birds, but on 1 Aug 1983, 56 birds were counted in cut alfalfa fields near Intercourse in Lancaster Co. (Morrin et al. 1991). Recent high counts are far fewer than the counts containing several hundred birds recorded at the turn of the twentieth century and before. Stragglers may remain to the third week of Oct.

History: Upland Sandpipers quickly colonized newly created agricultural areas in the East. They were included on Audubon's list of the Philadelphia area in the early 1800s and were called "abundant" by Baird (1845) in Cumberland Co. Bird lists from the southeastern counties in the late 1800s described this species as "common" (Pennock 1886 in Chester Co.) or "frequent" (Thomas 1876 in Bucks Co.; Mombert 1869 in Lancaster Co.). At that time, Upland Sandpipers were common breeding birds throughout most of the northeastern states.[2] Market hunting of migrants and persecution on winter grounds greatly reduced their population (Harlow 1913; Todd 1940). Upland Sandpipers were listed most widely and commonly during the period of greatest agricultural acreage in the state, during the late 1800s. Beck (1924) said that it was not uncommon to find 300–400 birds in Manheim or Warwick townships in Lancaster Co. as recently as 1900. Declining agricultural acreage in the early 1900s reduced available habitat, and changing agricultural practices since the 1950s contributed to making farmland less suitable. By 1960 nowhere in the state was the Upland Sandpiper considered common.

Comments: Upland Sandpipers are listed as Threatened by the PGC.

[1] Dean, B. 1989. 104 Upland Sandpipers in Mercer County. PB 3:54.

[2] Carter, J.W. 1992. Upland Sandpiper, *Bartramia longicauda*. Pages 235–252 *in* Migratory nongame birds of management concern in the Northeast, ed. K.J. Schneider and D.M. Pence. U.S. Fish and Wildlife Service, Newton Corner, Mass.

Eskimo Curlew *Numenius borealis*

General status: Probably extinct, Eskimo Curlews formerly bred in northwestern Mackenzie, possibly west to western Alaska. Migration in spring was primarily through the Mississippi River valley and west of the Great Lakes and Hudson Bay. Migration in fall was west of the Hudson Bay and from Labrador and the Gulf of St. Lawrence, casually to the Great Lakes, and along the Atlantic coast to the West Indies. The inclusion of this species is based upon a single specimen that was shot by James Thompson at Erie on 17 Sep 1889 (Todd 1940). This specimen is currently at the CMNH.

History: Once abundant in North America, they were greatly reduced in numbers by shooting between 1850 and 1890 and possibly by habitat change. By 1929, Eskimo Curlews were nearly extinct, but they were still occasionally seen in migration in the 1980s (Hayman et al. 1986). Warren (1890) said, "A few of these birds seen every year about the shores of Erie bay, where, in October 1889, two were shot by Mr. James Thompson, of Erie City." This statement probably refers to the same incident noted by Todd, but with Warren citing the wrong month. Todd (1940) stated that "its occurrence in our region could scarcely have been more than accidental."

Whimbrel *Numenius phaeopus*

General status: In North America, Whimbrels breed in northern Alaska, northern Yukon, and northwestern Mackenzie and the western side of Hudson Bay. They winter along the Pacific cost from Oregon south through Mexico and along the Atlantic coast from South Carolina south and along the Gulf coast. Whimbrels are known in Pennsylvania primarily from the Lake Erie Shore at Gull Point on PISP. At this site they are uncommon to rare regular migrants, but in some years they may be observed in rather large flocks. They are casual along the lower Susquehanna River. Elsewhere in the state they are accidental; they have been re-

Whimbrel at Presque Isle State Park, Erie County, July 1992. (Photo: Rick Wiltraut)

corded in Berks, Butler, Centre, Crawford, Delaware, Indiana, Juniata, Lehigh, Luzerne, Montour, Philadelphia, and Westmoreland cos.

Habitat: Whimbrels prefer open sandy beaches, mudflats, shorelines of lakes, and large level fields.

Seasonal status and distribution

Spring: A few historical sightings of Whimbrels were made in Apr, but most occur after the second week of May. They appear every year at Gull Point on PISP, usually within a three- or four-day period within the third and fourth weeks of May. Distant flocks occasionally are seen flying past PISP over Lake Erie. The number of birds varies from year to year, from a few individuals to flocks containing 50 or more. Very unusual away from the Lake Erie Shore were flocks of 270 birds seen in an evening flight north of Bowmansville in Lancaster Co. on 23 May 1995[1] and of 81 birds that were observed on 28 May 1984 at BESP in Centre Co.[2] On 19 May 1986 at PISP, a flock of 331 Whimbrels was counted.[3] Stragglers have been recorded until the third week of Jun.

Fall: The large numbers seen in spring are not seen during fall migration. Most sightings are of single birds or small groups containing fewer than five or six. Whimbrels are recorded as early as the first week of Jul. Their usual migration period is from the third or fourth week of Jul to the second week of Sep. Stragglers remain to the third week of Oct.

History: Whimbrels were recorded at least once in Bucks, Lebanon, and Tioga cos. prior to 1960.

[1] Haas, F.C., and B.M. Haas, eds. 1995. Birds of note—April through June 1995. PB 9:87–88.

[2] Schiefer, T.L. 1984–1985. Large number of Whimbrel and late spring birding in Centre Co., Pa. Cassinia 61:78.

[3] Hall, G.A. 1986. Appalachian region. AB 40:470.

[Long-billed Curlew] ***Numenius americanus***

General status: Long-billed Curlews breed from British Columbia and Saskatchewan south to Texas. They winter from Washington south to northern Mexico, east along the Gulf coast to Florida, and north along the coast to South Carolina. They are listed as hypothetical in Pennsylvania.

Comments: This species was apparently a rare transient before 1900. Libhart (1869) listed this species as rare in Lancaster Co. Frey (1943) listed Long-billed Curlews as accidental on the lower

Susquehanna River (three shot in the 1890s). Beck (1924) wrote of specimens taken by sportsmen, and Poole (unpbl. ms.) stated that "these are probably the specimens now in the North Museum at Lancaster." Currently there is one specimen in the North Museum; it may have been taken in Pennsylvania, but data are lacking. It is likely that some reports of Long-billed Curlews were misidentified Whimbrels, since none of the above authors mentioned those, then known as Hudsonian Curlews, in their reports.

Black-tailed Godwit ***Limosa limosa***

General status: Black-tailed Godwits breed in grassy wetlands from Iceland, Britain, France, the Netherlands, and Denmark west through central Russia. They are vagrants to eastern North America (Chandler 1989). One record of this species has been reported in Pennsylvania.

Seasonal status and distribution

Fall: One was observed repeatedly by J. C. Miller and colleagues around Tinicum and the Philadelphia Sewage Ponds from 16 to 26 Oct 1979. Unfortunately, photographs taken of the bird proved to be inconclusive.[1] However, written descriptions clearly confirmed the identification of the bird as a Black-tailed Godwit.

[1] Paxton, R.O., K.C. Richards, and D.A. Cutler. 1980. Hudson-Delaware region. AB 34:146.

Hudsonian Godwit ***Limosa haemastica***

General status: Hudsonian Godwits breed locally in Alaska, Mackenzie, northwestern British Columbia, and the west side of Hudson Bay. They winter along the coasts of South America. In Pennsylvania, Hudsonian Godwits are almost always found during the fall, when they are rare but regular vagrants. They are accidental in spring. Hudsonian Godwits do not occur regularly at any site in Pennsylvania. Most sightings are at Tinicum, at several sites in the Piedmont, and at PISP.

Habitat: Hudsonian Godwits prefer open sandy beaches, mudflats, and shorelines of lakes.

Hudsonian Godwit at John Heinz National Wildlife Refuge, Philadelphia County, October 1986. (Photo: Franklin C. Haas)

Seasonal status and distribution

Spring: Two records are from this season, both from PISP in Erie Co. One was present on 17–18 May 1982[1] and one was seen on 15 and 17 May 1983.[2]

Fall: Fall dates range from the second week of Jul to the second week of Nov. Most sightings are of single birds during Sep and Oct. However, a high count of 32 birds was at PISP on 2 Oct 1977,[3] and 24 passed through Tinicum during the fall of 1986, when the impoundment was drained.[4] Fall records from the Coastal Plain and Piedmont have come from Berks, Bucks, Chester, Cumberland, Delaware, Lancaster, Lebanon, Lehigh, Montgomery, Philadelphia, and York cos. In the Ridge and Valley, Hudsonian Godwits have been recorded in fall in Centre and Clinton cos. In Centre Co. they have been recorded in fall at BESP in 1971, 1972, and 1992 and at the Penn State sludge ponds in 1975.[5] In Clinton Co., 12 birds were at Lock Haven on 1 Oct 1975.[6] West of the Allegheny Front they have been recorded at only two sites away from Lake Erie. One was photographed at YCSP in Indiana Co. on 2 Nov 1991.[7] They have been recorded during 1971, 1978, 1986, and 1991 in the Pymatuning area in Crawford Co.[8]

History: All historical records are from the Coastal Plain and Piedmont. Before 1900 there were about three reports from the vicinity of Philadelphia[9] (Warren 1890; Stone 1894). Between 1900 and 1960 they have been recorded about 10 times (Poole, unpbl. ms.).

[1] Hall, G.A. 1982. Appalachian region. AB 36:852.

[2] Hall, G.A. 1983. Appalachian region. AB 36:870.

[3] Hall, G.A. 1978. Appalachian region. AB 32:204.

[4] Paxton, R.O., W.J. Boyle Jr., and D.A. Cutler. 1987. Hudson-Delaware region. AB 41:65.

[5] J. Peplinski and B. Peplinski, pers. comm.

[6] Hall, G.A. 1976. Appalachian region. AB 30:69.

[7] Higbee, M., and R. Higbee. 1992. County reports—October through December 1991: Indiana County. PB 5:174.

[8] J. Barker, pers. comm.

[9] Wood, C.D. 1879. Bulletin Nuttall Ornithological Club 1879:235.

Marbled Godwit *Limosa fedoa*

General status: Marbled Godwits breed in Alaska, south-central Canada, the southern end of James Bay, and in the north-central U.S. They winter along the Pacific coast from Oregon south through Mexico and along the Atlantic coast from South Carolina south along the Gulf coast. In Pennsylvania, Marbled Godwits are casual at Tinicum, along the lower Susquehanna River, and at PISP. They are accidental elsewhere in the state. Most sightings are of one or two birds in the fall.

Habitat: Marbled Godwits are found on open sandy beaches and mudflats, along the shoreline of lakes, and occasionally in open wet fields.

Seasonal status and distribution

Spring: There are at least six spring records, all in recent years. Two birds were in the Lancaster Co. section of the MCWMA on 27 Apr 1980 (Morrin et al. 1991), and the same two birds (probably) were seen in the Lebanon Co. section of the MCWMA on 4 May 1980.[1] One was at Long Arm Dam in York Co. on 21 Apr 1990.[2] At Prince Gallitzin State Park in Cambria Co., a single bird was observed on 24 Apr 1995.[3] Two were at PISP on 18 May 1990,[4] and three were present there on 12 May 1993.[5]

Fall: Marbled Godwit observations range from the first week of Jul to the first week of Nov, but most sightings are from Jul through Sep. Five fall records of single birds have been accepted away from the Coastal Plain, the lower Susquehanna River, and the Lake Erie Shore: at South Avis from 11 to 13 Oct 1976,[6] at Lock Haven in Clinton Co. from 9 to 17 Aug 1981,[7] in Butler Co. on 27 Aug 1972,[8] near Scandia in Warren Co. on 9 and 10 Oct 1971,[9] and at the

Linesville Fish Hatchery, Crawford Co. on 18–20 Aug 1978.[10]

History: Authorities have always considered this species to be very rare. Warren (1890) reported Marbled Godwits at many of the same sites where they have been recorded in recent times. He reported that they were occasionally taken from Crawford Co.: at least from Conneaut Marsh, and two birds were taken from Pymatuning Swamp on 2 Oct 1929 (Todd 1940).

[1] Paxton, R.O., W.J. Boyle Jr., D.A. Cutler, and K.C. Richards. 1980. Hudson-Delaware region. AB 34:760.
[2] A. Spiese, pers. comm.
[3] Haas, F.C., and B.M. Haas, eds. 1995. Birds of note—April through June 1995. PB 9:88.
[4] Hall, G.A. 1990. Appalachian region. AB 44:426.
[5] Haas, F.C., and B.M. Haas, eds. 1993. Rare and unusual bird reports. PB 7:59.
[6] Pulcinella, N. 1996. Seventh Report of the Pennsylvania Ornithological Records Committee (PORC) June 1996. PB 10:49.
[7] Hall, G.A. 1982. Appalachian region. AB 36:177.
[8] Hall, G.A. 1973. Appalachian region. AB 27:61.
[9] T. Grisez, pers. comm.
[10] Hess, P.D. 1978. Area bird summaries for August. ASWPB 43:6.

Ruddy Turnstone *Arenaria interpres*

General status: In North America Ruddy Turnstones breed from northern Alaska across the Canadian Arctic islands, and they winter along the Pacific coast from Oregon south through Mexico and along the Atlantic coast from southern New England south along the Gulf coast. The greatest number of Ruddy Turnstones in the state is found along the lower Susquehanna River and at the Lake Erie Shore, where they are uncommon to rare regular migrants. Ruddy Turnstones are irregular to casual over the remainder of the state except in the High Plateau, where they are accidental. Most sightings away from the Lake Erie Shore are of birds put down by severe storms. One winter record exists in Pennsylvania.

Habitat: Ruddy Turnstones are found on sandy beaches, mudflats, gravel bars, and edges of ponds, lakes, and rivers. They are occasionally seen in migration at hawk watches along the Kittatinny Ridge in fall.

Seasonal status and distribution

Spring: Early migrants may appear the first week of May. Their typical migration period is from the third week of May to the first week of Jun. Migration peaks during the third or fourth week of May. Concentrations of over 30 birds have been recorded. Forty were on the Conejohela Flats on 30 May 1958 (Morrin et al. 1991), and 25 birds were counted at YCSP on 1 Jun 1993.[1] At PISP, a total of 55 was counted on 19 May 1986.[2] Stragglers have been observed until the fourth week of Jun.

Fall: Adults may appear the first week of Jul, but they usually arrive the third week of Jul. Juvenile Ruddy Turnstones are seen as early as the third week of Aug. There are two distinct peak migration periods: the first, during the first and second weeks of Aug, consists of adults, and the second, usually during the first and second weeks of Sep, consists of juveniles. Concentrations of 30 or more birds have been recorded: 70 birds were counted on 17 Sep 1981 migrating past the Bake Oven Knob hawk watch in Lehigh Co.,[3] and 132 were counted on 3 Aug 1992 flying past Gull Point on PISP in Erie Co.[4] Most Ruddy Turnstones have left the state by the third week of Sep, but stragglers may remain to the first week of Nov. A very late Ruddy Turnstone was seen on the Glenolden CBC in Delaware Co. on 26 Dec 1970.[5]

Winter: The only winter record was of one bird observed feeding among gulls along the edge of the bay ice on 26 Jan 1991 at PISP in Erie Co.[6]

History: Several authors considered Ruddy Turnstones to be rare and irregular. Todd (1940) considered this species to be a regular, but not common, transient in the fall along the beaches of Presque Isle. An unusual inland count was 75 birds seen on 24 May 1937 at Pymatuning Lake (Todd 1940). In eastern Pennsylvania, records came from the lower Susquehanna River, Lake Ontelaunee in Berks Co., Lebanon Co., Tinicum in Delaware and Philadelphia cos., and Pocono Lake in Monroe Co. (Poole, unpbl. ms.). Historical records from western Pennsylvania came from Conneaut Lake in Crawford Co., at Warren in Warren Co., and in Fayette Co. (Todd 1940).

[1] Haas, F.C., and B.M. Haas, eds. 1993. Rare and unusual bird reports. PB 7:59.
[2] McWilliams, G.M. 1986. International Shorebird Survey (Erie, Pa. site). Files of Manomet Bird Observatory, Mass.
[3] Heintzelman, D.S. 1980–1981. More autumn bird records from Bake Oven Knob, Pennsylvania (1976–1981). Cassinia 59:85.
[4] Hall, G.A. 1993. Appalachian region. AB 47:93.
[5] Rigby, E.H. 1971. Glenolden CBC. AFN 25:227.
[6] McWilliams, G.M. 1991. County reports—January through March 1991: Erie County. PB 5:38.

Surfbird ***Aphriza virgata***

General status: Surfbirds breed on rocky tundra above treeline in the mountains of central Alaska and the Yukon Territory, and they winter along the coast from Alaska south to southern Chile. They are extremely rare away from their breeding and wintering range in North America but have been recorded in central Alberta, interior California, the Gulf coast of Texas, Florida, and western Pennsylvania. This species has been observed once in Pennsylvania.

Seasonal status and distribution

Fall: On the morning of 18 Aug 1979 at 7:15, a Surfbird was discovered with a small group of Black-bellied Plovers by G. M. McWilliams and S. Stull at Gull Point on PISP.[1] For the next five hours, heavy rain from a passing storm front forced down over 400 shorebirds from which 20 species were identified.[2] This is the only known record of Surfbird in eastern North America away from Texas and Florida (AOU 1998). A full written description with sketches of the Surfbird is in the files of PORC.

[1] Hall, G.A. 1980. Appalachian region. AB 34:160.
[2] G.M. McWilliams, pers. obs.

Red Knot ***Calidris canutus***

General status: The largest *Calidris* regularly occurring in North America, Red Knots are best known for their long-distance flights. They breed in northern Alaska and the Canadian Arctic islands. They winter along the Pacific coast from northern California south through Mexico and along the Atlantic coast from Massachusetts south along the Gulf coast. Huge concentrations gather in Delaware Bay in spring to feed on the eggs of horseshoe crabs. In Pennsylvania most records are from the Lake Erie Shore, where they are uncommon to rare regular migrants. Red Knots are casual migrants along the lower Susquehanna River and are accidental in the remainder of the state. Most sightings are in the fall.

Habitat: Red Knots prefer sandy beaches, mudflats, and the muddy shorelines of lakes.

Seasonal status and distribution

Spring: Nearly all spring sightings are from Gull Point on PISP, where birds are seen in the third and fourth weeks of May, with stragglers to the first week of Jun. A rather early record of one was at Emmaus in Lehigh Co. on 7 May 1984.[1] Most spring sightings are of single birds, but on 1 Jun 1996, more than 200 were at Green Lane Reservoir in Montgomery Co.[2] On 19 May 1986 an unusually large flight of Red Knots, totaling 485 birds, was observed at PISP.[3] Red Knots have also been recorded in

Red Knot at Presque Isle State Park, Erie County, September 1992. (Photo: Gerald M. McWilliams)

spring at Bethlehem Steel Lake in Lebanon Co. (Poole 1964), BESP in Centre Co.,[4] Plains in Luzerne Co.,[5] and the Union City Dam in Erie Co.[6]

Fall: Most sightings of Red Knots in Pennsylvania are recorded in the fall at the Conejohela Flats in Lancaster Co. and on PISP in Erie Co. A few adults are observed in fall, appearing as early as the second week of Jul but usually not until the third or fourth week of Jul. Early juveniles, making up the majority of birds observed at this season, arrive the second week of Aug. Flocks containing up to 26 birds have been reported. On 3 Aug 1992, a total of 52 was counted on PISP.[7] The majority of birds are recorded from the second week of Aug to the second week of Sep. East of the Allegheny Front, fall records come from Centre, Chester, Cumberland, Delaware, Lehigh, Luzerne, Montgomery, Northampton, Northumberland, and York cos. West of the Allegheny Front, they have been reported in fall from Allegheny, Butler, Crawford, Indiana, Mercer, and Somerset cos. By the first week of Oct most Red Knots have left the state, with lingering birds to the last week of Oct.

History: Their status and distribution have changed little throughout history. Warren (1890) reported that Red Knots were somewhat common visitors along the shore of Lake Erie and at Erie Bay but were seldom seen in other parts of the state. Poole (1964) reported Red Knot as a common species of the outer beaches, with all specimens and all but two of the sightings in fall. Sutton (1928b) reported two shot on the late dates of 22 and 28 Nov 1912.

[1] Boyle, W.J., Jr., R.O. Paxton, and D.A. Cutler. 1984. Hudson-Delaware region. AB 38:891.
[2] Haas, F.C., and B.M. Haas, eds. 1996. Birds of note—April through June 1996. PB 10:90.
[3] McWilliams, G.M. 1986. International Shorebird Survey (Erie, Pa. site). Files of Manomet Bird Observatory, Mass.
[4] Schiefer, T.L. 1984–1985. Large number of Whimbrel and late spring birding in Centre County, Pa. Cassinia 61:78.
[5] Reid, B. 1996. Local notes: Luzerne County. PB 10:102.
[6] McWilliams, G.M. 1980. International Shorebird Survey (Erie, Pa. site). Files of Manomet Bird Observatory, Mass.
[7] Hall, G.A. 1993. Appalachian region. AB 47:93.

Sanderling *Calidris alba*

General status: Sanderlings are familiar sandpipers of open sandy beaches, especially along the Atlantic and Pacific coasts. They are usually seen in tight flocks actively foraging along the edges of water, running back and forth to catch food washed in by the waves. In North America they breed in northern Alaska and the Canadian Arctic islands, and they winter along the Pacific, Atlantic, and Gulf coasts. In Pennsylvania, Sanderlings are common regular migrants only at the Lake Erie Shore, primarily at PISP, on extensive sandy beaches. They are rare regular migrants along the lower Susquehanna River at the Conejohela Flats, and irregular to casual at most other widely scattered sites in the state. Sanderlings are accidental in the High Plateau. Most sightings away from Lake Erie occur in the fall after storms.

Habitat: The majority of Sanderlings are found on sandy beaches (especially along Lake Erie) and river sandbars. They are occasionally found on mudflats and muddy shorelines of ponds, lakes, and rivers.

Seasonal status and distribution

Spring: Fewer birds are seen in spring than in fall. Spring arrivals have been recorded as early as the third week of Apr. Their typical migration period is from the second week of May to the first week of Jun. Migration peaks during the third and fourth weeks of May and occasionally into the first couple of days in Jun. Daily highs have reached 70 birds at PISP.[1] Away from Lake Erie, Sanderlings are casual, with one to four typically recorded. On 1 Jun 1996 more than 40 were at Green Lane Reservoir in Montgomery Co.[2] Stragglers linger to the second week of Jun.

Fall: Early southbound adults may appear the first week of Jul, but most are seen in the third week of Jul. They may suddenly appear at this time in great numbers. Early juveniles appear the third week of Aug. Adult peak migration ranges from the third week of Jul to the first week of Aug, and juveniles usually peak some time within the third week of Aug to the first week of Sep. However, there is often a later influx of Sanderlings ranging from the third week of Sep to the first week of Oct. Flocks of over 100 birds have been seen at PISP. On 3 Aug 1992 a total of 317 birds was counted there.[3] Single individuals, to small flocks of fewer than 10, to highs of up to 15 are reported at sites away from Lake Erie. Most birds have moved south by the third or fourth week of Oct, with stragglers to the third week of Nov. A very late bird was seen on 2 Jan 1972 at PISP.[4] In the High Plateau, there was one sighting: 2 birds were at the Bradford sewage treatment plant in McKean Co. in Jul and Aug 1987.[5]

[1] McWilliams, G.M. 1993. International Shorebird Survey (Erie, Pa. site). Files of Manomet Bird Observatory, Mass.

[2] Freed, G.L. 1996. Local notes: Montgomery County. PB 10:90.

[3] Hall, G.A.1993. Appalachian region. AB 47:93.

[4] Hall, G.A. 1972. Appalachian region. AB 26:602.

[5] Anderson, L. 1987. McKean County. PB 1:89.

Semipalmated Sandpiper *Calidris pusilla*

General status: Semipalmated Sandpipers breed along the northern Alaska coast, northern Mackenzie and Manitoba, the Canadian Arctic islands, and east around Hudson Bay to Labrador. In North America they are known to winter only in southern Florida and the Bahamas; most winter along the coast of Central and South America. One of the most common *Calidris* in Pennsylvania, Semipalmated Sandpipers are fairly common to common regular migrants over most of the state. In the Glaciated Northeast and the northern and central High Plateau they are casual visitors in suitable habitat.

Habitat: Semipalmated Sandpipers prefer sandy beaches, mudflats, and muddy shorelines of ponds, marshes, lakes, and rivers. They are occasionally found on wet fields that have sparse vegetation.

Seasonal status and distribution

Spring: Early migrants may appear the third week of Apr. Their normal migration time is from the second week of May to the first week of Jun. Peak migration usually occurs during the third and fourth weeks of May. Concentrations during or after storms have numbered in the hundreds or thousands at sites including Tinicum (where they are more common than anywhere else in the state), the Conejohela Flats along the lower Susquehanna River, various other sites in the Piedmont, and PISP. The 194 counted in Montour Co. on 1 Jun 1974 (Schweinsberg 1988) was a very high count for the Ridge and Valley region. Stragglers may remain to the last week of Jun. Some stragglers may remain later, overlapping southbound migrants.

Fall: The earliest southbound adults may appear the first week of Jul, but most arrive in the state the second week of Jul. Juveniles first appear during the first week of Aug, and by the third week of Aug flocks are dominated by juveniles. Peak migration ranges from the fourth week of Jul to the first week of Sep. The first peak, usually in the last week of Jul or first week of Aug, consists mostly of adults. Juvenile birds dominate the peak period during the last week of Aug or the first week of Sep. Like the status and distribution of most species of shorebirds, the fall status and distribution of Semipalmated Sandpipers vary from year to year, depending upon water levels. These sandpipers are most common during drought years, when lakes, ponds, rivers, and manmade impoundments drop well below normal levels. They are least common during seasons of excessive rainfall, when lakes, ponds, and rivers are at their highest levels. Most birds have departed from the state by mid-Oct, with stragglers to the first week of Nov.

History: The tidewaters of the Delaware attracted more Semipalmated Sandpipers in the past than anywhere else in the state. This site once was one of the major staging areas in North America. E. Manners reported flocks containing 25,000 birds and indicated that flocks containing 2000 or 3000 birds were common in Jul and Aug.[1] Probably one of the most outstanding high counts of Semipalmated Sandpipers in recorded history in Pennsylvania was made by P. Schwalbe and G. Schwalbe at Tinicum in the fall of 1952. The silt from a major dredging operation along the Delaware River was dumped in a portion of Tinicum, forming a large mudflat that attracted an estimated 3 million Semipalmated Sandpipers in a single day.[2] Todd (1940) stated that outside of Erie in western Pennsylvania "the species can scarcely be considered common" and had "escaped the notice of most observers outside the Erie-Crawford district." He suggested that the species may have been confused with Least Sandpipers.

[1] Manners, E.R. 1945. Shorebirds along the Delaware. Cassinia 35:23–34.
[2] P. Schwalbe and G. Schwalbe, pers. comm.

Western Sandpiper ***Calidris mauri***

General status: In North America, Western Sandpipers breed only on islands in the Bering Sea and along the coasts of western and northern Alaska. They winter along the Pacific coast from southern Oregon south through Mexico and along the Atlantic coast from North Carolina south along the Gulf coast. In Pennsylvania, Western Sandpipers are rare but regular migrants at Tinicum in the Coastal Plain, the Conejohela Flats in the Piedmont, and PISP at the Lake Erie Shore. They are irregular to casual in the Ridge and Valley, Glaciated Northeast, and Glaciated Northwest and accidental in the High Plateau. Apparently this species has not been recorded in the Southwest in recent years. Western Sandpipers are usually found with flocks of Semipalmated Sandpipers. They are accidental in winter in Pennsylvania.

Habitat: Western Sandpipers are found on sandy beaches, mudflats, muddy shorelines of ponds, marshes, lakes, and rivers, and occasionally on open level wet fields with sparse vegetation.

Seasonal status and distribution

Spring: Western Sandpipers are less frequently seen in spring than in fall. The earliest arrivals have been recorded in the third week of Apr, but most birds appear in the second week of May and all have departed by the first week of Jun.

Fall: Most records are at this season. The first southbound adults may arrive during the first week of Jul. Most adults begin passing through the state the third week of Jul. Juveniles are first seen in the third week of Aug. Migrants continue to be seen regularly until the last week of Aug or the first week of Sep. Stragglers may remain until the third week of Dec. In the High Plateau, Western Sandpipers have been recorded only in McKean,[1] Somerset,[2] and Westmoreland[3] cos.

Winter: The single winter record in the state is that of 2 birds that wintered at Tinicum in Philadelphia Co. in 1968.[4]

History: The first record in the state was of a bird that Todd identified. It was in a mixed flock of Semipalmated, Least, and Baird's sandpipers on 21 Aug 1907. Todd collected the bird along the outer beach of the Ohio River at Beaver (Todd 1940). This species was not reported again until 1928, when a small flock was seen on Presque Isle in Erie Co. (Todd 1940). All other sightings were from eastern Pennsylvania, primarily from the southeastern corner. Western Sandpipers were reported as common fall migrants at Tinicum at least during the early 1950s (Poole, unpbl. ms.). The absence of nineteenth-century records suggests that Western Sandpipers were probably overlooked because of their similarity to Semipalmated Sandpipers.

Comments: Many of the smaller *Calidris*, especially long-billed female Semipalmated Sandpipers, have been misidentified as Western Sandpipers. Poole (unpbl. ms.) mentioned that even specimens have been misidentified. Conversely, short-billed male Western Sandpipers have been misidentified as Semipalmated Sandpipers. Since most Semipalmated Sandpipers winter outside of the U.S. and Western Sandpipers winter within the U.S., probably most, if not all, peeps identified as Semipalmated Sandpipers in Pennsylvania in late fall are Western Sandpipers.

[1] Anderson, L. 1987. McKean County. PB 1:89.
[2] Haas, F.C., and B.M. Haas, eds. 1994. Rare and unusual bird reports. PB 8:163.
[3] R.C. Leberman, pers. comm.
[4] J.C. Miller, pers. comm.

Least Sandpiper *Calidris minutilla*

General status: The breeding range of Least Sandpipers extends from Alaska across northern Canada east to Newfoundland. They winter along the Atlantic coast from North Carolina south to Florida and west across the southern states to California and north along the Pacific coast to Washington. The most widespread and smallest *Calidris* in Pennsylvania, Least Sandpipers are fairly common to common regular migrants throughout most of the state. In the northern and central High Plateau they are uncommon to rare. They are often seen mixed with Semipalmated Sandpiper flocks. Least Sandpipers are accidental in early winter.

Habitat: They are less confined to large open areas than are other *Calidris*. Most are found on sandy beaches, on mudflats, along the muddy or grassy shorelines of ponds, marshes, lakes, and rivers, and on wet fields.

Seasonal status and distribution

Spring: The earliest arrivals may appear the third week of Apr. Their typical migration period is from the first week of May to the second week of Jun. Peak migration is usually during the second and third weeks of May. Concentrations of Least Sandpipers have numbered in the hundreds during or after storms at shorebird hot spots, such as Tinicum, where they are most common; on the Conejohela Flats along the lower Susquehanna River; at various other sites in the Piedmont; and at PISP. Stragglers remain to the last week of Jun, sometimes overlapping with southbound migrants.

Fall: The earliest southbound adults consistently appear during the first week of Jul, with the number of birds increasing through Jul. Juvenile Least Sandpipers appear in Pennsylvania during the last week of Jul, earlier than juveniles of any other species of shorebirds that nest outside of the state. Peak migration ranges from the third week of Jul to the first week of Sep. The first peak period, usually from the third week of Jul to the first week of Aug, consists mainly of adults. The peak period during the third week of Aug to the first week of Sep is dominated by juvenile birds. Adults are rarely observed by the third week of Aug. Most birds have departed south by mid-Oct. Stragglers may remain well into Dec, very rarely until the first week of Jan. In 1983 two birds lingered until 2 Jan at the Octoraro Reservoir in Lancaster Co.[1]

Winter: A Least Sandpiper was observed, just into the winter period, on 12 Jan 1960 at Tinicum.[2]

Comments: On 14 Jan 1987, during a period of unseasonably warm weather, a small brown *Calidris* believed to be this species was seen at PISP.[3]

[1] Boyle, W.J., Jr., R.O. Paxton, and D.A. Cutler. 1983. Hudson-Delaware region. AB 37:285.

[2] Scott, F.R., and D.A. Cutler. 1960. Middle Atlantic Coast region. AFN 14:297.

[3] G.M. McWilliams, pers. obs.

White-rumped Sandpiper *Calidris fuscicollis*

General status: White-rumped Sandpipers breed from northern Alaska east across the Canadian Arctic islands and northwestern Hudson Bay. They winter only in South America, primarily east of the Andes. White-rumped Sandpipers are uncommon to rare regular migrants in Pennsylvania except in the Glaciated Northeast and in the north and central High Plateau, where they have not been recorded. White-rumped Sandpipers are frequently seen during or after storms.

Habitat: They are found on sandy beaches, on mudflats, and along muddy or grassy shorelines of ponds, marshes, lakes, and rivers.

Seasonal status and distribution

Spring: Early arrivals have been recorded from the third week of Apr to the first week of May. Migration continues much later than for other *Calidris*, often through the first week of Jun, with stragglers to the third week of Jun. Most sightings are during the third and fourth weeks of May. Remarkable counts of 150 White-rumped Sandpipers on 20 May 1978[1] and 250 birds on 18 May 1979[2] were made at Tinicum.

Fall: White-rumped Sandpipers are most frequently observed in the state during fall, but their distribution varies from year to year, depending upon water levels. The first southbound birds, mostly adults, are recorded in the third week of Jul. Juveniles begin to appear during the first or second week of Sep. The peak number of White-rumped Sandpipers may appear any time from early Sep through the end of Oct. Fall migration continues to the fourth week of Oct, with stragglers to the third week of Nov.

History: Their status and distribution have changed little throughout history. Todd (1940) stated that this species was quite rare in western Pennsylvania and that they were common in the Erie region only after "the great storm of August 29, 1893." White-rumped Sandpipers were reported to be fairly common at Tinicum in the springs of 1950 and 1951 and in the autumns of 1955 and 1957 (Poole, unpbl. ms.).

[1] Paxton, R.O., P.A. Buckley, and D.A. Cutler. 1978. Hudson-Delaware region. AB 32:985.

[2] N. Pulcinella, pers. comm.

Baird's Sandpiper ***Calidris bairdii***

General status: Baird's Sandpipers breed from western and northern Alaska, northwestern British Columbia, and east across the Canadian Arctic islands. They winter only in South America, locally in the Andes. Most Baird's Sandpipers migrate through the central portion of the continent. They are not common anywhere in the eastern U.S. In Pennsylvania, most sightings are in fall at PISP, where they are uncommon to rare regular migrants. Baird's Sandpipers are rare irregular fall migrants at Tinicum and in the Piedmont and casual to accidental in fall in the Ridge and Valley, Southwest, and Glaciated Northwest. No record exists from the Glaciated Northeast or the High Plateau. They are rarely reported in spring anywhere in the state.

Habitat: Baird's Sandpipers prefer sandy beaches, mudflats, and muddy or grassy shorelines of ponds, marshes, lakes, and rivers. They are occasionally found on wet level fields with short grass, especially sod farms.

Seasonal status and distribution

Spring: Baird's Sandpipers are very rare in spring, when most records are from PISP. Observations, mostly of single

Baird's Sandpiper at Presque Isle State Park, Erie County, September 1991. (Photo: Gerald M. McWilliams)

birds, range from the third week of Apr to the first week of Jun. Reports at this season come from the Conejohela Flats in Lancaster Co. (Morrin et al. 1991), 12 birds on Brunner's Island in York Co. on 11 Jun 1993,[1] the Parnell Fish Hatchery in Franklin Co.,[2] Montour Preserve in Montour Co. (Schweinsberg 1988), BESP in Centre Co.,[3] and MSP in Butler Co.[4]

Fall: Adult Baird's Sandpipers are observed rarely in Pennsylvania during fall, when their distribution varies from year to year, depending upon water levels. They first appear the second week of Jul. Juveniles, which make up the majority of sightings, appear the second week of Aug. Peak migration ranges from the third week of Aug to the third week of Sep. A total of 21 was counted on the Conejohela Flats on 3 Sep 1959 (Morrin et al. 1991). Fall migration has usually ended by the second week of Oct, with stragglers to the first week of Dec.

History: The first Pennsylvania record was of a bird collected by Todd on 16 Sep 1889, at the mouth of the Beaver River in Beaver Co. (Todd 1940). Todd (1940) also listed the species as moderately common in fall along the outer beaches of Presque Isle. Baird's Sandpipers were not reported in eastern Pennsylvania until 1950, after which they had been observed several times by 1957 at the Conejohela Flats, Lancaster Co.

Comments: In recent years there have been more sightings of Baird's Sandpipers, perhaps because of better field guides and more observers in the field. It is likely that this species has been overlooked, especially in the past, since it resembles several other small *Calidris* sandpipers.

[1] Spiese, A. 1993. Notes from the field: York County. PB 7:64.

[2] Haas, F.C., and B.M. Haas, eds. 1993. Rare and unusual bird reports. PB 7:60.

[3] Hall, G.A. 1984. Appalachian region. AB 38:910.

[4] Hall, G.A. 1986. Appalachian region. AB 40:471.

Pectoral Sandpiper *Calidris melanotos*

General status: In North America the breeding range of Pectoral Sandpipers is in western and northern Alaska east across the Canadian Arctic islands and the western end of Hudson Bay. They winter in South America but are casual in Florida and on the Gulf coast. One of the most widespread *Calidris* species in the state, Pectoral Sandpipers are fairly common to common regular migrants over most of the state. They are uncommon to rare in the northern and central High Plateau and at the Lake Erie Shore.

Habitat: Pectoral Sandpipers use a wide variety of wet habitats such as flooded fields, mudflats, and grassy or muddy shorelines of ponds, marshes, lakes, and rivers. They can also be common in wet fields with short grass. Pectoral Sandpipers are less confined to large open areas than are other *Calidris* species.

Seasonal status and distribution

Spring: Pectoral Sandpipers are among the first shorebirds to appear in spring, sometimes arriving as early as the second week of Mar. Four birds discovered on 2 Mar 1991 at BESP represent an early record.[1] Their usual migration period is from the last week of Mar or the first week of Apr to the first week of May. Peak migration usually occurs from the first to the third week of Apr. In flooded fields, flocks containing 100 or more birds are not unusual after spring thaws. Stragglers have been recorded until the first week of Jun. At Tinicum a count of 30 birds on 4 Jun 1968 was an exceptionally high count for such a late date.[2]

Fall: At most sites they are more common in fall than in spring, especially in the Coastal Plain and the Piedmont. Pectoral Sandpipers are among the first shorebirds to return south from their breeding grounds, occasionally returning to Pennsylvania as early as the last week of Jun. However, most adults arrive the third or fourth week of Jul. First juveniles arrive the first week of Aug. Peak migration may occur any time from the second week of Aug to the fourth week of Sep. Concentrations of 170 or more birds have been recorded in the Coastal Plain and the Piedmont. Migration has usually ended by the last week of Oct. Stragglers remain to the second week of Nov. A very late bird was found on 15 Dec 1979 in the Tinicum area.[3]

History: Poole (unpbl. ms.) listed this species as an uncommon or irregular spring transient but often common in fall. He also listed the Pectoral Sandpiper as an abundant migrant at Tinicum, including one observed in 1953 on the exceptionally late date of 26 Dec.

[1] Hall, G.A. 1991. Appalachian region. AB 45:445.
[2] Scott, F.R., and D.A. Cutler. 1968. Middle Atlantic Coast region. AFN 22:512.
[3] N. Pulcinella, pers. comm.

Purple Sandpiper ***Calidris maritima***

General status: Purple Sandpipers breed in North America only in the East from Melville to Baffin Island south to James Bay. They are familiar wintering sandpipers of rocky coastlines and jetties of the Atlantic coast from New Brunswick south to Virginia. They are rare but regular visitors to a few inland sites on their southbound journey from breeding grounds in the Arctic to the Atlantic coast in late fall. In Pennsylvania they are known to occur only at PISP and only in the fall. They were recorded in only seven years between 1958 and 1978.[1] Since then they have been rare but regular fall migrants. They are frequently found with Dunlins.

Habitat: Purple Sandpipers are found exclusively on the outer beaches and on the breakwaters of PISP. They have also been seen in migration flying past the east end of the peninsula.

Seasonal status and distribution

Fall: The first confirmed record of Purple Sandpiper in Pennsylvania was on 16 Nov 1958, when J. G. Stull and J. H. Stull found two birds with a small group of Dunlins on the outside beach of Gull Point.[2] Purple Sandpipers are one of the last shorebirds to migrate through

Purple Sandpiper at Presque Isle State Park, Erie County, November 1986. (Photo: Franklin C. Haas)

Pennsylvania. Observations range from 23 Oct to 19 Dec. Most sightings are of single birds in Nov during days of strong cold north winds. They often accompany the passage of the first severe cold front in late Oct or early Nov. All birds of known age have been in first-winter plumage.

[1] G.M. McWilliams, pers. recs.
[2] Stull, J.H., and J.G. Stull. 1958. Field notes. Sandpiper 1:21.

Dunlin ***Calidris alpina***

General status: Dunlins breed in North America from Alaska, northern Mackenzie, and to the south coast of Hudson Bay. In winter they are found along the Pacific coast and along the Atlantic coast from Massachusetts south along the Gulf coast. Dunlins are common to abundant regular migrants at Tinicum in the Coastal Plain and at PISP at the Lake Erie Shore, and they are fairly common to common regular migrants in the Piedmont and the Glaciated Northwest. They are uncommon to rare regular migrants in the Ridge and Valley and in the Allegheny Mountain section of the High Plateau. Elsewhere in the High Plateau, Dunlins have been reported only in McKean Co. at the Bradford Sewage Treatment Plant.[1] They are uncommon regular migrants in the Southwest.

Habitat: Dunlins prefer sandy beaches; mudflats; muddy or grassy shorelines of ponds, marshes, lakes, and rivers; and flooded fields.

Seasonal status and distribution

Spring: The earliest migrants appear in the second or third week of Apr, but there are a few records in late Mar. One on 9 Mar 1991 at PISP was more likely a bird that had wintered somewhere nearby than a northbound bird from a coastal wintering site.[2] Their typical migration period is from the first week to the last week of May. Peak migration occurs during the third or fourth week of May. Areas such as the Tinicum and Conejohela Flats have recorded concentrations numbering over 200 birds. The 218 counted at the Pennsylvania Power and Light Montour Fly Ash Basin on 13 May 1979 (Schweinsberg 1988) was a very high count for the Ridge and Valley region. At PISP, concentrations of 60 to 90 are not uncommon, and as many as 350 in single day have been recorded.[3] Stragglers may remain to the first week of Jun.

Summer: All three midsummer records are from PISP and are more likely to be lingering nonbreeding birds than very early fall migrants. They were reported about 12 Jul and 19 Jul before 1975 (Stull et al. 1985), and one was seen on 1 Jul 1989.[3]

Fall: At most sites in Pennsylvania, Dunlins are most frequently encountered in fall. Their distribution at this season varies from year to year, depending on water levels. They are among the latest shorebirds to migrate south in the fall; the earliest migrants have appeared the second week of Aug. Most begin to appear during the third or fourth week of Sep. Arrival times of the different age groups have not been fully documented, but juveniles have been identified in the second week of Sep. Peak migration usually occurs within Oct, when daily highs may reach several hundred at Tinicum and PISP. Daily highs at other sites are usually fewer than 20 birds, but flocks containing 100 or more are not unusual. Migration extends well into the fall, with birds still present to the second or third week of Nov during most years. CBCs in Delaware Co. have recorded Dunlins nine times, with a high count of 40 birds.[4] Stragglers remain to the first week of Jan if there has not been a severe freeze.

History: Poole (unpbl. ms.) listed this species as a "rather irregular" transient throughout the state and most common in fall, but rather rare in spring, especially along the shores of Lake Erie. He listed the species as fairly common in both spring and fall at Tinicum, on the Conejohela Flats, and on Pymatuning Lake. Elsewhere in the state, Todd (1940) considered Dunlins to be very uncommon to rare. Todd (1904) quoted S. E. Bacon as saying that in "former years" great flights of shorebirds were noted in Nov along the outside beach of Presque Isle, from which bushels of birds could have fallen to a single gun. Bacon also claimed to have killed 53 birds out of two flocks (Todd 1940).

[1] Anderson, L. 1987. McKean County. PB 1:89.
[2] G.M. McWilliams, pers. obs.
[3] McWilliams, G.M. 1990. International Shorebird Survey (Erie, Pa. site). Files of Manomet Bird Observatory, Mass.
[4] Rigby, E.H. 1971. Glenolden CBC. AB 25:227.

[Curlew Sandpiper] *Calidris ferruginea*

General status: Curlew Sandpipers breed mainly on the tundra in northern Siberia and rarely in North America in Alaska at Barrow. They winter in the Mediterranean region south and east to Africa, India, Australia, and New Zealand. In eastern North America these sandpipers are casual during migration from southeastern Canada south to Florida and west along the Gulf coast to Texas. Curlew Sandpipers are listed as hypothetical in Pennsylvania.

Comments: There are four sight records of Curlew Sandpipers in Pennsylvania, three of which were in the Tinicum area. The first was one in breeding plumage on 16 May 1953, and two birds were on the Conejohela Flats after the hurricane of 14 Aug 1955 (Poole 1964). More recently at Tinicum, this species was reported on 12–14 Jul 1981[1] and in Jul 1982.[2] Because there was no documentation with photographs or written descriptions, it was not included on the official list of the birds of Pennsylvania.[3]

[1] Boyle, W.J., Jr., R.O. Paxton, and D.A. Cutler. 1981. Hudson-Delaware region. AB 35:924.
[2] Paxton, R.O., W.J. Boyle Jr., and D.A. Cutler. 1982. Hudson-Delaware region. AB 36:960.
[3] Pulcinella, N. 1995. Official list of the birds of Pennsylvania. PB 9:118–123.

Stilt Sandpiper *Calidris himantopus*

General status: Stilt Sandpipers breed from northern Alaska along northern Mackenzie and the west side of Hudson Bay. They winter from the West Indies south through Central and South America and casually in the U.S. in California, along the Gulf coast, and in Florida. Stilt Sandpipers normally migrate through the central part of the U.S. in spring, so they are casual to accidental at this season in Pennsylvania. They are uncommon to rare regular migrants in fall over most of the state and accidental in the High Plateau north of the Allegheny Mountain section.

Habitat: Stilt Sandpipers prefer sandy beaches, mudflats, and muddy or grassy shorelines of ponds, marshes, lakes, and rivers.

Seasonal status and distribution

Spring: Stilt Sandpipers are casual spring migrants at PISP and accidental elsewhere in the state at this season. They are very rare in spring, with nearly all records consisting of single birds. All but two records are between 12 and 21 May: one was on the Conejohela Flats on 4 Jun 1960, and two birds were at the MCWMA in Lancaster Co. on 13 Jun 1986 (Morrin et al. 1991).

Fall: Most Pennsylvania records are during this season, but birds are usually

Stilt Sandpiper at Presque Isle State Park, Erie County, August 1983. (Photo: Gerald M. McWilliams)

not very numerous at any one site. Adults appear the third or fourth week of Jul, and the first juveniles are seen the third week of Aug. Most migrants are seen in Aug and Sep. Daily concentrations have totaled over 40 birds. At Green Lane Reservoir in Montgomery Co., 48 birds were counted on 2 Sep 1981 (Morris et al. 1984); at Pymatuning in Crawford Co., 50 or 60 were counted on 3 Sep 1962.[1] The only record of a Stilt Sandpiper in the High Plateau outside of the Allegheny Mountain section was one found in a drained pond at Lander, Warren Co., on 7 Sep 1988.[2] Most birds have left the state by the last week of Sep, with stragglers to the last week of Oct. Their distribution varies from year to year at this season, depending upon water levels.

History: Warren (1890) listed this species as a "very rare spring and fall migrant in Pennsylvania," perhaps as a result of overhunting in the nineteenth century. Poole (1964) upgraded the Stilt Sandpiper's status to uncommon regular transient.

Comments: Stilt Sandpipers, especially in molt, are occasionally confused with other species of shorebirds such as Lesser Yellowlegs and Short-billed Dowitcher.

[1] Leberman, R.C. 1962. Field notes. Sandpiper 5:29.
[2] Hall, G.A. 1989. Appalachian region. AB 43:102.

Buff-breasted Sandpiper ***Tryngites subruficollis***

General status: Buff-breasted Sandpipers breed from northern Alaska east across the Canadian Arctic coast south to Victoria and King Williams islands and they winter in South America. Like Baird's Sandpipers, most Buff-breasted Sandpipers migrate through the central portion of the continent. In Pennsylvania all but one record of Buff-breasted Sandpipers is in the fall. They are rare irregular migrants in the Coastal Plain and Piedmont. Buff-breasted Sandpipers are accidental in the Ridge and Valley and High Plateau and casual in the Glaciated Northwest. No record exists from either the Glaciated Northeast or the Southwest region. Buff-breasted Sandpipers are rare regular migrants only at PISP.

Habitat: They inhabit sandy beaches, mudflats, recently plowed fields, grassy or muddy edges of lakes or ponds, sod farms, and airports (e.g., the Philadelphia airport).

Seasonal status and distribution

Spring: A Buff-breasted Sandpiper observed in spring at the Philadelphia

Buff-breasted Sandpiper at Salunga, Lancaster County, September 1991. (Photo: Randy C. Miller)

sewage ponds on 15 May 1980 is the only modern record.[1]

Fall: Buff-breasted Sandpipers are most likely to be found during years when lakes and ponds, especially manmade, drop below normal levels. Especially in the Glaciated Northwest, they are not reported when lakes and ponds are at their highest levels. Adults are rarely reported, with the earliest reports the third week of Jul. Juveniles first appear about the third week of Aug. Most birds are reported from the third week of Aug to the third week of Sep. As many as eight Buff-breasted Sandpipers were seen on 20 Sep 1964 at the Philadelphia airport.[2] Buff-breasted Sandpipers rarely linger in Pennsylvania past the fourth week of Sep. Two records are after Sep: of eight birds at the Philadelphia airport, one remained from 20 Sep 1964 to 3 Nov 1964,[2] and at the Conejohela Flats one was reported on 11 Oct 1959 (Poole, unpbl. ms.). In the Coastal Plain, all records are from the Philadelphia airport and Tinicum areas.[3] In the Piedmont, records come from Berks, Chester, Franklin, Lancaster, Lehigh, and Montgomery cos. They have been seen at four sites in the Ridge and Valley: in the State College area on 28 Aug 1974[4] and 24 Jul 1976,[5] at BESP in Centre Co. on 18–19 Sep 1982,[6] at Beltzville Lake in Carbon Co. 11 Sep 1991,[7] and at Plymouth in Luzerne Co. on 21 Aug 1994.[8] West of the Allegheny Front, Buff-breasted Sandpipers have been recorded in the counties of Crawford, Erie, Mercer, and Somerset. The one found at Somerset on 22 Sep 1980 is the only record in the High Plateau.[9]

History: Before 1940, Todd (1940) listed only the following two records of Buff-breasted Sandpipers for western Pennsylvania. One was shot near Chambersburg in Franklin Co. on 31 May 1899, and two were seen on Presque Isle on 1 Sep 1928. Poole (unpbl. ms.) wrote of only one record in eastern Pennsylvania before 1940: one taken in Lancaster Co. on 26 Sep 1900, which was placed in the North Museum. Between 1940 and 1960 records came from Berks, Crawford, Erie, Lancaster, and Philadelphia cos. (Poole, unpbl. ms.).

[1] Paxton, R.O., W.J. Boyle Jr., D.A. Cutler, and K.C. Richards. 1980. Hudson-Delaware region. AB 34:760.

[2] Scott, F.R., and D.A. Cutler. 1965. Middle Atlantic Coast region. AFN 19:36.

[3] J.C. Miller and E. Fingerhood, pers. comm.
[4] Hall, G.A. 1975. Appalachian region. AB 29:59.
[5] Hall, G.A. 1976. Appalachian region. AB 30:955.
[6] Hall, G.A. 1983. Appalachian region. AB 37:180.
[7] Paxton, R.O., W.J. Boyle Jr., and D.A. Cutler. 1992. Huson-Delaware region. AB 46:69.
[8] Reid, W. 1994. Notes from the field: Luzerne County. PB 8:161–162.
[9] Hall, G.A. 1981. Appalachian region. AB 35:183.

Ruff ***Philomachus pugnax***

General status: This Eurasian species breeds from western and northern Europe east across Russia and Siberia and winters from the British Isles through southern Europe to Africa, China, India, and Australia. They are rare but regular vagrants throughout North America. Ruffs were regular rare vagrants in Pennsylvania before 1980, but fewer than 10 records have been reported statewide since then. From 1960 to about 1973 they were nearly annual at Tinicum. From 1973 to 1986 they became irregular, and since 1987 they have not been recorded there. About 50 birds were seen in a 24-year span from 1956 to 1980 at Tinicum.[1] All sightings were of single birds. Away from Tinicum, all other records are from the Piedmont, Ridge and Valley, and Lake Erie Shore.

Habitat: Ruffs are found on mudflats and grassy or muddy edges of lakes, ponds, and marshes.

Seasonal status and distribution

Spring: Records are from the last week of Mar to the third week of May. Many of the sightings occurred during the first half of May. Both male and female (known as a reeve) have been reported at this season. All spring records away from Tinicum have been reeves. Reports have come from Allentown in Lehigh Co., where one was found in a flooded field from 14 to 15 May 1983;[2] MCWMA in Lancaster Co., where a bird was seen on 7 May 1987;[3] Lake Ontelaunee in Berks Co. on 4 Jun 1994;[4] State College in Centre Co., where one was identified on 27–29 Mar 1972 (Wood 1983); and PISP in Erie Co., where one was identified on 5 May 1990.[5]

Fall: Male, female, and juvenile birds have been identified in fall. Records are scattered from the third week of Jul to the first week of Nov. Away from Tinicum, Ruffs have been reported from Lake Ontelaunee in Berks Co. on 1 Oct 1965,[6] Green Lane Reservoir in Montgomery Co. on 14 Sep 1983,[7] the Conejohela Flats in Lancaster Co. on

Ruff at Tinicum, Philadelphia County, October 1986. (Photo: Franklin C. Haas)

16 Jul 1985,[8] Sunbury in Northumberland Co. on 20 Jul 1981,[9] and PISP in Erie Co. on 9 Aug 1973 and again on 24 Sep 1989.[10]

History: Before Ruffs were recorded regularly at Tinicum, they were regarded as accidental in fall. They were first reported in Pennsylvania by T. Hake and colleagues on the Conejohela Flats in Lancaster Co. on 1 Oct 1950 (Poole, unpbl. ms.). The first bird recorded at Tinicum was one seen on 30 Aug 1954 and later was joined by another.[11] One bird remained at Tinicum for the CBC on 1 Jan 1955.[12] Apparently the only record, except the Lancaster Co. record, away from Tinicum was one was at Churchville in Bucks Co. in 1951 (Thomas 1955).

[1] Paxton, R.O., W.J. Boyle Jr., D.A. Cutler, and K.C. Richards. 1980. Hudson-Delaware region. AB 34:760.
[2] Boyle, W.J., Jr., R.O. Paxton, and D.A. Cutler. 1983. Hudson-Delaware region. AB 37:852–853.
[3] Boyle, W.J., Jr., R.O. Paxton, and D.A. Cutler. 1987. Hudson-Delaware region. AB 41:408.
[4] R. Wiltraut, pers. comm.
[5] Hall, G.A. 1990. Appalachian region. AB 44:426.
[6] Scott, F.R., and D.A. Cutler. 1966. Middle Atlantic Coast region. AFN 20:24.
[7] Paxton, R.O., W.J. Boyle Jr., and D.A. Cutler. 1984. Hudson-Delaware region. AB 38:183.
[8] Paxton, R.O., W.J. Boyle Jr., and D.A. Cutler. 1985. Hudson-Delaware region. AB 39:892.
[9] Hall, G.A. 1981. Appalachian region. AB 35:940.
[10] Hall, G.A. 1974. Appalachian region. AB 28:50.
[11] Potter, J.K., and J.J. Murray. 1955. Middle Atlantic Coast region. AFN 9:16.
[12] Rigby, E.H. 1955. Glenolden CBC. AFN 9:106.

Short-billed Dowitcher *Limnodromus griseus*

General status: Short-billed Dowitchers breed primarily along the eastern coast of Alaska, central Canada, Quebec, and Labrador. They winter along the Pacific coast of California and Mexico and along the Atlantic coast from South Carolina south along the Gulf coast. In Pennsylvania, Short-billed Dowitchers are uncommon to fairly common regular migrants in the Coastal Plain, Piedmont, and Glaciated Northwest and at the Lake Erie Shore. They are rare regular migrants in the Ridge and Valley and Southwest and are accidental in the High Plateau north of the Allegheny Mountain section.

Habitat: They inhabit mudflats; sandy beaches; grassy and muddy edges of ponds, lakes, marshes, and rivers; and flooded fields and meadows.

Seasonal status and distribution

Spring: Earliest migrants have been recorded the first week of Apr. Their typical migration time is from the first week of May to the last week of May. Peak migration occurs during the second and third weeks of May. In spring, Short-billed Dowitchers are usually seen in greater numbers at PISP than anywhere else in the state. During storms it is not unusual to see flocks containing 100 or more birds forced to the ground by heavy rain. Storms forced down an estimated 880 on 18 May 1984[1] and 750 on 15 May 1990[2] at Gull Point on PISP. Stragglers remain to the first week of Jun.

Fall: Short-billed Dowitchers are more widely distributed in the state in fall than in the spring. However, their distribution varies from year to year, depending on water levels. They are among the first southbound shorebirds to arrive. The earliest adults arrive the last week of Jun. Most begin to appear the second week of Jul. Juveniles are seen the second week of Aug and replace adults by the third week of Aug. Peak migration for adults usually occurs during the second or third week of Jul and for juveniles during the third and fourth weeks of Aug. Short-billed Dowitchers are among the first shorebirds to leave the state. Most have moved south by the second week of Sep, with stragglers to the third week of Oct.

History: Poole (1964) listed this species as a regular transient that was rare in spring and fairly common in the fall at favorable locations.

Comments: Short-billed and Long-billed dowitchers were once considered to be the same species, so some of the historical records may have been of the latter species. It is also likely that many juvenile dowitchers after about mid-Sep may have been misidentified as Short-billed, since this is the time when juvenile Long-billed appear in the U.S. Better field guides have reduced the number of misidentifications, so most recent reports of Short-billed Dowitchers are probably correct. Two subspecies of Short-billed Dowitchers have been identified in Pennsylvania. At PISP both *L. g. griseus* and *L. g. hendersoni* have been identified in spring. The two subspecies are frequently observed in the same flock.[3]

[1] Hall, G.A. 1984. Appalachian region. AB 38:911.
[2] McWilliams, G.M. 1990. International Shorebird Survey (Erie, Pa site). Files of Manomet Bird Observatory, Mass.
[3] E. Kwater and G.M. McWilliams, pers. obs.

Long-billed Dowitcher *Limnodromus scolopaceus*

General status: Long-billed Dowitchers breed along coastal western and northern Alaska, northern Yukon, and northwestern Mackenzie. They winter along the Pacific coast of the U.S. and Mexico and east along the Gulf coast to Florida. The status of Long-billed Dowitcher in Pennsylvania is somewhat uncertain because of the difficulty in separating this species from Short-billed Dowitcher. They are irregular (perhaps regular) fall migrants in the Coastal Plain at Tinicum, Piedmont, and Glaciated Northwest, casual at PISP, and accidental in the Southwest.

Habitat: Long-billed Dowitchers are found on mudflats, on sandy beaches, and along grassy and muddy edges of ponds, lakes, rivers, and marshes as well as on flooded fields and meadows.

Seasonal status and distribution

Spring: Apparently the only reliable records of Long-billed Dowitchers in spring are from Gull Point at PISP, on 26 May 1979,[1] 11 and 18 May 1980,[2] and 22 May 1982.[3]

Fall: Nearly all recent sightings are of one or two birds, but as many as eight have been recorded. Adults may first appear the third week of Jul, but usually not until the first week of Aug. The majority of Long-billed Dowitchers seen in Pennsylvania are juveniles, which appear the second week of Sep and have been recorded until the third week of Nov. They peak from the third week of Sep to the second week of Oct. Away from Tinicum and PISP, they are most likely to be found during drought years when lakes and ponds, especially manmade, drop well below normal levels. In the Piedmont they have been recorded from Berks, Dauphin, Lancaster, Lebanon, Lehigh, Montgomery, and York cos. The single record in the High Plateau is of one photographed at Lake Somerset in Somerset Co. on 11 Sep 1997.[4] Two records are from the Southwest: a bird was identified at Greensburg in Westmoreland Co. on 28 Oct 1979,[5] and three were found at YCSP in Indiana Co. on 25 to 31 Oct 1997.[6] In the Glaciated Northwest they have been recorded at only four sites: in Erie Co. at the Union City Dam[7] and Siegel Marsh,[8] in Crawford Co. at Pymatuning (at least five times),[9] and near Volant in Lawrence Co.[10] One very late bird remained to 20 Dec 1958 at Tinicum.[11]

History: Poole (unpbl. ms.) listed this species as hypothetical because there was no specimen in existence and the difficulty in separating them from Short-billed Dowitchers made sight records questionable. He did, however,

Long-billed Dowitchers at Somerset Lake, Somerset County, November 1997. (Photo: Walter Shaffer)

mention sight records from Tinicum, the Conejohela Flats, and Lake Ontelaunee in Berks Co.

Comments: The longer bill of the Long-billed Dowitcher has in the past, and even recently, been used to separate this species from the shorter billed Short-billed Dowitcher. Most authorities agree, however, that this characteristic alone is not diagnostic. Call notes, especially of birds in breeding plumage, are probably the safest method to separate the two species. In addition, the plumages of juveniles of the two species are distinctly different. Juvenile dowitchers seen before the second week of Sep are most likely to be Short-billed. Kaufman stated that Long-billed Dowitchers arrive in the Pacific Northwest in early Sep and are extremely scarce farther south and east until the last third of the month (Kaufman 1990).

[1] Hall, G.A. 1979. Appalachian region. AB 33:771.
[2] Hall, G.A. 1980. Appalachian region. AB 34:776.
[3] Hall, G.A. 1982. Appalachian region. AB 36:852.
[4] Haas, F.C., and B.M. Haas, eds. 1998. Photographic highlights. PB 11:217.
[5] Hall, G.A. 1980. Appalachian region. AB 34:161.
[6] Higbee, M., and R. Higbee. 1997. Local notes: Indiana County. PB 11:239.
[7] Hall, G.A. 1981. Appalachian region. AB 35:183.
[8] G.M. McWilliams, pers. obs.
[9] J. Barker, pers. comm.
[10] Butcher, S. 1996. Local notes: Lawrence County. PB 10:165.
[11] Rigby, E.H. 1959. Glenolden CBC. AFN 13:119.

Common Snipe ***Gallinago gallinago***

General status: Common Snipe have suffered from the draining of wetlands throughout their breeding range in North America, which extends from Alaska throughout most of Canada and the northern and western U.S. They winter primarily across the southern half of the U.S. In Pennsylvania they are best known as fairly common to common and regular migrants statewide. They also breed in the state, primarily in the Glaciated Northwest. They are regular winter residents, mainly in the Coastal Plain and Piedmont. Though there is a hunting season for snipe, most hunters do not consider them to be favored targets or table fare.

Habitat: They inhabit almost any open wet area during migration, such as flooded fields and pastures; marshes; swamps; beaver dams; shorelines of ponds, streams, and lakes; and mudflats. In the breeding season they use a range of wetland conditions, including wet pastures and brushy marshes. In winter, their habitat is similar to that used in migration but confined more to creek edges and open springs in fields, pastures, marshes, or beaver ponds.

Seasonal status and distribution

Spring: Birds appear soon after the first spring thaw. Their usual migration period is from the first or second week of Mar to the second week of May. Migration may peak any time from the third week of Mar to the second or third week of Apr. Along the mudflats of Darby Creek at Tinicum snipe are more common in spring than anywhere else in the state, with some flocks containing as many as 100 birds.[1]

Breeding: Snipe are rare, generally irregular breeding birds in a variety of wetland habitats across Pennsylvania. They are most regular in the northwestern counties of Crawford, Erie, Lawrence, and Mercer, but they also have been found during the breeding season in Armstrong, Bradford, Centre, Lackawanna, Luzerne, McKean, Potter, Somerset, and Tioga cos. A half-grown young snipe was found in a wet field near Elderton in Armstrong Co. on 10 May 1991.[2] Nests in recent years were found in wet pastures in Mercer Co.[3] and a degraded shrub-wetland in Centre Co.,[4] indicating that snipe may be more widespread and adaptable than was previously suspected. Snipe populations vary with local conditions; they may be absent during dry years at regular locations (Sutton 1928b) and opportunistic by nesting in unusual sites during wet years. The nest is usually a cupped structure built of grasses in a wet area. Eggs are more darkly spotted than are those of woodcock and most often are found from late Apr to mid-May.

Fall: Southbound adults become evident about the second or third week of Jul, with an increasing number of birds (possibly juveniles) appearing the second or third week of Aug. In the Coastal Plain and Piedmont most birds have departed by late Nov, with stragglers remaining into winter. In areas such as Tinicum and southern Lancaster Co., over 20 or 30 birds have been recorded on CBCs in recent years. Birds depart earlier from the High Plateau and Glaciated Northwest. Most have left these regions by the fourth week of Oct, but stragglers may remain to late Dec or until the first severe freeze.

Winter: Common Snipe rarely winter successfully anywhere across the northern half of the state. Few remain anywhere in the state later than the fourth week of Jan. Wintering birds are found mainly in the Coastal Plain and Piedmont, but numbers vary from year to year, depending on availability of open water. In the Ridge and Valley and in the Southwest, winter records are widespread but local and not found every winter. In the Glaciated Northeast they were found for several winters where water flowed from a barn into a swampy area in Wyoming Co.[5] A group of 37 found on 19 Jan 1979 in a wet spot near Limestoneville in Montour Co. (Schweinsberg 1988) was a very high number for a site this far north. Except for southern Somerset Co., where they have been recorded, the only winter record from the High Plateau is of four birds on 27 Jan 1991, discovered in a pasture seep at Wellsboro in Tioga Co.[6] In the Glaciated Northwest they have been recorded at least once in winter in Butler, Crawford, southern Erie, and Lawrence cos.

History: Few nests or other confirmed breeding evidence were reported historically, but most summer records were from Crawford and Erie cos. (Warren 1890; Todd 1904). Warren (1890) also listed Bradford and Susquehanna cos. as areas where snipe nested rarely. Scattered breeding records from Susquehanna and possibly Bucks Co. (Poole 1964) indicate the potentially widespread breeding distribution of this species. Populations varied even in primary locations, such as the southern end of the Pymatuning Swamp, where Sutton (1928b) estimated "probably a dozen pairs nested" some years but were absent in others.

Comments: The OTC lists this species as Candidate–Rare.

[1] N. Pulcinella, pers. comm.
[2] M. Higbee and R. Higbee, pers. comm.
[3] S. Butcher, pers. comm.
[4] C. Bier, pers. comm.
[5] W. Reid, pers. comm.
[6] Ross, B. 1991. County reports—January through March 1991: Tioga County. PB 5:49.

[Eurasian Woodcock] *Scolopax rusticola*

General status: Eurasian Woodcocks breed across Europe and central Asia. Reports of vagrants in North America are from Quebec and Newfoundland, New Jersey, Pennsylvania, Virginia, and Alabama. This species is listed as hypothetical in Pennsylvania on the basis of birds reportedly shot before the beginning of the twentieth century.

Comments: Warren (1890) cited a large female that was shot on the barrens in East Nottingham Township in Chester Co. in Nov 1886. Stone (1894) mentioned a specimen taken in Northampton Co. before 1890. Apparently the specimens are no longer in existence.

American Woodcock *Scolopax minor*

General status: American Woodcocks are nocturnal and secretive, spending most of the day hidden in damp woodlands and thickets. They are most easily observed at dusk in spring when males perform elaborate flight displays. They breed in the eastern half of the U.S. north to Newfoundland and winter primarily in the southern half of their breeding range. Woodcocks are fairly common regular migrants, breed statewide, and are regular winter residents or visitors in Pennsylvania. Arrival and departure dates of woodcocks at most sites depend on thawing and freezing of the ground. They are hunted as game.

Habitat: Unlike other species of shorebirds, woodcocks prefer damp open or semi-open woodlands with thickets or damp brushy areas bordering woods or fields. Areas containing hawthorne, alder, birch, or aspen are especially favorable habitats. Open patches are used as singing grounds.

Seasonal status and distribution

Spring: Woodcocks may arrive as soon as the ground has thawed, sometimes as early as the second or third week of Feb, but the majority do not appear until the first week of Mar. Birds are most frequently encountered during spring, when they are most active and vocal. They become more difficult to find after the first or second week of May as calling decreases.

Breeding: Woodcocks are uncommon breeders in Pennsylvania. They have nested in every Pennsylvania county, but severe declines are a concern. Although woodcocks are not well documented on BBS routes, hunting records show a significant decline in harvest since 1970 (Diefenbach 1996). A reduction in brushy forest land is a major factor. Woodcocks are uniquely equipped to probe deep in the soil for earthworms; woodcock populations may vary with soil moisture. They also eat a variety of insects. Woodcocks may be best known for the male's dawn and dusk courtship, heard on singing grounds as early as late Feb but peaking in Mar and Apr. The "peent" call is made on the ground, and the wings produce chirps in flight. Females nest close (within 150 yards) to singing grounds, either in old fields or within adjacent woods.[1] The nest is a simple depression on the ground in the open, lined with dead plant material, typically leaves. Egg sets are found from late Mar through May.

Fall: It is difficult to determine exactly when migration begins, since they are not often reported during the summer or early fall. Hunters frequently see them in Oct. Woodcocks are typically reported until the ground freezes. In the northern portions of the state, most have departed by the last week of Oct, and in the southern portions of the state they move south about the second week of Nov. A few may remain to late Dec or later during mild winters in the southern portion of the state.

Winter: Woodcocks regularly winter in the Coastal Plain in the vicinity of Tinicum, but they leave the area during years of harsh weather.[2] They may be found almost anywhere across the southern portion of the state in suitable habitat during mild winters. The only winter record of woodcocks from the northern portion of the state is from McKean Co., where they have been reported in wetlands during mild winters in Jan and Feb.[3]

History: American Woodcocks were described by ornithologists in the nineteenth century as abundant, at least as migrants (Baird 1845). Local hunting regulations were established for woodcocks in 1839.[4] Populations declined precipitously in the 1900s in response to overhunting, spring cold snaps, and maturation of young forests (Sutton 1928b; Todd 1940).

[1] Fergus, C. undated. Woodcock. Wildlife notes, no. 175–21. Pennsylvania Game Commission, Harrisburg.
[2] N. Pulcinella, pers. comm.
[3] J.P. Dzemyan, pers. comm.
[4] J. Kosack, pers. comm.

Wilson's Phalarope *Phalaropus tricolor*

General status: Wilson's Phalaropes are less pelagic than the other two species of phalaropes. They would rather feed by picking along muddy shorelines than by spinning in circles over open water. Wilson's Phalaropes breed from southern Yukon east across southern Canada to Quebec and south to New York and locally in the western U.S. They winter primarily in South America and are casual in California, Utah, New Mexico, Texas, Louisiana, and Florida. They are rare but regular migrants over most of the state but do not occur regularly at any one site. Most records are from Tinicum, several sites in the Piedmont, and at PISP. No record of Wilson's Phalarope exists from the Glaciated Northeast or the High Plateau. They are accidental in the Southwest.

Wilson's Phalarope at Middle Creek Wildlife Management Area, Lancaster County, August 1991. (Photo: Randy C. Miller)

Habitat: Wilson's Phalaropes prefer sandy beaches, mudflats, flooded fields, and shorelines of ponds, lakes, and rivers.

Seasonal status and distribution

Spring: There are fewer sightings of this species in spring than in fall. Records range from the last week of Apr to the last week of May. All sightings are of single birds or pairs. Away from the Coastal Plain, Piedmont, and Lake Erie Shore, Wilson's Phalaropes have been recorded in spring in the Ridge and Valley from Centre, Clinton, Dauphin, Lycoming, Luzerne, and Northumberland cos. In the Southwest there is one spring record: one was found at the Imperial grasslands in Allegheny Co. on 14 May 1994.[1] In the Glaciated Northwest they have been observed in spring in Crawford, Erie, and Lawrence cos. Stragglers have been found until the first week of Jun.

Summer: Wood (1983) listed extreme dates of 18 Jun to 1 Aug but did not cite any specific records. There was a female Wilson Phalarope on 18–20 Jun 1997 at the Landingville Dam spillway mudflat in Schuylkill Co.[2]

Fall: Most records are from the last week of Jul to the second week of Sep. The majority of sightings in fall are of one or two birds, but up to four or five have been recorded. On 3 Sep 1976 and 6 Sep 1977 as many as 13 were present at Tinicum.[3] East of the Allegheny Front and outside the Piedmont and Coastal Plain, they have been recorded from the Ridge and Valley only in Centre and Luzerne cos. West of the Allegheny Front, Wilson's Phalarope has been recorded in the Southwest only in Allegheny Co.; in the Glaciated Northwest only from Crawford, Erie, and Mercer cos.; and at the Lake Erie Shore from PISP. Stragglers have lingered to the second week of Nov.

History: Poole (1964) listed Wilson's Phalaropes as casual, occurring irregularly both in spring and fall. In his manuscript, Poole listed records from Tinicum in Philadelphia or Delaware Co., the Conejohela Flats and Oregon Pond in Lancaster Co., Lake Ontelaunee and Moselem Springs in Berks Co., and Pocono Lake in Monroe Co. In the western part of the state, Poole (unpbl. ms.) listed Wilson's Phalaropes at Lake Pymatuning in Crawford Co. and Presque Isle in Erie Co.

[1] Haas, F.C., and B.M. Haas, eds. 1994. Rare and unusual bird reports. PB 8:105.

[2] Clauser, T. 1997. Local notes: Schuylkill County. PB 11:100.

[3] N. Pulcinella, pers. comm.

Red-necked Phalarope *Phalaropus lobatus*

General status: In North America, Red-necked Phalaropes breed from Alaska east across northern Canada to Labrador, and they winter in the Pacific and Atlantic oceans off southern South America. Of the three species of phalaropes they are the most widely distributed in Pennsylvania, with scattered records represented in all physiographic regions. They are rare irregular migrants in both spring and fall. Only two records are from the northern and central portion of the High Plateau.

Habitat: Red-necked Phalaropes are found on mudflats, in marshes, and on muddy shorelines of lakes, ponds, and rivers. They are often seen swimming in open water just off shore. Red-necked phalaropes have also been recorded on the water several miles from shore in Lake Erie.

Seasonal status and distribution

Spring: Birds have been recorded from the third week of Apr to the first week of Jun, but most are recorded during the third and fourth weeks of May. Sightings are usually of one or two birds. However, 11 male Red-necked Phalaropes were observed at Antes Fort in Lycoming Co. on 18 May 1980,[1] and 10 were found at Marysville in Dauphin

Red-necked Phalarope at Shenango Reservoir, Mercer County, August 1993. (Photo: Mike Fialkovich)

Co. on 15 May 1997 with 4 remaining to 16 May 1997.[2] Apparently the only recorded spring sightings west of the Allegheny Front are from Lake Arthur in Butler Co. on 27 May 1972,[3] the Union City Dam in Erie Co. on 19 May 1982,[4] and at Powdermill in Westmoreland Co., where three were found on 18 May 1984.[5]

Fall: More sightings are in fall than in spring. Red-necked Phalaropes have been seen from the third week of Jul to the last week of Nov, with most observed from the second week of Aug to the third week of Sep. As many as 7 birds were counted on 25 Aug 1981 at Green Lane in Montgomery Co. (Morris et al. 1984). There are two records from the Glaciated Northeast: one at Tunkhannock in Wyoming Co. from 17 Aug to 1 Sep 1975 and one at Rummerfield in Bradford Co. on 28 Oct 1993.[6] Records from the High Plateau include one found on Kinzua Dam in Warren Co. on 13 Oct 1967[7] and one at Greenman Hill in Potter Co. on 20 to 22 Sep 1995.[8]

History: Their status and distribution have remained relatively unchanged through history. Poole (1964) listed this species as uncommon in fall and very rare in spring. One collected near Reading on 10 Apr 1890 was very early. It was placed in the Reading Museum (Poole, unpbl. ms.). Todd (1940) listed no spring record for western Pennsylvania.

Comments: Caution is advised when identifying Red-necked Phalaropes in late fall, especially in Oct and Nov when the later-arriving Red Phalaropes are most likely to occur. Juvenile Red Phalaropes molting into first-winter plumage, usually Sep birds, can resemble Red-necked Phalaropes molting into winter plumage.

[1] P. Schwalbe and G. Schwalbe, pers. comm.
[2] Williams, R. 1997. Local notes: Dauphin County. PB 11:94.
[3] Hall, G.A. 1972. Appalachian region. AB 26:761.
[4] Hall, G.A. 1982. Appalachian region. AB 36:852.
[5] Mulvihill, R.S., and R.C. Leberman. 1986. Birdbanding at Powdermill, 1984. Research report no. 45. Powdermill Nature Reserve, Carnegie Museum of Natural History, Pittsburgh.
[6] W. Reid, pers. comm.
[7] Stull, J.H., and J.G. Stull. 1967. Field notes. Sandpiper 10:15.
[8] Hauber, D. 1995. Local notes: Potter County. PB 9:150.

Red Phalarope *Phalaropus fulicaria*

General status: In North America, Red Phalaropes breed from western and northern Alaska across the Canadian Arctic coast and islands east to Baffin Island. They winter in the Pacific Ocean off the coast of South America and in the Gulf of Mexico and Atlantic Ocean from Florida south. Red Phalaropes are the most pelagic of the three phalarope species. In Pennsylvania they are frequently observed on the water rather than along the shore. Nearly all reports in the state are during the fall. They are rare fall migrants and regular only at the Lake Erie Shore. Red Phalaropes are casual in the Piedmont and accidental elsewhere in the state.

Habitat: They are most often seen swimming in open water just off shore. They have also been reported on mudflats and along muddy shorelines of lakes, ponds, rivers, and marshes. On Lake Erie most birds are observed migrating past Gull Point far from shore.

Seasonal status and distribution

Spring: Two Red Phalaropes have been recorded at this season: a female at Van Sciver Lake at Penn Manor in Bucks Co. on 5–7 Jun 1971 was photographed.[1] A dead bird molting into breeding plumage was found on a suburban lawn at Blue Bell in Montgomery Co. on 22 Apr 1984.[2]

Fall: The earliest southbound birds appear the last week of Aug, but there is one rather early date: one on 10 Aug 1961 at the Conejohela Flats in Lancaster Co. (Morrin et al. 1991). Most Red Phalaropes are seen after the second week of Oct. Usually only singles are seen, but up to four have been recorded. Away from the Piedmont and Lake Erie Shore, they have been recorded from Allegheny, Bucks, Cambria, Chester, Centre (two sites), Crawford, Erie, Greene, Indiana, Northumberland, and Westmoreland cos. Birds have departed by the second week of Dec, with the exception of a late bird appearing at PISP on 5 Jan 1983.[3]

History: The first documented Red Phalarope record in Pennsylvania was made in 1899 of a bird shot by J. Thompson at Erie (Todd 1940). The first record in eastern Pennsylvania came from a specimen collected about 15 Dec 1918 near Lenape in Chester Co.[4] Thereafter, all but one sighting was from the Piedmont in the counties of Berks, Bucks, Chester, and Lancaster or at the Lake Erie Shore. Apparently the only record away from those areas before 1960 was one reported at Centre Furnace Pond in Centre Co. in 1946 (Wood 1983).

[1] Sehl, R.H. 1971–1972. Red Phalarope in Bucks County. Cassinia 53:46–47.

[2] Boyle, W.J., Jr., R.O. Paxton, and D.A. Cutler. 1984. Hudson-Delaware region. AB 38:891.

[3] Hall, G.A. 1983. Appalachian region. AB 37:302.

[4] Ehinger, C.E. 1919. General notes. Cassinia 23:31.

Family Laridae: Jaegers, Gulls, and Terns

Birds of the family Laridae are usually associated with water. Their wings are long and narrow, and their body is generally long and streamlined. Their bill is short, either hooked at the tip or straight and daggerlike. Their legs are short, and feet webbed. This family is divided into three major subfamilies (jaegers, gulls, and terns), each with distinctive behaviors and feeding strategies.

Jaegers are pirates of large lakes and rivers, chasing gulls and terns and stealing what they have consumed by forcing them to regurgitate food. Jaegers breed across the tundra and winter in the Atlantic and Pacific oceans. The two species that occur in Pennsylvania are listed as rare vagrants.

Most gulls feed on fish or animals washed ashore or on garbage at landfills. They also catch live fish by picking them from the surface or occasionally by plunge-diving. Larger gulls are occasionally predatory, especially during the winter; in summer they frequently raid nests of smaller species of gulls, terns, and shorebirds. Some species, such as Ring-billed Gulls, are opportunists, frequenting mall parking lots for handouts or discarded food. Of the 16 species of gulls recorded in Pennsylvania, only two, Ring-billed and Herring gulls, have been known to nest here, and those only recently. Ring-billed Gulls have yet to hatch eggs or to brood young in their nesting attempts here. Most jaegers and gulls reported in the state are migrants or winter visitors.

Of all the birds in the world, terns are considered to be the most graceful flyers. Eleven out of the 20 species of terns found in North America have been recorded in Pennsylvania. Terns are sleeker, thinner versions of gulls and have a straight rather than hooked bill. They spend more time in the air or feeding over water than most species of gulls, and they feed exclusively on fish that they capture from the surface or by plunge-diving. Common and Black terns, the only two species that have bred in the state, are in serious decline. Common Terns have not successfully nested in the state for over 30 years.

Pomarine Jaeger ***Stercorarius pomarinus***

General status: In North America, Pomarine Jaegers breed in western and northern Alaska and east across the Canadian Arctic islands to Greenland south to northeastern Quebec. They winter over the open sea in the Pacific Ocean from California south, in the Gulf of Mexico, and in the Atlantic Ocean from Florida south. They are rare but are regularly observed from shore along the East Coast during migration, but they are rarely reported inland from the Atlantic Ocean during migration. They are accidental in Pennsylvania, with about 9 or 10 modern records of single birds from three sites in summer or fall. Most birds have been in immature plumage.

Seasonal status and distribution

Summer to fall: In Lancaster Co. two reports are from the Susquehanna River. A bird identified as an immature Pomarine Jaeger at the Conejohela Flats remained from 29 Aug 1993 until 6 Sep 1993, when it was joined by an adult.[1] On 15 and 16 Jul 1969 an adult was photographed at Colyer Lake in Centre Co.[2] and remained to 27 Jul 1969 (Wood 1983). A Pomarine Jaeger in near-adult plumage remained at Hinckston Run Dam in Cambria Co. from 24 Jun to 3 Jul 1987.[3] One adult was reported at MSP in Butler Co. during the passage of Hurricane Fran on 7 Sep 1996.[4] Pomarine Jeagers have been reported five times from the Lake Erie Shore at PISP: on 27 Nov 1965 (Stull et al. 1985), 24 Sep 1966,[5] 20 Dec 1980,[6] one photographed on 5 Dec 1996,[7] and one on 7 Dec 1997.[8]

History: Baird (1845) wrote, "adult procured on [the] Susquehanna [River] in Summer." Warren (1890) mentioned two sightings: one collected on the Susquehanna River in Lancaster Co. and another shot in the winter of 1885 or 1886 near Eagles Mere in Sullivan Co. A Pomarine Jaeger was collected opposite Andulusia in Bucks Co. in Oct 1898 and was placed in the ANSP. The last three specimens can no longer be found.

Comments: A few distant jaegers, possibly Pomarines, have been seen migrating far from shore over Lake Erie.

[1] Haas, F.C., and B.M. Haas, eds. 1993. Rare and unusual bird reports. PB 7:104.
[2] Hall, G.A. 1970. Appalachian region. AFN 24:653.
[3] Georg, K., and B. Mulvihill. 1987. The first Pomarine Jaeger record for Cambria County. PB 1:30–33.
[4] Hess, P. 1996. Local notes: Butler County. PB 10:162.
[5] Hall, G.A. 1967. Appalachian region. AFN 21:31.
[6] Hall, G.A. 1981. Appalachian region. AB 35:300.
[7] G.M. McWilliams, pers. obs.
[8] Haas, F.C., and B.M. Haas, eds. 1997. Photographic highlights. PB 10:217.

Parasitic Jaeger ***Stercorarius parasiticus***

General status: In North America the breeding range of Parasitic Jaegers is from Alaska east across the Canadian Arctic islands to Greenland and south through Mackenzie and along the south side of Hudson Bay. They winter in the Pacific Ocean from California south, in the Atlantic Ocean from Maine south, and in the Gulf of Mexico south. Of the three species of jaegers, Parasitic Jaegers are more often observed on the eastern Great Lakes than either of the other two species. In Pennsylvania, most Parasitic Jaegers are observed flying offshore along Lake Erie, where they are rare irregular fall migrants. Elsewhere in the state, they are accidental in fall in the lower Susquehanna River in Lancaster Co. and in the Southwest.

Habitat: Parasitic Jaegers are observed over Lake Erie and on the lower Susquehanna River. In the Southwest they have been found on large lakes or reservoirs.

Seasonal status and distribution

Fall: On days of cold northerly winds this species may be seen over Lake Erie primarily from the outer beaches of PISP. Most Parasitic Jaegers are observed pursuing gulls, but occasionally they are found resting on the outer beaches of Gull Point. All sightings have been of immatures, and all three color morphs have been reported. They have been seen from the first week of Aug to the fourth week of Dec, with most sightings in Oct and Nov. There are three records from Lancaster Co.: a light-morph bird at Muddy Run on the early date of 2 Aug 1973 (Morrin et al. 1991), an immature at the same site on 27–30 Aug 1975 (Morrin et al. 1991), and a dark immature at the Conejohela Flats on 2–3 Nov 1991.[1] In the Southwest there are only two records both during the passage of Hurricane Fran on 7 Sep 1996. One was at MSP in Butler Co.,[2] and one was at Keystone State Park in Armstrong Co.[3]

History: Most historical records have been in the fall from the waters surrounding PISP. Todd (1940) cited a spring record of a jaeger believed to be a Parasitic chasing a Common Tern on 22 May 1904. Todd also reported a female Parasitic Jaeger captured at Shintown, near Renova in Clinton Co. on the unusual date of 18 Jun 1911. The bird was placed in the Carnegie Museum. Another rather late record was of a light-morph bird on the Conejohela Flats on 1 Jun 1952 (Morrin et al. 1991). Poole (1964) listed this species as accidental, with records outside Lake Erie coming from the Susquehanna River below Columbia, one or two from the Delaware River around Philadelphia, and one over Hawk Mountain on 29 Oct 1963.

Parasitic Jaeger at Presque Isle State Park, Erie County, November 1991. (Photo: Gerald M. McWilliams)

Comments: In recent years, frequent observations from PISP over Lake Erie during ideal conditions have changed the status of Parasitic Jaeger in this region. Once believed to be an accidental vagrant, this species is now expected almost annually. There have been many sightings of unidentified jaegers from the Lake Erie Shore that probably refer to this species. Two records are outside of the time period when Parasitic Jaegers are usually reported: a jaeger reported from Tinicum on 22 May 1960 may have been a Parasitic,[4] and a rather late unidentified jaeger was seen on 4 Jan 1983 at PISP.[5] Distinguishing jaegers in the field, especially Pomarine and Parasitic, is very difficult. It is very likely that there have been some misidentifications.

[1] Paxton, R.O., W.J. Boyle Jr., and D.A. Cutler. 1992. Hudson-Delaware region. AB 46:69.

[2] Hess, P. 1996. Local notes: Butler County. PB 10:162.

[3] Higbee, M., and R. Higbee. 1996. Local notes: Armstrong County. PB 10:161.

[4] Scott, F.R., and D.A. Cutler. 1960. Middle Atlantic Coast region. AFN 14:375.

[5] Hall, G.A. 1983. Appalachian region. AB 37:302.

Laughing Gull *Larus atricilla*

General status: Laughing Gulls breed along the Atlantic coast from Nova Scotia south and along the Gulf coast, and they winter in the southern part of their breeding range. In Pennsylvania they are common to abundant regular migrants and summer residents along the Delaware River in the Coastal Plain. Laughing Gulls are uncommon regular fall and irregular spring migrants and are accidental in winter along the lower Susquehanna River in Lancaster Co. They are regular visitors in spring and are casual in summer and fall at the Lake Erie Shore. These gulls are irregular to casual visitors from spring through fall in the Piedmont away from the lower Susquehanna and Delaware rivers and their tributaries. This species is casual elsewhere in the state, where records have come from Armstrong, Butler, Cambria, Centre, Clinton, Indiana, Montour, Northumberland, and Warren cos. Laughing Gulls were not recorded anywhere west or north of the Allegheny Front until 1971, when the first record came from the Lake Erie Shore. No known record of Laughing Gull exists from the Glaciated Northeast.

Habitat: They are found near rivers and lakes, especially the lower Delaware River.

Seasonal status and distribution

Spring: Along the lower Delaware River, Laughing Gulls arrive the first or second week of Apr, with numbers gradually increasing through May. Daily concentrations reach 100 or more birds at Tinicum.[1] Away from the Delaware River they have been recorded as early as the first week of Apr. The greatest number does not appear until the first week of May and leaves about the third week of May. Usually no more than one or two birds are seen at each sighting, but up to seven were recorded at Wales in Montgomery Co.[2] Stragglers may remain to the third week of Jun.

Summer: Laughing Gulls regularly summer along the lower Delaware River north to Bucks Co., where hundreds can sometimes be found. Although not expected to nest locally, Laughing Gulls were observed copulating at the landfills in southern Bucks Co. during the 1980s. In 1986 west of Tulleytown in Bucks Co. an estimated 800 birds in breeding plumage were observed on the ground. At least 30 copulating birds were counted in this concentration by G. Dewaghe in May, and in Jul five or six juveniles were present. Summer sightings are rare elsewhere.[3] They have been observed along the lower Susquehanna River, in Montgomery

Co. at Green Lane Reservoir[4] and at PISP.

Fall: As summer progresses, Laughing Gull numbers begin to build into the hundreds or thousands along the lower Delaware River. The number of birds peaks from about the third week of Aug to the third week of Sep. On the Delaware River at the Philadelphia airport on 17 Sep 1985, 3000 were counted.[5] On the Lower Susquehanna River at the mouth of Fishing Creek in Lancaster Co., 26 were seen on 30 Aug 1987 (Morrin et al. 1991). Away from these areas, usually only one or two birds are seen at each sighting, but up to six have been recorded. Beginning the second week of Aug 1985, there was a notable Laughing Gull invasion on Lake Erie, most of which were immature birds. At least four different immature birds were at PISP from 10 Aug to 8 Sep.[6] This was only a token number compared with the numbers at surrounding sites such as along Lake Erie in Ohio, where numbers of birds grew to flocks of 18 at Cleveland.[7] In the Ontario region, 20 birds were reported from sites on Lake Erie.[8] There were also inland Pennsylvania Laughing Gull reports in Centre and Montgomery cos. around this time that were likely related to this event. The unusually large concentration of gulls at the Philadelphia airport noted above was probably also related to this event. The number of birds begins to decline through Oct. The majority have left the state by the third week of Nov, with stragglers to the first week of Jan.

Winter: In the winter there is one record. Two immature Laughing Gulls in first-winter plumage were at the Holtwood Dam on the Susquehanna River in Lancaster Co. from 11 to 20 Jan 1984 (Morrin et al. 1991).

History: Laughing Gulls were formerly considered irregular along the lower Delaware River. Poole (1964) listed this species as an irregular spring and fall transient along the lower Delaware and Susquehanna rivers. There were only four historical sightings of this species away from the Coastal Plain and Piedmont: three from the West Branch of the Susquehanna River in Clinton Co. (Todd 1940) and one in Monroe Co. at Pocono Lake (P. B. Street 1954).

[1] N. Pulcinella, pers. comm.
[2] Boyle, W.J., Jr., R.O. Paxton, and D.A. Cutler. 1987. Hudson-Delaware region. AB 41:408.
[3] G. Dewaghe, pers. comm.
[4] Paxton, R.O., W.J. Boyle, Jr., and D.A. Cutler. 1985. Hudson-Delaware region. AB 39:892.
[5] Paxton, R.O., W.J. Boyle, Jr., and D.A. Cutler. 1986. Hudson-Delaware region. AB 40:89.
[6] G.M. McWilliams, pers. obs.
[7] Peterjohn, B.G. 1986. Middlewestern Prairie region. AB 40:120.
[8] Weir, R.D. 1986. Ontario region. AB 40:107.

Franklin's Gull ***Larus pipixcan***

General status: Native to the prairie Midwest, Franklin's Gulls breed in south-central Canada and the north-central U.S. and winter in South America and casually to California and Florida. They were recorded only once in Pennsylvania before 1970. Since then, they have been found here nearly every year. Franklin's Gulls are consistently reported only at PISP, where they are rare but regular visitors. Though still not annual, the frequency of sightings along the lower Susquehanna River has increased in recent years. Franklin's Gulls are accidental elsewhere in the state. Nearly all sightings are of single birds.

Habitat: Franklin's Gulls typically are seen on lakes and rivers, especially Lake Erie and the Susquehanna River.

Seasonal status and distribution

Spring: Most records are at this season, from the last week of Mar to the third week of Jun. Away from Lake Erie, Franklin's Gulls have been reported only since 1982 from the counties of Dauphin, Lancaster, and York along the Susquehanna River; and once in Somer-

Franklin's Gull at Edinboro Lake, Erie County, June 1992. (Photo: Gerald M. McWilliams)

set on Somerset Lake on 8 Apr 1984;[1] in Warren on 12 May 1984;[2] in Luzerne on Harvey's Lake on 17–18 Apr 1992;[3] in Erie on Edinboro Lake from 1 Apr to 14 Jun 1992;[4] and in Cambria near Patton in the spring of 1997, where two birds were seen together.[5]

Summer: The report of two birds discovered with Ring-billed Gulls at Pymatuning Lake on 6 Jul 1984[6] is the only record away from PISP, where they have been observed several times in summer.

Fall: The first modern record of Franklin's Gull in Pennsylvania was on 14 Nov 1971, when D. Bollinger and colleagues discovered one at PISP.[7] They may appear almost any time during the fall season. They are frequently associated with Bonaparte's Gull migration periods in Oct and through the second week of Nov. Away from PISP, there was a rather early record from 27 Aug to perhaps 17 Sep 1988 at the West Fairview boat launch in Cumberland Co.[8] They have been recorded at least four times on the Conejohela Flats in Lancaster Co. during the period from 15 Sep (Morrin et al. 1991) to 3 Nov.[9] Two were at Codorus State Park in York Co. on 11 Nov 1991,[10] and one was at YCSP in Indiana Co. from 6 to 10 Nov 1990[11] and on 30 Sep 1997.[12] A late fall straggler was recorded at YCSP on 5 Jan 1993.[13] Late fall stragglers have also been seen through Dec and early Jan on PISP in Erie Co.[7]

Winter: The only winter sightings are of one bird at Brunner's Island in York Co. on 27 Feb 1992[14] and two from PISP during the third and fourth weeks of Jan in 1975 and 1979.[7]

History: Todd (1940) cited one record within the boundaries of Pennsylvania: an adult in transition from breeding to winter plumage was collected by L. E. Hicks at Presque Isle on 13 Sep 1937. The same observer collected a specimen in Ashtabula, Ohio, on the same day.

[1] Hall, G.A. 1984. Appalachian region. AB 38:911.

[2] Hess, P.D. 1984. Area bird summaries for May. ASWPB 49:11.

[3] Boyle, W.J., Jr., R.O. Paxton, and D.A. Cutler. 1992. Hudson-Delaware region. AB 46:400.

[4] McWilliams, G.M. 1992. County reports—April through June 1992: Erie County. PB 6:80.

[5] Lamer, G. 1997. Local notes: Cambria County. PB 11:93.

[6] Hall, G.A. 1984. Appalachian region. AB 38:1020.

[7] G.M. McWilliams, pers. recs.

[8] Hoffman, D. 1988. County reports—July through September 1988: Cumberland County. PB 2:103.

[9] Haas, F.C., and B.M. Haas, eds. 1996. Birds of note—October through December 1995. PB 9:208.

[10] Spiese, A. 1992. County reports—October through December 1991: York County. PB 5:186.

[11] Higbee, M., and R. Higbee. 1991. County reports—October through December 1990: Indiana County. PB 4:158.
[12] Higbee, M., and R. Higbee. 1997. Local notes: Indiana County. PB 11:155.
[13] M. Higbee and R. Higbee, pers. comm.
[14] A. Spiese, pers. comm.

Little Gull ***Larus minutus***

General status: Formerly restricted to the Old World, Little Gulls were first known to nest in North America in 1962 at Oshawa in Ontario, Canada (Godfrey 1986). Since then they have nested sporadically in Manitoba, Minnesota, Wisconsin, Michigan, Ontario, and Quebec. They winter on the Great Lakes and along the Atlantic coast from Newfoundland to Virginia. Little Gulls have been regular only since 1975 in Pennsylvania. Little Gulls are now rare spring and fall migrants in Pennsylvania but are regular only at the Lake Erie Shore. They are accidental in summer and casual in winter at PISP and accidental in spring elsewhere in the state. However, an influx of recent sightings along the lower Susquehanna River may soon change the status of this species. Most records away from Lake Erie have been during the 1990s. Little Gulls are nearly always associated with Bonaparte's Gulls.

Habitat: They occur on lakes, especially Lake Erie, and larger rivers, especially the lower Susquehanna River.

Seasonal status and distribution

Spring: Little Gulls nearly always arrive with flocks of Bonaparte's Gulls. Sightings begin as early as the second week of Mar. Their typical migration period is from the last week of Mar to the fourth week of May. Away from Lake Erie they have been recorded only during spring. Along the lower Delaware River there is a record from the Philadelphia International Airport in Philadelphia Co. on 10 May 1980,[1] and two birds were seen flying upriver with Laughing Gulls on 29 Apr 1993 at Marcus Hook in Delaware Co.[2] Spring sightings have come from several sites on both sides of the Susquehanna River in York and Lancaster cos. from the Maryland border north to Fort Hunter in Dauphin Co. Away from the Susquehanna River one was at Long Arm Creek Reservoir in southeastern York Co. on 13 Apr 1994,[3] at Lake Ontelaunee in Berks on 25 Apr 1995,[4] and at Hibernia Park in Chester Co. on 20 Apr 1996.[5] Isolated spring records include one at Stone Valley Recreation Area in Huntingdon Co. on 18 Apr 1972 (Wood 1983) and along the North Branch Susquehanna River at Plymouth, Luzerne Co.

Little Gull at Presque Isle State Park, Erie County, October 1997. (Photo: Gerald M. McWilliams)

on 2 Apr 1994.[3] Only two spring records are known from West of the Allegheny Front away from Lake Erie: one was at Red Bank Creek in New Bethlehem in Clarion Co. on 4 Apr 1995[4] and one was at Lake Lebouef in Erie Co. on 28 Apr 1985.[6]

Summer: The only summer record is from PISP, where one was observed on 26 Jul 1976.[7]

Fall: Early fall migrants are seen on Lake Erie, especially at PISP. Migration begins as early as the fourth week of Aug. Juvenile-plumaged birds have been seen on three occasions on PISP: 23 Aug 1987,[8] 5 Sep 1989,[9] and 29 Aug 1992.[7] Their usual migration time is from the third or fourth week of Oct to about the first week of Dec. A few may linger through Dec into the beginning of winter, but they depart with Bonaparte's Gulls as soon as lakes and rivers freeze.

Winter: Little Gulls usually leave by the third week of Jan. When open water and food are available they have been recorded with Bonaparte's Gulls through the winter. On 8 Jan 1983 a total of 13 birds was counted at PISP.[7] In recent years, Little Gulls have become quite rare and irregular after the fourth week of Dec at Presque Isle Bay, at least in part as a result of the closing of an electric power plant in the early 1990s. This plant discharged warm water into Presque Isle Bay that prevented the water from freezing and attracted millions of gizzard shad that provided food for thousands of gulls.

History: The first sighting of a Little Gull in Pennsylvania was on 27 Dec 1933, when C. M. Beal and R. T. Peterson observed an adult at the foot of State Street in Erie.[10] This is the only record of this gull in Pennsylvania before 1960.

Comments: Quite remarkable was the discovery of the remains of a Little Gull at Beaver Run Reservoir in Westmoreland Co. in Jun 1996. This bird was banded as a chick north of Stockholm, Sweden, on 7 Jul 1995. It is the first banded bird from Sweden ever recovered in the U.S.[11]

[1] Miller, J.C. 1986–1987. Birds of the Tinicum Wildlife Refuge and adjacent areas Philadelphia, Pennsylvania 1970–1987. Cassinia 62:46.
[2] N. Pulcinella, pers. comm.
[3] Haas, F.C., and B.M. Haas, eds. 1994. Rare and unusual bird reports. PB 8:105.
[4] Haas, F.C., and B.M. Haas, eds. 1995. Birds of note—April through June 1995. PB 9:88.
[5] Blust, B. 1996. Local notes: Chester County. PB 10:98.
[6] Hall, G.A. 1985. Appalachian region. AB 39:300.
[7] G.M. McWilliams, pers. obs.
[8] Hall, G.A. 1988. Appalachian region. AB 42:75.
[9] Hall, G.A. 1990. Appalachian region. AB 44:90.
[10] Beal, F.E.L., and R.T. Peterson. 1934. Erie CBC. Bird-Lore 36:33.
[11] Leberman, R.C. 1996. Local notes: Westmoreland County. PB 10:167.

Black-headed Gull *Larus ridibundus*

General status: Originally vagrants from Eurasia, Black-headed Gulls were uncommon along the East Coast and casual elsewhere in the U.S. They were first discovered breeding in North America when, in 1977, at least two pairs of adults with three newly fledged young were photographed at Stephenville Crossing in Newfoundland (Godfrey 1986). Since then they have been reported breeding in Greenland, Quebec, Maine, and Massachusetts. They winter along the Atlantic coast from Labrador south to New York and casually in Ontario and on the Great Lakes. In Pennsylvania, Black-headed Gulls are currently listed as accidental. This species has been recorded in the state at least 10 times, with only one record before 1988.

Seasonal status and distribution

Spring: The first record in the state was in spring. An immature bird molting into first spring plumage was discovered by D. Snyder at PISP in Erie Co. on 28 May 1979.[1] Black-headed Gulls were

Black-headed Gull at Conowingo Pond, Lancaster County, April 1992. (Photo: Robert M. Schutsky)

not reported in Pennsylvania again until 1989, when one in breeding plumage was identified at PISP on 1 and 2 Apr.[2] Two were at the same site on 19 May 1990,[3] and one was there on 12 Apr 1992.[4] Also in 1992, a first spring bird was photographed on 22–23 Mar at Conowingo Pond on the lower Susquehanna River in Lancaster Co.:[5] one observer suspected that as many as six different immature birds may have been present those two days because of plumage variations. One in breeding plumage was also present at this site with a Franklin's Gull on 17–18 Apr 1992,[6] and one was reported from the Conowingo Pond on 8 Apr 1995.[7]

Fall: Only two records fall within this season: a bird in first-winter plumage was seen on 23 Dec 1988 at the foot of State Street in Erie,[8] and the most recent record was on 4 Nov 1997, when an adult Black-headed Gull was photographed at Riverfront Park in Pittsburgh, Allegheny Co.[9]

Winter: Black-headed Gulls have been reported twice in winter. At a sewage treatment plant near Oxford in Chester Co. one was discovered on 23 Jan 1993 and remained for several weeks.[10] The most recent record in the state was of at least one bird in Bucks Co. at Nockamixon State Park on 4 Jan 1997 and one at Peace Valley Park on 2 Mar 1997.[11]

[1] Hall, G.A. 1979. Appalachian region. AB 33:863.
[2] Hall, G.A. 1989. Appalachian region. AB 43: 1315.
[3] Hall, G.A. 1990. Appalachian region. AB 44:426.
[4] Hall, G.A. 1992. Appalachian region. AB 46:422.
[5] Witmer, E. 1992. County reports—January through March 1992: Lancaster County. PB 6:38.
[6] Witmer, E. 1992. County reports—April through June 1992: Lancaster County. PB 6: 82–83.
[7] Haas, F.C., and B.M. Haas, eds. 1995. Birds of note—April through June 1995. PB 9:88.
[8] Hall, G.A. 1989. Appalachian region. AB 43:313.
[9] Fialkovich, M. 1998. Local notes: Allegheny County. PB 11:234.
[10] Haas, F.C., and B.M. Haas, eds. 1993. Photographic highlights. PB 7:17.
[11] Kitson, K. 1997. Local notes: Bucks County. PB 11:27.

Bonaparte's Gull ***Larus philadelphia***

General status: Bonaparte's Gulls breed from Alaska south across Canada to southwestern Quebec. They winter along the Pacific coast of the U.S. and Mexico in the West. In the East they winter on the Great Lakes south through the Ohio and lower Mississippi valley, along the Gulf coast and north along the Atlantic coast to Massachusetts. In Pennsylvania, Bonaparte's Gulls are abundant migrants at the Lake Erie Shore and uncommon to fairly common migrants over the remainder of the state. They are casual summer and irregular winter visitors or residents at the Lake Erie Shore and accidental elsewhere in summer and winter, except in the Coastal Plain, and Piedmont. In the Coastal Plain Bonaparte's Gulls are regular winter residents, and in the Piedmont they are casual in winter.

Habitat: Bonaparte's Gulls are found on lakes, ponds, rivers, and marshes. They are occasionally seen in plowed wet fields or migrating overland far from water.

Seasonal status and distribution

Spring: The earliest migrants may appear in the third week of Feb after a mild winter. Their typical migration period is from the fourth week of Mar or the first week of Apr to mid-May. Peak migration ranges from the first week of Apr to the third week of Apr. Flocks containing 1000 or more birds are not unusual around PISP.[1] Flock sizes vary from year to year away from the Lake Erie Shore. At most sites in Pennsylvania daily highs are usually fewer than 50 birds, but concentrations in the hundreds have been recorded, and as many as 300 have been counted at Beltzville Lake in Carbon Co.[2] and at YCSP in Indiana Co.[3] Notable numbers have been recorded at Pymatuning in Crawford Co. and MSP in Butler Co., where concentrations of 400–500 birds have been counted.[4] Lingering birds, mostly immatures in first spring plumage, remain to the second week of Jun.

Summer: Bonaparte's Gulls found between the second week of Jun and the third week of Jul are considered summer records. Most birds during this period are one-year-olds. Away from the Lake Erie Shore summer records are few and lack precise dates.

Fall: The first evidence of southbound birds usually occurs at the Lake Erie Shore in the third or fourth week of Jul. Migration typically occurs from the third or fourth week of Aug through Dec. Most Bonaparte's Gulls move through the state with the first strong cold front from the last week of Oct to about the fourth week of Nov. Larger numbers pass along the Lake Erie Shore than at any other site in the state; as many as 15,000 birds have been counted in a day migrating by PISP.[5] Bonaparte's Gulls are casual to accidental in the Glaciated Northeast in fall.[6] Elsewhere they are usually less common in the fall than in the spring, with highs usually fewer than 35 birds, but up to 150 have been recorded.[7] The number of birds may occasionally build to several thousand by late Dec or into the first or second week of Jan. Hundreds have been recorded along the lower Susquehanna River in Lancaster Co. in Dec. On 18 Dec 1983, 700 were counted near Muddy Run on the CBC.[8] At Pymatuning, concentrations may build to thousands of birds by late Dec, with a high of up to 15,000 recorded in 1989.[9] An estimated 150,000 were attracted to abundant gizzard shad at PISP in late Dec 1988.[10] Stragglers remain to the second or third week of Jan.

Winter: Most birds are forced south by freezing water before the middle of Jan. During mild winters the number of birds may remain high until the last week of Jan. After the fourth week of Jan, Bonaparte's Gulls usually have left the state, except at PISP and at Tinicum, where fewer than five birds usually are recorded. Before the electric power plant at the foot of State Street at Erie was closed in the early 1990s, Bonaparte's Gulls were regular and far more common in winter. Since then, they have been absent nearly every winter.

[1] G. M. McWilliams, pers. recs.
[2] Boyle, W.J., Jr., R.O. Paxton, and D.A. Cutler. 1987. Hudson-Delaware region. AB 41:409.
[3] M. Higbee and R. Higbee, pers. comm.
[4] Hall, G.A. 1986. Appalachian region. AB 40:471.
[5] Hall, G.A. 1985. Appalachian region. AB 39:53.
[6] W. Reid, pers. comm.
[7] Hall, G.A. 1987. Appalachian region. AB 41:89.
[8] Boyle, W.J., Jr., R.O. Paxton, and D.A. Cutler. 1984. Hudson-Delaware region. AB 38:300.
[9] Hall, G.A. 1990. Appalachian region. AB 44:90.
[10] Hall, G.A. 1989. Appalachian region. AB 43: 313.

Mew Gull ***Larus canus***

General status: Of the three described subspecies of Mew Gulls, two occur in eastern North America. *Larus canus brachyrhynchus* breeds in North America from Alaska southeast to Manitoba and south along the coast to British Columbia and Northwest Canada. They winter along the Pacific coast and are casual east to New England and Delaware. *Larus canus canus* breeds in Eurasia from the British Isles east through Russia and Siberia. This subspecies winters from the Mediterranean east to Japan. In North America they have been confirmed only in Greenland and Nova Scotia. Mew Gulls are accidental in Pennsylvania, with only two accepted records.

Seasonal status and distribution

Fall through winter: On 2 Jan 1992 a Mew Gull described as *L. c. canus* was discovered by J. Heller at Lake Ontelaunee in Berks Co.[1,2] On 3 Jan it was found at Blue Marsh Lake in Berks Co. and remained there until 13 Jan. It was observed by many people and was photographed. A Mew Gull was reported again at Strausstown in Berks Co. on 22–23 Feb 1992.[3] One of undetermined subspecies was observed by D. Allison and H. Rufe on the Southern Bucks CBC on 18 Dec 1993[4] and was reported again until 11 Jan 1994.[5]

Comments: A bird reported from Nockamixon State Park in Bucks Co. on 27 Mar 1996[6] may have been a Mew Gull, but the description was not conclusive.

[1] Rich, D., H. Morrin, and J. Heller 1992. First Pennsylvania record of Mew Gull. PB 6:7.
[2] Kwater, E. 1992. Pennsylvania's first Mew Gull, with notes on its racial identification. PB 6:8–9.
[3] Keller, R. 1992. County reports—January through March 1992: Berks County. PB 6:29.
[4] Kitson, K. 1994. Notes from the field: Bucks County. PB 7:155.
[5] Haas, F.C., and B.M. Haas, eds. 1994. Rare and unusual bird reports. PB 8:42.
[6] Haas, F.C., and B.M. Haas, eds. 1996. Birds of note—January through March 1996. PB 10:14.

Ring-billed Gull ***Larus delawarensis***

General status: This species is probably the most common and widespread gull found inland in North America. Ring-billed Gulls breed from central Canada south locally in the western states to Colorado and from the Great Lakes north to Labrador. They winter along the Pacific coast of the U.S. south through Mexico and over most of the eastern U.S. They are common to abundant regular migrants throughout Pennsylvania. More birds are observed along the Lake Erie Shore in all seasons than in any other area of the state. In summer they are common to abundant in the Coastal Plain and at the Lake Erie Shore and uncommon to fairly common elsewhere in the state, except in the mountains, where they are rare or absent. They have nested several times (though unsuccessfully) at PISP along the Lake Erie Shore. In winter they are locally fairly common to abundant (during mild winters) around waterways throughout most of the state. They frequently associate with Herring Gulls, especially at major feeding sites. The number of birds has increased in all seasons in recent years. Ring-billed Gulls have become year-round residents at many sites.

Habitat: Ring-billed Gulls inhabit lakes, ponds, rivers, marshes, fields, cities (around parking lots), landfills, and in migration almost anywhere.

Seasonal status and distribution

Spring: Because this species winters over much of the state, the beginning of spring migration is not well defined. Ring-billed Gull numbers begin to build as soon as the ice starts to melt on larger bodies of water, usually with the first thaw in mid- to late Feb. Flock sizes continue to increase and usually peak before mid-Mar, often numbering in the tens of thousands in southeastern and northwestern Pennsylvania. At the Lake Erie Shore, counts may reach hundreds of thousands after winter mortalities of alewives and gizzard shad, which birds aggressively consume. Ring-billed Gulls are frequently seen in migration along the Delaware, Susquehanna, and West Branch Susquehanna rivers and the Lake Erie Shore. Birds rapidly disperse by late Mar when the ice has left. They may remain in the tens of thousands into May only at Lake Erie. Most of the birds that remain into summer are immatures.

Summer: Ring-billed Gulls have nested at Gull Point on PISP periodically since 20 nests, three of which contained an egg, were discovered by J. R. Hill in 1983.[1] Predators have destroyed most nests containing eggs, and to date no nest is known to have produced young. In 1995 at least 50 nests were found at this site but none produced young,[2] and in 1997 up to 50 nests contained 38 eggs, which again did not hatch. The latter nesting site may have been destroyed by a storm in early Jun.[3] Each nest contained from one to three eggs that varied in color from buff to light olive green; the eggs were heavily blotched and spotted with brown. Ring-billed Gulls may be found during summer in the same sites as in migration, but in lower numbers. In western Pennsylvania away from Lake Erie most nonbreeding birds are found along the Allegheny River north through Warren Co. and along the larger lakes, especially manmade impoundments. In eastern Pennsylvania most are found along the Susquehanna River north to Montour Co. and along the lower Delaware River to Bucks Co. They are also found along the larger lakes, especially manmade impoundments.

Fall: The number of birds begins to increase through early fall. A noticeable influx of birds arrives with the first cold front around the third or fourth week of Oct. Their numbers usually peak from about mid-Dec to mid-Jan before the first hard freeze. In late Dec 1988, 40,000 were on the southern Lancaster Co. CBC (Morrin et al. 1991). At the same time, abundant gizzard shad attracted an estimated 300,000 Ring-billed Gulls into Presque Isle Bay.[4]

Winter: As lakes and rivers freeze, Ring-billed Gulls become more local. The number of birds rapidly decreases by late Jan during most winters. Along Lake Erie they may become scarce or even absent for about 10 days to two weeks in late Jan or early Feb. Away from Lake Erie most birds are found on open large rivers such as the Allegheny, Susquehanna, and Delaware rivers, especially in areas of slow-moving water such as around dams. Ring-billed Gulls also occur on larger lakes that remain at least partially unfrozen, particularly where landfills or shopping centers are nearby.

History: Ring-billed Gulls were less common before 1960. Poole (unpbl. ms.) stated, "The Ring-billed Gull is a fairly common transient along the Delaware and Susquehanna rivers, as well as at Lake Erie and more rarely on lakes and ponds throughout the Commonwealth." Very few wintered or summered in the state away from these sites. Todd (1940) stated that he could find no winter records at Presque Isle before 1904 and that Ring-billed Gulls were not so numerous as Herring Gulls. Ring-billed Gulls had increased substantially soon after the beginning of the twentieth century.

[1] Hill, J.R., III. 1986. First recorded breeding attempt of the Ring-billed Gull in Pennsylvania. Colonial Waterbirds 9:117–118.

[2] McWilliams, G.M. 1995. Attempted nesting of three species of Laridae at Presque Isle State Park, 1995, Erie County. PB 9:79–80.

[3] G.M. McWilliams, pers. obs.

[4] Hall, G.A. 1989. Appalachian region. AB 43:313.

Herring Gull ***Larus argentatus***

General status: In North America, Herring Gulls breed from Alaska across Canada and south around the Great Lakes and along the Atlantic coast to South Carolina. They winter along the Pacific coast, in the Great Lakes region, and south from Newfoundland along the coast and inland. Like Ring-billed Gulls, Herring Gulls have spread rapidly around the Great Lakes. In Pennsylvania the number of birds has increased in all seasons since the 1960s. They have become year-round residents at many sites. Herring Gulls are abundant migrants along the lower Delaware and Susquehanna rivers and at the Lake Erie Shore and are fairly common migrants elsewhere in the state. In summer they are locally fairly common in the Coastal Plain, common only at the Lake Erie Shore, and uncommon to rare and local elsewhere. Herring Gulls have successfully nested in Pennsylvania since 1996. In winter, they are regular on larger bodies of water over most of the state. Herring Gulls frequently associate with Ring-billed Gulls, especially at major feeding sites.

Habitat: They prefer lakes, ponds, rivers, marshes, fields, and landfills, and in migration they are seen along mountain ridges. They are less frequently seen away from larger rivers and lakes than are Ring-billed Gulls.

Seasonal status and distribution

Spring: Because this species winters over much of the state, the beginning of spring migration is not well defined. The number of birds begins to build as soon as the ice starts to melt from the larger bodies of water, usually with the first thaw in mid- to late Feb. Flock sizes continue to increase and usually peak before mid-Mar; numbers reach the thousands in southeastern and northwestern Pennsylvania. At the Lake Erie Shore numbers of gulls may reach the thousands after winter mortalities of alewives and gizzard shad, on which they aggressively feed. They are frequently seen in migration along the lower Susquehanna and Delaware rivers and along the Lake Erie shore. The number of birds begins to diminish rapidly by late Mar when the ice has left. Most of the birds that remain into late spring or summer are immatures.

Breeding: The state's first nesting of Herring Gulls was confirmed in Mar 1994 by T. Floyd and P. Brown; two nests were on a navigation structure in the Allegheny River just upstream of the Highland Park Bridge in Pittsburgh.[1] Apparently both nests were unsuccessful, as no young were reported. Nesting has continued annually there, with at least two or three nests producing chicks in 1996[2] and four nests producing three chicks there in 1997.[3] Four nests of this species were found at Gull Point on PISP in 1995; at least one nest containing two eggs was found in a Ring-billed Gull colony, but the nest was abandoned before the eggs hatched.[4] The eggs were olive colored and heavily blotched and spotted with brown. A Herring Gull was seen sitting in an incubating posture on Turning Point Island west of Erie in 1990[5] during a aerial colonial waterbirds survey, but breeding was not confirmed and the site was not checked again. East of the Allegheny Front most summer nonbreeders are found along the Delaware River and along the Susquehanna River north through Bradford Co. and west to Centre Co. Away from the major rivers they have been recorded in Berks and Lehigh cos. West of the Allegheny Front nonbreeders are uncommon to rare and local in summer away from Lake Erie in Butler, Crawford, Mercer, and Somerset cos.

Fall: Returning Herring Gulls usually do not appear in Pennsylvania until the first cold fronts arrive in Oct. At this time the number of birds gradually builds to the thousands at some sites by mid- to late Dec. During mild falls they may be rather scarce until a very hard freeze occurs in early to mid-Jan. In late Dec 1988 about 25,000 were on the Southern Lancaster Co. CBC (Morrin et al. 1991). At the same time abundant gizzard shad attracted an estimated 50,000 Herring Gulls into Presque Isle Bay.[6]

Winter: Along Lake Erie they are abundant, but they may become scarce or even absent for about 10 days to a couple of weeks in late Jan or early Feb. Away from Lake Erie, Herring Gulls are fairly common to common in winter on open large rivers such as the Allegheny, Susquehanna, and Delaware, and especially on areas of slow-moving water such as around dams. They have a more limited winter distribution than Ring-billed Gulls, preferring larger bodies of water. They are absent in winter in the High Plateau except along the Allegheny River in Warren Co.[7]

Comments: Herring Gulls may be confused with the smaller more common Ring-billed Gulls, especially those birds in immature or transition plumages. Of the 11 or so described subspecies of Herring Gulls in the world (Grant 1986), only *L. a. smithsonianus* has been identified in the state. However, dark-mantled Herring Gulls resembling some Eurasian subspecies have been observed at PISP.[8] A dark-mantled bird was photographed at PISP on 10 Jan 1998.[9] Apparent Herring × Glaucous Gull hybrids have been identified periodically since 1983 at PISP, with usually only one or two sightings each year. All probable hybrids have been one- or two-year-old birds.[8]

[1] Floyd, T. 1994. First breeding colony of Herring Gulls in Pennsylvania. PB 8:34.
[2] Fialkovich, M. 1996. Local notes: Allegheny County. PB 10:161.
[3] Fialkovich, M. 1997. Local notes: Allegheny County. PB 11:91.
[4] McWilliams, G.M. 1995. Attempted nesting of three species of Laridae at Presque Isle State Park, 1995, Erie County. PB 9:79–80.
[5] Scharf, W.C., G.W. Shugart, and J.L. Trapp. 1991. Distribution and abundance of gull, tern, and cormorant nesting colonies in the U.S. Great Lakes, 1989 and 1990. U.S. Fish and Wildlife Service Bulletin No. 14-16-0009089-006.
[6] Hall, G.A. 1989. Appalachian region. AB 43:313.
[7] T. Grisez, pers. comm.
[8] G.M. McWilliams, pers. obs.
[9] Haas, F.C., and B.M. Haas, eds. 1998. Photographic highlights. PB 12:20.

Thayer's Gull *Larus thayeri*

General status: Formerly considered a subspecies of either Herring or Iceland gulls, Thayer's Gull is treated as a full species in accordance with the current AOU (1998) listing. They breed from Banks, Melville, Cornwallis, Axel Heiberg, and Ellesmere Islands south to Victoria Island Keewatin, Southampton and Baffin Islands, and on Greenland. Thayer's Gulls winter along the West Coast of North America, the Gulf of St. Lawrence, the southern Great Lakes and rarely inland as far south as Arizona east to Florida and north along the Atlantic coast. They have only recently been included in the avifauna of Pennsylvania. Thayer's Gulls have been recorded in only two areas of the state: they are regular but rare visitors only at the Lake Erie Shore and are accidental on the lower Susquehanna River. Little is known about this gull in Pennsylvania partly because of the difficulty and uncertainty of separating it from Iceland and Herring gulls. Most birds are found with concentrations of Herring Gulls.

Seasonal status and distribution

Fall through spring: The first record of Thayer's Gull in Pennsylvania was in 1982. A first- or second-year bird was discovered by S. Stull on 21 Dec and another in first-winter plumage by G. M. McWilliams on 24 Dec in the waters surrounding PISP.[1] This species has been recorded every year since at this site. It was not until the winter of 1991 that an adult were discovered,[2] and now nearly all sightings of Thayer's Gulls are adults. Sightings range from 8 Dec to 4 Apr. The partial remains of an adult were discovered at Gull Point on PISP on 22 Apr 1991. Part of a wing was salvaged and deposited in the bird collection at Edinboro University.[3] The only accepted record away from the PISP area was of a first-winter bird photographed at Safe Harbor along the Susquehanna River in Lancaster Co. It remained from 31 Dec 1989 to 9 Jan 1990.[4] Several photographs of both immature and adult birds have been taken in the surrounding waters of PISP and deposited in VIREO and in the files of PORC.

[1] Hall, G.A. 1983. Appalachian region. AB 37:302.
[2] McWilliams, G.M. 1991. County reports—January through March 1991: Erie County. PB 5:38.
[3] McWilliams, G.M. 1991. Partial remains of Thayer's Gull, *Larus thayeri*, from Presque Isle State Park, Erie County. PB 5:72–73.
[4] Boyle, W.J., Jr., R.O. Paxton, and D.A. Cutler. 1989. Hudson-Delaware region. AB 43:291.

Iceland Gull *Larus glaucoides*

General status: Iceland Gulls are part of a complex group of gulls commonly referred to as white-winged gulls. In North America they breed from Baffin Island south to northwestern Quebec east to Greenland. They winter from Newfoundland and the Gulf of St. Lawrence south along the Atlantic coast to Virginia and on Lake Erie and Ontario. In Pennsylvania they are uncommon to rare regular visitors in the Coastal Plain (occasionally locally fairly common), Piedmont, and Lake Erie Shore in late fall and winter. In spring Iceland Gulls are uncommon to rare regular migrants only at the Lake Erie Shore and are irregular in the Coastal Plain. They are accidental elsewhere in the state. Iceland Gulls frequently associate with Herring Gulls.

Iceland Gull at Strausstown, Berks County, January 1992. (Photo: Ed Kwater)

Habitat: They inhabit large lakes and rivers that are at least partially unfrozen, landfills, and, rarely, plowed fields.

Seasonal status and distribution

Spring: After the first major thaw in late Feb or early Mar, Iceland Gulls usually appear in singles. The arrival of these birds may indicate a northward return from areas south of Pennsylvania. High counts of from five to seven birds have been recorded in the PISP area at the Lake Erie Shore. Elsewhere in the state the highest daily count was four along the Schuylkill River in Philadelphia on 14 Mar 1969.[1] Only three spring records are away from the Coastal Plain and Lake Erie Shore. One was identified at Montour Preserve on 4 Mar 1974 (Schweinsberg 1988); one at MSP on 26 Mar 1977;[2] and one was at Raystown Dam in Huntingdon Co. on 21 Mar 1996.[3] Most Iceland Gulls leave the state by the fourth week of Mar with birds lingering to the first week of May. An exceptionally late individual remained at PISP to 30 May 1982.[4]

Fall: Single Iceland Gulls wander into Pennsylvania from the north after cold weather has arrived in late fall or early winter. They may rarely appear as early as the first week of Nov and usually arrive around the third week of Dec to the second week of Jan, especially along the lower Susquehanna River. If the weather remains mild through late fall they may not appear until after the second week of Jan. All but one fall record, a single bird at South Avis in Clinton Co. on 4 Dec 1996,[5] are from the Coastal Plain, Piedmont, or Lake Erie Shore.

Winter: The majority of winter records are along the Delaware River from Bucks Co. southward. Iceland Gulls are also frequently reported along the Delaware River north to central Northampton Co., the lower Susquehanna River north to Muddy Run in Lancaster Co., and the PISP area. They also are observed on lakes in central Berks Co. and in Lebanon Co. Usually winter totals are fewer than five birds at most sites. However, more than 30 have been recorded in a single day in winter at Tullytown in Bucks Co.,[6] and at least 14 birds were recorded during the winter of 1990 in the PISP area.[7] Only three winter records are away from these traditional sites: one at Pittston along the Susquehanna River in Luzerne Co. on 14 Jan 1995,[8] one on 17 Jan 1990 at BESP in Centre Co.,[7] and one on Cowanesque Lake in Tioga Co. on 13 Jan 1998.[9]

History: Iceland Gulls were not recorded in Pennsylvania until 1934, when one was seen by W. M. Guynes in Presque Isle Bay on 24 Mar and another on Presque Isle on 3 Apr 1937. From then until 1960 there were only eight records: from the lower Delaware and Schuylkill rivers and in and around Philadelphia (Poole 1964).

Comments: Except at the Lake Erie Shore, nearly all sightings are of one- or two-year-old birds.

[1] Scott, F.R., and D.A. Cutler. 1969. Middle Atlantic Coast region. AFN 23:464.

[2] Hall, G.A. 1977. Appalachian region. AB 31:1001.

[3] Grove, G. 1996. Local notes: Huntingdon County. PB 10:19.

[4] Hall, G.A. 1982. Appalachian region. AB 36:852.

[5] Haas, F.C., and B.M. Haas, eds. 1997. Birds of note—October through December 1996. PB 10:221.

[6] Boyle, W.J., Jr., R.O. Paxton, and D.A. Cutler. 1997. Hudson-Delaware region. NASFN 51:733.

[7] Hall, G.A. 1990. Appalachian region. AB 44:267.

[8] Reid, B. 1995. Local notes: Luzerne County. PB 9:34.

[9] G.M. McWilliams, pers. obs.

Lesser Black-backed Gull *Larus fuscus*

General status: This widely distributed species of Europe more closely resembles Herring Gulls in size and shape than the larger, more robust Great Black-backed Gulls. In North America they occasionally breed on Greenland. Lesser Black-backed Gulls winter in the Great Lakes region and in Labrador south along or near the Atlantic coast to Florida. In Pennsylvania they are rare regular visitors from late fall to early spring in the Coastal Plain, in the Piedmont, and around PISP, and they are accidental elsewhere in the state. They are accidental in summer. In winter Lesser Black-backed Gulls are regular in the Coastal Plain, where they may become quite numerous in some years. Lesser Black-backed Gulls have only recently been documented within Pennsylvania's boundaries; they were first recorded in the southeastern corner of the state in 1960[1] and were not found in the northwestern corner of the state until 1978 (Stull et al. 1985). The number of birds in the state has been steadily increasing since then. These birds are nearly always found with Herring Gulls.

Habitat: Lesser Black-backed Gulls prefer large lakes, rivers, and landfills and are occasionally found in fields.

Seasonal status and distribution

Spring: Many sightings are of birds that have wintered, but the arrival of some birds may indicate a northward return from areas south of Pennsylvania. After the first major thaw in late Feb or early Mar, they usually appear as single individuals, but up to 10 Lesser Black-backed Gulls were recorded on 25 Mar 1992 at the Octoraro Reservoir in Chester Co.[2] Most spring sightings are in Mar, with some birds lingering to as late as the second week of May. The only records away from the Coastal Plain, Piedmont, and Lake Erie Shore are in the spring. One was at Spruce Run Reservoir on 18 Mar 1982 in Union Co.,[3] and an adult was at Siegel Marsh in Erie Co. from 16 to 31 Mar 1991.[4]

Summer: The only summer sighting was of one on 15 Jul 1989 at Brunner's Island in York Co.[5]

Fall: Early fall (late summer) records have been observed from Aug through Nov, but most birds begin to appear no earlier than the second week of Dec, usually after Herring Gull numbers have increased substantially. Some of the largest concentrations of Lesser Black-backed Gulls have occurred during the latter half of Dec. On 26 Dec 1991 in Falls Township in Bucks Co., 32 birds were counted.[6]

Winter: The first winter sighting of this European species in Pennsylvania was one observed by A. Brady and R. Sehl on 7 Feb 1960 at Penn Manor in Bucks Co.[1] Most winter records are around the lower Delaware River, particularly in lower Bucks Co. near Tullytown, along the lower Susquehanna River, and at PISP. A landfill next to the Delaware River in Tullytown may host up to 100 birds of this species in some winters.[7] Most other sightings are of single overwintering birds, but as many as seven birds have been known to winter at PISP. Lesser Black-backed Gulls may be absent during mild winters or during extremely cold winters when there is very little open water.

Comments: All age classes have been recorded in Pennsylvania, with adult birds making up the majority of sightings. They are known to hybridize with Herring Gulls.[8] On 31 Jan 1998 an apparent Lesser Black-backed cross with a Herring Gull was studied at PISP.[9]

[1] Scott, F.R., and D.A. Cutler. 1960. Middle Atlantic Coast region. AFN 14:297.

[2] Pasquarella, J. 1992. County reports—January through March 1992: Chester County. PB 6:32.

[3] Boyle, W.J., Jr., R.O. Paxton, and D.A. Cutler. 1982. Hudson-Delaware region. AB 36:835.

[4] McWilliams, G.M. 1991. County reports—January through March 1991: Erie County. PB 5:38.

[5] Spiese, A. 1989. County reports—July through September 1989: York County. PB 3:158.

[6] French, R. 1992. County reports—October through December 1991: Bucks County. PB 5:166.

[7] Haas, F.C., and B.M. Haas, eds. 1997. Summary of the season—January through March 1997. PB 11:21.

[8] Brown, R.G.B. 1967. Species isolation between the Herring Gull *Larus argentatus* and Lesser Black-backed Gull *L. fuscus*. Ibis 109:310–317.
[9] G.M. McWilliams, pers. obs.

Glaucous Gull ***Larus hyperboreus***

General status: In North America, Glaucous Gulls breed from the western coast of Alaska east across the Canadian Arctic coast and islands to Greenland. They winter along the Pacific coast from Alaska south to California, on the Great Lakes, along or near the Atlantic coast from Labrador south, and along the Gulf coast. Glaucous Gulls are the largest of the "white-winged" gulls that occur in Pennsylvania—nearly equal in size to Great Black-backed Gulls. They are uncommon to rare regular visitors in the Coastal Plain, Piedmont, and Lake Erie Shore in late fall and early winter and again in late winter and early spring. They are accidental along the Susquehanna River north of Lancaster and York cos., and in the Ridge and Valley, Southwest, and Glaciated Northwest regions. Glaucous Gulls are accidental in summer.

Habitat: Glaucous Gulls are found on large lakes and rivers that are at least partially unfrozen, landfills, and, rarely, plowed fields.

Seasonal status and distribution

Spring: After the first major thaw in late Feb or early Mar, they are most evident at the Lake Erie Shore, where up to 10 or more may be found in a single day. Their arrival may indicate a northward return from areas south of the state. Most Glaucous Gulls leave by the first week of Apr, but a few occasionally linger until early May and very rarely until late May. All age classes have been documented in Pennsylvania, but most birds are one- or two-year-olds.

Summer: Pennsylvania records in this season are only from PISP, where one was observed on 24 Aug 1980[1] and another from 8 to 18 Jul 1981.[2]

Fall: Single Glaucous Gulls wander into Pennsylvania from the north soon after cold weather has arrived in late fall or early winter. They may first appear in early Nov, but in most years they are not found until mid- to late Dec. Glaucous Gulls have been recorded in fall along or near the lower Delaware River, on lakes in the Piedmont and along the lower Susquehanna River, and very rarely north to Snyder and Northumberland cos. In western Pennsylvania most records are from Lake Erie. Recent sightings away from Lake Erie in fall have been in Armstrong, Butler, Crawford, and Warren cos.

Winter: If there is open water or a readily available food source (at landfills or dumps) some birds will remain through the winter. Most records of Glaucous Gulls in the Coastal Plain along the Delaware River and in the Piedmont occur at this time. Along the Susquehanna River they have been observed as far north as Sunbury, Northumberland Co. (Schweinsberg 1988). They are regular winter residents at the Lake Erie Shore; the only other site with winter records west of the Allegheny Front is Pymatuning in Crawford Co.[3] Glaucous Gulls are very rare or absent during mild winters and during extremely cold winters when there is very little open water.

History: Before 1961 Glaucous Gulls had been reported only from the lower Delaware River and Lake Erie (Poole 1964). Apparently the first record in Pennsylvania was made by R. B. Simpson, who reported a Glaucous Gull with about 25 Herring Gulls at the mouth of Mill Creek in Erie on 22 Feb 1908 (Todd 1940). This species was not recorded in the lower Delaware River until 1 Jan 1918, when one was observed to 4 Jan at Philadelphia.[4] Glaucous Gulls became more frequently reported when Herring and Great Black-backed gulls also became more abundant. An increase in

Glaucous Gulls was also noted along the lakefront in Ohio during the 1960s and 1970s (Peterjohn 1989).

[1] Hall, G.A. 1981. Appalachian region. AB 35:183.
[2] Hall, G.A. 1981. Appalachian region. AB 35:940.
[3] Hall, G.A. 1984. Appalchian region. AB 38:317.
[4] Erskine, R. 1918. General notes. Cassinia 22:28.

Great Black-backed Gull ***Larus marinus***

General status: In North America, Great Black-backed Gulls breed primarily from Labrador to Newfoundland south along the St. Lawrence River to Ontario and Lake Huron and along the Atlantic coast to North Carolina. They winter on the Great Lakes and along the Atlantic coast from Newfoundland to Florida. Great Black-backed Gulls are the largest of the four gulls that most commonly occur in Pennsylvania. They are regular migrants and winter residents in Pennsylvania, primarily along the lower Delaware and Susquehanna rivers and at Lake Erie. A few, mainly immature birds, spend the summer in the state. In winter, when fish are scarce, they frequently prey on grebes, waterfowl, or other species of gulls, especially those in a weakened condition. They were rarely reported anywhere in the state during the first half of the twentieth century and have increased in number only since the 1950s or 1960s. Great black-backed Gulls are now found in the state in all seasons, and their range has been rapidly expanding. They are often seen with Ring-billed and Herring gulls.

Habitat: They are usually observed on large lakes and rivers. Great Black-backed Gulls are also frequently seen perched along shorelines, on islands, or on sand spits; at landfills; and, rarely, in plowed fields.

Seasonal status and distribution

Spring: Because Great Black-backed Gulls have become regular winter residents in the state, the beginning of spring migration is hard to detect. Numbers begin to build as soon as the ice begins to melt, usually with the first thaw in mid- to late Feb. At the Lake Erie Shore they may become abundant, the number reaching the hundreds during winter mortalities of alewives and gizzard shad, on which they gorge themselves. They are frequently seen in migration along the Delaware and lower Susquehanna rivers and the Lake Erie Shore. Birds begin to disperse rapidly by late Mar when ice has melted. Most of the birds that remain into late spring or summer are immatures.

Summer: During summer they are uncommon to rare in the Coastal Plain and Piedmont and are mostly confined to the wide, slow-moving portions of the lower Delaware River from southern Bucks Co. south and along the lower Susquehanna River from West Fairview, Cumberland Co., south. At the Lake Erie Shore, Great Black-backed Gulls are fairly common only around Gull Point on PISP in summer. There are two summer records well away from the Delaware and lower Susquehanna rivers and Lake Erie: in Centre Co. at BESP on 30 Jun 1980[1] and in Crawford Co. at Pymatuning on 31 Aug 1991.[2]

Fall: These gulls usually do not begin to return to Pennsylvania until the first cold fronts arrive in Oct. At this time the number of birds gradually builds to several hundred at some sites by mid- to late Dec. During mild falls the number may remain low until a very hard freeze occurs in early to mid-Jan, at which time concentrations may reach the hundreds.

Winter: Most birds are confined to the Delaware River from Bucks Co., the Susquehanna River primarily from Sunbury, Northumberland Co. south, and around PISP at Lake Erie. Hundreds may be found at these sites in some winters. At Tullytown in Bucks Co. as many as a thousand birds may be present at one time.[3] Most winter records from inland sites away from the Delaware and Susquehanna rivers are at manmade lakes in the Piedmont. The number of birds varies from year to year, depending on the amount of open water and the food supply. They occasionally appear in winter across the remainder of the state, primarily along the larger rivers where water is open.

History: Very little is mentioned of Great Black-backed Gulls before or shortly after 1900, but most authors considered them rare and local in winter in the vicinity of the lower Delaware River and Lake Erie. On the Delaware, Cassin (1862) and Krider (1879) reported them as early as the 1860s. No definite date is known when they were first recorded along the lower Susquehanna River, but Poole (unpbl. ms.) listed them as rare in that area. The first Great Black-backed Gulls to appear west of the Allegheny Front probably originated from the St. Lawrence River rather than from the Delaware River. Todd (1940) did not report the occurrence of this gull in western Pennsylvania until the fall of 1900, when he reported what he suspected was a Great Black-backed Gull at Erie. It was not verified in this vicinity until the spring of 1936. Sutton (1928b), however, mentioned its occurrence at Conneaut Lake in Crawford Co. as early as 1904. By 1957 they had increased dramatically along the lower Delaware River, where as many as 67 birds were counted in Delaware Co. (Poole, unpbl. ms.). There were also increased sightings of birds on interior lakes and rivers during this period.

[1] Hall, G.A. 1980. Appalachian region. AB 34:895.
[2] Leberman, R.F. 1991. County reports—July through September 1991: Crawford County. PB 5:128.
[3] Boyle, W.J., Jr., R.O. Paxton, and D.A. Cutler. 1997. Hudson-Delaware region. NASFN 51:733.

Sabine's Gull *Xema sabini*

General status: Sabine's Gulls breed in North America from coastal western Alaska east across the Canadian Arctic coast and islands to Baffin Island. They winter at sea primarily in the Pacific Ocean and less commonly in the tropical Atlantic. They are casual migrants across the interior of North America. In Pennsylvania they have been recorded about 16 times, with all but two being in juvenile plumage. All sightings have been of single birds in fall during or after the passage of cold fronts. They are usually observed with migrating Bonaparte's Gulls.

Seasonal status and distribution

Fall: Sabine's Gulls were first recorded in the state on 15 Oct 1979, when a moribund bird was picked up by A. Castele on the outer beach of Gull Point on PISP.[1] The bird died, was prepared as a study skin, and was placed in the bird collection at Edinboro University.[2] Observations have been recorded from 5 Sep to 3 Nov, with 10 of the records occurring within the four-week period of 17 Sep to 15 Oct. Four sightings were along the lower Susquehanna River in Lancaster Co., and six were from PISP. Other reports are from Sunbury in Northumberland Co. from 19 to 21 Sep 1981 (Schweinsburg 1988), an adult seen flying past HMS in Schuylkill Co. on 17 Sep 1987,[3] one at Prince Gallitzin State Park in Cambria Co. from 27 Sep to 2 Oct 1993,[4] and another at MSP in Butler Co. on 7 Sep 1996.[5]

[1] Hall, G.A. 1980. Appalachian region. AB 34:161.
[2] G.M. McWilliams, pers. obs.
[3] Clauser, T. 1987. Schuylkill County. PB 1:97.
[4] Haas, F.C., and B.M. Haas, eds. 1994. Rare and unusual bird reports. PB 7:162.
[5] Hess, P. 1996. Local notes: Butler County. PB 10:162.

Sabine's Gull at Sunbury, Northumberland County, September 1981. (Photo: Franklin C. Haas)

Black-legged Kittiwake ***Rissa tridactyla***

General status: In North America, Black-legged Kittiwakes breed along the coast of Alaska and from Somerset Island east to Baffin Island and Newfoundland south to Nova Scotia. They winter in the Great Lakes region and mostly offshore along the Pacific and Atlantic coast. Black-legged Kittiwakes are pelagic gulls that occasionally wander into Pennsylvania. There have been numerous reports, but few have been well documented. Most kittiwake reports have been from the Lake Erie Shore of birds in first-winter plumage. All are of one to three birds, usually during or after storms.

Habitat: Black-legged Kittiwakes are usually found on large lakes and rivers.

Seasonal status and distribution

Spring: The only accepted spring sighting in the state is of two adults observed on Lake Ontelaunee in Berks Co. on 31 Mar 1991.[1] Other unverified spring reports since 1960 have come from Berks, Crawford, and Somerset cos.

Fall: There have been 11 fall reports of kittiwakes from PISP in Erie Co., ranging in dates from 21 Sep to 3 Jan, with most sightings in Nov. Single accepted

Black-legged Kittiwake at Pennsylvania Power and Light Montour Power Plant, Montour County, November 1993. (Photo: Rick Wiltraut)

fall sightings have come from Washington Boro in Lancaster Co,[2] Marsh Creek State Park in Chester Co.,[3] Harvey's Lake in Luzerne Co.,[4] Pennsylvania Power and Light Montour Power Plant in Montour Co.,[5] and PISP in Erie Co.[6] Since 1960 they have also been reported in fall from Berks, Delaware, and Montgomery cos.

History: Apparently there were only three records before 1960. On 17 Oct 1900 a single bird was observed off the outer beach at Presque Isle, and a mounted specimen in the CMNH was purported to have been shot at McKees Rocks in Allegheny Co. on 15 Jan 1891 (Todd 1940). Poole (unpbl. ms.) cited a bird that he saw on Lake Ontelaunee in Berks Co. on 17 Nov 1935 during a severe storm.

[1] Keller, R. 1991. County reports—Janurary through March 1991: Berks County. PB 5:31.
[2] Haas, F.C., and B.M. Haas, eds. 1997. Birds of note—October through December 1996. PB 10:221.
[3] Paxton, R.O., W.J. Boyle Jr., and D.A. Cutler. 1984. Hudson-Delaware region. AB 38:183.
[4] Paxton, R.O., W.J. Boyle Jr., and D.A. Cutler. 1982. Hudson-Delaware region. AB 36:279.
[5] Brauning, D.W. 1994. Notes from the field: Montour County. PB 7:158.
[6] Haas, F.C., and B.M. Haas, eds. 1998. Birds of note—October through December 1997. PB 11:231.

Ross's Gull *Rhodostethia rosea*

General status: Ross's Gulls breed in northern Siberia and in northeastern Manitoba, and there is one record from west-central Greenland. Their winter range is unknown but is probably in open waters of the Arctic. They are accidental in Ontario, Quebec, along the Atlantic coast from Newfoundland south to Maryland and inland Pennsylvania and Tennessee. Only one record is known in Pennsylvania.

Seasonal status and distribution

Fall: From 10 to 13 Oct 1991 a Ross's Gull in first-winter plumage was observed and videotaped by K. Lippy and J. Wentz at Lake Marburg at Codorus State Park in York Co.[1]

[1] Spiese, A. 1992. County reports—October through December 1991: York County. PB 5:186.

Gull-billed Tern *Sterna nilotica*

General status: Gull-billed Terns breed locally from Long Island, New York, south along the Gulf coast. They winter in southern Florida and along the Gulf coast. Three sightings of this coastal species have been accepted in Pennsylvania.

Seasonal status and distribution

Spring: A bird in breeding plumage was observed by J. Book, T. Garner, and E. Witmer on 2 Jun 1996 on the Conejohela Flats in Lancaster Co.[1] Two were discovered by A. Fuller, T. Fuller, and B. Snyder on 25 May 1997 at BESP in Centre Co.; one of them was documented with a photograph.[2]

Fall: One was seen feeding among numerous gulls and terns by J. Book at the Conejohela Flats in Lancaster Co. on 11 Aug 1994.[3]

History: Warren (1890) mentioned a specimen that was captured in the fall in Chester Co. and two that were shot near Philadelphia. The species was also listed as occurring in Berks Co. (Poole 1964). Poole (1964) considered this species hypothetical because no specimen could be found.

Comments: One modern Gull-billed Tern record has been listed as hypothetical in Pennsylvania. Apparently a Gull-billed Tern was observed sitting with Ring-billed Gulls and in flight at Pittston on the Susquehanna River in Luzerne Co. on 21 May 1962.[4]

[1] Carl, B. 1996. Local notes: Lancaster County. PB 10:101.
[2] Peplinski, J., and B. Peplinski. 1997. Local notes: Centre County. PB 11:93.
[3] Pulcinella, N. 1994. Rare bird reports. PB 8:147.
[4] Scott, F.R., and D.A. Cutler. 1962. Middle Atlantic Coast region. AFN 16:392.

Caspian Tern ***Sterna caspia***

General status: The largest terns in the world, Caspians are the size of Herring Gulls and are easily recognized by their enormous blood-red bill. In North America Caspian Terns breed locally in Canada, along the Pacific coast of the U.S. and Mexico, and along the Atlantic coast at scattered sites from New Jersey south along the Gulf coast. In the U.S. they winter along the Atlantic coast from North Carolina south along the Gulf coast. At the Lake Erie Shore, Caspian Terns are common to abundant regular migrants, especially in spring, and are rare regular summer residents. They are rare in spring but uncommon to fairly common regular migrants in fall along the lower Delaware and Susquehanna rivers. They are uncommon to rare migrants in the remainder of the state, except in the northern and central portions of the High Plateau, where they are accidental during migration. Caspian Terns have become more widely reported in Pennsylvania in recent years than formerly. Although they have remained well into the breeding season at PISP at least since the early 1990s, there is no recent evidence of nesting.

Habitat: Caspian Terns are found on large lakes, rivers, and, occasionally, marshes.

Seasonal status and distribution

Spring: Caspian Terns may arrive in Pennsylvania as early as the first week of Apr. There was a report of two very early birds on 11 Mar 1997 at Lake Somerset in Somerset Co.[1] Their usual migration time is from the second week of Apr to the second week of May. Peak migration occurs during the second and third weeks of Apr. At PISP numbers have increased dramatically at this season in recent years; the east end of the peninsula has become a significant staging site. In the late 1970s maximum daily counts rarely exceeded 10 birds; by 1982 daily highs were reaching the mid-eighties, and from 1994 to present maximum daily highs reach 250–350 birds.[2] Stragglers may linger to the first or second week of Jun. Larger concentrations are seen at the south end of Gull Point at PISP than at any other site in the state. Although they have been reported recently at more sites in the state, they are still uncommon to rare in spring away from PISP.

Summer: Most birds at this season, from about the second week of Jun to the first week of Jul, probably are late spring or early fall migrants. They have been recorded during the summer from numerous counties in the state, especially in early Jul. More Caspian Terns are at PISP in summer than at any other single site in the state. Future nesting is possible here; copulating pairs and courtship behavior are observed almost annually where Ring-billed Gulls have recently made nesting attempts.[2]

Fall: Birds that have not summered in the state begin passing through by the second week of Jul and rarely as early as the last week of Jun. Peak migration ranges from the second week of Aug to the second week of Sep. More birds are seen along the lower Delaware and Susquehanna rivers, where they are common, in fall than in any other areas of the state. The majority have left the state by the second week of Sep. Stragglers have remained to the second week of Oct.

History: Poole (1964) listed this species as fairly regular but uncommon on larger rivers and lakes and casual elsewhere in the state. Todd (1940) found them to be rather common in fall at Presque Isle and reported records in spring. He stated that the Caspian Tern was rather common in summer on the Great Lakes but that most of the nesting colonies had ceased to exist. None of these nesting colonies were ever found in Pennsylvania, but on 6 Jul 1924 Todd discovered a nest containing two eggs along the outer beach of Presque Isle. He photographed the nest and left but upon returning three days later found no trace of the eggs or adults (Todd 1940).

[1] Haas, F.C., and B.M. Haas, eds. 1997. Birds of note—January through March 1997. PB 11:23.
[2] G.M. McWilliams, pers. obs.

Royal Tern ***Sterna maxima***

General status: Royal Terns breed along the Atlantic coast from New Jersey south along the Gulf coast. They winter along the coast of southern California and Mexico and from North Carolina south along the Gulf coast. Royal Terns are rarely observed away from saltwater but have been reported at least five times in Pennsylvania.

History: Inclusion of this species on the list of Pennsylvania birds is based on a specimen found dead by M. E. Bowers at Tinicum in Delaware Co. on 15 Aug 1955 after Hurricane Connie. Five years later one was observed by A. Conway and J. Conway at Tinicum in Sep 1960 (Poole 1964). Warren (1890) and Stone (1894) mentioned a Royal Tern captured in Berks Co. or Chester Co. in Sep 1879.

Comments: There have been about three reports of this tern in Pennsylvania in recent years; they are probably valid but were not adequately documented for acceptance. These reports are: a single bird seen in the Tinicum area on 11 Aug 1973,[1] one at Sunbury in Northumberland Co. on 23 Sep 1989 believed to have been carried inland by Hurricane Hugo,[2] and one at Memorial Lake in Lebanon Co. on 14 Apr 1991.[3]

[1] Scott, F.R., and D.A. Cutler. 1974. Middle Atlantic Coast region. AB 28:35.
[2] Hall, G.A. 1990. Appalachian region. AB 44:90.
[3] Kwater, E. 1992. Third report of the Pennsylvania Ornithological Records Committee, April 1992. PB 6:21.

Roseate Tern ***Sterna dougallii***

General status: Very rarely reported anywhere in North America away from the coasts, Roseate Terns breed locally from Nova Scotia discontinuously south in the U.S. to the Florida Keys. They winter in the West Indies south along the Atlantic coast of South America. Roseate Terns were reported three times in Pennsylvania before 1960.

History: The inclusion of this species on the list of birds of Pennsylvania is based on a specimen in the Reading Museum that was collected by C. H. Shearer on the Schuylkill River above Reading on 17 Aug 1895 (Poole 1964). One was reported by R. F. Miller from Holmesburg in Philadelphia Co. on 1 Jul 1938,[1] and one was reported at Green Lane Reservoir in Montgomery Co. on 17 May 1959 by E. Reimann (Poole 1964).

[1] Miller R.F. 1938–1941. General notes. Cassinia 31:45.

Common Tern ***Sterna hirundo***

General status: Like many other waterbirds that nest on sandy beaches and dunes, Common Terns have been severely adversely affected by human activity. In North America they breed from Alberta east across Canada and south to Illinois, along the Atlantic coast to North Carolina, and locally along the Gulf coast. They winter casually from southern California south through Central and South America. Common Terns have been known to breed in Pennsylvania only in a small area at the east end of PISP. Although they were a fairly common nesting species in the early part of this century, there have been no successful nests since the mid-1960s. The breeding population quickly declined when their nesting area became a popular area for unrestricted swimming. Common Terns attempted to nest again after the long-overdue protection of this valuable ecological site in the mid-1990s. The breeding population has declined throughout the Great Lakes and the Atlantic coastal regions, owing in part to the same problems. Common Terns are uncommon to fairly common regular migrants over most of the state; the greatest numbers are recorded along Lake Erie, where they are fairly common to abundant. They are casual to accidental in most of the High Plateau in spring, summer, and fall. Nonbreeding birds are occasionally found at the Lake Erie Shore in summer, but they are casual to accidental summer visitors elsewhere. They are often observed feeding or resting with Forster's Terns.

Habitat: Common Terns are usually seen either flying over lakes, slow-moving rivers, or, occasionally, marshes or perched on beaches, on sand spits, on mudflats, or on such structures as buoys or piers over water.

Seasonal status and distribution

Spring: They are fairly common to common only along the Lake Erie Shore, and they are uncommon to rare elsewhere. They may arrive in the state as early as the second week of Apr. Their typical migration period is from the last week of Apr or the first week of May to the fourth week of May. Peak migration occurs during the second or third week of May. Daily high counts may reach into the hundreds at PISP, but fewer than 100 are normal. Away from Lake Erie, single birds or groups of fewer than five birds are usually recorded. However, on one occasion 125 Common Terns were counted flying over PNR in Westmoreland Co. on 18 May 1968 (Leberman 1976), an unusually high number of birds for the Allegheny Mountains. Stragglers may remain to the first week of Jun.

Summer: The majority of sightings in Jun and Jul are probably of late spring or early fall migrants rather than summering birds. Most sightings at this time are at Gull Point on PISP, where they formerly nested until 1966. They did not nest at this site again until 11 May 1995, when a single nest containing three eggs was discovered by G. M. McWilliams. The nest consisted of a small depression in the sand; the depression was next to a large stick on a sand dune that had small, scattered patches of grass. On 1 Jun the nest was abandoned and the eggs were gone, but there was no evidence of predation. A second unsuccessful nest, containing one egg, was discovered near the original nest in Jul of the same year. Predation was believed to be the reason for the second nesting failure.[1] This time, human disturbance was not likely to have been a factor in nesting failure, because the area had recently been closed to entry.

Fall: Common Terns generally occur in the same abundance and in the same places in fall as in spring. Adults may begin returning to the state the second or third week of Jul, but typically by the second or third week of Aug, and leave by the third week of Oct. Peak migration ranges from the fourth week of Aug to the second week of Sep. Daily highs during this period may reach several hundred birds at the Lake Erie Shore. Migrant flocks are smaller now than they were 30 or more years ago. On 18 Sep 1960, 10,000 birds were estimated at PISP,[2] but a concentration of that magnitude has not been repeated. Lingering birds may remain until the first week of Dec.

History: Common Terns were never considered to be very common in the eastern half of the state. Poole (unpbl. ms.) mentioned that the largest count at any one time was 30 in 1935 and 1940. A flock of 75 was reported on the Churchville Reservoir, Bucks Co. in 1953 (L. S. Thomas 1955). In western Pennsylvania, they were most common at Pymatuning Lake, Conneaut Lake, and Erie. Even though birds were present all summer at Pymatuning and Conneaut lakes, no evidence of nesting was found. They were first suspected of nesting in the state some time in the late 1920s, when a colony was reported on a sandbar at the east end of Presque Isle. On 1 Jun 1930 Todd visited this site and counted at least 139 nests. In 1931 most nests were destroyed by bathers who walked among and even stepped on the eggs. In 1934 a party from the CMNH visited the colony and located at least 50 nests with birds still incubating eggs as late as 2 Aug (Todd 1940). Although there are no published data of the continuation of nesting at this site from 1935 until the early 1950s, nesting was observed again from the mid-1950s until 1966 (Stull et al. 1985).

Comments: Common Terns are listed as Endangered by the PGC in Pennsylvania.

[1] McWilliams, G.M. 1995. Attempted nesting of three species of Laridae at Presque Isle State Park, 1995, Erie County. PB 9:79–80.

[2] Hall, G.A. 1960. Appalachian region. AFN 15:37.

Arctic Tern ***Sterna paradisaea***

General status: Arctic Terns are well known for their remarkable long-distance migrations. They are circumpolar, breeding around the Arctic and in sub-Arctic regions of North America south in the West to British Columbia and in the East to Massachusetts, and wintering as far south as Antarctica. Arctic Terns normally migrate well offshore and are found only casually inland during migration. There is presently one accepted record in Pennsylvania.

Seasonal status and distribution

Spring: Arctic Terns have been included on the list of birds of Pennsylvania on the basis of two adults that were photographed by R. Wiltraut at Beltzville Lake in Carbon Co. on 16 May 1989. They were first discovered flying around the lake during a storm with strong easterly winds and were seen later perched on the beach.[1]

History: Warren (1890) reported several Arctic Tern records during the nineteenth century. Apparently two were

Arctic Tern at Beltzville State Park, Carbon County, May 1989. (Photo: Rick Wiltraut)

shot on the Delaware River below Philadelphia in Sep around 1886 and one was collected on the Lehigh River in the fall, but no remaining specimen could be found. He also reported the species as accidental in Berks Co.

Comments: There is a sight record of an adult at Gull Point on PISP observed periodically from 17 to 28 May 1983.[2] An Arctic Tern observed flying over Glen Morgan Lake in Berks Co. on 3 May 1998 is currently under review.[3]

[1] Wiltraut, R. 1989. Arctic Terns in Carbon County. PB 3:56.

[2] Hall, G.A. 1983. Appalachian region. AB 37:870.

[3] Pulcinella, N. 1998. Rare bird reports. PB 12:62.

Forster's Tern ***Sterna forsteri***

General status: Forster's Terns breed at scattered sites in the West south to Colorado, along the Atlantic coast from Massachusetts south locally to South Carolina, and along the Gulf coast from Louisiana west to Texas. They winter along the Atlantic coast from Virginia and south along the Gulf coast. Forster's Terns are fairly common to common regular migrants along the lower Delaware and Susquehanna rivers and at Lake Erie. They are the most frequently encountered terns away from these sites. Forster's Terns are uncommon to fairly common regular migrants elsewhere in the state except in the northern and central High Plateau, where they are casual to accidental. They are accidental in summer and winter in the state. Forster's Terns frequently associate with Common Terns.

Habitat: They prefer lakes, ponds, marshes, and slow-moving streams.

Seasonal status and distribution

Spring: The typical migration period of Forster's Terns is from the second week of Apr to the first or second week of May. Most are seen in early May. Stragglers may remain until the second week of Jun.

Summer: A few records in Jun and early Jul probably represent either very late spring migrants or very early fall migrants. Such birds have been reported at YCSP in 1989,[1] Pymatuning on 3 Jul 1988,[2] and PISP in mid-Jun 1985.[3]

Fall: Forster's Terns are most common at this season in the southeastern portion of the state. Their typical migration season is from the second or third week of Jul to the fourth week of Oct. Peak migration varies from year to year but frequently occurs in Aug or Sep. Along the lower Delaware and Susquehanna rivers, they may be abundant at times, with concentrations in the hundreds or even thousands. For example, as many as 100 were present at the mouth of Darby Creek along the Delaware River in Delaware Co. on 1 Aug 1991,[4] 1200 were counted on 25 Sep 1987 at the Conowingo Pond in Lancaster Co. (Morrin et al. 1991), and 500 were observed on 3 Sep 1988 near Washington Boro, Lancaster Co.[5] Birds may linger to the first week of Dec. A very late bird was recorded at Tunkhannock in Wyoming Co. on 14 Dec 1995,[6] and one was on the Lancaster CBC in Dec 1970 (Morrin et al. 1991).

Winter: The only winter record was of an individual standing on the ice with gulls at Presque Isle Bay in Erie on 13–19 Jan 1991.[7]

History: All other records were during fall migration, mostly from the Coastal Plain and Piedmont. Todd (1940) did not include this species in his *Birds of Western Pennsylvania*. It was not reported in the western part of the state until 28 Sep 1949 at Oneida Lake in Butler Co. (Parkes 1956). There were more sightings through the 1950s as stated by Poole (1964): "During the past decade the Forster's Tern has been appearing at a number of localities throughout the state where it was previously unknown." The number of birds reported also increased during the late 1950s and early 1960s.

Comments: Forster's Terns may easily be mistaken for Common Terns, especially when their primaries become worn and rather dark. Birds in their first spring plumage, which are rarely encountered in Pennsylvania, also look very similar to Common Terns. The confusion between these two species may be the reason why few Forster's Terns were reported in historical times. At PISP Forster's Terns were not confirmed until the late 1950s, probably because they were overlooked.[8]

[1] Hall, G.A. 1988. Appalachian region. AB 42: 1287.
[2] Hall, G.A. 1989. Appalachian region. AB 43: 1315.
[3] Hall, G.A. 1985. Appalachian region. AB 39:912.
[4] Guarente, A. 1991. County reports—July through September 1991: Delaware County. PB 5:129.
[5] Witmer, E. 1988. County reports—July through September 1988: Lancaster County. PB 2:107.
[6] W. Reid, pers. comm.
[7] McWilliams, G.M. 1991. County reports—January through March 1991: Erie County. PB 5:38.
[8] J.H. Stull, pers. comm.

Least Tern ***Sterna antillarum***

General status: Least Terns breed along the coast of California and Mexico, in the Mississippi and Ohio river systems, and along the Atlantic coast from Maine south along the Gulf coast. They winter along coastal South America. These, the smallest of North American terns, are casual vagrants in Pennsylvania. They may be from populations either in the Mississippi Valley or along the Atlantic coast. Most sightings are of single birds, and most are in the spring and early summer. Both adult and immature birds have been observed in the state. Least Terns have been recorded only in the Coastal Plain at Tinicum, in the Piedmont, and on the Lake Erie Shore at PISP. They have nested twice in Pennsylvania, both times at Tinicum.

Habitat: They have been observed on large lakes, rivers, marshes, and, rarely, ponds.

Seasonal status and distribution

Spring and summer: Most of the 16 or more sightings recorded statewide since 1960 fall between 1 May and 15 Jul. Away from Tinicum and PISP, Least Terns have been recorded in spring at Hamburg in Berks Co. on 1 May 1977,[1] Pine Run in Bucks Co. on 30 Jun 1992,[2] Holtwood Flyash Pond in Lancaster Co. from 8 to 10 Jul 1992,[3] and Green Lane Reservoir in Montgomery Co. on 1 Jun 1995.[4] The only recent nesting was on 14 Jun 1961, when a nest containing two eggs was found near Tinicum .[5]

Fall: The few sightings recorded in fall include one bird in Brecknock Township in Lancaster Co. on 21 Sep 1973 (Morrin et al. 1991), five on the Conejohela Flats in Lancaster Co. on 5 Sep 1975 (Morrin et al. 1991), and one at Tinicum on 15 Aug 1985[6] and again from 2 Aug to 2 Sep 1994 in both Delaware and Philadelphia cos.[7] One also was seen through much of Jul until 1 Aug 1994 along the lower Susquehanna River in Lancaster Co.,[7] and another at Green Lane Reservoir in Montgomery Co. from 15 to 25 Jul 1997.[8]

History: The first published record of Least Terns in Pennsylvania was from Libhart (1869), who reported spring and fall sightings in Lancaster Co. Warren (1890) stated that he had seen seven of these terns taken in the previous 10 years, one from the spring and the others from Aug and Sep in the counties of Chester, Delaware, Lancaster, and Montgomery. They were not reported again until Aug 1933 at Lake Ontelaunee in Berks Co. after a hurricane. In 1939 a total of 18 were found on 30 Jul at Tinicum (Poole, unpbl. ms.). The next known report was in 1950 when a nest was found on 15 Jun near the Tinicum marsh on a spoil bank created by nearby dredging; two young were banded there on 3 Jul.[9] In 1955 Hurricane Connie brought in 10 to the Conejohela Flats in Lancaster Co. and eight to Lake Ontelaunee in Berks Co. on 13–14 Aug.[10] In western Pennsylvania Todd (1940) cited two records: one shot on 10 Apr 1912 at Conneaut Lake, and two seen along the Raystown Branch of the Juniata River at Ardenheim in Huntingdon Co. on 16 Aug 1929.

[1] R. Keller, pers. comm.
[2] French, R. 1992. County reports—April through June 1992: Bucks County. PB 6:72.
[3] Haas, F.C., and B.M. Haas, eds. 1992. Rare and unusual bird reports. PB 6:129.
[4] Freed, G.L. 1995. Local notes: Montgomery County. PB 9:95.
[5] Meritt, J.K. 1962. Field notes—1961 (Least Tern). Cassinia 46:29.
[6] N. Pulcinella, pers. comm.
[7] Haas, F.C., and B.M. Haas, eds. 1994. Rare and unusual bird reports. PB 8:164.
[8] Haas, F.C., and B.M. Haas, eds. 1997. Birds of note—July through September 1997. PB 11:149.
[9] Hoy, N.D. 1951. First nesting record for the Least Tern in Pennsylvania. Cassinia 38:34.
[10] Poole, E., 1957. General notes. Cassinia 42:20.

Sooty Tern *Sterna fuscata*

General status: These pelagic terns breed primarily on the Dry Tortugas, Florida, and islands off Texas and Louisiana. They spend most of their life flying over tropical seas. Severe storms sometimes carry them many miles inland and as far north as coastal New England and the Maritime Provinces. Modern records of Sooty Terns have been documented in Pennsylvania only during or immediately after the passage of two hurricanes.

Seasonal status and distribution

Fall: During the passage of Hurricane Fran in Sep 1996 several birds were carried inland on 7 and 8 Sep. Sooty Terns were distributed over several areas in the southeastern corner of the state.[1] Presently the two accepted records from this storm are of two immatures seen flying up the Delaware River in Delaware Co. by N. Pulcinella and J. Lockyer[2] and an adult and an immature observed flying over the Susquehanna River at the Conejohela Flats in Lancaster Co. by B. M. Haas and colleagues.[3] These were the first Sooty Terns carried into Pennsylvania since Hurricane Connie in 1955.

History: Warren (1890) cited two birds collected in Delaware Co. Burns (1919) gave the date of one of these birds as "in the spring of 1878, found dead in a field." The other was apparently a male shot in West Goshen Township near West Chester on 20 Sep 1878 (Poole, unpbl. ms.). Warren (1890) mentioned birds collected in Lycoming Co. as well, but no data were given. Many Sooty Terns were brought in with the passage of Hurricane Connie in 1955. On 13 Aug three birds appeared on Lake Ontelaunee in Berks Co. (Poole, 1964), and on 14 Aug two were seen on the Conejohela Flats in Lancaster Co. (Poole, unpbl. ms.). On 16 Aug one was found at York (Poole, unpbl. ms.), and another was picked up at Athens on 17 Aug.[4] Both birds were given to the Philadelphia Zoo. The last bird, found on 18 Aug, was picked up and photographed at Wrightsville.[4]

Comments: Two other other reports from Hurricane Fran were of an adult and an immature seen over the Schuylkill River near Gibralter in Berks Co.[5] and one in Chester Co. at Chambers Lake.[6] Unlike Black-capped Petrels, which have been reported in or near the eye of the hurricane, Sooty Terns from Hurricane Fran were deposited several hundred miles east of the eye.[1]

[1] Pulcinella, N. 1996. Hurricane Fran's fallout. PB 10:138–142.
[2] Guarente, A. 1996. Local notes: Delaware County. PB 10:164.

[3] Blust, B. 1996. Local notes: Chester County. PB 10:163.
[4] Hoyt, S., D. Cutler, and E. Reimann. 1957. General notes. Cassinia 42:20.
[5] Carl, B. 1996. Local notes: Lancaster County. PB 10:165.
[6] Keller, R. 1996. Local notes: Berks County. PB 10:161.

Black Tern ***Chlidonias niger***

General status: In North America, Black Terns breed across the southern provinces of Canada east to Maine and south in the east to Pennsylvania, Ohio, and Illinois. They winter along the coasts of South America. Black Terns have recently decreased as migrants and breeders in Pennsylvania. They breed annually only in Crawford Co. and have nested at PISP. Black Terns are uncommon to rare regular migrants over most of the state, except in the Ridge and Valley and High Plateau, where they are rare irregular migrants.

Habitat: In migration Black Terns are seen over lakes, ponds, large slow-moving rivers, and marshes. During the breeding season they are found in deep-water marshes that have up to 75% emergent vegetation and at least some areas of open water.

Seasonal status and distribution

Spring: Black Terns are most frequently reported during this season but are still usually uncommon to rare. Spring migration begins the last week of Apr and continues to the first week of Jun. No apparent peak migration is evident, but most pass over the state during May. An unusually high concentration of 200 birds was at Tinicum in Delaware Co. on 14 May 1967;[1] no concentration close to that number has been reported since.

Breeding: Black Terns are rare colonial breeders, nesting only in Crawford and Erie cos. Most historical colonies had been either abandoned or reduced to a few pairs by 1980. Leberman believed it unlikely that any single site was occupied throughout the 1980s.[2] Hartstown Marsh, then the location of the largest known remaining colony in the state, supported three nests in 1988[3] and one pair in 1991 and 1992.[4] Up to three pairs were present at Hartstown in 1996, and at least two young fledged; it was the only breeding site that year.[5] A nesting attempt at PISP in 1993 was believed to be unsuccessful.[6] Summer sightings continue in the Pymatuning Upper Reservoir, but nesting has not been documented there since the mid-1980s. The crude nest, a slight depression in a pile of rotting vegetation, is built on a floating mat, usually associated with others in a loose colony. Nests with eggs are found in late May; young are flying by mid-Jun. The eggs are olive to buff and are heavily blotched or spotted with black or brown. The decline of Pennsylvania's breeding population corresponds to a severe continental decline. Nonbreeding terns recorded between the second week of Jun and the first week of Jul may be late spring or early fall migrants. These have appeared in many areas of the state away from the established breeding sites in northwestern Pennsylvania.

Fall: Their typical fall migration period is from the last week of Jul to the second week of Sep without an apparent peak. Daily highs are usually fewer than five birds, but as many as 16 have been recorded at the Pennsylvania Power and Light Fly Ash Basin in Montour Co.[7] and 18 on the Conejohela Flats in Lancaster Co. (Morrin et al. 1991). A flock of 65 Black Terns was seen after a heavy afternoon thunderstorm at Donegal Lake, Westmoreland Co. on 14 Aug 1994.[8] During the passage of Hurricane Fran on 8 Sep 1996 an unusual flight totaling 53 Black Terns was observed passing Gull Point on PISP.[9]

Stragglers have remained to the first week of Oct. A record from the Lehigh Valley area on 23 Oct 1976 was rather late (Morris et al. 1984).

History: Although the first breeding records are from the early 1900s, it is possible that the species nested earlier, perhaps sporadically, on Presque Isle.[10] J. W. Detwiller claimed to have obtained eggs of this species near Erie (Warren 1890), but Todd (1904) resoundingly discredited any of Detwiller's records, so the report remains in question. Colonies near Sandusky, Ohio, in the late 1800s supported the prospect that terns nested at Presque Isle before any rigorous surveys were conducted (the first of which was Todd's). The first undisputed Pennsylvania nesting record was from marshes near Conneaut Lake in Crawford Co. about 1910 (Sutton 1928b). Peak populations in Pennsylvania include about 50 pairs at the Pymatuning Upper Reservoir at Linesville in 1934,[11] 25–30 pairs in Hartstown Marsh in 1940 (Grimm 1952), and 15 nests at Presque Isle in 1958 (Stull et al. 1985). Additional nesting sites in Crawford Co. include Smith Marsh (also known as Meadville Junction), with up to 40 birds, and numerous individuals at Conneaut Lake Outlet, where Grimm (1952) reported regular breeding. Numbers appear to have declined from these peaks between 1940 and 1960. Todd (1940) cited a rather early arrival date at Conneaut Lake of 10 Apr 1912. During the same year and at the same site the latest record for the state was established when one was collected on 12 Nov. Concentrations of up to 50 birds were noted on Crystal Lake, Crawford Co. in 1922 (Sutton 1928b) and on Lake Ontelaunee, Berks Co. in 1935 (Poole, unpbl. ms.). Perhaps the most noteworthy tallies ever recorded in the state were of 500 birds near Hartstown Marsh on 21 May 1940 (Grimm 1952) and as many as a thousand on the Conejohela Flats in Lancaster Co. on 4 Sep 1957 (Poole, unpbl. ms.).

Comments: The Black Tern is listed by the PGC as State Endangered, with a breeding population at the brink of disappearing from the state.

[1] N. Pulcinella, pers. comm.

[2] Leberman R.C. 1992. Black Tern *(Chlidonias niger)*. BBA (Brauning 1992):144–145.

[3] Bush, W.L. 1989. Black Tern *(Chlidonias niger)* nesting platform and habitat study in Crawford and Erie cos., Pennsylvania. M.S. thesis. Allegheny College, Meadville, Penn.

[4] R.C. Leberman, pers. comm.

[5] Brauning, D.W. 1997. Black Tern nesting study. Annual report. Pennsylvania Game Commission, Harrisburg.

[6] McWilliams, G.M. 1993. Notes from the field: Erie County. PB 7:62.

[7] Brauning, D.W. 1994. Notes from the field: Montour County. PB 8:161.

[8] Leberman, R.C. 1994. Notes from the field: Westmoreland County. PB 8:162.

[9] McWilliams, G.M. 1996. Local notes: Erie County. PB 10:164.

[10] Kibbe, D. 1995. Black Tern management plan. Pennsylvania Game Commission, Harrisburg.

[11] Trimble, R. 1937. Some recent developments in the Pymatuning region, western Pennsylvania. Cardinal 4:102–108.

Black Skimmer ***Rynchops niger***

General status: Black Skimmers are saltwater birds that breed along the Atlantic coast from Massachusetts south along the Gulf coast. They winter along the coast from North Carolina south to Florida and along the Gulf coast to south Texas. There are six fall records of Black Skimmer in Pennsylvania since 1959.

Seasonal status and distribution

Fall: A Black Skimmer was collected by D. Snow near Essington, Delaware Co., in Sep 1976; the specimen is in the skeletal collection of the Delaware Museum of Natural History. G. Kopf and colleagues carefully studied an immature Black Skimmer at Tinicum from 23 Aug to 5 Sep 1980.[1] This bird was documented with a photograph.[2] E. Witmer,

H. Morrin, and R. M. Schutsky reported a Black Skimmer after a storm at Muddy Run on the Susquehanna River in Lancaster Co. on 20 Oct 1989.[3] Two days later a bird was reported flying overhead near Pipersville in Bucks Co. on 22 Oct 1989 by B. Webster.[4] An adult and a juvenile skimmer were found on 26 Sep 1992 at the Conejohela Flats in Lancaster Co. by R. Wiltraut; the juvenile remained until the next day and was photographed.[5] An adult was photographed by P. Schwalbe and G. Schwalbe during the passage of Hurricane Fran and remained from 7 to 9 Sep 1996 at BESP.[6]

History: Warren (1890) obtained a specimen near Philadelphia shortly after a severe storm in Sep 1880. The exact location was not given, so this bird may not have been within the political boundaries of the state. Subsequently, several sightings have been from the Conejohela Flats in Lancaster Co. One was seen on 1 Dec 1940 (Poole 1964), one on 31 Aug 1952 (Morrin et al. 1991), one on 14 Aug 1955 after a hurricane, and one on 31 May 1959 (Poole 1964). On 12 Sep 1954, one was at Tinicum, and on 16 Oct 1954 a bird visited Lake Ontelaunee in Berks Co. The latter bird was blown in with a hurricane that had passed over Reading the night before (Poole, unpbl. ms.).

[1] Paxton, R.O., W.J. Boyle Jr., and D.A. Cutler. 1981. Hudson-Delaware region. AB 35:164.
[2] R. Wiltraut, pers. comm.
[3] Witmer, E. 1989. County reports—October through December 1989: Lancaster County. PB 3:148.
[4] French, R. 1989. County reports—October through December 1989: Bucks County. PB 3:138.
[5] Kwater, E. 1994. Fifth report of the Pennsylvania Ornithological Records Committee, June 1994. PB 8:21.
[6] Peplinski, J., and B. Peplinski. 1996. Local notes: Centre County. PB 10:162.

Family Alcidae: Auks, Murres, and Puffins

Auks, Murres, and Puffins are black-and-white birds that fit the niche, in the Northern Hemisphere, that penguins do in the Southern Hemisphere. Except when they come ashore to breed, their entire life is spent on the open ocean. They have a short neck, stout body, short tail, and short narrow wings. Their legs also are short and are attached far back on their body, so they must stand very erect on land. When they fly, their wings beat rapidly. Their flight is rapid, direct, and usually low over the water. They are expert divers capable of reaching great depths in pursuit of their staple diet of fish and marine invertebrates. All species

Black Skimmer at Bald Eagle State Park, Centre County, September 1996. (Photo: Rick Wiltraut)

are accidental away from saltwater. Three species have been included on the list of birds of Pennsylvania: Dovekie, Thick-billed Murre, and most recently, Ancient Murrelet.

Dovekie ***Alle alle***

General status: In North America these chunky little alcids breed on eastern Baffin Island and in Greenland. They winter chiefly in the North Atlantic Ocean and are casual farther south offshore to North Carolina. Dovekies may be found at inland locations after fall storms. At least five modern sightings have been reported in Pennsylvania, all in the fall.

Seasonal status and distribution

Fall: A Dovekie was captured between 22 and 30 Nov 1966 in Hatboro in Montgomery Co. by L. Murphy and G. Murphy.[1] On 12 Nov 1970 one was found in Bethlehem (Morris et al. 1984), and another was found the next day on a road near Dingmans Ferry in Pike Co. (P. B. Street 1954). A northeast storm, possibly caused by tropical storm Gilda, during 28–29 Oct 1973, apparently was responsible for two inland Dovekie records: one picked up alive by D. Steffey at Oley in Berks Co. on 30 Oct 1973 and another found by J. E. Silagy at Pikeville in Berks Co. on 31 Oct 1973.[2]

History: Warren (1890) stated that between 1880 and 1890 he had seen two specimens that were captured in winter on the Delaware River near Philadelphia. He did not cite an exact location, so these birds may not have been collected within the boundaries of Pennsylvania. This species was recorded in Feb 1924 on the Lehigh River in Allentown (Morris et al. 1984). In Nov 1932 after a storm, many Dovekies were stranded around Philadelphia and as far inland as Birdsboro and Pottstown in Berks Co. (Poole, unpbl. ms.). They were reported at Bristol, Langhorne, Newtown, Norristown, and Oxford Valley in Bucks Co., within the city of Philadelphia and at Pottstown and Schwenksville in Montgomery Co. between 19 Nov and 3 Dec 1932.[3] The Birdsboro specimen was mounted and placed in the Reading Museum, and six others were preserved and placed in the collection at the ANSP. Two other sightings occurred on 28–29 Nov 1950 when one was found alive in the Pocono Mountains in Monroe Co. and another was found at Malvern in Chester Co.[4] The Pocono Dovekie was placed in the ANSP. Also on 28 Nov 1950, five were seen at Fairmont Park Reservoir in Philadelphia Co., and one was at Upper Darby in Delaware Co. (Poole, unpbl. ms.).

[1] Scott, F.R., and D.A. Cutler. 1967. Middle Atlantic Coast region. AFN 21:17.
[2] Scott, F.R., and D.A. Cutler. 1974. Middle Atlantic Coast region. AB 28:35.
[3] Murphy, R.C., and W. Vogt. 1933. The Dovekie influx of 1932. Auk 50:325–349.
[4] Gillin, J.R., and P.B. Street. 1951. General notes (other unusual records). Cassinia 38:38.

Thick-billed Murre ***Uria lomvia***

General status: Thick-billed Murres breed along the Alaska coast and offshore islands and in Greenland south to Hudson Bay and east to the Gulf of St. Lawrence. They winter in offshore waters within their breeding range and regularly south to British Columbia in the West and to New Jersey and inland to southern Ontario in the East. No modern sightings of this species have been recorded in Pennsylvania.

History: A specimen collected by C. H. Shearer in Dauphin Co. on 11 Dec 1893 (Stone 1894) was apparently the first valid record of this species in Pennsylvania. It was placed in the Reading Museum. Another was collected in the Barrens in Centre Co. in the fall of 1895 and was placed in the State College collection (Todd 1940). An unprecedented invasion of these birds occurred in the fall of 1896; single birds were collected at Lititz and on the Conestoga Creek in Lancaster Co. (Beck 1924), and at Erie Bay several were shot and one was mounted and placed in the CMNH. This species was found in Pennsylvania again in 1899, when a bird was collected from Erie Bay on 27 Nov and again on 2 Dec 1900 (Todd 1940). None has been recorded from this site since then. One was killed on the Delaware River at Byberry in Philadelphia Co. on 11 Jan 1901,[1] and another was captured at Williamsport in Lycoming Co. on 13 Dec 1902.[2] One shot at Conneaut Lake in Crawford Co. on 3 Dec 1907 was placed in the CMNH (Todd 1940). During the same year, one was collected on the Susquehanna River according to D. E. Kunkle (1951). The most recent records are of one collected at Eynon in Lackawanna Co. on 30 Nov 1950, and a live bird picked up on a street in Shillington on 4 Dec 1950 was taken to the Reading Museum (Poole, unpbl. ms.).

[1] Fowler, H.W. 1903. Water birds of the middle Delaware Valley. Cassinia 7:45.
[2] Koch, A. 1903. Abstract of the proceedings of the Delaware Ornithological Club. Cassinia 7:71.

[Razorbill] *Alca torda*

General status: In North America, Razorbills breed in Canada from southeastern Baffin Island to Newfoundland, northeastern Nova Scotia, and in southern New Brunswick south to eastern Maine. They winter offshore south to New York and inland to southern Ontario. Listed as hypothetical in Pennsylvania, this species has been reported once.

Comments: A Razorbill was reportedly shot on a lake near Pittston before 1902 and was placed in a private collection (Poole 1964). This collection later came to the Reading Museum. Unfortunately, the collecting data of the Razorbill could not be located after labels had been removed from all specimens.

[Black Guillemot] *Cepphus grylle*

General status: In North America, Black Guillemots breed in Canada, in the central and eastern Arctic, and south to Nova Scotia (Godfrey 1986). They winter within their breeding range and wander along the Atlantic coast as far south as New Jersey (AOU 1998). Black Guillemots are listed as hypothetical because records have not been confirmed within Pennsylvania.

Comments: Specimens were collected in the vicinity of Philadelphia in 1843–1844, before 1894, and on the Delaware River in Chester Co. in 1898, but they have not been proved to have been collected within the boundaries of Pennsylvania. The most recent report was of a bird in nonbreeding plumage observed with a flock of Buffleheads on 6 Apr 1957 at Lake Ontelaunee in Berks Co. (Poole 1964).

Ancient Murrelet *Synthliboramphus antiquus*

General status: Ancient Murrelets breed from southern Alaska south to British Columbia. They winter offshore from the Aleutian Islands south to central California and are occasionally found far inland in Canada and the U.S. (AOU 1998). One record is known from Pennsylvania.

Seasonal status and distribution

Fall: On 8 Nov 1992 at Lake Ontelaunee in Berks Co. M. Spence and colleagues discovered an Ancient Murrelet. It was documented with a photograph.[1]

[1] Wlasniewski, M. 1993. First Pennsylvania record of Ancient Murrelet. PB 6:143.

[Atlantic Puffin] *Fratercula arctica*

General status: Atlantic Puffins breed in eastern North America from Labrador south in coastal areas to eastern Maine.

They winter in Atlantic waters south to Massachusetts and casually to Virginia (AOU 1998). Atlantic Puffin is listed as hypothetical in Pennsylvania.

Comments: Warren (1890) and Stone (1894) cited a record of a bird shot in winter about 1876 along the Delaware River near Chester in Delaware Co. The specimen no longer exists.

Order Columbiformes
Family Columbidae: Pigeons and Doves

Pigeons and doves are recognized by their small round head in proportion to their plump body, very short legs, and their habit of bobbing their head as they walk. Approximately 290 species are found worldwide, most in the tropics. They feed primarily on seeds and fruit and frequently flock where food is plentiful. Six species have been found in Pennsylvania, including the Passenger Pigeon, which is now extinct. The White-winged Dove, Band-tailed Pigeon, Common Ground-Dove, and Eurasian Collared-Dove (a recently established bird of the southeastern U.S.) are accidental visitors. The Rock Dove and Mourning Dove are widespread residents, common around towns, cities, and farmlands.

Rock Dove *Columba livia*

General status: Rock Doves, widely called pigeons, were introduced from Europe and are now well established across southern Canada and the U.S. They are common to abundant residents statewide, often found close to human habitation. They depend on cultivated crops or trash for food. Rock Doves are considered pests, particularly around buildings. They are absent only from extensively forested areas, except where road cuts or quarries create vertical walls that provide nest sites.

Habitat: Rock Doves are found almost anywhere near human habitations, especially around farms with open barns or silos for nesting and in parks and urban areas. They also nest on bridges, on rock ledges, and, occasionally, in caves.

Seasonal status and distribution

Breeding: Rock Doves are well established in all Pennsylvania counties but are found only locally in extensively forested areas of the High Plateau. The Rock Dove's nest is placed on a ledge or in a cavity and normally contains two unmarked white eggs. Over time, accumulated fecal material builds around the nest into a sizable pile. Their nesting season is from about mid-Apr to early Jul; two or more broods are raised. Like the young of other altricial species, Rock Dove young are fully feathered and the size of adults when they leave the nest.

History: Rock Doves were introduced to North America with some of the earliest European colonists during the 1600s.[1] Considered domesticated, this species was not recorded in most ornithological accounts until the 1960s, when they were acknowledged as feral and were included on regional and statewide bird lists.[2]

Comments: The Rock Dove is one of three species, including European Starling and House Sparrow, that is afforded no protection by state or federal laws.

[1] Schorger, A.W. 1952. Introduction of the domestic pigeon. Auk 69:462–463.

[2] Schwalbe, P.W. 1992. Rock Dove *(Columba livia)*. BBA (Brauning 1992):146–147.

Band-tailed Pigeon *Columba fasciata*

General status: Band-tailed Pigeons breed in coniferous forests from British Columbia south to southern California and in oak or oak-conifer woodlands to the southwestern U.S. They winter throughout most of their breeding range. One accepted record of this species exists in Pennsylvania.

Seasonal status and distribution

Fall: A Band-tailed Pigeon was discovered along West Fisher Drive at PISP on 11 Dec 1991 by D. Snyder. The bird was observed at close range picking grit along the edge of the road.[1]

Comments: A Band-tailed Pigeon reported from Centre Co. on 18 Dec 1982 is listed as hypothetical.[2]

[1] McWilliams, G.M. 1992. County reports—October through December 1991: Erie County. PB 5:173.

[2] T. Schiefer, pers. comm.

Eurasian Collared-Dove ***Streptopelia decaocto***

General status: Eurasian Collared-Doves were evidently brought from the Netherlands to a pet shop in Nassau, Bahamas, in the early 1970s and later were delivered to a bird propagator. In 1974 fewer than 50 birds escaped from captivity and quickly began nesting in the wild. They rapidly spread throughout other islands in the Bahamas. It is not known whether they arrived on the mainland of Florida assisted or unassisted by humans, but they were nesting in Florida by the early 1980s.[1] Collared-doves rapidly expanded their breeding range throughout Florida and recently spread north to the Carolinas. It was only a matter of time before the first vagrant made an appearance in Pennsylvania.

Seasonal status and distribution

Summer: One was discovered by C. Nicholls in Spring Township near Conneautville in Crawford Co. on 28 Jul 1996 and remained to be photographed and seen by many birders until 1 Aug.[2] The call was also heard and well described. The origin of this bird is not known, but because of the bird's behavior and plumage condition, it was believed to be wild and was accepted as the first state record.[3]

[1] Smith, P.W. 1987. The Eurasian Collared-Dove arrives in the Americas. AB 41:1371–1379.

[2] Pulcinella, N. 1996. Rare bird reports. PB 10:155–156.

[3] Pulcinella, N. 1997. Eighth report of the Pennsylvania Ornithological Records Committee, June 1997. PB 11:129.

White-winged Dove ***Zenaida asiatica***

General status: White-winged Doves breed and winter across the Southwest from Baja California to southern Texas and Florida. They occasionally wander north along the Atlantic coast to Nova Scotia and inland to northern Ontario. White-winged Doves are accidental in Pennsylvania.

Seasonal status and distribution

White-winged Dove at Fairview Township, Erie County, May 1996. (Photo: Gerald M. McWilliams)

Spring: Two White-winged Doves have been recorded in Pennsylvania. One was identified in fresh plumage visiting a feeder in Fairview Township in Erie Co. from 1 to 10 May 1996 by G. M. McWilliams.[1] The dove was documented with photographs and placed in the files of PORC and VIREO. Another was seen and heard from 29 Jun to 3 Jul 1998 in Wayne Township in Schuylkill Co. and was photographed by S. Weidensaul.[2]

[1] Pulcinella, N. 1996. Rare bird reports. PB 10:95.
[2] Pulcinella, N. 1998. Rare bird reports. PB 12:63.

Mourning Dove ***Zenaida macroura***

General status: Mourning Doves breed across southern Canada and throughout the U.S. They winter over most of their breeding range. Mourning Doves are among the most common and widely distributed birds in Pennsylvania. They are common to abundant regular migrants and are breeding and wintering residents. Mourning Doves are probably permanent residents in some parts of the state, such as in the Piedmont.

Habitat: Mourning Doves are adaptable generalists, occupying a range of habitats including fields, farmlands, parks, towns, cities, open woods, and forest edges. They are less common, but present, even in openings of extensive forested areas.

Seasonal status and distribution

Spring: Migration is difficult to determine because Mourning Doves winter over most of the state, but numbers increase soon after the first major spring thaw in late Feb or early Mar. Migration may be delayed in the higher elevations and in the traditional snowbelt area of the Glaciated Northwest.

Breeding: The Mourning Doves' diet—almost exclusively seeds, including grain—dictates their habitat. They are most common in the Piedmont and Glaciated Northwest and least common through the High Plateau (see Map 7). Although flocking much of the year, they form pairs during the nesting season. The nest, a flimsy collection of sticks, is placed on a horizontal branch or near the trunk of a shrub, hedge, or small tree, often a conifer. It may be placed on an old nest of another species and is frequently 8–15 feet from the ground. Mourning Doves have a remarkably long nesting season, producing two or three broods each year. Two white, unmarked eggs usually are found from Apr through Aug, but nests occasionally are active in Mar or Sep.

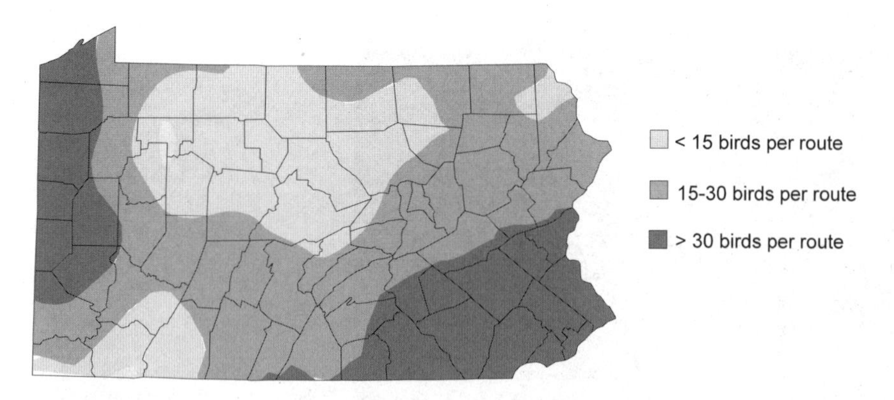

7. Mourning Dove relative abundance, based on BBS routes, 1985–1994

BBS data indicate a significant increase (2.6% per year) since 1966.

Fall: The number of birds begins to increase by mid- to late Jul. Peak migration ranges from mid-Aug to mid-Sep when farmers harvest their grain. Flocks containing several hundred birds gather in fields where grains have been harvested or in freshly plowed fields that are planted with winter wheat. Numbers begin decreasing as cold weather approaches in Oct, and by Dec many have moved to lower elevations or have migrated south.

Winter: The greatest number of Mourning Doves in winter is found around large open fields near streams and rivers in the Piedmont. Elsewhere most are found in towns and cities—especially at bird feeding stations—and around farms. They become uncommon to rare during winters with heavy snow cover, especially at higher elevations, in heavily forested areas, and in the snowbelt of the Glaciated Northwest.

History: Harlow's (1913) comment, "the Dove is one of the most common and generally distributed birds that nest within the state" reflects the state's deforested condition at the turn of the twentieth century. In sharp contrast, Todd (1940) stated that "more than forty years ago this dove was one of the most abundant birds in western Pennsylvania, but where there may once have been fifty pairs there is probably not one now." At that time, doves were less likely to remain in winter.

Passenger Pigeon ***Ectopistes migratorius***

General status: Extinct. Immense flocks of Passenger Pigeons once blackened Pennsylvania's skies. Accounts describe flocks many miles long taking as much as an hour to pass. Their habitat was destroyed and they were slaughtered in huge numbers. With unabated killing and habitat destruction over many years, Passenger Pigeons finally fell to extinction at the beginning of the twentieth century.

History: Much has been written on Passenger Pigeons, including lengthy reports from Pennsylvania.[1] Before 1880 these wild pigeons nested in great numbers in Cameron, Elk, Forest, McKean, Potter, and Warren cos., less commonly in Crawford and Erie cos. (Warren 1890). Huge nesting colonies, referred to as cities, extended over square miles of forest. One of the last "great cities" was in Potter Co., reportedly 2 miles wide and 40 miles long.[1] According to Warren (1890), a few pairs still bred in Luzerne Co. as late as 1889. F. L. Burns reported a nest and one egg taken in Lancaster Co. in 1889,[2] possibly Pennsylvania's last documented nest. Reports of the species continued until 1910 in Lancaster Co. (Beck 1924). The last Passenger Pigeon died in the Cincinnati Zoo in 1914.[3]

[1] French, J.C. 1919. The Passenger Pigeon in Pennsylvania: Its remarkable history, habits, and extinction. Altoona Tribune Company, Altoona, Pa.

[2] Burns, F.L. 1910. The status of the Passenger Pigeon in eastern Pennsylvania. WB 22:47–48.

[3] Schorger, A.W. 1955. The Passenger Pigeon. Univ. Wisconsin Press, Madison.

Common Ground-Dove ***Columbina passerina***

General status: Common Ground-Doves are resident across the southern U.S. from southern California, southern Texas across southern Louisiana, Mississippi, and Alabama, and throughout Florida north along the coast to South Carolina. Two sightings of this species have been documented in Pennsylvania.

Seasonal status and distribution

Fall through spring: A Common Ground-Dove appeared at G. P. Cole's feeding station on 14 Dec 1974 at Cardell's Corners near Newtown in Bucks

Co. The bird was observed daily at this site through 25 Apr 1975. It was photographed by A. Brady on 19 Jan 1975, and the photograph was published in *Cassinia*.[1] On 19 Oct 1986 one was discovered at PISP in Erie Co. by S. Stull.[2] It was seen feeding along the sand dunes at Gull Point and was well studied as close as 10 feet by the observer. The dove was seen twice on the ground and three times in the air.

History: Libhart (1869) reported that a Common Ground-Dove was shot in 1844 near Marietta in Lancaster Co. A specimen in the North Museum, without data, may be this bird (Poole, unpbl. ms.). Another bird was seen on the Conejohela Flats in Lancaster Co. on 14 Aug 1955 (Poole 1964) after Hurricane Connie, but details are lacking on this record.

1 Sparks, D., and A. Brady. 1974–1975. Ground Dove in Bucks County. Cassinia 55:41.
2 Hall, G.A. 1987. Appalachian region. AB 41:89.

Order Psittaciformes
Family Psittacidae: Parrots

Most species of parrot are found in the tropics. They are colorful, noisy birds with a hooked bill and a short, rounded to very long, pointed tail. Many species are kept as cage birds and occasionally escape and establish local colonies. Escaped Monk Parakeets nested in the Pittsburgh area from 1971 to 1973.[1] Only the now-extinct Carolina Parakeet occurred naturally in Pennsylvania. It is listed as hypothetical because no specimen or written description can be located.

1 Freeland, D.B. 1973. Some food preferences and agressive behavior by Monk Parakeets. WB 85:332–334.

[Carolina Parakeet] *Conuropsis carolinensis*

General status: The Carolina Parakeet formerly ranged from Nebraska and Oklahoma east to New Jersey and Florida. Most information about this species has been gleaned from anecdotal notes of early settlers or naturalists. The last wild bird was collected in Florida on 12 Mar 1913, and the last known living individual died at the Cincinnati Zoo on 21 Feb 1918 (AOU 1998). Very little is known about their status and distribution in Pennsylvania.

Comments: The oldest record of Carolina Parakeet in Pennsylvania comes from Barton (1799), who stated that "about the year 1760 large flocks of the Carolina Parakeet . . . were seen in Shareman's Valley, some twelve miles north of Carlisle." Turnbull (1869) listed this species as occurring at rare intervals in southern Pennsylvania. Libhart (1869) reported that a flock was seen many years previously near Willow Street in Lancaster Co., and a specimen was collected from this flock and deposited in the North Museum in Lancaster. Stone (1894) listed what he presumed to be the same specimen still in that collection, but without data. Apparently this specimen no longer exists. Beck (1924) stated that the last record was of a flock seen near Manheim, Lancaster Co. (probably before 1850). No known specimens from Pennsylvania exist today.

Order Cuculiformes
Family Cuculidae: Cuckoos

Of the 125 or more species of cuckoo found in the world, only the similar-looking Black-billed and Yellow-billed cuckoos are found in Pennsylvania. They are slender and have a long tail and a bill that curves slightly downward. Cuckoos have zygodactylous feet: that is, two toes in front and two toes in back. Uncommon and secretive but often quite vocal, cuckoos fly with agility, maneuvering through dense growth with ease. They eat a wide variety of insects but have a voracious appetite for caterpillars. During outbreaks of webworms, tent caterpillars, or gypsy moths, cuckoos may become locally common. The two species lay eggs in each other's nests, and Black-billed Cuckoos are known to lay eggs in the nests of other species of birds.

Black-billed Cuckoo ***Coccyzus erythropthalmus***

General status: Cuckoos are quite secretive and are more often heard than seen. They breed from Alberta east across southern Canada to Nova Scotia and south to Texas, Alabama, and South Carolina and winter in South America. Black-billed Cuckoos are uncommon to rare regular migrants in Pennsylvania. They breed statewide but are more common across the northern portion of the state than Yellow-billed Cuckoos. Populations fluctuate with outbreaks of hairy moth larvae, such as the gypsy moth and tent caterpillar.

Habitat: During migration Black-billed Cuckoos prefer forest edges, woodlands, and thickets. In the breeding season they prefer extensively wooded areas.

Seasonal status and distribution

Spring: They may arrive as early as the last week of Apr, but most appear the first week of May. A bird seen on 10 Apr 1977 at PISP was very early.[1] Their secretive habits make migration difficult to discern, but fewer are found after the third week of May.

Breeding: In Pennsylvania, Black-billed Cuckoos are uncommon breeders. They are most common in the mountains and are regularly more common there than are Yellow-billed Cuckoos. During caterpillar outbreaks, Black-billed Cuckoos may be found in any Pennsylvania county. BBA reports included confirmed records in most counties. The nest is a bulky collection of twigs, most often placed 4–10 feet from the ground. The clutch size, mostly three or four pale bluish green eggs, varies with abundance of caterpillars and with the extent of nest parasitism from conspecifics or Yellow-billed Cuckoos. Most eggs are found from late May through Jun, with apparently only a single clutch each year.

Fall: As during spring migration, their secretive habits make it difficult to discern the fall migration period. Cuckoos are not easy to find after the breeding season; birds seen from the first week to the third week of Sep are likely southbound birds. Stragglers have been recorded to the first week of Nov. A very late bird was recorded in Washington Co. on 15 Dec 1984.[2]

History: Alexander Wilson (1808–1814) first described this species, probably from a specimen taken near Philadelphia (AOU 1983). No major change in distribution or abundance is apparent in the literature.

[1] G.M. McWilliams, pers. recs.
[2] Hall, G.A. 1985. Appalachian region. AB 39:167.

Yellow-billed Cuckoo ***Coccyzus americanus***

General status: Yellow-billed Cuckoos breed across most of the U.S. and winter in South America and casually in the southern U.S. In Pennsylvania they are uncommon to rare regular migrants and breeding residents. Their breeding range is statewide, but they are more common in the southern portion of the state than are the Black-billed Cuckoos. Populations fluctuate with outbreaks of hairy moth larvae such as gypsy moths.

Habitat: Yellow-billed Cuckoos are found in forest edges, woodlands, and thickets. During the breeding season they prefer more-open, brushier areas than Black-billed Cuckoos and are more likely to be found at lower elevations and in suburbs.

Seasonal status and distribution

Spring: A few Yellow-billed Cuckoos arrive as early as the last week of Apr. More appear in southern counties in the first week of May and across the northern tier of the state in the second week of May, but many may not arrive until Jun. This species has been recorded as early as 7 Apr in Delaware Co.[1] In Erie Co. they have been recorded twice on 10 Apr, once in 1960 and once in 1987.[2] They are not usually encountered in numbers during migration. More birds are found after the third week of May.

Breeding: Yellow-billed Cuckoos have a more southerly distribution than Black-billed Cuckoos. They are fairly common in southern portions of the Ridge and Valley and Southwest regions but may be encountered statewide. No long-term population trend is apparent on BBS routes. The nest is sometimes placed low (3 feet from the ground but generally below 20 feet) and is loosely constructed and flat like a dove's nest. Clutch size varies with abundance of food and egg-dumping of conspecifics or Black-billed Cuckoos, but two to four pale bluish green eggs are normal. The extent of the nesting season, from the third week in May through mid-Jul, suggests that two broods are possible. Nests are occasionally active through Aug.

Fall: Yellow-billed Cuckoos are difficult to find after the breeding season, but a few birds make appearances from about the first to the third week of Sep. Stragglers have been recorded to the second week of Nov. One rather late bird was at Reading on 20 Nov 1971,[3] and one was at PISP on 18 Nov 1960.[2]

History: Harlow (1913) thought that this species was expanding its range north. Yellow-billed Cuckoos have probably benefited from human habitat alteration.

[1] N. Pulcinella, pers. comm.

[2] G.M. McWilliams, pers. recs.

[3] Scott, F.R., and D.A. Cutler. 1972. Middle Atlantic Coast region. AB 26:43.

Order Strigiformes
Family Tytonidae: Barn Owls

Barn Owls are the only species in the family Tytonidae that live in North America. Like typical owls, they are nocturnal birds of prey and have a large head with facial disk, immobile eyes directed forward, excellent vision, and a strongly hooked bill. They have powerful toes, with long curved talons used for catching and holding prey; the outer toe is reversible. Their light, soft plumage enables them to fly silently. Barn Owls differ from typical owls in having a long heart-shaped face, smaller eyes, long legs that are covered with short feathers, and toes that are covered with bristles.

Barn Owl *Tyto alba*

General status: Barn Owls are resident across the U.S. Little is known about the status of this secretive and nocturnal species in Pennsylvania. Where found in Pennsylvania, Barn Owls usually are very rare and local. Barn Owls may be permanent residents in the Coastal Plain and Piedmont but apparently move south for the winter from the north and west. They winter locally, primarily southeast of the Ridge and Valley region. They may breed in the state at any time of the year.

Habitat: Barn Owls inhabit open areas such as abandoned fields, pastures, or meadows near suitable nesting cavities, such as hollow trees, silos, barns, bridge girders, or abandoned buildings. They are occasionally found roosting in conifers during the migration period and in winter.

Seasonal status and distribution

Spring: Though Barn Owls are mainly sedentary, migration is evident at some sites. Barn Owls have been recorded at PISP, where they do not nest, ranging from mid-Apr and mid-May. They usually appear one day and are gone the next.[1]

Breeding: Barn Owls are rare to accidental; potentially found statewide. They are found most often in the western Piedmont and in broad valleys of the southern Ridge and Valley. Barn Owls are rare and irregular in the rest of the state. Barn Owl breeding populations are not monitored, but numbers apparently have dwindled since the late-1970s on CBCs. For example, they are absent from Greene and Washington cos., where they were formerly regular. The current population status is unclear, although birds remain at some traditional sites in south-central counties. The nesting season is potentially year-round; eggs or flightless young have been found from Jan through Dec, but more often in Mar and Apr. No nest is built; usually from five to seven unmarked white eggs are laid.

Winter: Most winter records are from the Coastal Plain and the Piedmont, with others scattered from the valleys of the Ridge and Valley. West of the Allegheny Front they have been recorded in winter only from Allegheny, Clarion, Crawford, Greene, and Washington cos. They may withdraw to river valleys and to the south from northern sites during severe winters with heavy snow cover. They are vulnerable to harsh winter conditions; severe winters in the mid-1990s eliminated them from some areas.

History: Long associated with human agriculture and buildings, Barn Owls were always most common in the southern counties. Bird lists of Chester, Delaware, and Lancaster cos. from 1860 to 1885 considered them rare residents, but they were not listed on Baird's (1845) thorough summary of Cumberland Co. An expanding range, or at least new records, followed the pattern of rapid agricultural expansion during the 1800s. At the beginning of the twentieth century Harlow (1913) and Sutton (1928a) thought that they were expanding and common, though local, in southern counties, with a few records to the north, even during winter in Crawford Co. (Sutton 1928b). Breeding records were reported statewide by the mid-1900s (Wood 1979). Barn Owl migration also was noted historically. According to Bent (1938), a young Barn Owl banded in Pennsylvania was later discovered in Georgia. A nestling that was banded near Kempton in Berks Co. on 10 Jun 1953 was recovered the following Dec at Key West, Florida (Uhrich 1997).

Comments: The value of Barn Owls to farmers has long been known. Barn Owls received protection (with other small owls) in the late 1800s even when bounties were placed on the larger owls (Kosack 1995). Although closely associated with humans, Barn Owls continue to be confused with other owls by rural residents. The OTC lists the species as Candidate–At Risk in Pennsylvania.

[1] G.M. McWilliams, pers. obs.

Family Strigidae: Typical Owls

The family strigidae includes all the owls except Barn Owls. Compared with Barn Owls, typical owls have more rounded and less triangular faces. Their eyes are usually larger, ear tufts are often present, and they have shorter legs. Most species are nocturnal, but some North American species, including Snowy, Short-eared, and Northern Hawk owls, also feed by day. Roosting owls are often mobbed by songbirds. They can also be located by finding regurgitated pellets of fur, feathers, and bones on the ground below roosts or nests. Ten species have been recorded in Pennsylvania. The Snowy Owl is a regular winter visitor, and three species are accidental vagrants: Northern Hawk, Great Gray, and Boreal owls. Owls rarely construct their own nests but occasionally add material to an existing nest, in which they lay up to eight unmarked white eggs. Eastern Screech-Owls and Northern Saw-whet Owls will use nest boxes.

Eastern Screech-Owl ***Otus asio***

General status: Eastern Screech-Owls are resident from Montana and Texas east across the U.S. They are permanent residents in Pennsylvania. Screech-owls are the most common owls of towns and cities that have large deciduous trees. Populations may fluctuate because of severe winters across the northern-tier counties and in the mountainous areas of the High Plateau.

Habitat: Eastern Screech-Owls are found in a variety of habitats including residential areas in towns, city parks, cemeteries, orchards, and deciduous and mixed woodlands.

Seasonal status and distribution

Breeding: These owls are fairly common across most of the state. They are most common across the southern half of the state and are rare (or at least infrequently reported) in the higher elevations and more extensively forested regions. Three to six eggs are placed in a natural cavity, old woodpecker hole, or nest box; but no nest is constructed. Screech-owls may be heard calling year-round, but most often during early spring in preparation for the breeding season. Most eggs are found in Apr, with some in Mar and May.

History: The screech-owl's status and distribution in early ornithological records differ little, if at all, from today. Harlow (1918) went as far as to state that screech-owls were "not found at all in the primeval forests." Where the primeval forest was cleared, this species became fairly common, even in northern counties such as Susquehanna (Cope 1898). These small owls are vulnerable to attacks by the larger species; Sutton (1928b) said that the screech-owl was very rare at Pymatuning Swamp, "where the Barred Owl constantly preys upon it."

Comments: Screech-owls also will use nest boxes as winter roosts. The box should be placed in the woods and have an entrance diameter of 3 inches. Screech-owls are found in two color morphs; the gray morph is most common, but red-morph screech-owls also are found in Pennsylvania.[1]

[1] Sutton, G.M. 1927. Mortality among Screech Owls in Pennsylvania. Auk 44:563–564.

Great Horned Owl ***Bubo virginianus***

General status: Great Horned Owls are resident throughout North America to the treeline in Arctic Canada. The northernmost populations are partially migratory, wintering south to southern Canada and the northern U.S. The largest and most widely distributed resident owls in Pennsylvania, Great Horned Owls are powerful birds of prey that feed on a wide variety of mammals and birds, including other species of owl. They are uncommon to fairly common permanent residents statewide. Bounties were paid before 1965 for Great Horned Owls killed in Pennsylvania.

Habitat: Great Horned Owls are found in a variety of habitats ranging from large tracts of forests to towns, city parks, and cemeteries, provided woodlots with large trees are available.

Seasonal status and distribution

Breeding: Great Horned Owls are most common in fragmented mixes of farmland and woodlots or at least are more easily found in open habitats.[1] They are most often detected when vocalizing during the winter breeding season. In a survey in the south-central counties, the density was estimated at one pair per 2–3 square miles.[2] The Great Horned Owl is one of the state's earliest-breeding birds; territorial hooting, often heard as a duet, begins in the fall. Horned owls are on eggs in Feb and early Mar, occasionally as early as late Jan. A nesting female may sit on eggs through the worst of winter weather and sometimes is seen covered with snow. Great Horned Owls build no nest but lay two or three eggs in an old crow or hawk nest or in a tree cavity. CBCs show an expansion of the population since around 1960, at least in the Piedmont.

History: Few birds have received humans' wrath with the fury directed toward Great Horned Owls. Listed by early authors as common residents, Great Horned Owls were considered a varmint for occasionally preying on domestic fowl and game. Bounties were paid for their "scalp" intermittently from 1885 until 1965, bringing $5 in 1949 (Kosack 1995). Even Sutton (1928a), the first official state ornithologist, called them "one of our most destructive birds of prey." Unregulated shooting affected the population, at least in the settled southeast. Harlow (1918) stated that Great Horned Owls were "becoming exterminated in all but the wilder sections" of Pennsylvania. These owls were not easily obtained by bounty hunters in the southeastern counties during the 1930s,[3] but they began to reappear in that area during the 1940s.[4] Great Horned Owls did not receive full protection until the Federal Migratory Bird Treaty was ratified with Mexico in 1972 and protection was extended to all native birds.

Comments: An extensive study of Great Horned Owl food habits in southeastern Pennsylvania documented a diversity of prey items, but opossum, cottontail rabbit, and Norway rat made up a majority of the diet.[5]

[1] Morrell, T. 1994. Status and habitat characteristics of the Great Horned Owl in south-central Pennsylvania. Ph.D. dissertation. Penn State Univ., State College.

[2] Morrell, T., and R. Yahner. 1990. Status and habitat characteristics of the Great Horned Owl in Pennsylvania. Final report. Pennsylvania Game Commission, Harrisburg.

[3] McDowell, R.D. 1940. The Great Horned Owl. Pennsylvania Game News 8:10–11, 29.

[4] Schwalbe, P.W. 1992. Great Horned Owl *(Bubo virginianus)*. BBA (Brauning 1992):158–159.

[5] Wink, J., L. Goodrich, and S. Senner. 1988. Research results: Raptors and their prey. Hawk Mountain News 70:25–30.

Snowy Owl ***Nyctea scandiaca***

General status: Snowy Owls breed from northern Alaska east across the Canadian Arctic south to the western side of Hudson Bay and east to Labrador. They winter irregularly south to California and east to Texas, the Gulf states, and Georgia. Snowy Owls are rare regular fall visitors or winter residents in the northwestern portion of the state and irregular to casual elsewhere in Pennsylvania. They are absent in the heavily forested areas of the High Plateau and Allegheny Mountains. A single record of a summering Snowy Owl exists in the state. Snowy Owls are irruptive and cyclical; numbers vary from year to year. During invasion years, as many as 10 or more may appear over the state. Usually single individuals, but as many as three together, have been recorded from various sites. Most owls observed are the heavily marked immatures.

Habitat: Snowy Owls are found in large open fields, at airports, along lake shores, and on buildings in towns and cities.

Seasonal status and distribution

Spring: Snowy Owls seen in spring usually are birds that arrived some time in late fall or winter and, with sufficient food, remained into spring. Most leave by mid-Mar, with stragglers to the first week of Apr. A very late bird was shot near Allentown on 9 May 1981.[1] Another, captured in mid-May 1985 at Canoe Creek State Park, Blair Co., was rehabilitated.[2]

Snowy Owl at Washington Township, Erie County, February 1991. (Photo: Gerald M. McWilliams)

Summer: Unexpected was the Snowy Owl that was discovered east of Erie on the roof of a building in midsummer. It was observed and photographed by many people from the third week of Jul to at least the third week of Sep 1997. It spent the summer perched either on the roof of a building, on equipment outside the building, or on nearby utility poles.[3]

Fall: Snowy Owls have been recorded as early as the third week of Oct. They often appear with the passage of a cold front after the first or second week of Nov. Some years they may not appear until the second week of Jan or even later.

Winter: Most winter reports are of birds that have arrived late in the fall. Some owls will remain into early spring. When food becomes scarce in rural areas Snowy Owls sometimes move into towns and cities, where they can be seen perched on rooftops or church steeples.

History: Snowy Owls were probably very rare before the great forests in Pennsylvania were cleared. The most notable invasion occurred during the winter of 1926–1927, when as many as 204 birds were reportedly shot and 39 others were seen in Pennsylvania.[4] Nineteen of those birds were killed in Bucks Co. (Poole, unpbl. ms.). Grimm (1952) reported a very late bird said to have been killed at the Pymatuning Fish Hatchery near Linesville on 8 May 1946. Otherwise, their status has remained relatively unchanged throughout recorded history.

[1] Paxton, R.O., W.J. Boyle Jr., and D.A. Cutler. 1981. Hudson-Delaware region. AB 35:807.
[2] Hall, G.A. 1985. Appalachian region. AB 39:300.
[3] J. Stull Cunningham, pers. comm.
[4] Sutton, G.M. 1927. The invasion of goshawks and Snowy Owls during the winter of 1926–1927. Cardinal 2:35–41.

Northern Hawk Owl *Surnia ulula*

General status: Northern Hawk Owls breed north to the treeline in Canada, from northern Yukon east to Newfoundland, and south to southern Ontario, northern Minnesota, Wisconsin, and Michigan. They winter throughout their breeding range and rarely south in the U.S. to Oregon, Nebraska, Pennsylvania, and Ohio. There are two accepted records in Pennsylvania, one modern and one historical, and several other historical and unverified reports.

Seasonal status and distribution

Winter to spring: A Northern Hawk Owl was identified by F. Haas and col-

Northern Hawk Owl at Lookout, Wayne County, February 1991. (Photo: Randy C. Miller)

leagues on 19 Feb 1991 near the town of Lookout in Manchester Township, Wayne Co. The owl remained for many birders to see until 17 Mar 1991 and was photographed by many.[1]

History: The only accepted historical record is from a specimen at the Reading Public Museum that was collected at Albany in Berks Co. on 26 Jan 1887. The first reported occurrence of this species in Pennsylvania was of one shot at Haddington, near Philadelphia in 1866 (Turnbull 1869). Gentry (1877) wrote, "In eastern Pennsylvania it is occasionally met during severe winters. . . . specimens have been taken by the writer and others, as early as the middle of October."

Comments: Although there is little documentation, this species has also been reported from Pittsburgh in the winter of 1963;[2] in 1963 at Meadville, Crawford Co. on the unusually late date of 13 May;[3] and at Erie from 13 to 18 Jan 1979 (Stull et al. 1985).

[1] Haas, F.C. 1991. First documented Northern Hawk Owl, *Surnia ulula*, in Pennsylvania in the twentieth century. PB 5:16–18.

[2] Hall, G.A. 1963. Appalachian region. AFN 17: 324.

[3] Hall, G.A. 1963. Appalachian region. AFN 17: 402.

Barred Owl *Strix varia*

General status: Like many other species of owls, Barred Owls are more often heard than seen, although they occasionally perch in the open along woodland edges. Barred Owls are resident from British Columbia south to Oregon, across central Canada to Nova Scotia, and south across the eastern U.S. and west to the Great Plains. Northern populations are partially migratory. They are uncommon permanent residents nearly statewide.

Habitat: Barred Owls prefer extensive mature forested swamps, wet mature woodlands, and wooded ravines. However, they have been found in urban, agricultural, and industrialized areas of the state.

Seasonal status and distribution

Breeding: Barred Owls have been expanding their range in recent decades and now probably are resident in every county except Philadelphia. The BBA project reported records in Chester Co., where they had long been scarce, but failed to detect this owl in Delaware, Bucks, and Montour cos., where they irregularly were found during the 1990s. Barred Owls nest in highest densities in old-growth forests of the High Plateau.[1] Two or three eggs are placed in a cavity of a large tree; on a broken treetop; or in an old crow, hawk, or squirrel nest (Sutton 1928b; H. H. Harrison 1975). Nests with eggs are expected during Mar or early Apr, although few have been documented. A Barred Owl nested in a box erected at Loyalville in Luzerne Co.[2]

Winter: These owls generally remain sedentary through the winter. However, some may retreat from the mountains of the northern and central counties to lower elevations or may wander into open areas where food is more plentiful. They are most likely to move during severe winters with substantial snow accumulations.

History: The infrequent reports of this species in historical accounts are typical of nocturnal birds. Warren (1890) stated that Barred Owls were the most common of the owl species in mountainous and heavily wooded sections of the state. He also elaborated on the persecution that owls suffered in the late 1800s because of the perception that they reduced populations of poultry and game. Harlow (1913) speculated that the species bred in every county except Allegheny, Chester, Greene, Montgomery, and Philadelphia, a distribution that is remarkably similar to modern patterns. Since Barred Owls generally rely on nest sites in cavities of large trees, the reduction of forest cover undoubtedly affected the species' distribution and abundance by the turn of the twentieth century.

[1] Haney, J.C. 1997. Spatial incidence of Barred Owl *(Strix varia)* reproduction in old-growth forest of the Appalachian Plateau. Journal of Raptor Research 31:241–252.

[2] W. Reid, pers. comm.

Great Gray Owl *Strix nebulosa*

General status: These birds of prey are easily approached and always attract attention when they occur south of their normal range. They are the largest owls in North America. Great Gray Owls breed in northern coniferous or mixed forests from Alaska south to California and Montana and east across Canada to southern-central Ontario. In winter they occasionally wander south and east of their breeding range from Canada to the north-central and northeastern U.S. to Pennsylvania and New Jersey. There are only two modern records in Pennsylvania.

Seasonal status and distribution

Winter to spring: In Feb 1979 a Great Gray Owl was found dead in a melting snowbank near the office at Nockamixon State Park in Bucks Co. The bird was mounted and placed on display at Tobyhanna State Park in the Poconos[1] but was transferred to the ANSP in 1994. A Great Gray Owl was reported on 28 Jan 1992, and confirmed by T. Grisez the next morning, on the grounds of the United Refining Company in Warren, Warren Co., and later along a nearby railroad track. This bird remained until 27 Mar 1992 and was documented with numerous photographs.[2] This was one of the most celebrated bird sightings in Pennsylvania in many years. At least 2500 birders and nonbirders visited the area to see this easily approached owl.[3]

History: The first authentic record for Pennsylvania was of a bird shot on 16 Feb 1898 about 6 miles west of Waynesburg in Greene Co. The bird was reported to have been wounded, kept alive in a box for about a week, and then killed. The wings and tail were re-

Great Gray Owl at Warren, Warren County, February 1992. (Photo: Gerald M. McWilliams)

moved for ornaments, but the head was taken to J. W. Jacobs, who identified the bird as a Great Gray Owl (Todd 1940).

[1] Boyle, W.J., Jr., R.O. Paxton, and D.A. Cutler. 1981. Hudson-Delaware region. AB 35:284.
[2] Grisez, T. 1992. First live record of Great Gray Owl in the twentieth century in Pennsylvania. PB 6:10.
[3] Hall, G.A. 1992. Appalachian region. AB 46:265.

Long-eared Owl ***Asio otus***

General status: Long-eared Owls breed primarily across the southern Canadian provinces south across the U.S., to New Mexico in the West and to Indiana and West Virginia in the East. They winter throughout the U.S. Long-eared Owls are uncommon to rare regular migrants and regular winter visitors or residents. They breed in the state, but their status and distribution are not well known. Long-eared Owls may be permanent residents in many parts of the state.

Habitat: They prefer pine forests or plantations during migration and in winter. Long-eared Owls inhabit pine plantations and mixed woodlands close to open fields during the breeding season.

Seasonal status and distribution

Spring: Migration is not well known. Evidence of transient birds has been noted in Lancaster Co. and at PISP in Erie Co. They are not known to breed at the latter site. The beginning of migration is difficult to determine because winter residents may overlap spring arrivals. Most transients are recorded the first to the third week of Apr, with daily highs from 4 to 10 birds. Lingering migrants have been recorded to the last week of May.

Breeding: Long-eared Owls are rare and local breeding residents, potentially found statewide in summer. They are rarely reported during the breeding season and are easily overlooked because of their strictly nocturnal habits. Although this owl is generally associated with conifers, Todd (1940) reported that most prey items were meadow and house mice, residents of open country. This report supports a general picture of Long-eared Owl habitat that includes a pine stand adjacent to a meadow or pasture. Nesting was confirmed during the BBA project in Beaver, Cumberland, Lancaster, Lebanon, and Tioga cos. and subsequently in Columbia[1] and Indiana cos.[2]

Clutches of four to six eggs are laid from late Mar through Apr, often in an old crow or squirrel nest in dense pines. Nesting sometimes occurs in groves that supported a winter roost.

Fall: Long-eared Owls first appear in areas where they do not breed about the second or third week of Nov but have been recorded as early as the third week of Oct. Most of these presumed migrants remain through the winter. Usually only one or two birds are recorded at each sighting.

Winter: Most Long-eared Owls are reported during winter, but sightings are usually rare. Many of the sightings are in the Piedmont and Lake Erie Shore, with fewer winter records in the remainder of the state. They have been reported at only two areas in the Glaciated Northeast; in the 1960s they regularly wintered near Wyalusing in Bradford Co.,[3] and they have been found in the northeastern corner of Tioga Co.[4] No winter record is known in the High Plateau. They are sometimes found in communal roosts of up to a dozen or more birds (Morrin et al. 1991). In one communal roost near Nockamixon State Park, 17 were counted on 30 Jan 1972 (Morris et al. 1984).

History: Warren (1890) was one of few historical authors to suggest that the Long-eared Owl was anything but rare. He stated that it was "one of the most abundant of all the owl tribe in this state." This was probably an overstatement, biased by winter observations of more than two dozen birds that were found at communal roosts. About 750 pellets—about a half-bushel—were collected from a colony in the barrens near Oxford in Chester Co. around 1940; the collection contained 1279 skulls of meadow mice (Poole, unpbl. ms.). B. H. Christy and G. M. Sutton (1928) thought that Long-eared Owls were "found generally over the region" of Cook Forest, Forest Co., and called the species a permanent resident. Nests were only rarely documented, historically attributed to at least 12 counties: Beaver, Berks, Blair, Butler, Chester, Crawford, Erie, Lehigh, Montgomery, Warren, Westmoreland, and York (Todd 1940; Poole, unpub. ms.). Poole thought that Long-eared Owls were persecuted in mistake for the widely despised Great Horned Owl and reported further that the Long-eared Owls were "formerly much more common" (Poole 1964).

Comments: The Long-eared Owl is listed by the OTC as a Species of Special Concern, Candidate–Undetermined.

[1] D.A. Gross, pers. comm.
[2] M. Higbee and R. Higbee, pers. comm.
[3] W. Reid, pers. comm.
[4] J. Stickler, pers. comm.

Short-eared Owl *Asio flammeus*

General status: Short-eared Owls breed throughout Canada except on most of the Canadian Arctic islands south primarily through the northern half of the U.S. They winter throughout the U.S. Short-eared Owls are uncommon to rare regular migrants, most evident at the Lake Erie Shore. They have recently been found nesting on reclaimed strip mines in Pennsylvania. In winter they are regular, but the number of birds may vary from year to year.

Habitat: Short-eared Owls were probably more confined to bogs, marshes, and scattered grassy glades when most of Pennsylvania was covered by forest. Now they inhabit reclaimed strip mines, open grassy fields, large meadows, and airports. Short-eared Owls are usually found on or near the ground, but they occasionally roost in shrubs and conifers.

Seasonal status and distribution

Spring: Lingering winter residents may overlap spring migrants. Their typical migration period is from the last week of Mar to the last week of Apr. The greatest number of migrants is usually observed during the second or third week of Apr. Stragglers may remain to the third week of May.

Breeding: Short-eared Owls are rare, irregular, and irruptive breeding residents in extensive grassy areas scattered across Pennsylvania. Summer residents were observed at nine locations across western Pennsylvania during 1997, all at reclaimed strip mines. Young were observed in Allegheny, Clarion, Jefferson, Lawrence, and Venango cos. that year.[1] Summer records came from reclaimed mineland in Clearfield Co. in Jun 1993,[2] and up to three pairs in a year nested on and off between 1988 and 1998 on the extensive reclaimed strip mine of Piney Township, Clarion Co., known locally as the Piney Tract.[1,3] Summering birds were observed again in 1998 at the Imperial grasslands, Allegheny Co. Nesting was reported irregularly in fields adjacent to the Philadelphia International Airport until 1989 (Brauning et al. 1994). These owls lay clutches of up to nine eggs in a depression on the ground in May and Jun, but rarely have more than four fledged owlets been observed at any nesting site.

Fall: The first evidence of southbound birds may be as early as the third week of Sep. Their usual migration period is from the third or fourth week of Oct to the third week of Nov. Some Short-eared Owls remain through the winter. Owls were unusually common in reclaimed coal strip mines near Curllsville, Clarion Co., in Oct 1986, when a total of about 75 was counted.[3]

Winter: Short-eared Owls are generally uncommon to rare but may be locally common in some areas. They are widely distributed over the state in winter and occur almost annually in some large open fields with large small-rodent populations. Short-eared Owls have not been recorded in the mountainous areas of the Ridge and Valley region, in the north-central portion of the High Plateau, or in the northeastern and southwestern corners of the state in winter. Some large wintering concentrations, with counts as high as 20 or more birds, have recently been discovered in several of the reclaimed strip mines in the Southwest region. The number of birds may increase through the winter in areas where the food supply remains high.

History: There have been fewer than 10 confirmed Short-eared Owl nesting events from the time of Audubon's observations in the early 1800s to the BBA project in the 1980s. Breeding reports came from Crawford, Berks, and Lehigh cos. before 1960, with additional summer records in Adams, Erie, and Philadelphia cos. Short-eared Owls have suffered, as have many other species associated with grasslands, from a decline in agricultural acreage and changes in farming practices.

Comments: Short-eared Owls are listed as Endangered by the PGC.

[1] J. Fedak and B.S. Butcher, pers. comm.

[2] Schwalbe, G., and P. Schwalbe. 1993. Breeding season record of a Short-eared Owl. PB 7:53.

[3] Buckwalter, M. 1988. Short-eared Owls in Clarion County. PB 2:55–56.

Boreal Owl *Aegolius funereus*

General status: Boreal Owls breed in North America from Alaska south and east across most of Canada in dense northern forests and locally at high elevations in the Rocky Mountains. They are an irruptive species that winter mostly within their breeding range and rarely wander south of the northern and New England states. This small owl has been reported once in Pennsylvania.

Seasonal status and distribution

Spring: A single Boreal Owl was shot near Wilkinsburg in Allegheny Co. on

12 Mar 1896; the specimen resides in the CMNH collection. Dr. D. A. Atkinson, who collected the owl, stated that this bird was killed at the edge of the woods in the afternoon (Todd 1940).

Northern Saw-whet Owl ***Aegolius acadicus***

General status: These tiny, rather tame owls breed from southern Alaska along the Pacific coast, south through the Rocky Mountains to New Mexico, east across Canada, south to Iowa, and in the Appalachian Mountains to North Carolina and Tennessee. Northern Saw-whet Owls winter within their breeding range and south in the East to the Gulf coast and Georgia. They are probably more common in Pennsylvania than records indicate. Saw-whet owls spend the day roosting in dense vegetation, where they are well concealed. They are regular migrants and winter visitors or residents and may be fairly common during invasion years. Saw-whet owls breed in the state, and some may be permanent residents within their breeding range.

Habitat: Northern Saw-whet Owls are found in a variety of habitats, from hedgerows and thickets in suburban areas of towns and cities to swamps, mixed woodlands, and pine plantations. They are found most frequently in extensive forests with dense understory or in conifers near a wetland. They usually prefer roosting in the densest portions of vegetation where they are well concealed, but they will occasionally perch in the open.

Seasonal status and distribution

Spring: Evidence of migration begins the first or second week of Mar and continues to the last week of Apr. Most birds are seen from the last week of Mar to the third week of Apr. Usually daily counts are from one to five birds, but as many as 10 were discovered on 3 Apr 1993 at PISP.[1] Stragglers may remain to the second week of May.

Breeding: Although their breeding status is poorly known, the saw-whet owl is probably a rare breeder in the forested mountains and an accidental or irregular summer resident in a variety of wooded settings almost statewide. Breeding has been confirmed in widely scattered locations such as in Bucks Co., at HMS, and in the city of Erie. In some years (possibly after migration irrup-

Northern Saw-whet Owl at Presque Isle State Park, Erie County, January 1993. (Photo: Dave Darney)

tions) saw-whet owls are locally fairly common in spring, but confirmed nesting records have not been clustered in any particular year. Numerous birds were heard in Centre Co. in 1986, but few were confirmed breeding during the BBA project. Saw-whet owls are cavity nesters, using an old woodpecker cavity or occasionally a nest box. Few nests have been documented in the state, but eggs are expected in Apr and birds may sing until early Aug.[2]

Fall: They have been found as early as the second week of Oct. Saw-whet owls are usually detected beginning the last week of Oct or the first half of Nov. Because they are nocturnal migrants, they are rarely, if ever, actually seen migrating in Pennsylvania. On 21 Nov 1995 one was observed, at 8:30 A.M., making landfall after its flight across Lake Erie from Canada.[3] Roosting birds, probably migrants, continue to be seen through Dec or early Jan. Many saw-whet owls remain through winter.

Winter: Northern Saw-whet Owls are widely distributed over the state in winter, but they are difficult to locate. They are usually found roosting as individuals, but occasionally several are found within a small area. A survey conducted during the winter of 1986 in the State College area revealed an unexpected high total of 92 birds.[4]

History: The modern pattern is not unlike that described in historical references. Actual documented records of breeding were few, and most involved recently fledged young in northern locations. Saw-whet owls have a history of appearing in unlikely places. Todd (1940) tells of two instances in which young birds were found alive in the basement of the Carnegie Museum (in Jun 1927 and Jul 1932), suggesting a nest was active in the Pittsburgh city limits.

Comments: Northern Saw-whet Owls are listed by the OTC as a Species of Special Concern, Candidate–Undetermined.

[1] G.M. McWilliams, pers. obs.
[2] D.A. Gross, pers. comm.
[3] Hall, G.A. 1996. Appalachian region. NASFN 50:53.
[4] Hall, G.A. 1986. Appalachian region. AB 41:283.

Order Caprimulgiformes
Family Caprimulgidae: Goatsuckers

Goatsuckers, also known as nightjars, are chiefly nocturnal birds with a very large mouth that helps them capture insects while in flight. They have a large flat head with a tiny bill, and a huge mouth that opens far back under their ears. Their soft plumage enables them to fly silently, and their large eyes provide nocturnal vision. Their plumage is dappled with somber browns and grays that camouflage them during the day, when they roost on the ground or perch lengthwise on a branch. At dusk and dawn they become vocal; they are more easily identified by their distinctive calls than by sight. Three species have been recorded in Pennsylvania: Common Nighthawk, Chuck-will's-widow, and Whip-poor-will. Whip-poor-wills and Common Nighthawks regularly breed in the state.

Common Nighthawk *Chordeiles minor*

General status: Common Nighthawks breed across the southern half of Canada and most of the U.S. and they winter in South America. In Pennsylvania they are regular migrants but are far more common during their southbound passage in fall than during spring. Nighthawks breed across the state. They have been declining both as migrants and as breeders in Pennsylvania.

Habitat: Nighthawks are usually observed at dawn, in late afternoon, or in the evening, feeding or migrating overhead. They often spend the daylight hours roosting on horizontal tree limbs, on the ground, or on manmade structures such as flat-topped roofs.

Seasonal status and distribution

Spring: Spring migration is less spectacular than fall migration. Nighthawks have arrived as early as the third week of Apr, but they usually appear the first or second week of May. Usually only individuals, pairs, or groups containing from 5 to 10 birds are seen in a day.

Breeding: Uncommon to fairly common breeding summer residents, Common Nighthawks are regularly heard calling from dusk to dawn over urban areas and towns statewide. No nest is built; two eggs are laid directly on the ground or on a gravel roof. Summer reports have come from every county. Away from towns, nighthawks summer but are rare in barrens and extensively forested areas where rock outcrops provide nest sites (e.g., Pine Creek Gorge, Tioga Co.). They are generally thought to be declining and have disappeared from some towns (particularly in southern counties) in recent years. Most eggs are laid from the last 10 days of May through Jun, but sets have been found in Jul. Only two cream-colored or pale gray eggs spotted with brown and gray are laid.

Fall: The greatest number of birds is seen in migration during this season. The peak migration occurs during a short period as nighthawks pass through rapidly. Migration may begin the third week of Aug. Their typical migration period is from the fourth week of Aug to the second week of Sep. Some years, hundreds and even thousands of birds have been counted in migration in a single evening. On 1 Sep 1972, between 5000 and 10,000 birds were estimated to move southwestward over southern Erie Co. across a broad front.[1] Stragglers have remained in the state to the first week of Nov. A rather late bird was seen on 20 Nov 1978 in Washington Co.[2]

History: Early ornithological records generally listed nighthawks as common or abundant. Turnbull (1869) apparently was the first to publish that nighthawks nest on buildings, when he wrote "its nest has frequently been found on the roofs of warehouses" (of Philadelphia). The use of human structures appears to have largely replaced natural sites on boulders, rock ledges, burned fields, and other open habitats in Pennsylvania.

Comments: A major change in roof construction may be reducing available nesting habitat, contributing to declines in local nesting populations.

[1] Hall, G.A. 1973. Appalachian region. AB 27:58.
[2] Hall, G.A. 1979. Appalachian region. AB 33:177.

Chuck-will's-widow *Caprimulgus carolinensis*

General status: The largest of North American nightjars, Chuck-will's-widows breed in the U.S. from Kansas and Texas east to New Jersey and Florida. They winter along the Gulf coast to central Florida. An exclusively nocturnal species, they are easily identified by their loud call, *chuck-will's-widow*. In Pennsylvania, they are casual during spring migration and into midsummer. Most birds are probably migrants overflying their normal range. Their breeding range has been expanding northward from the pine forests of the southern U.S. Most records of Chuck-will's-widows are of birds heard. All records, except four in Erie Co., are from the southern half of the state.

Habitat: Chuck-will's-widows are usually found in pine plantations, but some birds have been found in deciduous woodlots near fields.

Seasonal status and distribution

Spring: The first modern record was of one found by many observers in downtown Philadelphia on 16 Apr 1972.[1] The species was recorded again on

6 Jun 1974 at Wyncote in Montgomery Co.[2] Chuck-will's-widows were not reported again in the state until 1982, after which they were found nearly every year until 1992. Most records are during spring, and most birds are one-day visitors. Dates range from 26 Apr to mid-Jun. They have been recorded as probable migrant overflights in Berks, Bucks, Centre, Chester, Cumberland, Erie, Franklin, Fulton, Montgomery, Philadelphia, Schuylkill, and Somerset cos. The most recent spring record was at Penn Manor in Bucks Co. on 18–29 May 1996.[3]

Summer: Although their breeding in Pennsylvania has not yet been confirmed, there is good evidence that nesting occurred near the town of Pyrra in Armstrong Co. From late May to Jul in 1985 through 1988, as many as four calling birds were found in this area of deciduous woodlots and farmland by R. Higbee and M. Higbee.[4,5] In Berks Co., a Chuck-will's-widow was heard calling well into the summer of 1984 in a pine woods near Hamburg,[6] and more recently one was heard in 1996 on 18 Jun in Upper Tulpehocken Township in Berks Co.[7] A more likely location for a regular breeding location may be in southern Franklin and Fulton cos., adjacent to breeding records in Maryland (Robbins and Blom 1996), where a summer record was reported during the BBA project.

History: This species was reported at least three times before 1960. It was found at Chestnut Hill in Philadelphia on 9 Aug 1942 and at Morris Park in Philadelphia on 18 May 1943.[8] One was reported calling at Mingoville in Centre Co. from 13 to 24 May 1957 or 1958 (Wood 1983).

[1] Harty, S.T. 1973. Chuck-will's-widow in Center City, Philadelphia. Cassinia 54:26.

[2] Scott, F.R., and D.A. Cutler. 1974. Middle Atlantic Coast region. AB 28:888.

[3] Kitson, K. 1996. Local notes: Bucks County. PB 10:97.

[4] Santner, S. 1992. Chuck-will's-widow *(Caprimulgus carolinensis)*. BBA (Brauning 1992):170–171.

[5] M. Higbee and R. Higbee, pers. comm.

[6] Paxton, R.O., W.J. Boyle Jr., and D.A. Cutler. 1984. Hudson-Delaware region. AB 38:1003.

[7] Haas, F.C., and B.M. Haas, eds. 1996. Birds of note—April through June 1996. PB 10:91.

[8] Hebard, F.V., and Q. Kramer. 1943. General notes. Cassinia 33:33.

Whip-poor-will *Caprimulgus vociferus*

General status: As do Chuck-will's-widows, Whip-poor-wills identify themselves by calling out their name. They breed in Arizona, New Mexico, south through Mexico and from southeastern Canada south to Georgia and west to North Dakota and Minnesota. They winter from South Carolina to Florida and along the Gulf coast south through Mexico. Because of their nocturnal and secretive habits, their migration is not well known in Pennsylvania. They are probably uncommon to fairly common regular migrants. Whip-poor-wills breed primarily in the Ridge and Valley and Southwest regions and are less common and more local in other parts of the state. Most records of this species are of birds heard. Whip-poor-wills appear to be declining in Pennsylvania.

Habitat: They prefer open woodlands or areas of secondary growth with abandoned fields or cleared areas nearby. They also occur in open grassy pine plantations, at least during migration. When observed during the day they are usually roosting on the ground but will occasionally roost on a low horizontal branch. The eye-shine of Whip-poor-wills is frequently seen on rural roads at night.

Seasonal status and distribution

Spring: Birds first appear as early as the second week of Apr. Records suggest that their usual migration period is from the last week of Apr to the third week of May.

Breeding: Whip-poor-wills are uncommon to locally fairly common breeding residents in open wooded areas in the Ridge and Valley and Poconos, and they are rare to uncommon at forest edges and openings west and north of the Allegheny Front. Populations decreased drastically in Luzerne Co., but there has been an increase since 1995 near edges of power lines and at reclaimed strip-mine areas that have been reforested with birches, oaks, or maples.[1] They are no longer regular in the Piedmont, except in woodlands and pine barrens of southern Lancaster Co. The BBS shows a precipitous decline of 12.4% per year that confirms the general impression that Whip-poor-wills are not as common as formerly. Most egg sets were reported from mid-May through mid-Jun, but fresh eggs have been found as late as 10 Jul. Harlow's (1918) average date was 28 May. They lay two white eggs that are spotted and blotched with gray and brown.

Fall: Whip-poor-wills are rarely reported after calling stops, so fall migration is poorly known. Most sightings are of birds that have been disturbed from their day roosts. Records are in Sep and to the fourth week of Oct.

History: Whip-poor-wills were historically thought to be found nearly statewide. Populations were reported to be declining by Todd (1940), by Poole (unpbl. ms.), and in the BBA.[2]

[1] W. Reid, pers. comm.

[2] Santner, S. 1992. Whip-poor-will *(Caprimulgus vociferus)*. BBA (Brauning 1992):172–173.

Order Apodiformes
Family Apodidae: Swifts

Swifts are small, fast-flying birds with long narrow wings. Their primaries are long and stiff, and their wrists are set close to the body. Casual observers often confuse swifts for bats because of their dark color and erratic flight. Confusion with bats also occurs when they are seen exiting and entering caves, canyons, or chimneys, where they nest. The wing beats are shallow and rapid. They are usually seen flying overhead, constantly beating their wings and only occasionally making short glides on bowed wings. Swifts feed entirely on the wing, catching insects with their large mouth. Of the four species of swifts that breed in North America, only the Chimney Swift breeds in Pennsylvania.

Chimney Swift *Chaetura pelagica*

General status: Chimney Swifts spend most of their life in flight, feeding on insects in the air. They are often observed entering or exiting open chimneys, which they depend on for nesting and roosting. Chimney Swifts breed from southeastern Canada south across the eastern U.S. and winter in South America. They are fairly common to abundant regular migrants and breeding residents throughout Pennsylvania.

Habitat: Swifts are usually observed flying around buildings that have suitable chimneys for nesting. As diurnal migrants, they may be seen over almost any habitat during migration.

Seasonal status and distribution

Spring: Early migrants may appear in the second week of Apr. An unusually early swift was seen on 30 Mar 1980 in Pittsburgh,[1] and another was seen on 5 Apr 1971 in the central Susquehanna Valley (Schweinsberg 1988). Their typical migration period is from the third or fourth week of Apr to the third week of May.

Breeding: Chimney Swifts are common summer residents statewide around human habitation, including urban areas and small towns with open chimneys. They are uncommon in woodlands with large hollow trees adjacent to open

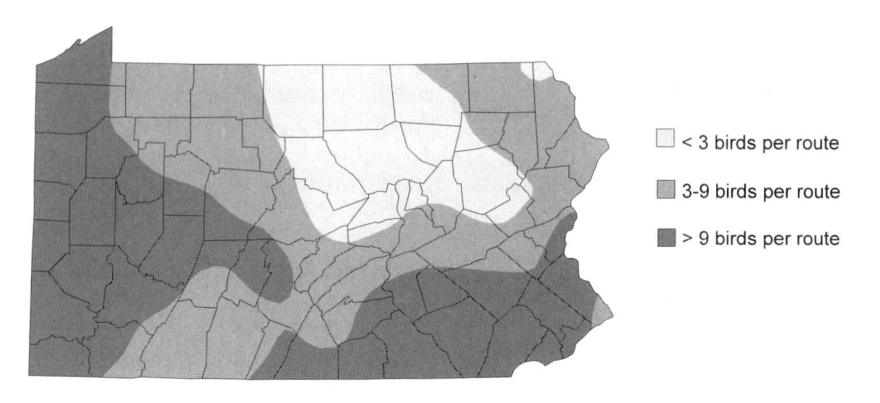

8. Chimney Swift relative abundance, based on BBS routes, 1985–1994

foraging areas. Swifts are most commonly reported on BBS routes in the Piedmont and the western-tier of counties (see Map 8). The breeding distribution is constrained by their specific nest site: the inside wall of a chimney or a hollow tree. Chimney Swifts are occasionally found far from human dwellings in summer, suggesting that nests in hollow trees may still be used. One of the best examples is at the Tionesta Scenic and Natural Area, McKean Co., where a tornado leveled hundreds of acres of old-growth forest in 1985. Swifts are fairly common there, presumably nesting in the remaining large snags. The nest is made of sticks and saliva, forming a half-cup glued to the wall of a dark place, usually inside a chimney. Nesting material is collected by breaking small twigs from treetops while in flight. Nests contain four or five white eggs and are most often found in Jun but have been reported as early as 10 May (Todd 1940) and as late as 15 Jul (Harlow 1918).

Fall: The first migrants are detected the second or third week of Aug; birds begin forming communal roosts in large chimneys by early Sep. Hundreds or even thousands may gather in the evening, swirling around the entrance of a chimney in unison before suddenly descending into the chimney at dusk. For example, over 8000 were counted at the Coopersburg School in Lehigh Co. in Sep (Morris et al. 1984). Loose flocks of migrating swifts may be seen high overhead almost anywhere, but particularly along the mountain ridges. Flocks are usually larger in fall than in spring, with daily highs containing loose flocks of from 5 to 20 or more birds. Swifts have been known to remain at these communal roosts as late as the second week of Oct, but the number of birds normally has declined by this time. Birds usually leave the state by the third week of Oct.

History: Before European settlement, Chimney Swifts nested in hollow trees and possibly caves. As their name denotes, this swift has long been associated with human structures. The earliest ornithologists (e.g., Wilson 1808–1814) identified chimneys alone as the nest site. Occasional instances of swifts nesting in dark barns and attics were reported long ago (Todd 1940). Harlow (1913) stated, "Here is one species for which civilization has little terror" and said that they bred abundantly in every county, "even in the heart of our great cities." This species has adapted to and survived in highly altered environments.

[1] Hess, P.D. 1980. Area bird summaries for March. ASWPB 44:10.

Apus Species

General status: About 17 species of swifts *(Apus apus)*, ranging throughout Europe, Africa, Asia, the East Indies, and Australia, belong to the genus *Apus*. The Common Swift *(Apus apus)* is the most common and wide-ranging swift throughout Europe and is a vagrant to Iceland and an accidental visitor to North America. A bird of this genus seen in Pennsylvania was believed to be a Common Swift. This is the first record for this taxon in Pennsylvania.

Seasonal status and distribution

Spring: On 10 May 1996 R. Ridgley, P. Ridgley, and L. Bevier saw a large dark swift with a grayish white throat and a long forked tail flying with Chimney Swifts at Lafayette Hill, in Montgomery Co. The bird was accepted as an *Apus* species although the species was not determined definitely.[1]

[1] Pulcinella, N., 1997. Eighth report of the Pennsylvania Ornithological Records Committee, June 1997. PB 11:129.

Family Trochilidae: Hummingbirds

The family Trochilidae is confined to the New World; most species are found in the tropics. Hummingbirds are among the smallest birds in the world. Most males have plumages of brilliant iridescent colors. Hummingbirds are easily recognized by their habit of hovering while feeding on nectar or insects found inside flowers and by their ability to fly backward. Most species have a long, thin, straight or decurved, bill and long narrow wings. When in flight, their wings beat so rapidly that they blur and produce a humming sound, hence the name. Only two species have occurred in Pennsylvania: Rufous Hummingbird, which is an occasional vagrant from western North America, and Ruby-throated Hummingbird, the only species of hummingbird that breeds in eastern North America.

Ruby-throated Hummingbird *Archilochus colubris*

General status: Ruby-throated Hummingbirds breed across southern Canada and the eastern half of the U.S. and winter along the Pacific slope of Mexico south through Central America. A few winter along the Gulf coast to Florida. Ruby-throated Hummingbirds are the only hummingbirds that regularly occur in Pennsylvania. They are fairly common regular migrants, except in urbanized and densely forested areas, where they are rarely observed. Ruby-throated Hummingbirds breed statewide.

Habitat: They are found in a variety of habitats including near wooded streams and in gardens, parks, woodlands with openings, and thickets.

Seasonal status and distribution

Spring: The first arrivals have been reported as early as the second week of Apr. A remarkably early Ruby-throated Hummingbird on 29 Mar 1997 was attracted to a red tag on a telephone pole at the Little Gap Raptor Research Station in Northampton Co.[1] The typical migration period is from the fourth week of Apr to the fourth week of May. Migration normally peaks between the first and the third weeks of May.

Breeding: Uncommon to fairly common over most of the state, Ruby-throated Hummingbirds breed statewide where woodlands are found. Adaptable, they frequent flower gardens and hummingbird feeders even in suburban habitats, provided woodland is available nearby for nesting. They are sparsely distributed in the Piedmont even in suitable habitat. The BBS indicates a steady, increasing population trend of 3.2% per year in Pennsylvania. Ruby-throated Hummingbirds build a tiny nest about 10–20 feet above the ground on a twig or small branch of a tree or shrub. Nests are made of bud scales and plant fibers and are lined with plant down; the outsides are covered with grayish green lichens. They are attached to a branch with spider silk. Two glossy white eggs are laid, sometimes before the nest has been completed. Most eggs are found from late May through Jun, with second clutches reported in early Jul. Males perform a rapid pendulum-like flight in courtship, sometimes near a feeder.

Fall: Migration is usually under way by the third week of Aug and continues to the second week of Sep. It peaks from the fourth week of Aug to the first week of Sep. Southbound birds are often observed from hawk watches along the mountain ridges. Stragglers may remain to the second week of Oct. An adult male seen on 6 Nov 1996 at Irwin in Westmoreland Co. was very late.[2]

History: The historical descriptions of Ruby-throated Hummingbirds in Pennsylvania were very similar to modern patterns. Harlow (1913) indicated that they were rare breeders in the immediate vicinity of Philadelphia.

Comments: It is questionable as to whether all immature *Archilochus* hummingbirds seen after the first week of Oct are Ruby-throated. One seen in Allegheny Co. on 2 Dec 1982 was believed to be a Ruby-throated Hummingbird, a remarkably late record for this species.[3] A hummingbird tentatively identified as a female Ruby-throated Hummingbird was seen on 16 Nov 1992 at Red Lion in York Co.[4] The very similar Black-chinned Hummingbird has been recorded east of the Appalachian Mountains, so all immature *Archilochus* hummingbirds, especially males, should be carefully examined.

[1] Wiltraut, R. 1997. Local notes: Northampton County. PB 11:99.
[2] Leberman, R.C. 1997. Local notes: Westmoreland County. PB 10:230.
[3] Hall, G.A. 1983. Appalachian region. AB 37:302.
[4] Haas, F.C., and B.M. Haas, eds. 1993. Rare and unusual bird reports. PB 6:180.

Rufous Hummingbird *Selasphorus rufus*

General status: They breed from coastal southern Alaska south along the Pacific coast to northern California and east to Wyoming, migrate chiefly along the ridges of the Rocky Mountains in fall, and are casual vagrants in the eastern U.S. They winter primarily in Mexico, but some winter along the Gulf coast and Florida and casually northward in the east. Rufous Hummingbirds are accidental in Pennsylvania. Seven known records of *Selasphorus* hummingbirds have been accepted as Rufous Hummingbirds in Pennsylvania. All of these records have been from late summer through late fall. Birds held in captivity have survived to early winter, and one survived to spring and was released.

Seasonal status and distribution

Fall: The first record in the state was of a male found in Devon, Chester Co. on 6 Nov 1975 by J. Donatone and colleagues. The bird was photographed and was captured a week later and taken to the Philadelphia Zoo.[1] The next record was not until 1990, when an immature male came regularly to sugar-water feeders between 17 Oct and 21 Nov about a mile north of Ligonier in Westmoreland Co. It was trapped for positive identification and was measured; a tail feather was taken as a voucher specimen.[2] Another adult male was discovered at a feeder at Gibsonia in Allegheny Co. and was photographed on 10 Nov 1991.[3] Still another Rufous Hummingbird was found in

Rufous Hummingbird at Gibsonia, Allegheny County, November 1991. (Photo: Gerald M. McWilliams)

late Oct 1991 near Latrobe in Westmoreland Co. It was captured and given to a rehabilitator. This one survived in captivity and was released on 14 Apr 1992.[3,4] An adult male was observed and photographed at a feeder west of Edinboro in Erie Co. from 27 Sep 1992 until cold weather arrived. It was then enticed into the host's house.[5,6] The bird survived inside the house until it escaped on 24 Jan 1993 and was not seen again.[7] An adult male was photographed at Cherry Valley, Monroe Co., and was present from 7 to 9 Aug 1994.[8] The most recent accepted record was of an adult male at Trexlertown in Lehigh Co. from 28 Sep 1996 to 12 Jan 1997.[9]

Comments: In addition to the identifiable Rufous Hummingbirds, there have been several other records of *Selasphorus* hummingbirds in Pennsylvania since 1985 that could not be definitively distinguished from the very similar Allen's Hummingbird *(Selasphorus sasin)*. These birds were immature males or females, whose plumage can be identified with certainty only by examining them in the hand. They have been reported in Adams, Allegheny, Berks, Bucks, Huntingdon, Lehigh, Somerset, and Westmoreland cos. A combination of more hummingbird feeders in backyards and gardens and more people keeping their feeders full through the fall, has probably contributed to the recent increase in sightings of western hummingbirds.

[1] Paxton, R.O., P.A. Buckley, and D.A. Cutler. 1976. Hudson-Delaware region. AB 30:44.
[2] Mulvihill, R., and R.C. Leberman. 1991. Rufous Hummingbird. PB 4:139–141.
[3] Hall, G.A. 1992. Appalachian region. AB 46:88.
[4] Hall, G.A. 1992. Appalachian region. AB 46:423.
[5] Hall, G.A. 1993. Appalachian region. AB 47:94.
[6] G.M. McWilliams, pers. recs.
[7] Hall, G.A. 1993. Appalachian region. AB 47:257.
[8] Wiltraut, R. 1994. Notes from the field: Monroe County. PB 8:160.
[9] Pulcinella, N., 1997. Eighth report of the Pennsylvania Ornithological Records Committee, June 1997. PB 11:129.

Order Coraciiformes
Family Alcedinidae: Kingfishers

Kingfishers are stocky, large-headed birds with a long, heavy bill and very short legs. Many species have a ragged crest accentuating the size of the head. In North America all species feed on fish, which they catch by plunge-diving from an exposed perch or by hovering above the surface of the water. One species is found in Pennsylvania, the Belted Kingfisher.

Belted Kingfisher ***Ceryle alcyon***

General status: Belted Kingfishers breed from Alaska south across much of Canada and most of the U.S. They winter from Alaska south along the Pacific coast and east across most of the U.S. They are uncommon to fairly common regular migrants in Pennsylvania. Belted Kingfishers are regular breeding and winter residents.

Habitat: Belted Kingfishers are found at almost any body of water such as lakes, ponds, and streams where fish and suitable nesting habitat are available. They prefer exposed perches such as trees, shrubs, wires, or utility poles above quiet water when feeding. Kingfishers are absent or very rare along small streams in large tracts of mature forests and in areas where nesting sites are not available.

Seasonal status and distribution

Spring: Nonwintering birds arrive as soon as the ice has melted, usually beginning the first or second week of Mar. Peak migration usually occurs from the first to the third week of Apr. Migration ends by the second week of May.

Breeding: Belted Kingfishers are uncommon to fairly common summer residents statewide. Although pairs are widely scattered, kingfishers are conspicuous; they have large territories and are vocal as they fly between feeding areas and their nest. Fairly tolerant of development pressures, they are found even in highly suburbanized areas of the Piedmont and Southwest region but are absent from urban areas where nesting habitat is lacking. The BBA project found kingfishers to be scarce in Clearfield Co. and surrounding areas, where acid mine drainage widely affects fish populations. The BBS results show a steady, long-term decline (3.1% per year) statewide. Nest burrows are excavated in steep dirt banks, sometimes a distance from water. The entrance is a circular hole 4 inches in diameter, usually placed near the top of a steep bank. The tunnel typically extends three or more feet into the bank. Nest cavities may be used in successive years. Six or seven white eggs are found in a chamber, often accompanied by fish bones and crayfish parts (Todd 1940). Most egg sets are found between early May and early Jun, with early clutches in late Apr.

Fall: Apparently, there is no defined migration period during this season. Most birds remain in the state to at least mid-Dec or to the first week of Jan, when lakes and streams freeze. If water remains open, some will stay through the winter.

Winter: During mild winters with open water, kingfishers are uncommon. They are found statewide but become rare and local in mountainous and northern portions of the state. During severe winters when open water is scarce, they are confined to areas of warm-water discharges from factories, sewage treatment plants, fish hatcheries, and open areas of the largest rivers. They have also been reported from lakes kept open by water circulators.[1]

History: Most of Pennsylvania's published ornithological summaries included this species, and most considered it common (probably meaning regular). Sutton (1928a) and Todd (1940) mentioned that kingfishers may be absent where water is polluted or acidic from mine seepage. Still, this species apparently never underwent dramatic declines in response to environmental degradation as had other fish-eating birds. They retreated to headwaters, abandoning larger rivers that were more affected by pollution (Todd 1940). At the height of deforestation, Harlow (1913) considered Belted Kingfishers a statewide breeder, although he astutely pointed out that their conspicuous habits make them appear more common than more secretive birds.

[1] W. Reid, pers. comm.

Order Piciformes
Family Picidae: Woodpeckers

Woodpeckers are found worldwide wherever trees are found. All the birds in this family are highly specialized for climbing and for extracting wood-boring adult insects and larvae. They range in size from crow-sized Pileated Woodpeckers to sparrow-sized Downy Woodpeckers. Their short legs, the position of their body in relation to the limb or trunk they are climbing, and their stiff tail adapt them for this niche. Woodpeckers have a bill that is straight and chisel-like and a tongue that is long and tipped with barbs for removing food from holes they have excavated by hammering or digging. Their skull is thick and heavy, enabling them to withstand the shock of hammering tree trunks in search of adult insects and larvae. They will also feed on various seeds and fruits, especially in winter, and are regular visitors to bird feeding stations that offer suet. Eight species are included on the list of birds of Pennsylvania, and two are considered hypothetical, Three-toed Woodpecker and Red-cockaded Woodpecker. Woodpeckers use dead or dying trees, and occasionally live ones; some species excavate holes in utility poles or use artificial nest boxes. Both sexes assist in the excavation of a nest cavity and in the rearing of young. Three to ten, though usually no more than seven, unmarked white eggs are laid in the cavity. European Starlings frequently compete for nest sites and will often take over a cavity immediately after a woodpecker has excavated a hole. Most species are permanent residents, but some are migratory.

Red-headed Woodpecker *Melanerpes erythrocephalus*

General status: Red-headed Woodpeckers breed across the southernmost edge of Canada south to Montana and Arizona east across the U.S. They winter primarily in the southern two-thirds of their breeding range. In Pennsylvania, Red-headed Woodpeckers are uncommon to locally fairly common regular migrants. They are locally distributed breeders and irregular winter residents throughout most of their breeding range. Red-headed Woodpeckers have disappeared from much of their former range because of competition with European Starlings for nest sites and removal of dead trees from woodlots (Bull and Farrand 1977).

Habitat: Red-headed Woodpeckers prefer open areas with scattered trees, especially deciduous woodlots with mature oaks and large dead trees or dead branches for nesting.

Seasonal status and distribution

Spring: Nonwintering birds may arrive as early as the second week of Apr or as late as the first week of May. Migrants are seen until the third week of May. Migrants may be seen as individuals or pairs flying past the hawk watches and along Lake Erie.

Breeding: Red-headed Woodpeckers are locally fairly common in the Piedmont from Lancaster Co. west through much of the southern Ridge and Valley, in the Glaciated Northwest, and along the Lake Erie Shore. They are rare and irregular elsewhere, with summer records from almost every county. Red-headed Woodpeckers are casually reported in the extensive wooded regions of the High Plateau. They have been reported on 19 of 122 state BBS routes; on the basis of data from those routes, their populations have been declining at a rate of 11% per year since 1966. As a migrant, they nest later than most other woodpeckers (except sapsuckers). Nests are active from late May through Jun, with possible second clutches in mid-Jul. Harlow's (1918) average date of 28 clutches was 3 Jun. Nests are usually placed in the dead branch or trunk of an isolated tree or at a woodlot edge, at various heights from the ground. This is the prime situation for starlings, which usurp this woodpecker's nest and have undoubtedly contributed to its decline. Red-bellied Woodpeckers also have been found to take over Red-headed Woodpeckers' nests.

Fall: Red-headed Woodpeckers are more frequently seen during fall than spring migration. Most are seen migrating as individuals or sometimes in small groups along the mountain ridge hawk watches in the Ridge and Valley. Migration extends through the fall, with birds moving as food becomes less available. At PNR most fall sightings are in Sep (Leberman 1976). Few birds remain after the last week of Dec or the first week of Jan in some years, while during other years many birds remain to winter. At PISP, where there is a large breeding population of Red-headed Woodpeckers, they usually depart by late Dec or early Jan. However, on 27 Dec 1980 a total of 84 was counted in the park.[1]

Winter: Winter populations fluctuate from year to year. During some winters they are uncommon to rare within their breeding range, while in other winters they are absent. The only area in the High Plateau where they have been reported in winter is in the Allegheny Mountains at PNR and around Ligonier, Westmoreland Co. (Leberman 1976).

History: A review of the literature would suggest that Red-headed Woodpeckers are perpetually declining. Even before the introduction of starlings, they were reported to be declining (Libhart 1869 [in Lancaster Co.]; Warren 1890 [in eastern Pennsylvania]). Todd (1940) described them as one of the most erratic of all North American breeding birds. Several early authors, generally reporting from western counties, called Red-headed Woodpeckers "abundant" (e.g., Baird 1845 [Cumberland Co.]; Townsend 1883 [Westmoreland Co.]) shortly after the era when land-clearing expanded habitat. Harlow (1913) considered Red-headed Woodpeckers to be one of the most abundant species in Centre Co. and said that they "undoubtedly nest in every county of the state." This prospect was rapidly reversed as starling populations exploded. By the 1960s, Poole (1964) called them increasingly rare and local. Through this era they were irregular as winter residents.

[1] Hall, G.A. 1981. Appalachian region. AB 35:300.

Red-bellied Woodpecker ***Melanerpes carolinus***

General status: Red-bellied Woodpeckers are resident in the eastern U.S. from North Dakota and Texas east through southern Ontario to Massachusetts and Florida. Vocal and conspicuous, Red-bellied Woodpeckers have expanded their range since the 1940s or 1950s from southwestern Pennsylvania to areas north and east across the state. They are now permanent residents throughout the state, with the densest populations east of the Allegheny Front. They are uncommon to rare and local elsewhere in the state, but their range is rapidly expanding northward, especially along the river valleys.

Habitat: Red-bellied Woodpeckers are found in deciduous woodlands or in fields with scattered trees, mostly at lower elevations and river valleys, and occasionally in coniferous woodlands.

Seasonal status and distribution

Spring: A slight northward movement of Red-bellied Woodpeckers is noted along the Lake Erie Shore in spring. Single birds are occasionally seen flying over Gull Point at PISP from the fourth week of Apr to the second week of May. An increase in the number of birds in the park during this period and then a decrease toward the end of May also suggest some northward movement.

Breeding: Red-bellied Woodpeckers are fairly common permanent residents across most of Pennsylvania, except in extensively forested sections in the north-central High Plateau, where they are rare and irregular. They are most abundant and uniformly distributed (found in 92% of BBA blocks) in the Piedmont, southern Ridge and Valley, and areas south of Pittsburgh (see Map 9). Red-bellied Woodpeckers are common in suburban areas that support large trees. They have been recorded in all counties and are uncommon but regular in river-bottom woodlands along the North Branch of the Susquehanna and Allegheny rivers north to the New York State line. The nest cavity is excavated in the dead trunk or branch of a large tree, generally 20–50 feet from the ground. The BBS documented a long-term expansion of 11% per year, increasing most sharply since 1980. The population expanded from both the southeastern and southwestern corners of the state.[1] Their range continued to expand during the 1990s, for example into Carbon and Pike cos. (P. B. Street and Wiltraut 1996) and into Clarion Co.[2]

History: The early ornithological records list Red-bellied Woodpeckers in the Piedmont region, sometimes as abundant, but not farther north. In eastern Pennsylvania they apparently became rare and irregular during the early 1900s (Poole 1964). West of the Alleghenies they were "common" in Greene Co. in the late 1800s (Jacobs 1893), but north of Pittsburgh they were not reported breeding (Harlow 1913). However, four specimens collected during fall and winter in Erie Co., between 1874 and 1876 (Todd 1904), reflected an erratic winter occurrence well north of their established range. Red-bellied Woodpeckers expanded dramatically during the twentieth century. They were documented breeding in Beaver Co. in 1902 (Todd 1940) and north to Crawford Co. by 1923 (Sutton 1928b).

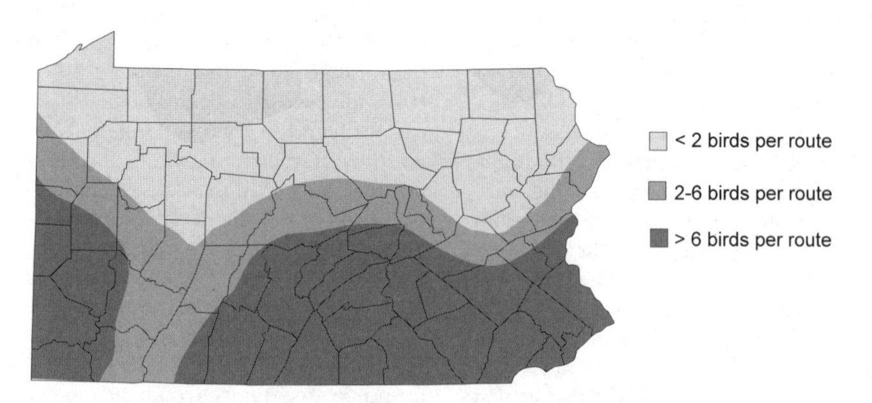

9. Red-bellied Woodpecker relative abundance, based on BBS routes, 1985–1994

Red-bellied Woodpeckers expanded northward in the eastern half of the state after the 1950s, nesting sporadically north to Berks Co. by the 1950s (Poole 1964), regularly around State College, Centre Co. by 1960 (Wood 1983), and reaching some of the northeastern counties by the 1980s.[3]

Comments: Red-bellied Woodpeckers are regular visitors to feeders, taking sunflower seeds and suet.

[1] Haas, F.C. 1987. Recent range expansion and population increase of the Red-bellied Woodpecker, *Melanerpes carolinus* (Linnaeus) in Pennsylvania. PB 1:107–110.

[2] M. Buckwalter, pers. comm.

[3] Ickes, R. 1992. Red-bellied Woodpecker *(Melanerpes carolinus).* BBA (Brauning 1992):182–183.

Yellow-bellied Sapsucker ***Sphyrapicus varius***

General status: Yellow-bellied Sapsuckers' presence is most often revealed by their very distinctive drumming. They breed from the southern half of Canada south through the Rocky Mountains in the West and to Illinois and in the Appalachian Mountains to North Carolina and Tennessee in the East. Sapsuckers winter from southern California across the Southwest and Mexico east and north to Pennsylvania and New Jersey. In Pennsylvania they are fairly common regular migrants at the Lake Erie Shore and regular uncommon to rare migrants elsewhere in the state. Sapsuckers breed primarily across the northern-tier counties. They winter mainly in the southern half of the state.

Habitat: Sapsuckers are found almost anywhere in migration where softwoods such as birches, aspens, pines, or apple trees are found. During the breeding season, they are found in northern hardwood forests, primarily in moist woodlands at elevations above 1500 feet. In winter they inhabit mixed woodlands but prefer conifers and will also visit bird feeding stations.

Seasonal status and distribution

Spring: Early individuals may appear the third or fourth week of Mar. Their usual migration time occurs over a very short period from the first to the fourth week of Apr. Peak migration is around the third or fourth week of Apr. A few individuals may linger until the second week of May away from known breeding areas.

Breeding: Sapsuckers are uncommon to locally fairly common in extensive forested areas of the High Plateau north of the Allegheny Mountains, uncommon to rare in woodlands of the Glaciated Northeast and Glaciated Northwest regions, and accidental in summer in forested mountains elsewhere. They are strongly associated with northern hardwood forests, where they may be the most common summer woodpeckers in a few northern counties. A few nests were confirmed in widely scattered locations in the Ridge and Valley and in the Allegheny Mountains during the BBA project[1] and in 1992 in Butler Co.,[2] but these records are outside the normal breeding range. Nesting begins in May. Nest cavities are excavated in the trunk of a living or dead tree and may be found at various heights above ground. The entrance hole is about 1.5 inches in diameter. Young, almost ready to fledge, are boisterous in mid-Jun. The BBS shows a strong increase (15% per year) since 1966.

Fall: Individuals may appear away from breeding sites as early as the third week of Aug. Their typical migration period is from the third week of Sep to the fourth week of Oct. Peak migration usually occurs during the fourth week of Sep and the first week of Oct. If the weather remains mild, a few may linger to late Dec or early Jan at nonwintering sites.

Winter: Sapsuckers are uncommon to rare winter residents in the valleys and lowlands east of the Allegheny Front.

In the Glaciated Northeast they have been reported in winter from Wyoming and Pike cos. The only winter report in the High Plateau north of the Allegheny Mountains is a sighting on 15 Jan 1996 in Potter Co.[3] In the Southwest region they are uncommon to rare along the Allegheny River valley north to Venango Co. and east across the southern portion of the Allegheny Mountains. In the Glaciated Northwest they have been recorded in winter only south of the snowbelt in Mercer and Lawrence cos. There are few winter records from the Lake Erie Shore.

History: The Yellow-bellied Sapsucker's breeding distribution in Pennsylvania has changed little from that of historical accounts. Some authors (e.g., Warren 1890) mention declines, probably the result of forest clearing. A major difference with modern patterns is in the Alleghenies, where they were formerly common breeders in Somerset Co. (Todd 1940). By the early 1900s they had virtually disappeared from the southern mountains. Scattered nestings outside the primary range, such as those reported in the BBA, were reported by Todd (1940).

Comments: They were considered "vulnerable" by Genoways and Brenner (1985), but the BBA project demonstrated that they were much more widespread than would warrant Special Concern status.

[1] Gross, D.A. 1992. Yellow-bellied Sapsucker *(Sphyrapicus varius)*. BBA (Brauning 1992):184–185.

[2] Wilhelm, G. 1993. Yellow-bellied Sapsucker nesting in Butler County. PB 6:146.

[3] Haas, F.C., and B.M. Haas, eds. 1996. Seasonal occurrence tables—January through March 1996. PB 10:36.

Downy Woodpecker *Picoides pubescens*

General status: Downy Woodpeckers are resident from southern Alaska south across Canada and throughout most of the U.S. They are the smallest and most familiar of the woodpeckers in Pennsylvania. Downy Woodpeckers are common permanent residents statewide but are more conspicuous in winter than in any other season.

Habitat: They inhabit a variety of habitats wherever trees are found, including cities, towns, and parks, but they prefer deciduous woods. Downy Woodpeckers can be found in open fields in winter, especially in fields with goldenrod or corn, where they search for larvae.

Seasonal status and distribution

Breeding: The BBS shows Downy Woodpeckers' highest densities (up to five birds per route) in the south-central Ridge and Valley and lowest (down to one bird per route) in the heavily forested High Plateau. A population decline of 2.1% per year reported on state BBS routes since 1966 has not resulted in a change in their status. Most egg sets have been collected during May, with a few in late Apr and in Jun. The frequency of double broods is not known. Their cavity nest is usually in dead wood. The entrance hole, 1.25 inches in diameter, is often placed on the underside of an exposed limb (H. H. Harrison 1975). Downy Woodpeckers are regular visitors at suet feeders and may come to other feeders for seeds.

History: Todd's (1940) comment, "with the exception of the Flicker, none of our woodpeckers is so common and familiar as the little Downy," applies equally today. The distribution and abundance of this familiar species have changed little since the time of the earliest writers.

Hairy Woodpecker *Picoides villosus*

General status: Hairy Woodpeckers' North American breeding and wintering range is very similar to that of Downy Woodpeckers. These larger versions of Downy Woodpeckers are less approachable, more reclusive, and less common, and they usually feed and nest higher in the trees. They are fairly common permanent residents statewide but are more conspicuous in winter than during the rest of the year.

Habitat: They are more habitat-specific than Downy Woodpeckers, preferring large tracts of woodlands with mature deciduous or coniferous trees. Hairy Woodpeckers also may be found in cities, towns, and parks where some mature trees are nearby.

Seasonal status and distribution

Breeding: Hairy Woodpeckers nest in every county. They are more closely associated with large, extensive timber than are the more universal Downy Woodpeckers and are generally most common in extensively forested regions of northern Pennsylvania and least common in the Piedmont. Rare in extensive open areas, agriculture, and urban centers, Hairy Woodpeckers will use scattered woodlots and groves of large trees even in generally open areas. As a result, they were well represented in every county, even Philadelphia, during the BBA project, but were reported at one-fifth the frequency of Downy Woodpeckers on BBS routes. Egg sets have been observed from 11 Apr through 2 Jun, but most were in late Apr. The entrance hole is about 2.5 inches high by 1.5–2 inches wide.

Winter: Hairy Woodpeckers may wander at this season in search of food, frequently entering suburban backyards where at least a few large trees are present. They are readily attracted to suet feeders.

History: Hairy Woodpeckers' distribution has changed little from that described by early authors. A few accounts described them as "abundant" (e.g., Baird 1845), but generally they were considered common and widely distributed. Apparently Hairy Woodpeckers did not decline noticeably even during the height of timber cutting; they were considered regular if not common breeders statewide at the end of the 1800s, even in cut-over Lancaster Co. (Libhart 1869) and Delaware Co. (Cassin 1862).

[Red-cockaded Woodpecker] *Picoides borealis*

General status: Red-cockaded Woodpeckers are resident locally across the southeastern states from Texas east to Florida and north to Oklahoma and North Carolina. Populations have decreased with the loss of their highly specialized habitat in mature pine or pine-oak woodlands. No twentieth-century record is known from Pennsylvania. This species is listed as hypothetical in Pennsylvania.

Comments: Stone (1894) stated that a specimen was shot near Philadelphia in 1861. The specimen was formerly housed in the ANSP but was missing in 1958 (Poole 1964). Cassin (1862), Michner (1863), and Gentry (1877) mentioned that the species also had been reported in Delaware and Chester cos.

[Three-toed Woodpecker] *Picoides tridactylus*

General status: In North America, Three-toed Woodpeckers are resident in Alaska, across the boreal forests of Canada to Newfoundland, south in the West through the Rocky Mountains to northern Arizona and New Mexico, and south in the East to northern New York. There have been four reports of this woodpecker in Pennsylvania but none has been accepted because they lacked sufficient detail. As a result, the species is listed as hypothetical in the state.

Comments: The first sighting was in 1974 of a female observed and heard calling on 3 Apr at Longwood Gardens in

Chester Co.[1] On 15 Dec 1974 at PISP in Erie Co., one was found removing bark from a dead larch tree and at one point was observed being chased by a Hairy Woodpecker.[2,3] One was at Tidioute in Warren Co. on 28 Oct 1979, and another at Tussey Mountain in Bedford Co. from 23 to 26 Nov 1987 (Leberman 1988).

[1] Scott, F.R., and D.A. Cutler. 1974. Middle Atlantic Coast region. AB 28:787.
[2] Hall, G.A. 1975. Appalachian region. AB 29:689.
[3] G.M. McWilliams, pers. recs.

Black-backed Woodpecker ***Picoides arcticus***

General status: Black-backed Woodpeckers are resident in the boreal forests of Alaska south in the mountains to central California and Idaho and east across much of boreal Canada to Newfoundland and south to northern New York. They are more common than Three-toed Woodpeckers in the eastern part of their range. Despite the numerous reports of this species in Pennsylvania since 1960, at least 14, only one was adequately documented. Black-backed Woodpeckers are likely to be casual in the state but, because of the lack of accepted records, their occurrence in Pennsylvania should be listed as accidental. In some years there are frequent reports, and then several years pass without any, suggesting that this woodpecker may be irruptive, perhaps moving into Pennsylvania only during years of food shortages farther north. Godfrey (1986) stated that "in certain years, irregularly spaced, relatively large numbers move southward." Most reports were during the 1960s, especially during 1961–1962, when at least eight different individuals were reported. The most recent sighting of this species in Pennsylvania was in Philadelphia on 2 May 1992, but it was not adequately documented.

Seasonal status and distribution

Fall through spring: Black-backed Woodpeckers have been reported from 30 Sep to 19 May. All sightings have been east of the Allegheny Front. In the Coastal Plain they have been found in Philadelphia Co., but in the Glaciated Northeast only in Wayne Co. In the Piedmont,

Black-backed Woodpecker at State College, Centre County, November/December 1981. (Photo: Rick Wiltraut)

they have been reported from Berks, Bucks, Chester, Delaware, Lehigh, and Montgomery cos. In the Ridge and Valley, they have been recorded from Carbon, Centre, Franklin, Huntingdon, Monroe, Pike, and Schuylkill cos. The only accepted sighting was of an adult male that was discovered by C. Schmidt and H. Henderson near State College in Centre Co. on 24 Nov 1981. The bird was photographed and remained until 19 May 1982.[1,2]

History: Turnbull (1869) referred to the species as occasionally seen in the northern counties of Pennsylvania. This species was not mentioned again in Pennsylvania until one was reported by Kunkle (1951) at Lost Farm on Shamokin Mountain near Lewisburg in Union Co. in Nov 1928, and another in Bear Meadows in Centre Co. in the winter of 1929 (Wood 1983). A female Black-backed Woodpecker was observed within 10 feet, flaking bark from a red pine tree on 1 Nov 1953 near Reading, and a male was seen at close range by E. Poole on 10 Mar 1956 at Lake Ontelaunee in Berks Co. (Poole, unpbl. ms.). They have also been reported from Hartstown in Crawford Co. on 23 Mar 1957[3] and at Waggoner's Gap in Perry Co. on 13 Sep 1957 (Poole, unpbl. ms.). One was shot on 1 Dec 1958 about 22 miles north of Penn State University and supposedly was mounted by M. Wood, but the specimen cannot be located.

Comments: All woodpeckers with a black back should be carefully studied because some melanistic Downy Woodpeckers with a black back have been recorded. Well-written descriptions and photographs are needed to verify the frequency of occurrence of this boreal visitor in Pennsylvania.

[1] Hall, G.A. 1982. Appalachian region. AB 36:178.
[2] Hall, G.A. 1982. Appalachian region. AB 36:852.
[3] Harrison, H. 1957. Black-backed Woodpecker in western Pennsylvania. Redstart 24:22.

Northern Flicker ***Colaptes auratus***

General status: Flickers breed throughout Canada and breed and winter throughout the U.S. Northern Flickers are unusual among woodpeckers in that they spend much of their time hopping on the ground in search of ants. They are common to abundant regular migrants in Pennsylvania and breed across the entire state. Flickers are regular winter residents, mainly at lower elevations.

Habitat: Northern Flickers are found in a variety of habitats with at least some mature trees, including woodlands, towns, and cities. They are conspicuous, feeding on the ground, often around suburban and rural yards and in forest openings. Their habitat preferences are similar throughout the year, except in winter, when by necessity they spend more time in trees.

Seasonal status and distribution

Spring: Birds that have not wintered in Pennsylvania may begin returning as early as the second week of Mar. Their typical migration period is from the fourth week of Mar or the first week of Apr to the fourth week of Apr. Peak migration usually occurs during the second or third week of Apr.

Breeding: Northern Flickers are fairly common breeders statewide. Using a range of deciduous wooded settings, they are the most widespread and abundant of the woodpeckers in Pennsylvania. The BBS reports flickers to be most common in the Piedmont and from Pittsburgh south and least common in the extensive forests of the High Plateau. Although they still retain the distinction of most common woodpecker statewide, the BBS reported a sizable long-term population decline of 5.5% per year, down from 10 birds per route in the 1960s to about 3 birds per route statewide in the early 1990s. Most of that decline occurred in the first 14 years of the BBS (1966–1979). Contributing to this decline are starlings, which compete for nest sites, sometimes taking newly excavated flicker cavities. Most eggs have been found from 9 May to 5 Jun, with extreme dates from 24 Apr to 24 Jun.

Fall: Migration is poorly defined in fall but is usually evident along the mountain ridges from the second week of Sep to the second or third week of Nov. The largest number of birds is recorded from the third week of Sep to the second week of Oct. However, large numbers have been recorded on CBCs, such as the 234 recorded during the Southern Lancaster Co. CBC on 16 Dec 1984 (Morrin et al. 1991). Away from wintering sites, some birds may linger to early Jan.

Winter: Northern Flickers have been recorded over most of the state in winter, but numbers vary with the severity of the weather. Winter residents in the Ridge and Valley and High Plateau are rare and local but at times may be locally fairly common in the lower elevations such as in river valleys.

History: Possibly because they are so conspicuous, flickers have generally been considered the most abundant of the woodpeckers in Pennsylvania (e.g., Baird 1845; Harlow 1913; Todd 1940). Supporting the idea that they have benefited from forest clearing, Harlow (1913) stated that they were "always more abundant in the open farming regions than in the wilder forested sections."

Comments: The western (Red-shafted) subspecies *(C. a. cafer)* has been reported at least twice in Pennsylvania; once in Pittsburgh on 27 and 31 Dec 1969[1] and once at State College on 3 Apr 1982.[2] There is apparently no specimen or photograph of this form from Pennsylvania. Numerous Yellow-shafted × Red-shafted hybrids have been reported, mostly from banding stations. Wood (1979) cautions that "occasionally yellow-shafted birds have some pink, but are not necessarily hybrids."

[1] Hall, G.A. 1970. Appalachian region. AFN 24: 504.

[2] Hall, G.A. 1982. Appalachian region. AB 36:852.

Pileated Woodpecker *Dryocopus pileatus*

General status: Their large size and loud ringing calls identify these woodpeckers. They are quite wary and are more often heard than seen. They can, however, sometimes be cautiously approached when they are intently tearing apart a rotting log in search of food. Pileated Woodpeckers are resident from California north to Montana and across southern Canada to Nova Scotia and south through the entire eastern U.S. In Pennsylvania they are uncommon to fairly common permanent residents over most of the state. Pileated Woodpeckers are less common and more local in agricultural areas where woodlots are sparse, especially in the Piedmont and in the urban areas of the larger cities. The BBS shows that they may be increasing in many parts of the state.

Habitat: Pileated Woodpeckers prefer mature deciduous or mixed woodlands but can also be found in second-growth woodlands if at least a few mature trees are present in the area. Formerly confined to large tracts of forest, Pileated Woodpeckers have recently accepted parks and suburban areas of towns and cities where mature trees are present.

Seasonal status and distribution

Breeding: Pileated Woodpeckers are most likely to be found in extensively wooded areas (more than 400 acres; Robbins et al. 1989). Birds range over several hundred acres. For reasons not clear, they are more common on BBS routes in the southern Ridge and Valley than elsewhere in the state. Pileated Woodpeckers are probably more common now than at any time in the past 150 years, recouping their range even in Philadelphia city parks during the late 1980s.[1] Reflecting a general maturation of Pennsylvania's woodlands and reduced persecution, Pileated Woodpeckers are increasing at 3.3% per year on BBS routes statewide. A nest cavity is excavated annually in an 8-inch or larger diameter tree trunk. Eggs have been reported during the month of May.

Winter: They usually remain within their breeding territory most of the year, except in winter, when they wander into smaller woodlots in search of food. Pileated Woodpeckers occasionally visit bird feeding stations where suet is provided.

History: Pileated Woodpecker populations declined precipitously with the felling of the great forests of presettlement Pennsylvania. Originally reported around Philadelphia (Barton 1799) and other southern localities (Baird 1845), this grand species disappeared before the axe and gun. By the mid-1800s they were "much more rare than formerly" and had already retracted their range into the mountainous areas (Turnbull 1869). Authors in the southeast considered them rare (e.g., in Lancaster Co.; Libhart 1869). Timbering statewide reduced numbers even across mountainous regions (Warren 1890). Harlow (1918) noted that they still bred rarely in wilder sections, across the northern tier of the High Plateau and south through the Allegheny Mountains, but thought they were still declining. Sutton (1928b) suggested that shooting contributed to the Pileated's decline. Regenerating forests during the early 1900s and new bird protection laws set the stage for recovery of these woodpeckers. By the 1920s and 1930s, bird lists suggested that Pileated Woodpeckers were increasing (e.g., Cumberland Co. area; Frey 1943). Poole (1964) considered them fairly common in heavily forested areas but still rare or absent in heavily settled areas.

[1] D.W. Brauning, pers. obs.

Order Passeriformes
Family Tyrannidae: Flycatchers

The Tyrannidae are a large family of birds of the Western Hemisphere of which most species live in the tropics. Flycatchers are small to medium in size and have a rather flat, broad-based bill with bristles at the base. The *Empidonax* flycatchers are among the most difficult genera to distinguish in the field; most are rather drab, and many can be separated only by call. They typically perch in an upright posture, snap insects out of the air in quick flights, and then return to their perch to wait for another insect to pass. Of the 16 species that have been recorded in Pennsylvania, 9 breed here. They occur in a variety of habitats, from deep woodlands and woodland margins to open fields. Those that breed in the state build a cupped nest of grasses, rootlets, mosses, plant downs, hair, or bark and line the nest with soft fibers. Phoebes build nests predominantly of mud. Flycatchers usually lay three or four white eggs, most of which are blotched or spotted with brown.

Olive-sided Flycatcher *Contopus cooperi*

General status: Olive-sided Flycatchers breed from Alaska south across Canada to Newfoundland and south in the west to California and through the Rocky Mountains. In the East they breed locally from the New England states south through the Appalachian Mountains to North Carolina and Tennessee. Olive-sided Flycatchers winter primarily in South America. These boreal birds are the rarest of the flycatchers that regularly migrate through Pennsylvania. They are nearly always observed as individuals perched on a high exposed tree branch, often near water. Olive-sided Flycatchers are uncommon to rare regular migrants. Though they have been recorded during the breeding season in suitable habitat, no nest or confirmed breeding evidence has been found since the 1930s.

Habitat: During migration, Olive-sided Flycatchers are usually seen in open woodlands, bogs, and swamps and, in fall, along mountain ridges. During the breeding season Olive-sided Flycatchers have been reported in lakeside wetlands, swamps, wooded ravines with conifers, and burned-over (or windthrown) woodlands with standing dead trees.

Seasonal status and distribution

Spring: They are rather late migrants, often not seen until passerine migration is nearly over. Olive-sided Flycatchers have been recorded as early as the first week of May, but most pass through the state the third or fourth week of May. A few late stragglers may not pass through Pennsylvania until the first or second week of Jun.

Summer: Olive-sided Flycatchers are accidental in summer at widely scattered sites in northern counties in areas that have standing dead trees or open forests along streams. They were last confirmed nesting in the 1930s (Poole 1964), but recent records of summering adults continue to raise the possibility of sporadic nesting in Pennsylvania. Summer records include a territorial male in the tornado blowdown of the Tionesta Natural Area of the Allegheny National Forest on the Warren-McKean county line during the summers of 1995 and 1996.[1] Records from late Jun or early Jul during the BBA project include territorial males near Mehoopany, Wyoming Co., in 1988 and 1989; near Arnot, Tioga Co. in 1987; and near Thornhurst, Lackawanna Co. in 1987.[2] The BBS indicates that Olive-sided Flycatcher populations nationwide are declining. Nests would be expected in Jun, high on the horizontal branch of a conifer.

Fall: Olive-sided Flycatchers are most frequently reported at hawk-watching sites in the Ridge and Valley. Earliest fall migrants have been recorded beginning the fourth week of Jul. They are typically observed from the fourth week of Aug through the third week of Sep. Stragglers have been recorded until the third week of Oct.

History: Authors of the late nineteenth century (e.g., Warren 1890) considered Olive-sided Flycatchers a predominantly northern bird but among the rarest of the flycatchers in the state in summer. They were widely reported as migrants but were never considered common. The reference to this species, "Breeds in [Lancaster] county" (Libhart 1869), seems improbable from our vantage, but parts of the lower Susquehanna River corridor have a Canadian feel that may have attracted the species. Characteristic of the bird lists from that era, there is no further discussion and no evidence supplied. As a late spring migrant, however, early Jun transients might have been mistakenly considered breeders by early ornithologists. Misidentification cannot be ruled out. Olive-sided Flycatchers remain an enigma of ornithological history. Pennsylvania's first record of a nest was in 1895 near Hazleton (Young 1896). Stone (1900) stated that they were common near Lopez, Sullivan Co. The most extensive reference to nesting was from Warren Co., where Simpson reported six nests (Todd 1940). The breeding distribution apparently ranged across the northern tier at the end of the nineteenth century. Olive-sided Flycatchers were considered rare summer residents in swamps and cut-over forest edges from the Pymatuning area (Sutton 1928b) east, where extensive forests remained, through Elk, Forest, Warren, Lycoming, Sullivan, Monroe, Pike, and Wayne cos. (Harlow 1913).

Comments: Olive-sided Flycatchers are listed by the OTC as Extirpated to reflect their current breeding status.

[1] D.W. Brauning and D.A. Gross, pers. obs.
[2] Gross, D.A. 1992. Olive-sided Flycatcher *(Contopus cooperi)*. BBA (Brauning 1992):194–195.

Eastern Wood-Pewee *Contopus virens*

General status: Eastern Wood-Pewees usually spend the day hidden in the dense foliage of the woodland canopy and choose open perches from which to feed. Their breeding range extends across the eastern half of North America from New Brunswick and Manitoba south, and they winter in South America. They are regular and fairly common migrants in Pennsylvania. Pewees breed widely across the state.

Habitat: They are found in a variety of forest types and situations, including woodlots, urban parks, and extensive forests.

Seasonal status and distribution

Spring: Early migrants have been recorded in the third week of Apr. Their typical migration period is from the second week of May to the first week of Jun. Peak migration usually occurs during the third or fourth week of May.

Breeding: Pewees are the most common nesting flycatchers on BBS routes statewide, but they were reported in fewer BBA blocks than phoebes. Although almost ubiquitous in distribution, Eastern Wood-Pewees are less common in the extensively forested High Plateau than in southern and western regions of the state. Relative abundance ranges from 1.5–3 birds per route in northern Warren to Potter cos. to 6–9 birds per route over the southern Ridge and Valley and the Blue Ridge west across the Southwest region. A declining population trend of 2.8% per year is reported on state BBS routes. The nest is a neat cup of moss and grass, often decorated with lichens, that usually saddles a horizontal limb away from the trunk at a height of 15 feet and higher. The nest tree is usually near a forest opening or woodland edge. Eggs have been found as late as early Aug, but most occur from 30 May to 2 Jul. This long nesting season suggests that two broods may be reared.

Fall: Migration may begin as early as the fourth week of Jul. Their usual migration time is from the fourth week of Aug to the second week of Oct. Stragglers remain to the third week of Oct. A pewee on 4 Nov 1992 at Prince Gallitzin State Park in Cambria Co. was unusually late.[1]

History: Most previous statewide summaries (e.g., Sutton 1928a; Todd 1940) considered Eastern Wood-Pewees to be common summer residents. A few considered them rare in heavily forested northern sections, such as at North Mountain, Wyoming Co. (Dwight 1892) and Cook Forest State Park, Forest Co. (Christy and Sutton 1928). But contradictory impressions can be found; Cope (1902) in Clinton and Potter cos. considered them particularly abundant in the "deep woodland," and Stone (1891) in Luzerne Co. said they were very common "on the edge of the forest." Pewees fared well even when forests were heavily cut. The adaptability of this species is reflected by Burleigh's (1931) summary of Centre Co.: "This species is also a plentiful breeding bird here, occurring in many of the scattered short stretches of woods about the town."

[1] Haas, F.C., and B.M. Haas, eds. 1993. Rare and unusual bird reports. PB 6:180.

Yellow-bellied Flycatcher ***Empidonax flaviventris***

General status: Yellow-bellied Flycatchers breed across Canada from British Columbia and Mackenzie east to Newfoundland and south to Minnesota and Pennsylvania, with isolated breeding in West Virginia. They winter in Central America. Of all the regularly occurring *Empidonax* flycatchers found in Pennsylvania, Yellow-bellied Flycatchers are the easiest to identify by plumage. However, they are very shy and difficult to find at any season. They usually remain hidden in dense thickets and are usually silent during migration, when they are most common. Yellow-bellied Flycatchers are uncommon to rare regular migrants statewide and local breeders in the northern portion of the High Plateau.

Habitat: During migration Yellow-bellied Flycatchers are found in thickets or the understory of deciduous, mixed, or coniferous woodlands. The breeding habitat is at elevations above 1700 feet in poorly drained areas such as bogs or open swamps overshadowed by scattered canopy trees including conifers and underlain with sphagnum moss.

Seasonal status and distribution

Spring: They are more frequently observed west than east of the Allegheny Front in spring. Rather late spring migrants, the earliest they have been recorded in Pennsylvania is the second week of May. Most migrants are seen the third and fourth weeks of May. Stragglers may remain to the second week of Jun.

Breeding: Yellow-bellied Flycatchers are rare in northern forested wetlands and are usually found above 1900 feet elevation. Nesting birds, at least territorial males, have been found in isolated wetlands in Lycoming, McKean, Monroe, Somerset, Sullivan, Tioga, Warren, and Wyoming cos. since 1980. The largest, most stable populations (involving fewer than a dozen pairs each) apparently are in a complex of swamps in Wyoming Co., including Coalbed Swamp, and in the tornado blowdown area of Tionesta Natural Area, Allegheny National Forest, on the McKean-Warren county line. Wind-thrown old-growth trees formed the swampy habitat at the Tionesta site. Nests are placed on the ground, usually in a sphagnum hummock or among free roots, protected overhead by a fern or a shrub. Some smaller wetlands (of a few acres) within the breeding range are not occupied every year. Males arrive on territory as early as 14 May and sing *kil-lik* (or *che-bunk*) frequently from prominent perches. Females initiate nest building in late May or early June while migrants are still passing through.[1] Nests with eggs have been found in Jun and as late as 21 Jul, suggesting at least occasional double broods. From 1990 to 1998, 17 nests were found in Pennsylvania,[2] adding substantially to our knowledge of the species.

Fall: Most Yellow-bellied Flycatchers are recorded at this season. Fall migrants have been observed as early as the third week of Jul. Their typical migration period is from the third week of Aug to the first week of Oct. The greatest number of birds is recorded from the fourth week of Aug to the third week of Sep. Stragglers may remain to the fourth week of Oct.

History: The Baird brothers first distinguished Yellow-bellied Flycatchers from other *Empidonax* flycatchers on the basis of migrants near Carlisle; they published their description in the first

Proceedings of the Academy of Natural Sciences in Philadelphia.[3] Some subsequent reports would appear to be misidentifications, such as Turnbull's (1869) report of breeding birds in Trenton, New Jersey, and Libhart's (1869) statement: "frequent, breeds in the county" [Lancaster]. Todd (1940), in characteristic conservative fashion, stated that "few of our local observers have dared to record this flycatcher in their lists, and even those who have done so seem sometimes to have confused it with related species." He did not regard them as an uncommon migrant. Since the early twentieth century, much of their breeding habitat has been flooded or mined for peat. Nesting pairs were found in the Pocono bogs that were subsequently destroyed (P. B. Street 1954). Likely breeding records of Yellow-bellied Flycatchers had been reported from eight counties between 1915 and 1980, adding only Clearfield, Forest, and Pike to the recent list. Confirmed nesting records were few.

Comments: Two reports during Dec are believed to be of this species. One flew into a truck window at Kutztown on 25 Dec 1969,[4] and an apparent Yellow-bellied Flycatcher was discovered and photographed at MCWMA on 15 Dec 1985.[5] But both of these birds could have been one of the western *Empidonax* species that may be more likely on such a late date. Owing to their rarity and restricted and fragile habitat, Yellow-bellied Flycatchers are on Pennsylvania's Threatened species list.

[1] Gross, D.A. 1991. Yellow-bellied Flycatcher *Empidonax flaviventris* nesting in Pennsylvania; with a review of its history, distribution, ecology, behavior, and conservation problems. PB 5:107–113.

[2] D.A. Gross, pers. comm.

[3] Baird, W.M., and S.F. Baird. 1843. Description of two species supposed to be new, of the genus *Tyrannula* Swainson, found in Cumberland County, Pennsylvania. Proceedings of the Academy of Natural Sciences of Philadelphia 1: 283–285.

[4] Scott, F.R., and D.A. Cutler. 1970. Middle Atlantic Coast region. AFN 24:491.

[5] Boyle, W.J., Jr., R.O. Paxton, and D.A. Cutler. 1986. Hudson-Delaware region. AB 40:263.

Acadian Flycatcher ***Empidonax virescens***

General status: Acadian Flycatchers breed from Nebraska and Texas east to Massachusetts and central Florida. They winter in Central and South America. Acadian Flycatchers are rarely observed during migration away from areas where they breed. Most field identifications during migration away from banding stations are made in spring when they are calling. Acadian Flycatchers are regular breeders, primarily across the southern half of the state.

Habitat: They prefer lowland areas in old-growth woodlands near streams. They are also found in narrow hemlock-lined ravines, especially in the northern counties.

Seasonal status and distribution

Spring: They have been recorded as early as the fourth week of Apr, but most arrive the second week of May. Migration is difficult to discern because most birds are found only on territory. Banding records suggest that the greatest numbers may be passing through during the second and third weeks of May.

Breeding: Acadian Flycatchers are fairly common to rare during summer. Greene and Washington cos. fall within the BBS physiographic region (the Ohio Hills) that supports the highest density of Acadian Flycatchers on BBS routes rangewide. These flycatchers are uncommon in the southeastern counties and are rare in the north, where they are generally associated with hemlock trees along streamsides in extensive woodlands. The population is increasing on BBS routes at 1.4% per year in Pennsylvania and at 0.5% per year nationwide. Since the 1960s, Acadian Flycatchers have expanded their range north of the Piedmont through the Ridge and Valley. Most nests with eggs are found during the first three weeks of Jun, with early dates of 22 May and late nesters into mid-Jul. The nest is slung between the arms of a fork in a narrow horizontal branch, generally less than 15 feet from the ground. Long streamers of vegetation often hang from the nest, presenting an untidy but distinctive appearance from below.

Fall: After singing stops, field identification becomes difficult, and they are rarely identified except at banding stations where they can be examined in the hand. Most of what we know about this species' fall migratory status in Pennsylvania is based on banding records. Records suggest that migration begins as early as the fourth week of Jul. The number of birds increases by the third week of Aug with a steady flow of birds to the third week of Sep. Banding records also suggest that the number of birds begins falling by the fourth week of Sep. Acadian Flycatchers have left the state by the first week of Oct, with stragglers banded through the third week of Oct. One was netted at PNR on 25 Oct 1989.[1]

History: Early authors considered Acadian Flycatchers common breeding residents in southern parts of Pennsylvania and rare in the north, where these flycatchers have recently expanded. At the end of the 1800s they were rather common summer residents in Erie Co., "one of the northernmost localities where this relatively southern species is known to breed regularly" (Todd 1904). The species had not been listed on breeding bird summaries in the High Plateau counties (e.g., Clinton and Potter cos.; Cope 1902), where the BBA project subsequently identified them. Loss of forested habitat, particularly in the southeastern counties, contributed to local declines. A bird was found singing at Indiantown Gap in Lebanon Co. on the very late date of 25 Nov. 1954 (Poole, unpbl. ms.).

[1] R.C. Leberman, pers. comm.

Alder Flycatcher ***Empidonax alnorum***

General status: Alder Flycatchers, recognized as a distinct species in 1973,[1] (AOU 1983) were formerly considered conspecific with Willow Flycatchers. Their habitat preferences and plumage subtleties may aid in separating them from Willow Flycatchers, but confident identification can be made only by their vocalizations. Very little is known about the status and distribution of this species during the periods after singing has stopped. They breed from Alaska south across Canada to Newfoundland and south to Pennsylvania, Ohio, and through the Applachian Mountains to North Carolina and Tennessee. Alder Flycatchers winter apparently exclusively in South America. In Pennsylvania they are uncommon to fairly common regular migrants but are rarely reported away from sites where they breed. Alder Flycatchers breed locally across the northern half of the state and south through the Allegheny Mountains.

Habitat: Alder Flycatchers prefer brushy swamps, fens, bogs, and wet meadows interspersed with larger trees—especially areas associated with beaver activity. They may also be found in uplands overgrown with brush, often far from water. Where their ranges overlap, Alder and Willow flycatchers can occasionally be found in the same habitat.

Seasonal status and distribution

Spring: They have been recorded the second week of May, but most are found from the third week of May to the first week of Jun. As they are rarely found away from their breeding grounds, determining the migration season is difficult. Migrants may remain until the second week of Jun.

Breeding: Alder Flycatchers are uncommon to locally fairly common in the Glaciated Northeast and Northwest and are rare across the central High Plateau south through the Allegheny Mountains. They are absent in extensively forested areas and occur rarely in larger clearcuts. They are rare but regular and local during summer in northern parts of the Ridge and Valley and the Southwest regions. Birds singing the Alder song have been heard in early summer at scattered locations almost statewide. Alder Flycatchers are reported irregularly and in low numbers on about one-quarter of the BBS routes, but sufficiently to document an increasing trend of 7.7% per year. The nest is built in Jun within a few feet of the ground in the upright fork of a shrub. Alder Flycatchers have been heard singing on their breeding grounds as late as the second week of Aug.[2]

Fall: It is very likely that Alder Flycatchers make up a certain percentage of the "Traill's" flycatcher complex that are recorded after singing has stopped. Most of what we know about Alder/Willow migratory status in Pennsylvania at this season is based on banding records. These records suggest that Alder/Willow Flycatchers begin leaving their breeding grounds as early as the fourth week of Jul. Numbers gradually increase to reach a maximum from about the third week of Aug to the second week of Sep. Banding records suggest that the number of birds decreases toward the end of Sep, with stragglers to the second week of Oct.

History: The history of Alder Flycatchers must be unraveled from records of "Traill's" Flycatchers before they were split. "Traill's" (often called Alder) was regarded strictly as a migrant in Pennsylvania until Warren (1890) reported possible nesting in "mountainous parts of the state." The generally northern distribution and wet habitat described in early accounts seem to refer to what we now know as Alder Flycatchers,[3,4] but nesting was rarely confirmed. They were considered locally common and were confined almost entirely to alder swamps by Sutton (1928a), who considered them among "the most characteristic birds of Pymatuning" Swamp (Sutton 1928b). The breeding distribution described by Todd (1940) was similar to that currently known for the Alder Flycatcher. Nesting ecology helped distinguish the Alder from the closely related Willow Flycatcher.[4]

Comments: Alder and Willow flycatchers should be distinguished in the field only by voice.

[1] Eiseman, E. 1973. 32nd supplement to the American Ornithological Union Check-list of North American Birds. Auk 90:411–419.

[2] G.M. McWilliams, pers. obs.

[3] Mulvihill, R.S. 1992. Alder Flycatcher *(Empidonax alnorum)*. BBA (Brauning 1992):202–203.

[4] Stein, R.C. 1963. Isolating mechanisms between populations of Traill's Flycatchers. Proceedings of the American Philosophical Society 107:21–50.

Willow Flycatcher *Empidonax traillii*

General status: Very little is known about the status and distribution of Willow Flycatchers after singing has stopped. At this time they become virtually indistinguishable from Alder Flycatchers. However, their very different call notes, when heard, may be useful in separating the two species. Willow Flycatchers breed primarily across the northern two-thirds of the U.S., and they winter in Central America. They are probably fairly common regular migrants statewide. Willows breed over most of the mountainous regions, but rarely above 1800 feet elevation.

Habitat: Willow Flycatchers may occur in somewhat drier habitats than Alder Flycatchers. They are usually found in brushy swamps and streams, wet meadows, old fields, and uplands such as pastures surrounded by or interspersed with brush or on brushy dry hillsides. They may be found with Alder Flycatchers where their breeding ranges overlap.

Seasonal status and distribution

Spring: Willow Flycatchers generally migrate earlier than Alder Flycatchers and they are seen in slightly greater numbers. The migration period probably occurs between the second and fourth weeks of May. However, since they are almost never found away from their breeding grounds, determining the migration season is difficult. Fewer birds are observed after the first week of Jun, suggesting the end of the migration period.

Breeding: Willow Flycatchers are uncommon to locally common statewide and rare at higher elevations of the High Plateau. Although they may be found in drier sites than Alder Flycatchers, Willow Flycatchers are generally not found in regenerating clearcuts or power line rights-of-way but may be common in shrub swamps in southern counties. The continued expansion of Willow Flycatchers, borne out by the BBS, which shows an increase of 3.1% per year, is contrary to the pattern of many species of brushy habitat. This species has expanded during the century into the Alder's range, and now the two species overlap regularly in the Northeast and Northwest Glaciated regions, sometimes occurring in the same fields. Nesting data are generally lacking, but the few nests definitely attributed to this species were active in mid- to late Jun. The nest is a compact cup placed several feet above the ground on a limb or in the forked branch of a shrub or small tree. Willow Flycatchers have been heard singing on their breeding grounds as late as early Sep.[1]

Fall: Data received from banding stations constitute the majority of Alder/Willow Flycatcher migrant records; these suggest that Alder/Willow Flycatchers begin leaving their breeding grounds as early as the fourth week of Jul. The number of birds gradually increases to reach a maximum from about the third week of Aug to the second week of Sep. Banding records suggest that the numbers decrease toward the end of Sep, with stragglers until the second week of Oct.

History: Alder/Willow Flycatchers were strictly regarded as migrants until the late nineteenth century, when scattered nestings of birds known as "Traill's" flycatchers were first reported. R. S. Mulvihill considered the first Traill's nest in Pennsylvania, in 1894, most likely to belong to the Willow Flycatcher, but thought most early records in northern counties were of Alder Flycatchers.[2] The scarcity of "Traill's" flycatchers in southern Pennsylvania during the first half of the twentieth century is evidence of the Willow Flycatcher's general scarcity in the state at that time. An early nest, best attributed to the Willow Flycatcher, was documented in Montgomery Co. near Holmesburg in 1939.[3] The expansion of the "Traill's" complex eastward and north across the state during the 1940s and 1950s primarily reflects the expansion of Willow Flycatchers into Alder Flycatcher's range. By the mid-1900s in Centre Co., Wood (1958) described "Traill's" as rare in "brush-covered fields," the habitat characteristic of Willow Flycatchers, and by 1958 they were common at the Tinicum refuge in Philadelphia (J. C. Miller and Price 1959). The expansion of Willow Flycatchers was concurrent with widespread reversion of old farms and regrowth of timbered areas, both of which provided extensive habitat for this species.

Comments: A "Traill's" Flycatcher banded on 12 Sep 1971 at PNR was caught and released alive on 8 Oct 1971 in Belize City, British Hondorus (Belize). This was the first record for that country.[4]

[1] W. Reid, pers. comm.
[2] Mulvihill, R.S. 1992. Willow Flycatcher *(Empidonax traillii)*. BBA (Brauning 1992):204–205.
[3] Miller, R.C. 1941. Field notes and reports—October 1937 to September 1941. Cassinia 31:47.
[4] Leberman, R.C., and M.H. Clench. 1972. Birdbanding at Powdermill, 1971. Research report no. 30. Powdermill Nature Reserve, Carnegie Museum of Natural History, Pittsburgh.

Least Flycatcher *Empidonax minimus*

General status: Least Flycatchers' persistent vocalizations make them easy to find and identify. Unlike others in this family they can usually be identified when they are not singing. Least Flycatchers breed from Yukon across Canada and the northern U.S. to Pennsylvania and south through the Appalachian Mountains to North Carolina and Tennessee. Their primary wintering range is from Mexico south through Central America. They are the earliest to arrive in spring and in many parts of Pennsylvania, the most common *Empidonax* flycatcher. Least Flycatchers are uncommon to fairly common regular migrants over most of the state except in the Coastal Plain, where they are rare. They regularly breed in the Glaciated Northeast and west and north of the Allegheny Front.

Habitat: Least Flycatchers are found in woodlands with an open understory, on woodland edges, and in small woodlots in suburbs or in parks.

Seasonal status and distribution

Spring: Some birds arrive in the state as early as the third week of Apr, but their typical migration period is from the fourth week of Apr to the first week of Jun. The greatest number of birds during migration is seen or heard during the second and third weeks of May.

Breeding: Least Flycatchers are fairly common to uncommon across much of the High Plateau and Glaciated Northeast and Northwest. They were most frequently reported (BBA) and most common (BBS) in the Glaciated Northeast and are the dominant flycatchers in the northern mountains. However, they are uncommon in the Glaciated Northwest and south through higher elevations of the Allegheny Mountains. Least Flycatchers are rare and irregular breeding birds in the Ridge and Valley and southern parts of the Southwest region. They are accidental in the Piedmont and through the central Susquehanna Valley. Statewide they are the most common (average of 2.6 birds per BBS route) and most widespread (1816 BBA blocks) of the five *Empidonax* flycatchers breeding in Pennsylvania. The BBS shows no significant change in the state, but birds have retracted from the southern extent of their range since the 1960s. In the 1960s they were widely distributed in Berks Co. but were last confirmed breeding there in 1986 (Uhrich 1997). The nest is a neat, deep cup, frequently placed less than 20 feet high, perched on a branch or between several horizontal twigs of a deciduous tree. Most eggs are found in late May and early Jun, with dependent young found into early Jul.

Fall: Banding records suggest fall movement beginning as early as the third week of Jul, but most begin migration the second week of Aug. Banding records as well as field observations point to the greatest movement of birds occurring from about the fourth week of Aug to the second week of Sep. The number of Least Flycatchers gradually falls through the first week of Oct, with stragglers to the fourth week of Oct. One banded at PNR on 29 Oct 1988 was recaptured on the very late dates of 23 and 27 Nov 1988.[1]

History: Along with the Yellow-bellied Flycatcher, the Least Flycatcher was first named by the Baird brothers on the basis of specimens taken in Cumberland Co.[2] Historical accounts described this common species as extending south as a breeding species even into Bucks and Lancaster cos. (e.g., Libhart 1869). The nesting range once included southern Montgomery Co., where Harlow (1913) called them infrequent but regular. Frey (1943) found them to be a local breeding resident in the same area where the Bairds identified the species. A general retraction of their range northward appears to have occurred primarily since 1950.

[1] Hall, G.A. 1989. Appalachian region. AB 43:103.
[2] Baird, W.M., and S.F. Baird. 1843. Description of two species supposed to be new, of the genus *Tyrannula* Swainson, found in Cumberland County, Pennsylvania. Proceedings of the Academy of Natural Sciences of Philadelphia 1: 283–285.

[Hammond's Flycatcher] *Empidonax hammondii*

General status: These *Empidonax* flycatchers breed across much of western North America from southern Alaska south and east to Colorado and northern New Mexico. They winter from Arizona and New Mexico south through Central America. In eastern North America they have been reported in Louisiana, Alabama, Michigan, Pennsylvania, Massachusetts, Maryland, and Delaware (AOU 1998). The only specimen collected in Pennsylvania was damaged, and the identity is in question; therefore, the Hammond's Flycatcher is listed as hypothetical in the state.

Comments: One was collected on 23 Dec 1966 at Schnecksville in Lehigh Co. The bird was sent to N. Johnson of the Museum of Vertebrate Zoology at Berkeley, who pointed out that the bird's color and the size and shape of its bill were reasonable for a Dec specimen of *E. hammondii.* He further stated that the tail could not be measured because of the damage sustained when the bird was collected, but the two outer rectrices were of the juvenile generation, "hence this bird would be aged as an immature, or a first-winter individual." Measurements were taken of the wings, and Johnson concluded that the bird was a typical female Hammond's Flycatcher.[1] The specimen is now at the State Museum in Harrisburg.

[1] Heintzelman, D.S. 1968. *Empidonax hammondii* in Pennsylvania. Auk 85:512.

[Dusky Flycatcher] *Empidonax oberholseri*

General status: Dusky Flycatchers breed across much of western North America from southern Yukon south to southern California and east to western South Dakota and New Mexico. They winter in Arizona, New Mexico, western Texas and south through Mexico. In eastern North America they have been recorded only in southern Ontario, Pennsylvania, and Delaware (AOU 1998). The only record of this flycatcher in Pennsylvania is listed as hypothetical.

Comments: On 25 Dec 1969 a bird identified as a Dusky Flycatcher was found near Kutztown in Berks Co.[1,2] Initially the bird was identified as a Yellow-bellied Flycatcher. Apparently it flew into the rear window of a pickup truck and was taken to A. Nagy, curator at HMS. F. Wetzel, assistant curator at HMS sent the bird to Dr. N. Johnson, an authority on the *Empidonax* genus of flycatchers, for identification.[3] Johnson wrote, "I am fairly certain that your bird is an immature . . . and I also think it is very likely a female rather than a male as labelled. This assumption is based on wing shape. . . . In any event a winter specimen of *Empidonax oberholseri* is of considerable interest from the eastern United States."[4] C. Robbins stated that this bird was the first record east of the Mississippi and perhaps east of the states where the species breeds.[5] A skin of the flycatcher was prepared, but apparently the specimen cannot now be located.

[1] Scott, F.R., and D.A. Cutler. 1970. Middle Atlantic Coast region. AFN 24:491.
[2] Scott, F.R., and D.A. Cutler. 1970. Middle Atlantic Coast region. AFN 25:39.
[3] Nagy, A.C. 1970. Hawk Mountain Sanctuary, 1970 curator's report no. 3.
[4] Letter to F. Wetzel, Assistant Curator at HMS from N. K. Johnson, Curator of Birds, Museum of Vertebrate Zoology, 7 October 1970.
[5] Letter to F. Wetzel, Assistant Curator at HMS from C. Robbins, Chief of Non-game Bird Studies, U.S. Fish and Wildlife Services, 17 December 1970.

Pacific-slope Flycatcher ***Empidonax difficilis***

General status: Pacific-slope Flycatchers and Cordilleran Flycatchers *(E. occidentalis)* until recently were considered conspecific [Western Flycatcher] (AOU 1998). Pacific-slope Flycatchers breed from southeastern Alaska to Baja California and winter from southern California to Baja California. They have been recorded in eastern North America only from Louisiana and Pennsylvania (AOU 1998). There is one accepted record of this species in Pennsylvania.

Seasonal status and distribution

Fall: A Pacific-slope Flycatcher was discovered by J. Meloney and colleagues on 16 Dec 1990 during the southern Lancaster Co. CBC in East Drumore Township, Lancaster Co. and was present until 26 Dec 1990. The flycatcher was photographed by R. Miller, and the voice was clearly recorded by F. C. Haas.[1,2] This was the first record east of Louisiana (AOU 1998).

Comments: On 15 Dec 1991 one was reported in Fulton Township in Lancaster Co. about 7 miles from the previous year's location. This may have been the same bird as the one found in 1990.[3] It also was audiotaped and photographed, but not clearly enough to be verified. A bird previously reported as a Yellow-bellied Flycatcher on 15 Dec 1985 at MCWMA[4] was later evaluated

Pacific-slope Flycatcher at East Drumore Township, Lancaster County, December 1990. (Photo: Randy C. Miller)

as a possible Pacific-slope Flycatcher, but details could not rule out Yellow-bellied or Cordilleran flycatchers.[5]

[1] Meloney, J., and H. Morrin. 1991. First record of a Pacific-slope Flycatcher *(Empidonax difficilis)* east of the Mississippi. PB 4:135.
[2] Haas, F.C. 1991. Pacific-slope Flycatcher. PB 4:142.
[3] Witmer, E. 1992. Second state record of Pacific-slope Flycatcher, *Empidonax difficilis*. PB 5:156.
[4] Boyle, W.J., Jr., R.O. Paxton, and D.A. Cutler. 1986. Hudson-Delaware region. AB 40:263.
[5] Kwater, E. 1993. Fourth report of the Pennsylvania Ornithological Records Committee, April 1992. PB 7:11.

Eastern Phoebe ***Sayornis phoebe***

General status: Eastern Phoebes breed from central Canada south through the eastern half of the U.S. They winter in the east from Tennessee west of the Appalachian Mountains south to Texas and Virginia south to Florida. Eastern Phoebes are fairly common regular migrants and breeding residents throughout Pennsylvania. They are irregular and local winter residents in the Coastal Plain and Piedmont. In the Ridge and Valley, Southwest, and Lake Erie Shore they are accidental in winter.

Habitat: During migration Eastern Phoebes inhabit a variety of habitats, from woodlands and brushy areas to suburban parks, towns, and cities. In the breeding season they are observed wherever suitable ledges for nesting exist, such as under bridges or culverts, in open buildings that provide beams or posts, and on rock ledges. In winter they prefer brushy areas or woodland edges, especially where there is open water near active barns.

Seasonal status and distribution

Spring: Arrival time usually depends on the extent and continuation of winter conditions. In years of persistent cold weather, they may not appear in Pennsylvania until the fourth week of Mar. After mild winters they may appear as early as the third week of Feb. They generally do not move into the northern parts or higher elevations of the state before the third week of Mar. The greatest number of migrants is usually recorded from the first to the third week of Apr. Migration usually ends by the second week of May.

Breeding: Eastern Phoebes are fairly common where nesting structures are available. The nest is bulky and distinct, composed primarily of moss and mud. Neither habitat nor elevation appears as important a factor in selection of territories as do nest sites; phoebes will nest on the crossbeam of a cabin on a forested ridge or under a bridge in farmland. They are least common in the Piedmont. In Pennsylvania their association with water is probably secondary to the occurrence of proper nest sites. Phoebe numbers on BBS routes declined by about half between 1966 and 1977, increased until 1995, and then declined again somewhat in 1996. The nesting season is early, with ample time for two, or potentially three, broods. Most egg sets were found from 21 Apr to 11 Jun, with peaks that reflect first clutches in late Apr and second clutches in Jun. The close association with human structures results in a bias on BBS routes, particularly in the northern mountain regions, where roads are limited and often follow streams. Higher densities of phoebes in these areas do not necessarily reflect their abundance across the region. As a result, a map is not included for this common bird.

Fall: The beginning of fall migration is difficult to discern, but there is a noticeable increase in the number of birds seen away from breeding sites by the second or third week of Sep. In the last week of Sep or the first week of Oct, substantial numbers of birds are banded at banding stations, suggesting the time of heaviest southerly movement. The number of Eastern Phoebes steadily drops through Oct. The majority of birds depart by the first week of Nov in the northern counties or at high elevations. Some may remain as late as the fourth week of Dec, with lingering birds until the first or second week of Jan, especially in the southeastern counties. Some may winter if the weather remains mild.

Winter: Whether birds remain through winter depends on weather conditions. They are rare wherever they occur at this season. Most records during the winter season are of lingering fall migrants or very early spring records that fall within our defined winter period. During mild winters most records come from the Coastal Plain and Piedmont, but they have been recorded in winter in the Ridge and Valley in Blair, Huntingdon, Juniata, Luzerne, and Perry cos. West of the Allegheny Front they have been recorded in winter only in the Southwest region in Armstrong, Beaver, Butler, and Greene cos. A Lake Erie Shore winter record of a bird at PISP on 13 Feb 1993[1] probably represents a very early spring migrant, since this species is not likely to have survived the severe winter conditions in this area.

History: The degree in which phoebes exploit human structures for nesting and the general scarcity of natural sites suggests that phoebes were rare and locally distributed before European settlement. Aside from rock ledges, a historical habitat still occasionally used is upturned roots of large trees.[2] These may have provided widespread sites in old forests but generally are not available in modern, younger forests. Nineteenth-century bird lists, including those for the southeastern portion of the state, all consider the species to be common or abundant. Eastern Phoebes were considered "a great favorite" by Cassin (1862) in Delaware Co. A decline of 85% in the southeastern counties was reported between 1945 and the early 1970s (Richards 1976). There was a single record of one in winter as far north as Scranton in Lackawanna Co. on 7 Feb 1937 (Poole, unpbl. ms.).

[1] Hall, G.A. 1993. Appalachian region. AB 47: 257.

[2] D.A. Gross, pers. comm.

Say's Phoebe ***Sayornis saya***

General status: Say's Phoebes are western birds that breed from Alaska south through British Columbia and in the western half of the U.S. They winter primarily in the southwestern U.S. Say's Phoebes are casual visitors to the East Coast in fall and winter. There have been at least 10 reports in Pennsylvania since 1957, all from east of the Allegheny Front. Four of these birds were close to HMS.

Seasonal status and distribution

Fall to spring: The first modern record of Say's Phoebe in Pennsylvania was of one found by R. Bubb and J. Falkenstein on Brunner's Island in York Co. on 21 Dec 1957; the bird remained to be photographed until 12 Jan 1958 (Poole 1964; photo with PORC). Near Eckville in the Berks-Schuylkill co. area one was present for about two weeks in early Apr 1975,[1] and another from 29 Dec to at least 6 Jan 1986;[2] one appeared from 27 Dec 1987 to 2 Jan 1988;[3,4] and one that was there from 4 to 15 Jan 1993 was photographed.[5] One was carefully described at Media in Delaware Co. on 24 Nov 1979.[6] On 8 Oct 1982 one was at BESP in Centre Co. (Wood 1983), and on 31 Oct 1982 one was at Mifflinburg in Union Co. (Schweinsberg 1988). In 1985, one was at MCWMA from 7 to 24 Dec and was photographed. The most recent Say's Phoebe in Pennsylvania was discovered on a CBC on 3 Jan 1998 at Newville in Cumberland Co. This

Say's Phoebe at Eckville, Berks County, January 1993. (Photo: Rick Wiltraut)

bird successfully wintered and was last observed on 10 Apr 1998.[7]

History: Poole (1964) listed a Say's Phoebe seen by six observers, including himself, at Lake Ontelaunee in Berks Co. from 22 to 28 Dec 1946.

[1] Scott, F.R., and D.A. Cutler. 1975. Middle Atlantic Coast region. AB 29:835.
[2] Boyle, W.J., Jr., R.O. Paxton, and D.A. Cutler. 1986. Hudson-Delaware region. AB 40:263.
[3] Slater, M. 1987. Berks County. PB 1:122.
[4] Slater, M. 1987. Berks County. PB 2:19.
[5] Haas, F.C., and B.M. Haas, eds. 1993. Rare and unusual bird reports. PB 7:17.
[6] Paxton, R.O., K.C. Richards, and D.A. Cutler. 1980. Hudson-Delaware region. AB 34:146.
[7] Hoffman, D. 1998. Local notes: Cumberland County. PB 12:70

Vermilion Flycatcher ***Pyrocephalus rubinus***

General status: Vermilion Flycatchers breed in the southwestern U.S. from southern California east to central Texas. They winter near the U.S. border south through Mexico and South America. One accepted record exists in Pennsylvania.

Seasonal status and distribution

Vermilion Flycatcher at Jobs Corners, Tioga County, October 1991. (Photo: Franklin C. Haas)

Fall: An immature male found by A. Brown at Jobs Corners in Jackson Township, Tioga Co., on 24–25 Oct 1991 was photographed.[1,2]

[1] Brown, A. 1992. Pennsylvania's first record of Vermilion Flycatcher. PB 5:156–157.
[2] PORC files.

Ash-throated Flycatcher ***Myiarchus cinerascens***

General status: This is a flycatcher of deserts and chaparrals with a breeding range in the U.S. extending from Oregon south to California and east to Colorado and Texas. They winter from southern California and Arizona south through Central America. Ash-throated Flycatchers are casual fall and winter visitors to the eastern U.S. One Pennsylvania record has been accepted.

Seasonal status and distribution

Fall: A single Ash-throated Flycatcher was discovered by A. Koch in her front yard in Willliams Township in Northampton Co. from 24 Nov 1997 to 16 Dec 1997.[1,2] It was photographed and videotaped, and its call notes were recorded.

Comments: Other *Myiarchus* flycatchers reported may have been Ash-throated Flycatchers, but there was inadequate documentation: one on 2 Dec 1959 near Center Point in Montgomery Co.[3] and another on 22 Nov 1989 at Toughkenamon in Chester Co.[4]

[1] Koch, A. 1998. The flycatcher in the honeysuckle bush. PB 11:222–223.
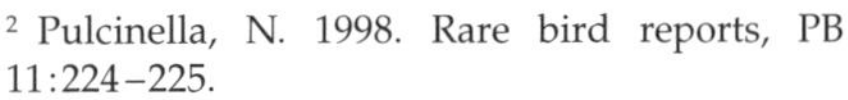
[2] Pulcinella, N. 1998. Rare bird reports, PB 11:224–225.
[3] Scott, F.R., and D.A. Cutler. 1960. Middle Atlantic Coast region. AFN 14:297.
[4] Paxton, R.O., W.J. Boyle Jr., and D.A. Cutler. 1990. Hudson-Delaware region. AB 44:65.

Great Crested Flycatcher ***Myiarchus crinitus***

General status: Great Crested Flycatchers breed from Alberta east across southern Canada and south through the eastern half of the U.S. They winter from central Florida south to South America. A canopy-dwelling woodland species, the Great Crested Flycatcher is the only cavity-nesting flycatcher in Pennsylvania. It is a fairly common regular migrant and breeding resident statewide.

Habitat: Great Crested Flycatchers prefer open deciduous or mixed woodlands or woodlots in parks and suburbs.

Seasonal status and distribution

Spring: The earliest migrants may appear in the fourth week of Mar, but most begin passing through the state in the third or fourth week of Apr in southern counties. In the northern counties and at higher elevations of the High Plateau most appear the first week of May. The greatest number of birds is noted during the second and third weeks of May. Migration has usually ended by the last week of May.

Breeding: Great Crested Flycatchers are fairly common throughout the Ridge

Ash-throated Flycatcher at Williams Township, Northampton County, November 1997. (Photo: Franklin C. Haas)

and Valley, Glaciated Northeast, Pocono section, and Blue Ridge section and in the Southwest region. Elsewhere they are uncommon. They are least common in the mountains and extensively forested sections of the High Plateau and most common east of the Allegheny Front within the wooded slopes of the mountain ridges (see Map 10).[1] These flycatchers nest in a natural cavity or old woodpecker hole, most often in the dead limb of a live tree, generally less than 25 feet from the ground (Sauffer and Best 1980). Nest boxes and other assorted manmade cavities are sometimes used. They will use a standard bluebird nest box with an entrance hole of 1.5–2 inches, placed at a deciduous woodland edge or in an orchard. Most eggs are found from late May to the third week of Jun. The nest cavity usually contains a snake skin, strip of birch bark, cellophane, or plastic wrapper.[2] No long-term population change is apparent, although BBA results suggest that they have become more widespread than historically reported in the High Plateau.

Fall: Great Crested Flycatchers are silent and inconspicuous at this season and are less frequently seen in fall than in spring. Migrants are first detected the second or third week of Aug. Their usual migration time is from the last week of Aug to the second week of Sep. Stragglers remain to the first week of Oct. Two very late fall sightings have been recorded: one was found from 18 to 20 Nov 1990 near Slippery Rock in Butler Co. and was photographed,[3] and another was seen and heard on the Bernville CBC in Berks Co. on 1–6 Jan 1992.[4]

History: Early ornithologists considered them to be a common resident of southern Pennsylvania but rare or absent in the north-central counties (Cope 1902). Great Crested Flycatchers appear to have expanded their range into northern counties during the twentieth century. For example, Todd (1940) stated that he "never encountered them in Potter, McKean, Cameron, or Elk cos., nor have others," where they were being reported in the late 1940s in Clinton Co. (Reimann 1947) and were considered rare by Poole (1964).

[1] Ickes, R. 1992. Great Crested Flycatcher *(Myiarchus crinitus)*. BBA (Brauning 1992):210–211.

[2] Lanyon, W.E. 1997. The Great Crested Flycatcher *(Myiarchus crinitus)*. BNA (A. Poole and Gill 1992), no. 300.

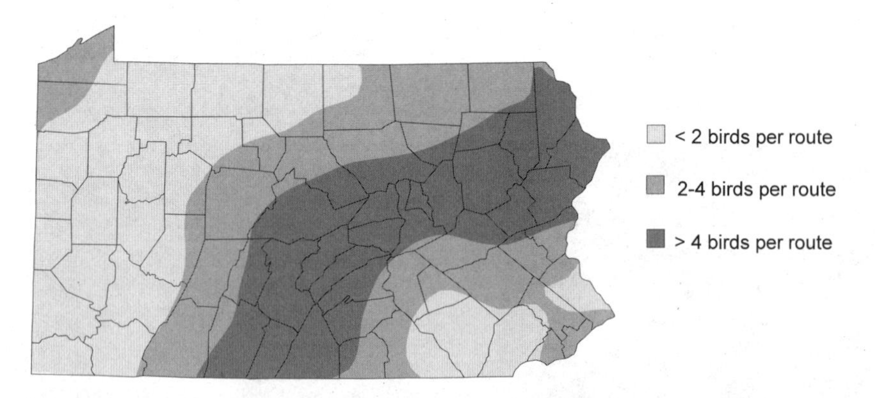

10. Great Crested Flycatcher relative abundance, based on BBS routes, 1985–1994

[3] Hess, P. 1991. County reports—October through December 1990: Butler County. PB 4:150.
[4] Keller, R. 1992. County reports—January through March 1992: Berks County. PB 6:29.

Western Kingbird *Tyrannus verticalis*

General status: Western Kingbirds breed from southern Manitoba west to British Columbia and south through the western half of the U.S. They winter primarily from Mexico through Central America, but a few winter from South Carolina south along the Atlantic and Gulf coasts. In Pennsylvania they are casual visitors in the Piedmont and accidental elsewhere. Most Western Kingbird sightings in Pennsylvania have been in the fall.

Habitat: Western Kingbirds are usually observed perched on wires surrounded by open fields.

Seasonal status and distribution

Spring: There have been three reports during this season. One was a very late spring record on 16 Jun 1978 at Upper Strasburg in Franklin Co.;[1] one was at Liberty in Tioga Co. on 31 May 1981;[2] and a single bird was at Allentown in Lehigh Co. on 30 May 1991.[3]

Summer: A bird that remained from spring into the summer season was photographed on 27 Jun 1997 at Beltzville Lake in Carbon Co.[4]

Fall: Sightings during this season are from 13 Aug to 18 Dec, with most occurring during the second and third weeks of Sep. All reports are of one or two birds at each site, and no more than four birds were recorded in the state in any single year. Away from the Piedmont, Western Kingbirds have been recorded at only a few other sites in Pennsylvania. In the Coastal Plain, they have been recorded at the Philadelphia International Airport on 3 Dec 1961[5] and at Tinicum on 15 Sep 1990.[6] Five records are known in the Ridge and Valley: in Centre Co. at BESP on 22 Aug 1970 and 31 Aug 1981 (Wood 1983),[7] and at Colyer Lake on 5 Sep 1986;[8] two at Beach Haven in Luzerne Co. on 28 Sep 1984;[9] and one at Wilkes-Barre in Luzerne Co. on 19 Sep 1987.[10] The only record in the Southwest was one discovered at Rector on 17 Sep 1961.[11] At the Lake Erie Shore one was banded and photographed at PISP on 26 Sep 1992.[12]

History: The first record of Western Kingbird in Pennsylvania was in about 1928, when one was observed by J. Cope and E. Cope in Uwchlan Township in Chester Co. in the fall (Conway 1940). Poole (1964) stated that "there have been at least 17 observations of this western species reported in Pennsylvania." All of these sightings were in the fall from 25 Aug to 16 Dec and in the Piedmont.

[1] Hall, G.A. 1978. Appalachian region. AB 32: 1160.
[2] Hall, G.A. 1981. Appalachian region. AB 35:824.
[3] Boyle, W.J., Jr., R.O. Paxton, and D.A. Cutler. 1991. Hudson-Delaware region. AB 45:423.
[4] Hawk, D. 1997. Local notes: Carbon County. PB 11:53.
[5] Scott, F.R., and D.A. Cutler. 1962. Middle Atlantic Coast region. AFN 16:318.
[6] Paxton, R.O., W.J. Boyle Jr., and D.A. Cutler. 1991. Hudson-Delaware region. AB 45:82.
[7] Hall, G.A. 1982. Appalachian region. AB 36:178.
[8] Hall, G.A. 1987. Appalachian region. AB 41:89.
[9] W. Reid, pers. comm.
[10] Paxton, R.O., W.J. Boyle Jr., and D.A. Cutler. 1988. Hudson-Delaware region. AB 42:51.
[11] Hall, G.A. 1962. Appalachian region. AFN 16:32.
[12] Haas, F.C., and B.M. Haas, eds. 1992. Rare and unusual bird reports. PB 6:129.

Eastern Kingbird *Tyrannus tyrannus*

General status: Eastern Kingbirds are flycatchers of open spaces. They may be quite aggressive, especially during the breeding season, when they will defend their territory fearlessly by driving away birds as large as Red-tailed Hawks. Eastern Kingbirds breed across Canada south across all but the West

Coast and southwestern U.S. They winter in South America. In Pennsylvania, Eastern Kingbirds are fairly common regular migrants; their breeding range extends across the state.

Habitat: Eastern Kingbirds are found in open country with scattered trees or hedgerows. They are frequently observed perched on exposed tree branches or wires. Kingbirds avoid heavily forested areas but may be found around open lots or parks in some urban areas.

Seasonal status and distribution

Spring: Earliest spring arrivals may be seen by the third week of Apr. Their typical migration period is from the last week of Apr to the first week of Jun. Spring migration usually peaks during the second or third week of May.

Breeding: Eastern Kingbirds are fairly common to locally common breeding residents statewide. They are among a handful of native birds that are regular summer residents around large shade trees in urban areas, at least in the eastern half of the state. The absence of birds in the Pittsburgh metropolitan area (most of Allegheny Co.) contrasts with the regular occurrence of birds in Philadelphia. Kingbirds are more common at Pennsylvania's corners: in the Piedmont and in the Northeast and Northwest Glaciated regions (see Map 11). They are expected to be absent from extensively forested areas, although the reason for their lower abundance in western counties is unclear. They are conspicuous, even pugnacious, and were reported on more than 70% of BBA blocks in most areas,[1] even where uncommon. The nest is an unkempt cup of grass and twigs that may be placed in a wide array of places and heights above ground, often on a horizontal branch out from the trunk of a small tree. Most eggs are reported between 16 May and 30 Jun, with extremes in early May and mid-Jul. The population appears stable on BBS routes.

Fall: Southbound movement may be detected beginning the first week of Aug, and most birds are migrating by the third week of Aug. The largest number of Eastern Kingbirds is recorded during the last week of Aug or the first week of Sep. Across the northern counties most depart by the second week of Sep, and stragglers may remain to the first week of Oct. In the lowland valleys, especially east of the Allegheny Front, they may still be present until the third or fourth week of Sep, with stragglers to the third week of Oct. Very late Eastern Kingbirds have been recorded in the

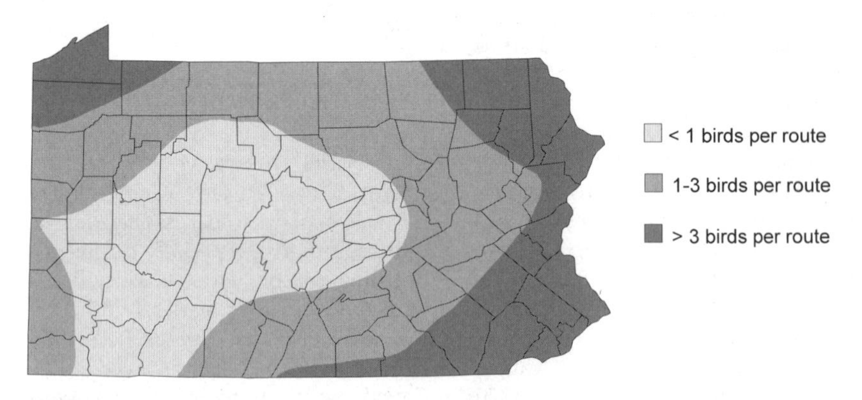

11. Eastern Kingbird relative abundance, based on BBS routes, 1985–1994

Piedmont: two birds remained at Riegelsville in Bucks Co. from 15 Dec 1971 to 2 Jan 1972,[2] and one was found in mid-Nov and remained to be recorded on the Lancaster Co. CBC in Dec 1984.[3]

History: The earliest ornithologists, primarily reporting on settled areas, recorded this species as common to abundant (e.g., Baird 1845; Libhart 1869). The few historical accounts of northern forested regions reflect the modern distribution of kingbirds. They were absent in forested areas but "more abundant in the valley region" (Dwight 1892) of northern counties. Todd (1940) considered them a common summer resident throughout western Pennsylvania.

[1] Master, T.L. 1992. Eastern Kingbird *(Tyrannus tyrannus)*. BBA (Brauning 1992):212–213.
[2] Scott, F.R., and D.A. Cutler. 1972. Middle Atlantic Coast region. AB 26:587.
[3] Boyle, W.J., Jr., R.O. Paxton, and D.A. Cutler. 1985. Hudson-Delaware region. AB 39:152.

Scissor-tailed Flycatcher ***Tyrannus forficatus***

General status: These long-tailed flycatchers breed in the south-central portion of the U.S. from southeastern New Mexico and Colorado north to Nebraska and Indiana and east to Tennessee, South Carolina, and Georgia. They winter in central and south Florida south to Central America. Scissor-tailed Flycatchers are casual wanderers north to southern Canada, with sight reports as far north as southeastern Alaska (AOU 1998). They are listed as accidental in Pennsylvania. All but three records have been from east of the Allegheny Front.

Seasonal status and distribution

Spring: East of the Allegheny Front one was observed by J. Pusey over Chillisquaque Creek in Northumberland Co. on 14 May 1987 (Schweinsberg 1988). West of the Allegheny Front two records exist from the Southwest region: one in Middlesex Township in Butler Co. on 13 May 1983 found by A. Elwood[1] and a very late spring vagrant seen east of Stahlstown in Westmoreland Co. on 18 Jun 1970 by C. Stringer and R. Stringer (Leberman 1976).

Fall: A Scissor-tailed Flycatcher was studied by J. C. Miller at Tinicum on 21 Oct 1972.[2] One was photographed at Ulster in Bradford Co. on 17 Nov 1973.[3] One found by K. Grim in Lehigh Co. near Bake Oven Knob from 10 Aug to 12 Sep 1986 was photographed.[4] On 15 Oct 1987, a bird was discovered by J. Brown and E. Brown near Washingtonville, Montour Co. (Schweinsberg 1988). One near Myerstown in Lebanon Co. on 1 Jan 1988[5] may have been the same bird seen at Hershey in Dauphin Co. in Jan 1988.[6]

History: The first specimen of this species in Pennsylvania, a bird collected on 22 Nov 1942 at Rothsville in Lancaster Co., was placed in the Reading Museum (Poole 1964). Scissor-tailed Flycatchers were not reported in the state again until 1955, when one was observed on 5 Jun at PISP in Erie Co. (Stull et al. 1985). Another was found on 22–23 May 1957 at Richboro in Bucks Co. (Poole 1964).

Comments: Additional sightings of long-tailed flycatchers have been reported that failed to eliminate the Fork-tailed Flycatcher.

[1] Hall, G.A. 1983. Appalachian region. AB 37:870.
[2] Scott, F.R., and D.A. Cutler. 1973. Middle Atlantic Coast region. AB 27:39.
[3] T. Gerlach, pers. comm.
[4] Paxton, R.O., W.J. Boyle Jr., and D.A. Cutler. 1987. Hudson-Delaware region. AB 41:67.
[5] Santner, S. 1987. County reports—January to March 1987: Lebanon County. PB 1:20.
[6] Church, J. 1987. County reports—January to March 1987: Dauphin County. PB 1:15.

[Fork-tailed Flycatcher] ***Tyrannus savana***

General status: The Fork-tailed Flycatcher is a species of Central and South America, breeding as far south as the

Falkland Islands. It is casual in North America, with records in the east from New Brunswick south to Florida. One single historical report of this species is known in Pennsylvania. It is listed as hypothetical because the specimen can no longer be found.

Comments: A specimen was reported to have been collected near Erdenheim, Philadelphia in the autumn of 1873 (Poole 1964). It was placed in the Princeton Museum but has not been found.

Family Laniidae: Shrikes

Nicknamed butcherbirds for their habit of impaling prey on thorns or barbed wire, shrikes are hunters of open areas, preying on insects, small birds, rodents, and snakes. They are solitary hunters most often seen perched on utility wires, on fence posts, or on the highest limbs of trees overlooking fields. The shrike's bill is heavy, with a hooked tip; the upper mandible is notched and has a toothlike projection used for capturing and holding prey. Shrikes are robin-sized birds with a large head and distinctive black mask, short neck and legs, and moderate-sized tail. They usually perch with their body held in a horizontal position. Two species regularly occur in North America, and both are found in Pennsylvania: Northern Shrikes are rare winter residents from the high Arctic regions of North America, and Loggerhead Shrikes are rare breeding birds whose breeding range barely extends into the state from the south.

Loggerhead Shrike *Lanius ludovicianus*

General status: Loggerhead Shrikes breed from south-central Canada and southern Ontario and Quebec south across the U.S. except in the Northeast, where they formerly nested as far north as Nova Scotia. They winter primarily across the southern two-thirds of the U.S. Loggerhead Shrikes are irregular to casual migrants in most areas of Pennsylvania, except in Adams and Franklin cos., where they are rare year-round residents. They often use lower perches than Northern Shrikes and hunt from hedges, brush piles, or small trees. Loggerhead Shrikes were formerly regular migrants, but very few acceptable records have been reported anywhere in Pennsylvania since the mid-1980s except in the south-central portion of the state. There are no records from the High Plateau, except in the southeastern corner of Westmoreland Co., where the last report was in 1969 (Leberman 1976). Loggerhead Shrikes formerly bred west of the Allegheny Front.

Habitat: Loggerhead Shrikes are found in hedgerows and scattered trees and shrubs in open fields and pastures, especially in agricultural areas.

Seasonal status and distribution

Spring: The timing of migration has been difficult to determine because of the scarcity of reports since 1970. The most recent records indicate that they are northbound beginning the second or third week of Mar and continue to be seen to the third or fourth week of Apr. Stragglers remain to the second week of Jun. All migrant records are of individual birds. Some nonbreeding individuals may remain into summer, especially in the south-central counties and eastward.

Breeding: Breaking a streak of no breeding records in 50 years in Pennsylvania, a pair was observed carrying nesting material by A. Kennell in 1990, and juvenile shrikes were observed in 1992 in Adams Co. by R. Chambers and A. Kennell.[1] Nests or fledged young were found at various points in Freedom and adjoining townships in Adams Co. during each subsequent year (1992 to 1998) and in scattered locations in southern

Franklin Co. in 1993, 1994, and 1996.[2] Nesting success varied; a high of 11 young fledged from three nests in 1997, but many nests fail. Summer records in which nesting was not confirmed came from Cumberland and Huntingdon cos. during the BBA project. Nests during the 1990s have been found in eastern red cedar, hawthorn, Norway spruce, and multiflora rose. Nest building generally begins in Apr, with birds on eggs from late Apr through May. From four to six white, gray, or buff eggs blotched with brown or gray are laid. Replacement clutches have been observed in Jun. A pair apparently raised two broods in Adams Co. during 1997.[2]

Fall: Little is known of their fall migration period in the state, but before 1970 in the Glaciated Northeast Loggerhead Shrikes appeared occasionally in the last week of Aug or the first week of Sep.[3] Since then, there have been few fall records, but single birds have been recorded from the third week of Aug to the fourth week of Dec.

Winter: The only acceptable recent winter records of Loggerhead Shrike are from Adams Co., where nesting pairs are basically permanent residents and have been reported at least three out of four years.[4] One was banded near Chambersburg in Franklin Co. in the winter of 1967,[5] and one was photographed at Montour Preserve in Montour Co. in the winter of 1973.[6] Other reports have come from Delaware Co. at Tinicum until 1978[7] and in Bucks Co. They were last reported at this season at Nockamixon State Park in 1979 (Kitson 1998).

History: Nesting historically was concentrated in the Glaciated Northwest. Loggerhead Shrikes were considered regular and common by Warren (1890) and Todd (1940) in Erie, Crawford, and Mercer cos. In Erie Co., Warren (1890) described finding three nests on 20 May 1889 and three additional nests the following day—an unimaginable feat today. Other breeding records include a nest in Beaver Co. in 1887, in Allegheny and Greene cos. in 1884, and a more easterly record in Blair Co. in 1934 (Todd 1940). The last western Pennsylvania nesting record was noted in 1937 in Crawford Co. (Grimm 1952). From 1950 to 1954 they were found in summer near Camptown in Bradford Co.,[3] where they probably nested. Loggerhead Shrikes were rare and irregular migrants, primarily observed during the fall across the rest of the state before 1960.

Comments: When active nests were discovered in 1993, Loggerhead Shrikes were upgraded from Extirpated to Endangered.[8] The Adams Co. pairs represent one of few nesting shrike populations in the northeastern U.S., possibly the farthest north outside Ontario and Quebec, where a declining population is being monitored. Northern Shrikes are far more likely to occur in winter than Loggerhead Shrikes.

[1] Kennell, A. 1992. First confirmed shrike nesting in Pennsylvania since 1934. PB 6:65–66.
[2] Brauning, D.W. 1998. Loggerhead Shrike nesting study. Annual report. Pennsylvania Game Commission, Harrisburg.
[3] W. Reid, pers. comm.
[4] A. Kennell and E. Kennell, pers. comm.
[5] Hall, G.A. 1967. Appalachian region. AFN 21:420.
[6] F.C. Haas and B.M. Haas, pers. comm.
[7] N. Pulcinella, pers. comm.
[8] Brauning, D.W., M.C. Brittingham, D.A. Gross, R.C. Leberman, T.L. Master, and R.S. Mulvihill. 1994. Pennsylvania breeding birds of special concern: A listing rationale and status update. Journal of the Pennsylvania Academy of Science 68:3–28.

Northern Shrike *Lanius excubitor*

General status: In North America, Northern Shrikes breed from Alaska south to British Columbia and east across Mackenzie, northern Ontario, and northern and central Quebec. They winter across southern Canada south to New Mexico in the West and to North Carolina and Tennessee in the East. Nearly always observed on an exposed perch at the top of the highest tree overlooking an open field, Northern Shrikes are rare regular spring, fall, and winter visitors or residents. They are more often reported in the northern portion of the state, especially the northwest corner, and are irregular to casual farther south. During invasion years they may become more widespread.

Habitat: Northern Shrikes prefer open brushy fields and agricultural areas and strip mines, where they use isolated trees or shrubs for perches.

Seasonal status and distribution

Fall through spring: The earliest Northern Shrikes may appear the first week of Oct, but they usually move into the state with a strong cold front during the fourth week of Oct or the first week of Nov. Nearly always found alone, they are aggressive toward other birds and are highly territorial, requiring large areas of open country for feeding. Fewer than 5 to 10 are observed statewide in most years; 20 or more are reported in the state during invasions. They are harder to find after the first or second week of Jan, especially during years with heavy snow cover. Northern Shrikes begin returning north after the fourth week of Feb or the first week of Mar. At this time, there may be a sudden increase in sightings. By the third week of Mar most have left the state, with stragglers lingering to the second week of Apr. The most recent invasion was during the winter of 1996, when they were recorded in 25 counties.[1]

History: There were few reports of Northern Shrikes historically, except during invasion years. Northern Shrikes were considered by most authorities to be rare and irregular throughout the state. There were several years in a row when this species was not recorded in the state (Poole, unpbl. ms.). Bent (1950) stated that since 1900, periodic invasions reached at least to New England and adjacent states at intervals of about four years. Reports were widespread historically, reaching as far south as Tinicum in Philadelphia and Delaware cos. in the east and Washington Co. in the west.

Comments: Northern Shrikes are frequently confused with Loggerhead Shrikes, especially during spring and fall migration, when both species may be present.

Northern Shrike at Beltzville State Park, Carbon County, February 1992. (Photo: Rick Wiltraut)

[1] Haas, F.C., and B.M. Haas, eds. 1996. Summary of the season—January through March 1996. PB 10:13.

Family Vireonidae: Vireos

Most vireos are rather plain birds with few prominent markings. They average 5 to 6 inches in length and have a comparatively short tail. They resemble some of the dull-plumaged warblers but are easily separable from them by their behavior and by the shape of their bill, which is longer, stouter, and slightly hooked. Vireos are less active than warblers, and, unlike most warblers, have a distinctive feeding behavior of alternately hopping from side to side along the length of a horizontal branch in search of insects and larvae. Their song, often monotonous, may continue for long periods before the bird is observed; vireos have a habit of sitting still and singing from a concealed location in the foliage of a tree or shrub. Six species are found in Pennsylvania, and all except the Philadelphia Vireo breed in the state. Their cup-shaped nests are suspended from a horizontal fork in a shrub or on an outer limb in a tree at a height of a few feet to over 50 feet above the ground. Nests are constructed of grass, plant fibers, mosses, lichens, rootlets, and pine needles (Blue-headed Vireo) woven together with spider webbing and insect silk. Vireos usually lay four white to creamy-colored eggs that are lightly spotted throughout, but more heavily spotted on the large end. Their nests are frequently parasitized by Brown-headed Cowbirds.

White-eyed Vireo *Vireo griseus*

General status: White-eyed Vireos breed from extreme southern Ontario, New York, and Massachusetts south to Florida and west to Texas and Kansas. They winter from Florida west across the Gulf coast to Texas. In Pennsylvania White-eyed Vireos are uncommon regular migrants across the southern half except at above about 1500 feet elevation, where they are rare. They are rare, irregular migrants across most of the northern half of the state. Migrants are observed primarily in spring. White-eyed Vireos breed primarily across the southern portion of the state. Their range is expanding northward.

Habitat: White-eyed Vireos inhabit thickets, brushy woodland understory, wet brushy areas, overgrown fields, and alder-lined streams.

Seasonal status and distribution

Spring: They usually arrive the second or third week of Apr at most southern sites but not until the fourth week of Apr or the first week of May in northern areas. Migration peaks during the first or second week of May. Stragglers away from nonbreeding sites may remain to the fourth week of May.

Breeding: White-eyed Vireos are uncommon and regular breeders in the Southwest (north to Armstrong, Butler, Indiana, and Lawrence cos.) and throughout the Piedmont. They are uncommon and local breeders in the Ridge and Valley and accidental and irregular to the north. Summer records have come from most counties except in the northern tier. White-eyed Vireos are most common on BBS routes in southwestern counties, where a population increase of 3.3% per year since 1966 has been noted. New records during the 1980s in western Cumberland Co. probably reflect survey effort by the BBA project more than a change from the historical rarity in this area. Most egg sets are found from 20 May to 17 Jun.

Fall: They become rather secretive and scarce after the breeding season is over, making migration difficult to discern. Where they do not breed, they have been observed from the second week of Aug to the second week of Oct. Most records are from the second week of Sep to the first week of Oct. A few stragglers have remained as late as the second week of Nov. A very late bird was discovered on 15 Dec 1990 on a CBC at Warren in Warren Co.[1]

History: The historical distribution was similar to that described above, except in the southwestern portion of the state, where they were historically not as common. Todd (1940) called them "unaccountably rare in the western part of the state." White-eyed Vireos were described variously from common in eastern Pennsylvania (Turnbull 1869) to abundant in Delaware Co. (Cassin 1862). They were notably absent from Baird's (1845) list of birds in Cumberland Co. However, Frey (1943) listed White-eyed Vireos as an "uncommon and extremely local summer resident breeding only in the eastern part [of Cumberland Co.]." The longstanding rarity in the western Piedmont (Adams, Cumberland, and Franklin cos.) is not paralleled by other southern or early succession species, so the cause is unclear.

[1] Grisez, T. 1991. County reports—October through December 1990: Warren County. PB 4:170.

[Bell's Vireo] *Vireo bellii*

General status: The Bell's Vireo breeds across the southwestern U.S. north through Texas to North Dakota, east to Ohio, and south along the Mississippi River to Louisiana. They winter primarily in Central America. Vagrants have been recorded east to southern Ontario, New York, New Jersey, and South Carolina (AOU 1998). There are two reports of this species in Pennsylvania, both from Erie Co.

Comments: A Bell's Vireo was found in Erie on 15 Jul 1957 (Stull et al. 1985). The observer stated that it was studied at a distance of 10–12 feet in his backyard from his bedroom window.[1] The second bird was discovered at PISP on 17 May 1959 (Stull et al. 1985). The observer consulted J. L. Baillie of the Royal Ontario Museum of Toronto, who felt certain that it was a Bell's Vireo. A Bell's Vireo had been reported at Point Pelee in Ontario, Canada, the weekend before.[1]

[1] J. Stull Cunningham, pers. comm.

Yellow-throated Vireo *Vireo flavifrons*

General status: Yellow-throated Vireos breed in the eastern half of the U.S. and winter from southern Florida to Central and South America. They are uncommon to rare migrants and regular breeders almost statewide, but they are unevenly distributed across the state. In the High Plateau, they are quite scarce or absent above 1800 feet elevation.

Habitat: During migration Yellow-throated Vireos prefer open woodlands and brushy woodland understory and edges. During the breeding season they prefer open mature deciduous woodlands, forest edges, and tall trees along riparian edges with little understory. They spend much of their time in the upper canopy.

Seasonal status and distribution

Spring: Migrants are more often observed during spring than fall. They consistently arrive during the fourth week of Apr in the southern portions of the state but may be delayed until the first or second week of May at northern sites. No apparent peak migration is evident, but most sightings away from their breeding grounds are reported during the second and third weeks of May. Migration usually ends by the fourth week of May. Migrants may linger to the first or second week of Jun.

Breeding: Nesting statewide, Yellow-throated Vireos are fairly common south of Pittsburgh and uncommon to fairly common in the Glaciated Northeast, the southern Ridge and Valley (north to Huntingdon Co.), the Northwest, and locally in the Piedmont. They are rare to accidental elsewhere, particularly in the High Plateau. The pattern of distribution does not follow any known physical features. Few nests have been collected, but the breeding season is typical of Neotropical migrant songbirds. Egg sets have been collected from 13 May to 7 Jul.

Fall: Yellow-throated Vireos are rarely reported during this season, but banding records suggest that migration is under way by the fourth week of Aug and has usually ended by the first week of Oct. Most birds are reported in Sep, with daily highs of no more than one or two birds. Stragglers have been recorded to the fourth week of Oct. One was banded on the late date of 19 Nov 1962 at Norristown, Bucks Co.[1]

History: Historical descriptions were variable through the years, with conflicting assessments in some localities. They differ markedly from recent patterns. Warren (1890) supposed (probably erroneously) that this vireo nested more commonly in the northern, mountainous parts of the state. Harlow (1918) more accurately considered Yellow-throated Vireos more frequent in the south. In contrast, and opposite recent patterns, Todd (1940) thought they were more common in the northwest than in the southwest portion of the state and considered them regular in Somerset Co., where the BBA project found them poorly represented.

[1] Scott, F.R., and D.A. Cutler. 1963. Middle Atlantic Coast region. AFN 17:21.

Blue-headed Vireo ***Vireo solitarius***

General status: Until recently, *V. s. solitarius*, of eastern North America, was considered to be conspecific with two western subspecies. The three subspecies were collectively called Solitary Vireo but are now recognized as distinct species. Blue-headed Vireos breed from British Columbia east across Canada and the north-central states to Newfoundland and south through the New England states and the Appalachian Mountains to Alabama and Georgia. They winter from Virginia south along the Atlantic coast to Florida and west across the Gulf states to Texas. In Pennsylvania Blue-headed Vireos are uncommon to fairly common regular migrants and breed primarily in northern forested regions.

Habitat: During migration Blue-headed Vireos are found in mature and second-growth woodlands of all types. During the breeding season they prefer mixed woodlands and conifer forests, especially where hemlocks predominate, but they will use deciduous woodlands at elevations of about 1800 feet and higher in northern counties.

Seasonal status and distribution

Spring: Blue-headed Vireos are usually our first vireos to arrive in spring. One was found as early as 7 Mar 1998 at Upper Wissahickon in Philadelphia Co.,[1] but the earliest migrants usually do not appear until after the fourth week of Mar. Their typical migration period is from the third and fourth weeks of Apr to the second week of May. Stragglers remain to the fourth week of May.

Breeding: Blue-headed Vireos are uncommon to fairly common breeders in extensive forests of the northern regions south in the Allegheny Mountains to the Maryland line; they also are uncommon to rare in Ridge and Valley forests and in the Blue Ridge section. Breeding was confirmed during the 1990s in Berks Co., where the BBA project failed to document the species (Uhrich 1997). They are several times more abundant in old-growth forests containing pine and hemlock than in typical second-growth woodlands, even more common than the Red-eyed Vireo in some old-growth sites (Haney and Schaadt 1996). Their populations have experienced one of the largest long-term increases (8.6% per year) on BBS routes of any forest bird in the state. Egg sets have been reported from 15 May through 2 Jul, but earlier nests are suspected.

Fall: Their usual migration period is from the second week of Sep to the first week of Nov. The greatest number of birds is recorded from the fourth week of Sep to the second week of Oct. Stragglers have been recorded to the first week of Jan.

History: Early accounts list Blue-headed Vireos farther south than currently. The species was listed as a breeder south to Lancaster Co. (Libhart 1869) and in Cumberland Co., where they were considered "rare in summer" (Baird 1845). In the few surveys available of historical old forests, Blue-headed Vireos were considered common, even as common as Red-eyed Vireos in Clinton and Potter cos. (Cope 1902) much as recent old-growth studies have reported. By the beginning of the twentieth century, when forest cover was at its lowest extent, Blue-headed Vireos apparently had retracted their range. They bred rarely or went unreported in Centre Co. (Burleigh 1931), in the northeast Ridge and Valley (Young 1896), and on Chestnut Ridge, Fayette Co. (Todd 1940). Sutton (1928b) found Blue-headed Vireos at Pymatuning before the flooding of the swamp in 1934. With the regrowth of Pennsylvania's forests, their numbers recovered and their range expanded south to many historical locations, although not yet to Lancaster Co.

[1] Floyd, T. 1998. Local notes: Philadelphia County. PB 12:25.

Warbling Vireo *Vireo gilvus*

General status: Warbling Vireos are nondescript and most easily observed in spring before the leaves have fully emerged. Much of their time is spent among the tops of the tallest trees. Warbling Vireos breed from Alaska south across Canada to southern Ontario and Quebec and south primarily across the northern two-thirds of the U.S. They winter from Mexico through Central America. Warbling Vireos are uncommon to fairly common regular migrants and regular breeders in Pennsylvania. The greatest concentrations of breeding birds are in the four corners of the state.

Habitat: Typical habitat is a row of tall deciduous trees, often sycamores, along streamsides or scattered tall shade trees. Warbling Vireos seem more closely associated with water in eastern counties than in the west. Also, they can be found in isolated tall trees in open areas such as along country roadways and in suburban areas of towns and cities.

Seasonal status and distribution

Spring: Their migration period is from the fourth week of Apr or the first week of May to the fourth week of May. Warbling Vireos' migration peaks during the second or third week of May.

Breeding: Warbling Vireos are unevenly distributed across the state, generally rare to fairly common. They are fairly common breeders throughout the Glaciated Northeast and Glaciated Northwest regions and south of Pittsburgh. They are uncommon and local in the Piedmont and rare elsewhere. Similar to the Yellow-throated Vireo, they are more regular and common in the corners of the state and are scattered and local in the central counties. Warbling Vireos, however, are more common to the north than are Yellow-throated Vireos. Most nests with eggs are found from 18 May to early Jul. The BBS reports a long-term increase of 4.1% per year statewide.

Fall: The beginning of migration is difficult to detect in fall, but movement is usually observed by the third week of

Aug. Migration begins early and does not last long. Most birds are reported from the fourth week of Aug to the second week of Sep, and by the third week of Sep most have left the state. Stragglers have remained to the second week of Oct. There are two late dates: one on 28 Nov 1988 at Rector in Westmoreland Co. (Leberman 1988) and one photographed on 28 Nov 1992 at Roderick Wildlife Reserve in Erie Co.[1]

[1] F.C. Haas and B.M. Haas, pers. comm.

Philadelphia Vireo ***Vireo philadelphicus***

General status: Philadelphia Vireos breed from British Columbia east to Newfoundland and south across northern North Dakota, Minnesota, Michigan, and New England south to New York. They winter from Mexico south to South America. The similarity of the Philadelphia Vireos' obscure plumage to that of other more common species probably allows many to pass through Pennsylvania undetected. They are regular but uncommon to rare spring and fall migrants statewide, more common west of the mountains than to the east. Philadelphia Vireos frequently associate with migrant flocks of warblers.

Habitat: They are found in a variety of habitats, but most are observed in deciduous woodlands and brushy woodland understories and edges.

Seasonal status and distribution

Spring: Most spring records are from west of the Allegheny Front. East of the Allegheny Front they are rare and irregular. Earliest arrivals may appear the last week of Apr. Their typical migration is from the second week of May to the first week of Jun. Most Philadelphia Vireos are recorded during the second and third weeks of May. Stragglers may remain to the second week of Jun.

Fall: East of the Allegheny Front, Philadelphia Vireos are more frequently seen in fall than in spring. The earliest migrants arrive the third week of Aug. Their typical migration period is from the first week of Sep to the first week of Oct. Peak migration occurs during the second and third weeks of Sep. Stragglers remain to the fourth week of Oct.

History: Before about 1935 Philadelphia Vireos were considered rare or unknown in the spring in eastern Pennsylvania (Poole, unpbl. ms.). Since then, Poole (unpbl. ms.) stated that observations had increased during spring and that there was always a possibility of misidentification, probably as Warbling Vireos before 19 May. Todd (1940) listed this species as "one of the rarer species" in western Pennsylvania. Even though Todd found them to be more difficult to find in fall, he believed they were just as numerous then as in the spring. Unusual in summer was a bird captured on 6 Jul 1895 at Waterford in Erie Co. (Todd 1940). In regard to this capture, Todd believed the bird may have been a nonbreeding bird, but he also mentioned that it may have actually been breeding at that time.

Comments: Philadelphia Vireos are frequently confused with Warbling Vireos. It is likely that most birds reported before the second week of May and later than the first week of Jun are Warbling Vireos. In fall Warbling Vireos may be strongly tinged with yellow and may be easily misidentified as Philadelphia Vireos, especially those found in Aug.

Red-eyed Vireo ***Vireo olivaceus***

General status: Long after nearly all species of birds have stopped singing, Red-eyed Vireos continue singing into the summer, even during the hottest afternoons. They breed across Canada south through the eastern U.S. and west to Washington, Colorado, and Texas. Red-eyed Vireos winter in South America. They are perhaps the most numerous and widespread woodland-nesting species in Pennsylvania. These vireos usually remain hidden among the leaves of trees well above the ground as they continue to sing, often without moving, for extended periods of time. They are the most common vireos during migration and the breeding season.

Habitat: Red-eyed Vireos are found in a variety of habitats during migration. During the breeding season they are found in almost all wooded habitats, including scattered woodlots, wooded riparian strips, connecting rows of suburban trees, and the most extensive forests.

Seasonal status and distribution

Spring: Common migrants, they have been recorded in the state as early as the second week of Apr but most typically appear the second week of May. Arrival times may vary somewhat from year to year. A very early Red-eyed Vireo was recorded on 20 Mar 1966 at HMS.[1] Some years they do not arrive in the northern portions of the state and in the mountains until the third week of May. Peak migration occurs during the second and third weeks of May, when daily highs reach 20 or more birds. The end of migration is difficult to discern, but fewer birds are found after the first week of Jun.

Breeding: Red-eyed Vireos are common breeding birds in the heavily forested sections of the High Plateau (see Map 12) and fairly common to common elsewhere statewide. Absent from treeless urban centers (e.g., downtown Philadelphia) and large areas of intensive agricultural activity (e.g., northern Lancaster Co.), they are ubiquitous woodland birds. Most nests with eggs have been found from 27 May to 28 Jun, with replacement (or second) clutches occurring sometimes into Aug, late for a vireo. A population increase of 1.4% per year is reported statewide on BBS routes.

Fall: Because Red-eyed Vireos are so common and widespread in the state, it is difficult to determine when fall migration begins. Local family groups begin wandering in late Jul. By late Aug the number of birds increases, indicating that migration has begun. The greatest number of birds is found between the first and the third weeks of Sep. Migration continues until the first week of Oct, with stragglers to the first week of Nov. There are two late dates: one sighting on 21 Nov 1969 at Indiana in Indiana Co.[2] and another exceptionally late bird seen on 21 Dec 1975 on the Reading CBC in Berks Co.[3]

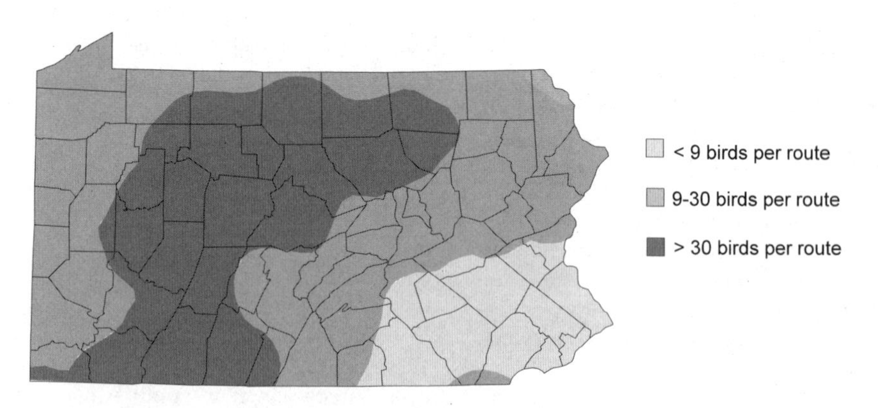

12. Red-eyed Vireo relative abundance, based on BBS routes, 1985–1994

History: Little difference can be found between historical accounts and the modern distribution. Even at the peak of agricultural activity in Pennsylvania's most intensively farmed county (Lancaster), Red-eyed Vireos were listed as "Frequent. Breeds in the county" (Libhart 1869). Harlow's (1913) assessment, "one of the most characteristic of our woodland birds," would stand today.

[1] Scott, F.R., and D.A. Cutler. 1966. Middle Atlantic Coast region. AFN 20:408.
[2] Hall, G.A. 1970. Appalachian region. AFN 24:47.
[3] Spence, M. 1976. Reading CBC. AB 30:294.

Family Corvidae: Jays and Crows

A large family of birds with worldwide distribution, the Corvidae, jays and crows, are conspicuous and often gregarious. They include the largest of passerine birds. They have a stout bill with stiff bristles covering the nostrils. The powerful bill is well suited for a variety of foods. Jays and crows often become pests to farmers by eating their cultivated crops of fruits and grains. Their voices are harsh and their manners are often aggressive. Jays differ from crows in being smaller and having a longer tail; some jays have a crest, and many have brightly colored plumage. Crows and ravens generally are more somber-colored, being entirely black; they walk rather than hop when on the ground.

Four species in this family are residents in Pennsylvania: Blue Jays, American Crows, Fish Crows, and Common Ravens. Blue Jays and American and Fish crows gather in huge flocks during migration and form spectacular flights along mountain ridges and (except Fish Crows) along the south shore of Lake Erie. Both species of crows are also known for their huge communal roosts. Common Ravens are more sedentary, usually stay at higher elevations in the mountains, and are less gregarious than others in this family. In addition to these native species, two, Black-billed Magpie and Eurasian Jackdaw, were introduced and briefly nested in the state. Those two and the Gray Jay and Clark's Nutcracker are listed as hypothetical.

[Gray Jay] *Perisoreus canadensis*

General status: The Gray Jay is a resident across the boreal forests of Canada south in the west through the Rocky Mountains to northern Arizona and New Mexico and south in the East to northern New York. They have been reported south of their breeding range to Pennsylvania, central New York, Connecticut, and Massachusetts (AOU 1998). No specimens exist, and all sightings have been reported with only brief notes.

Comments: The validity of the reports of Gray Jays in Pennsylvania is questionable. Todd (1940) listed this species in *Birds of Western Pennsylvania* on the basis of a letter to O. E. Jennings briefly describing a Gray Jay at Clarington in Forest Co. in Feb 1923. According to Warren (1890) and Stone (1894), one was collected in Lancaster Co. in Feb 1889. Turnbull (1869) listed Gray Jays as rare stragglers to the northern counties. A bird was reported at Stony Creek Mills near Reading from 21 May to 12 Jul in 1960 (Poole, unpbl. ms.). During the same year, another bird was reported at Grove City in Mercer Co. on 30 Dec,[1] and one was reported on 29 Dec 1961 near Somerset in Somerset Co.[2]

[1] Hall, G.A. 1961. Appalachian region. AFN 15:330.
[2] Grom, J. 1962. Gray Jay. Redstart 29:58.

Blue Jay *Cyanocitta cristata*

General status: Blue Jays breed across southern Canada from British Columbia east to Newfoundland and are resident south through the U.S. east of the Rocky Mountains. Northern populations are partly and variably migratory to the southern portions of their breeding range (AOU 1998). In Pennsylvania, Blue Jays are evident throughout most of the year nearly everywhere trees are found, but they become inconspicuous during the breeding season. They are abundant regular migrants and winter residents statewide. Apparently, a complex of populations of Blue Jay occurs in Pennsylvania. Some populations are migratory, passing through the state to more-northern areas in spring and south to winter outside the state in fall. Other populations are essentially permanent residents.

Habitat: Blue Jays are found in a variety of habitats, from forests and woodlands to parks and yards in residential areas of towns and cities, especially where evergreens or dense hedgerows are present.

Seasonal status and distribution

Spring: Northward movement begins the second or third week of Apr, accompanying a steady flow of warm southerly air. Peak migration occurs during the first and second weeks of May. Daily highs include loose migrating flocks containing hundreds of birds. Spectacular flights are recorded along the Lake Erie Shore. On a typical day during peak migration over 5000 jays can be seen flying overhead.[1] Northward movement has ended by the second week of Jun.

Breeding: Blue Jays are fairly common to common breeding residents. They were among the top 10 most widely distributed species in terms of the number of BBA blocks, in part because they are so well known and conspicuous. Blue Jays are predominantly forest birds; they are more commonly observed in fragmented and mixed open habitats but will nest in isolated trees in suburban yards. They were most common on BBS routes in the state's eastern third and northwestern corner; least common in the Laurel Highlands (see Map 13). Usually four or five buff-colored eggs spotted with brown or gray are laid in a nest constructed of twigs, bark, and leaves. The nest is usually constructed less than 25 feet above the ground in the branches of a shrub or tree. The long-term statewide BBS population trend is slightly negative (–1.4% per year). Most

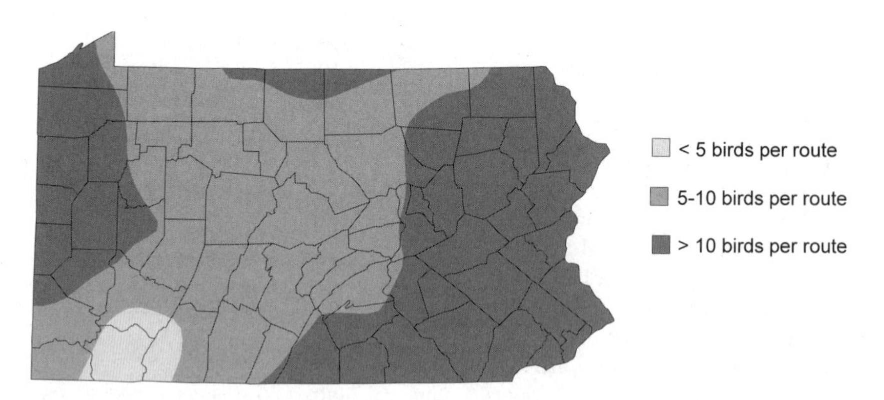

13. Blue Jay relative abundance, based on BBS routes, 1985–1994

egg sets have been found from 4 May to 16 Jun, but some birds start nesting in late Apr when northbound migrants are still passing through the state. Early nests are usually in a conifer.

Fall: Migration occurs mainly along the mountain ridges in fall, especially east of the Allegheny Front. Their typical migration period is from the third week of Aug to the second week of Nov; the greatest numbers are observed between the third week of Sep and the second week of Oct. Even though daily totals at hawk watches reach the thousands, they never reach the numbers recorded along Lake Erie in spring.

Winter: Blue Jays winter in every county and are most numerous around human habitation, where they are familiar visitors to bird feeding stations. Jays prefer areas with dense cover such as is provided by evergreen hedgerows or stands of conifers.

History: In the 1930s and early 1940s Blue Jays were uncommon in western Pennsylvania and were local in distribution.[2]

Comments: Banding records show evidence that young birds migrate but older birds do not.[2] Before it became illegal to kill Blue Jays, shooting them for sport during the spring flights along the Lake Erie Shore and fall flights on the ridges was still popular well into the mid-1960s. On 6–7 May 1967, at least 500 were shot near Girard in Erie Co.[3]

[1] G.M. McWilliams, pers. obs.
[2] G.A. Hall, pers. comm.
[3] Hall, G.A. 1967. Appalachian region. AFN 21: 505.

[Clark's Nutcracker] *Nucifraga columbiana*

General status: Nutcrackers are resident in the Rocky Mountains from British Columbia south to Arizona and New Mexico. They have been reported east of their breeding range to Minnesota, Michigan, Iowa, Wisconsin, Ontario, Pennsylvania, Missouri, Arkansas, Louisiana, and Texas (AOU 1998). One hypothetical record exists in Pennsylvania.

Comments: One was reported at Tuscarora Mountain in Fulton Co. on 4 Nov 1979 (Leberman 1988).

[Black-billed Magpie] *Pica pica*

General status: Black-billed Magpies are permanent residents from Alaska south to New Mexico and east to Kansas and Nebraska. This species occasionally escapes from captivity and on one occasion in Pennsylvania became established for at least 15 years.

Comments: Before 1955, several magpies escaped from the Pittsburgh Zoo and a small nesting colony resulted (Poole, unpbl. ms.). At least three nests were found.[1] The Black-billed Magpie colony survived in the wild in the Pittsburgh area until at least 1969 (Leberman 1988). Others found in the state during this period were probably birds wandering from the Pittsburgh colony. Poole (unpbl. ms.) mentioned magpies near Maxatawny, Berks Co. in 1953 and in Fayette Co. on 15 Nov 1959. In 1955 one was reported on 23 Jul and from 10 to 17 Sep at PISP (Stull et al. 1985).

[1] Hall, G.A. 1959. Appalachian region. AFN 13:370.

[Eurasian Jackdaw] *Corvus monedula*

General status: The Eurasian Jackdaw has a widespread breeding range across Europe and Asia. It is listed as hypothetical in Pennsylvania because the birds reported apparently did not arrive in North American unassisted.

Comments: On 23 May 1985 one was identified at the U.S. Penitentiary at Lewisburg in Union Co. (Schweinsberg 1988). Soon thereafter, it became apparent that there was a pair. There were signs of nesting that spring, but it was not until 5 Jun 1987 that a nest and a fledgling were found. The fledgling died two days later and was placed in the ANSP. They attempted to nest again in 1988 and 1990[1] but were unsuccessful; then, after six years of failed nesting two fledged young were found on 10 Jun 1991.[2] These birds may have been vagrants from a presumed ship-assisted flock discovered in Nov 1984 on the north shore of the Gulf of St. Lawrence.[3]

[1] Brauning D.W. 1992. Eurasian Jackdaw *(Corvus monedula)*. BBA (Brauning 1992):234–235.
[2] Schweinsberg, A. 1991. County reports—April through June 1991: Union County. PB 5:102.
[3] Yank, R., and Y. Aubry. 1985. Quebec region. AB 39:149.

American Crow ***Corvus brachyrhynchos***

General status: American Crows breed from Alaska south across Canada and throughout the U.S. but are local in the southwestern U.S. They winter across their breeding range in the U.S. One of the most common and widespread birds in Pennsylvania, these intelligent birds have been persecuted relentlessly as agricultural pests, yet they have managed to survive and have successfully adapted to life among humans. Like that of the Blue Jay, the American Crow's distribution patterns are fairly complex. Some populations in the state are migratory, and others are permanent residents. They are abundant migrants and regular winter residents statewide. American Crows are observed in greatest numbers at communal winter roosts.

Habitat: They frequent a variety of habitats at all seasons, from large tracts of forest to small woodlots, as well as parks and suburban areas of towns and cities wherever mature trees are available for perches and to roost.

Seasonal status and distribution

Spring: Migration may begin as early as the second week of Feb while snow covers the ground and ice is still on the lakes. Their typical migration period is from the third or fourth week of Feb to the first week of Apr. Peak migration ranges from the fourth week of Feb to the second week of Mar. Larger numbers of birds—thousands in a single day—are observed migrating along the shores of Lake Erie than in any other area in the state.

Breeding: American Crows are abundant and conspicuous, the second most widespread species in the BBA project and the seventh most abundant species on BBS routes. State BBS routes report crows in lower numbers in several irregular areas of central and southwestern counties (see Map 14), which may reflect observer biases more than an ecological pattern. The BBS shows a modest but consistent increase of 1.2% per year since 1966. American Crows are becoming more common in cities and towns. They nest early; some are on eggs as early as the last week in Mar, but most clutches date from 7 Apr through 27 May. They are solitary nesters. The nest is a large coarse structure, with an outside diameter of about 2 feet. It is usually well concealed in the crotch, or near the trunk, of a coniferous or deciduous tree as high as 80 feet above the ground. From four to six greenish or bluish eggs spotted with brown and gray are laid.

Fall: Migrant American Crows are usually detected along the mountain ridges during the third or fourth week of Sep. More migrants are observed at hawk watches along the mountain ridges than at any other area in the state during this season. The greatest number of birds is observed during Oct, with migrants diminishing through Nov. Daily highs along the mountain ridges reach

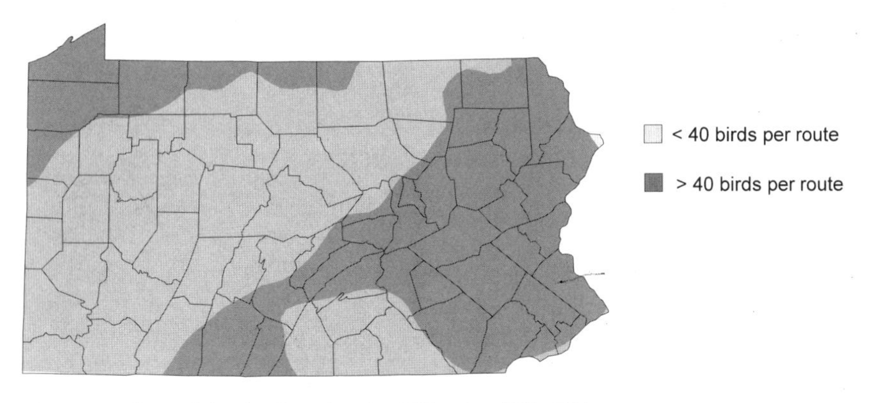

14. American Crow relative abundance, based on BBS routes, 1985–1994

the thousands during peak migration, especially east of the Allegheny Front. As winter approaches American Crows gather in communal roosts, and by mid-Dec congregations may number from 50,000 to over 100,000 birds at some sites in the Piedmont. A roost near Carlisle in Cumberland Co. was thought to have 200,000–300,000 birds in early Jan 1982.[1] A high of 300,000 was counted on the Lititz CBC in Lancaster Co. on 2 Jan 1984 (Morrin et al. 1991).

Winter: Populations may fluctuate from year to year depending on the severity of the winter. Many crows retreat to lowland valleys from areas of heavy snow cover, especially from the snowbelt areas of the Glaciated Northwest, the northern counties, and the mountains. Communal roosts reach maximum size by mid-Jan, and as spring approaches the roosts begin to decrease in size.

History: Despite direct persecution as an agricultural pest and by sportsmen as a target, American Crows have always been widespread and common. Many of the largest historical roosts in southeastern counties were dispersed because of conflicts with human residents.

Comments: Crows are still hunted as vermin, but hunting is now regulated.

[1] Hall, G.A. 1982. Appalachian region. AB 36:294.

Fish Crow ***Corvus ossifragus***

General status: Fish Crows are essentially birds of the Atlantic coast, but they have been expanding inland along major rivers. They are resident locally from New York and Massachusetts and south along the Atlantic and Gulf coasts. Fish Crows also breed inland along major river systems such as the Mississippi and Ohio rivers as far north as West Virginia. They have followed the Delaware, Susquehanna, and Potomac river valleys into Pennsylvania and are now breeding regularly as far west as Centre Co. Fish Crows are regular breeding residents in the Coastal Plain, Piedmont, and Ridge and Valley. A few birds have recently been reported in the Glaciated Northeast, but breeding is yet to be confirmed there. Some may winter within their breeding range, especially along the lower Delaware and Susquehanna rivers, but most retreat out of the state in fall. Central Pennsylvania is the most northwesterly portion of the species' normal breeding range. Fish Crows associate with American Crows most of the year, but less so during the breeding season.

Habitat: They prefer riparian habitats, especially along large rivers and their tributaries. They are also found in agricultural areas, parks, and residential areas of towns around mature trees.

Seasonal status and distribution

Spring: Where they have not wintered, Fish Crows may return to their northernmost breeding sites as early as the first or second week of Feb. Range expansion usually occurs in the spring as they continue to push farther north and west. The greatest number of birds is usually recorded along the lower Delaware and Susquehanna rivers. Concentrations have been noted as far north as Union Co., such as the flock of about 100 seen there on 19 Mar 1987 (Schweinsberg 1988).

Breeding: Fish Crows are uncommon to fairly common residents through most of the Piedmont and north along the major river corridors through the Ridge and Valley. They are also fairly common along the West Branch Susquehanna River northwest to Clinton Co. and into Centre Co. along Bald Eagle Creek. Fish Crows extend northeast along the North Branch to Scranton and even into the Glaciated Northeast in Wyoming Co. Since the 1980s they have expanded into scattered sites in Monroe and Pike cos. (P. B. Street and Wiltraut 1996) and have been reported in Clearfield Co.[1] They are most frequently reported in Cumberland, Franklin, Lancaster, and Lebanon cos., but even there American Crows vastly outnumber Fish Crows. Since 1980, the BBS indicates a substantial population increase (15% per year), but this species began expanding its range long before then. Few egg sets from Pennsylvania are in collections, but Harlow's (1918) range was 19 Apr to 24 May. Nests and eggs are similar to those of American Crows, but Fish Crows nest in small colonies, often in riparian trees.

Fall: Because Fish Crows begin to wander and mix with American Crows after the nesting season, their movements are difficult to follow. In late fall they frequently join roosts with American Crows. At one northern site near State College on 30 Nov 1986 flocks of just Fish Crows have numbered up to at least 100 birds.[2]

Winter: Fish Crows winter within their breeding range. The number of birds recorded may vary from year to year depending on the amount of available food and snow cover. Their late fall and early spring movements make it difficult to discern wintering individuals at any given site. Near winter roosts they may be seen in mixed groups (with American Crows) flying to and from feeding areas, mornings and evenings. A rather large winter roost containing 500 Fish Crows was located at Tinicum on 11 Feb 1976,[3] and hundreds gather daily along the Delaware River to feed at the Tullytown (Bucks Co.) landfill.[4]

History: Early historical records are ambiguous. Audubon (fide Warren 1890) referred to Fish Crows migrating north along the Delaware River "almost to its source," but returning south in the winter. Warren's (1890) description is of a very southern distribution, Lancaster Co. Fish Crows evidently were expanding their range in 1900 along the Delaware River into Philadelphia (Harlow 1918) and north into Cumberland Co. (Harlow 1913). Although Todd (1940) discredited the record, it is likely that J. F. Street (1921) was accurate in his listing of Fish Crows in Adams Co. An extralimital record in 1918 in Centre Co.[5] foreshadowed the current distribution. Poole (1947) listed a set of eggs taken in Berks Co. in 1933. By the mid-twentieth century Kunkle (1951) listed Fish

Crows as "fairly common" near Lewisburg, Union Co. By 1960 Fish Crows had attained a distribution not unlike the current pattern.

Comments: Fish Crows' status remains an enigma, overlooked by some observers, mistaken for young American Crows, and discernible only by voice. A tradition of skepticism follows reports of Fish Crows. Poole (unpbl. ms.) stated that Harlow and Burleigh questioned R. F. Miller's 1918 report of Fish Crows in the State College area, but that Miller's observations were later verified.[5] M. Wood discounted Fish Crow records in the State College area 50 years later, but the BBA project documented the species there in the 1980s. The range of Fish Crows has probably been accurately mapped in the state, but silent scouts are likely to move undetected with northbound flocks of American Crows outside their expected range.

[1] Gross, D.A. 1992. Fish Crow *(Corvus ossifragus)*. BBA (Brauning 1992):232–233.
[2] Hall, G.A. 1987. Appalachian region. AB 41:89.
[3] Buckley, P.A., R.O. Paxton, and D.A. Cutler. 1976. Hudson-Delaware region. AB 30:700.
[4] P.B. Street, pers. comm.
[5] Miller, R.F. 1941. Fish Crow in Center County, Pennsylvania. Auk 58:263.

Common Raven *Corvus corax*

General status: Common Ravens are resident in North America from the high Arctic regions of Alaska, Canada, and Greenland south primarily through and west of the Rocky Mountains. East of the Great Plains they are found from Minnesota north and east across Canada and south through the New England states and in the Appalachian Mountains to Georgia and Alabama. Pennsylvania probably supported a large population of Common Ravens when its landscape was dominated by wilderness. Ravens declined as the forests were cleared, but with forest regrowth in recent years populations have increased and ravens have been rapidly spreading south and east. They are no longer just a bird of the wilderness as they were previously. Adaptable and intelligent, they have exploited the human environment in numerous ways, including nesting on human-made structures (e.g., the stadium at Penn State University, State College) and feeding in dumpsters and landfills. Today they are uncommon to locally fairly common permanent residents, primarily in the Ridge and Valley and High Plateau. Considerable dispersal is noted along mountain ridges, especially in fall.

Habitat: Common Ravens prefer sparsely populated, forested areas at high elevations most of the year. They nest in quarries and mining high-walls during the breeding season and along mountain ridges during migration. In winter they are frequently found in open valleys and agricultural areas near extensive forests, and they even come into towns when winter snows bury natural foods.

Seasonal status and distribution

Spring: Movement is occasionally detected along the mountain ridges, primarily along the Kittatinny Ridge in Mar and Apr.

Breeding: Ravens are widely distributed in the High Plateau of Pennsylvania, including the Allegheny Mountains south through Somerset and Fayette cos. They are well established in the Blue Ridge section and the western and northern portions of the Ridge and Valley, but they occur sparingly east to Blue Mountain and Kittatinny Ridge. The population has expanded outward from the wildest portions of the state a century ago into more developed areas recently. The extensive raven distribution reported by the BBA project represents a significant discovery. The ex-

Common Raven at Canadensis, Monroe County, April 1993. (Photo: Rick Wiltraut)

pansion continued in the 1990s. The first confirmed breeding record this century in the Poconos was obtained in 1993,[1] and a nesting attempt in York Co. on a transmission tower[2] demonstrated the species' adaptability to expand away from the mountains. Ravens may be seen patrolling roads and farmland for carrion within their large territories. A large deer herd and extensive road network combine to provide a steady supply of carrion for this scavenger. Common Ravens are solitary nesters in remote areas. Nesting begins in Feb. Most eggs are reported from 3 Mar through 6 Apr. The nest, a bulky stick structure lined with deer hair, is sometimes placed high in a tall tree (often a white pine) or more typically on a cliff, rock outcrop, tower, or mine high-wall. Four or five greenish eggs spotted or blotched with brown or olive are laid.

Fall: Usually a few ravens are seen with migrating hawks along the mountain ridges mainly from the third week of Sep to the second week of Dec. Most birds are recorded in Oct and Nov.

Winter: When deep snow covers the ground in the mountains, ravens retreat to the valleys and roadways to search for carrion. Two records are far outside their breeding range: at Tyler Arboretum in Delaware Co. on 20–29 Jan 1974[3] and at MCWMA in Lancaster Co. on 12 Jan 1988.[4] They may become locally common at landfills or even forage at dumpsters in towns during severe winters when food is scarce. Over 50 ravens were observed circling over a landfill near Denton Hill State Park in Potter Co. in late Feb 1985.[5]

History: The range of ravens contracted during the 1800s and early 1900s as forests were cleared and unrestricted shooting took its toll. Warren's (1890) brief summary described a decidedly north-central distribution. Harlow (1918) stated that it "is nearing extinction in Pennsylvania" and then listed several central Ridge and Valley counties where ravens were "making a last stand." Protection in 1923[6] and regrowth of the forest paved the way for the raven's recovery. By the mid-1920s, they were "only in the wildest mountain gorges, chiefly in the central mountains" (Sutton 1928a); but 40 years later Poole (1964) described a broader range, including the Blue Ridge section.

[1] Wiltraut, R. 1993. First confirmed nesting of Common Raven in the Pocono Mountains; Monroe County. PB 7:53.
[2] Lippy, K. 1996. Raven nest discovered in York County. PB 10:45–46.
[3] Scott, F.R., and D.A. Cutler. 1974. Middle Atlantic Coast region. AB 28:624.
[4] E. Witmer, pers. comm.
[5] G.M. McWilliams, pers. obs.
[6] J. Kosack, pers. comm.

Family Alaudidae: Larks

Members of the family Alaudidae are small ground-dwelling birds of open fields that walk rather than hop. Larks are primarily an Old World family; only one species, the Horned Lark, is native to North America. They have a short, slender, pointed bill and a single hornlike tuft of feathers on each side of the crown, hence the name Horned Lark. Horned Larks breed and winter in Pennsylvania in large open fields. Sky Larks *(Alauda arvensis)*, released in the state of Delaware in 1852, were reported irregularly in Chester and Delaware cos. for two years, but they have not been seen since (Warren 1890).

Horned Lark *Eremophila alpestris*

General status: Horned Larks breed from the high Arctic regions of Alaska and Canada south throughout Canada and most of the U.S. They winter across southern Canada and throughout their breeding range in the U.S. A prairie species that was probably not very common in the state before forests were cleared for farming, Horned Larks moved into Pennsylvania from the west in the late 1800's and quickly established themselves in agricultural areas and more recently in reclaimed strip mines. They are common to abundant regular migrants and breed over most of the state. Larks are regular winter residents statewide, except during winters of heavy snow cover, when they are absent from the High Plateau and in the snowbelt area of the Glaciated Northwest.

Habitat: During migration Horned Larks are found in plowed fields, grain stubble fields, reclaimed strip mines, and airports. During the breeding season they inhabit extensive grassy fields, untrimmed lawns, golf courses, airports, open agricultural areas, reclaimed strip mines, and even large gravel parking lots. In winter Horned Larks are frequently found in agricultural areas spread with fresh manure.

Seasonal status and distribution

Spring: Horned Larks are fairly common to common migrants in spring. Northward movement usually begins with the first warm southerly wind, sometimes as early as the fourth week of Jan. Their typical migration period is from the second or third week of Feb to the second week of Apr, but the timing of peak migration varies from year to year. High numbers of birds may be recorded in several periods during spring. Arrival may be delayed until the first week of Mar at higher elevations and in the snowbelt after severe winters with deep snow cover. Stragglers may continue to pass northward until the last week of May.

Breeding: Horned Larks are local and generally uncommon during the breeding season statewide. They are most common on reclaimed strip mines in the western half of the state, where sparse vegetation provides ideal conditions. Larks are less frequently encountered, but regular, in agricultural fields in the eastern half of the state and sometimes occupy patches of grassland within extensively forested settings. The male Horned Lark's courtship displays are quite elaborate; he flies several hundred feet in the air and then suddenly plummets to the earth to land near his mate. Horned Larks are among the earliest songbirds to nest in the spring (Todd 1940). Eggs have been found in early Mar. Early nesting, an advantage over many grassland species, allows larks to raise a brood before fields are cultivated; yet state BBS data indicate a significant Pennsylvania decline of 8.9% per year. Nests in Jun and Jul probably involve second or replacement clutches. The nest is a bulky loose structure placed on the ground or in grass in an exposed position. Usually four grayish eggs blotched or spotted with brown are laid.

Fall: Fall migration is less defined than spring migration, with southbound birds detected beginning as early as the second or third week of Aug. Migration continues through Nov into Dec and early Jan. During some winters it is difficult to distinguish migrant birds from wintering birds.

Winter: East of the Allegheny Front most birds winter in wide-open agricultural areas of the Piedmont and the Ridge and Valley, where numbers of birds may vary from year to year. West of the Allegheny Front most wintering birds are found in reclaimed strip mines of the Southwest. In the Glaciated Northeast, High Plateau, and Glaciated Northwest, Horned Larks are rather local, and the number of wintering individuals depends on the amount of snow cover.

History: Before 1888 the Horned Lark was unknown as a breeding bird in Pennsylvania. Breeding records occurred widely across the state within a few years, such as Lycoming Co. in 1892 (Stone 1894) and Berks and Fulton cos. by the early 1900s (Harlow 1913).

Comments: Many subspecies have been described in North America. Because their plumages are similar and vary widely, these subspecies are not safely separable in the field.

Family Hirundinidae: Swallows

Known for their graceful and acrobatic flight, swallows are found in both the Southern and Northern hemispheres. They superficially resemble swifts, but their wrists are set farther away from their body and their flight is more fluid. Swallows have long pointed wings, more or less forked tail, short neck and legs, flat bill broadening at the base, and large mouth for catching insects on the wing. During migration they congregate in huge flocks and may be seen spiraling above a marsh or lined shoulder to shoulder on utility wires. Several species nest in colonies of hundreds of pairs. Seven species have been recorded in Pennsylvania and all breed here except the Violet-green Swallow, which is a vagrant from western North America. An assortment of habitats and structures (for nest sites) is used by the six breeding species. Four or five unmarked white eggs form a normal clutch of most species.

Purple Martin *Progne subis*

General status: Purple Martin populations are known to fluctuate primarily in response to their susceptibility to periods of prolonged wet weather. They breed locally from British Columbia south through the West, where populations are widely scattered, and east across southern Canada to Nova Scotia and south throughout the eastern U.S. In Pennsylvania they are fairly common to abundant regular migrants and breed statewide where nest boxes are placed in suitable habitat.

Habitat: During migration Purple Martins are observed overhead, along mountain ridges, at communal roosts in trees near water, or on islands of vegetation in lakes or marshes. During the breeding season they are found primarily around martin houses placed in open areas near water or, when feeding, over ponds, lakes, and marshes.

Seasonal status and distribution

Spring: Migrants have been observed as early as the third week of Mar. Their usual migration time is from the first or second week of Apr to the fourth week of May. The greatest number of migrants is usually recorded between the fourth week of Apr and the second week of May. Stragglers remain to the first week of Jun.

Breeding: In Pennsylvania and across the eastern U.S. martins nest only in man-made structures such as gourds or specially designed nest boxes. Nesting colonies have been found in every Pennsylvania county but are concentrated in the agricultural and rural areas of the western tier and the Piedmont. The fewest colonies reported during the BBA project were in mountains of the High Plateau, including the Allegheny Mountains. Although martins return from their South American winter residence in Apr, eggs are not laid until late May. Martins face numerous obstacles to survival; they cannot obtain adequate food (flying insects) during early frosts or extended periods of rain, such as occurred after Hurricane Abby in 1968 and Hurricane Agnes in 1972 (Robbins et al. 1986). They also face competition for their nest cavities from Tree Swallows, House Finches, and House Sparrows. Martins are slow to colonize vacant or new nest sites.

Fall: Flocks of martins begin gathering the fourth week of Jul; numbers gradually increase through Aug and peak between the third week of Aug and the first week of Sep. Larger concentrations are recorded at PISP than at any other single site in the state. During peak migration, communal roosts at PISP frequently consist of more than 20,000 birds.[1] Most birds leave the state by the third week of Sep, but stragglers remain to the last week of Sep.

History: Martin populations have been assisted by humans since Native Americans placed hollowed gourds around their villages to attract them (Wilson and Bonaparte 1870). Martins were probably not as common when the state was covered with forest but increased when the forests were cleared for farming.[2] Todd (1940) wrote, "There was a time when almost every town of importance, and many smaller ones as well, could boast of one or more colonies of purple martins." He went on to state that most of those colonies were no longer active. Many colonies had been abandoned by 1960, in part as a result of poor upkeep of martin houses (Uhrich 1997).

Comments: Erecting martin houses to attract these colonial birds has become very popular. Many sales were made after martin house distributors erroneously stated that Purple Martins feast on mosquitoes. The Purple Martin Conservation Association, based at Edinboro University of Pennsylvania, is an international organization dedicated to research and conservation of Purple Martins.

[1] Hill, J.R., III, pers. comm.

[2] Master, T.L., and J.R. Hill III. 1992. Purple Martin *(Progne subis)*. BBA (Brauning 1992):216–217.

Tree Swallow *Tachycineta bicolor*

General status: Tree Swallows breed from Alaska south across Canada and the northern two-thirds of the U.S. They winter from southern California southeast along the southern border of the U.S. and north along the Atlantic coast to New Jersey. Hardiest of the swallows, the Tree Swallow arrives earlier and remains later than any other species. They are the only species of swallow in Pennsylvania that currently nests in tree cavities. Tree Swallows readily occupy bird houses for nesting and frequently compete for nest sites with the Eastern Bluebird and the introduced European Starling and House Sparrow. They are common to abundant regular migrants, and they breed statewide.

Habitat: During migration Tree Swallows are observed most commonly over ponds and marshes and along rivers, mountain ridges, and Lake Erie. During the breeding season they occupy any open areas including upland agricultural areas, forest openings, and suburbs where there are at least a few scattered trees with cavities or artificial nesting boxes. Tree Swallows may be locally abundant in wetlands that have standing dead trees, such as beaver ponds.

Seasonal status and distribution

Spring: Tree Swallows have been noted as early as 17 Feb in 1972,[1] at Gifford Pinchot State Park and 23 Feb 1992 at JHNWR.[2] The earliest arrivals usually do not appear before the second week of Mar in the southern counties. It is usually the third or fourth week of Mar or later before they arrive in the northern counties and at higher elevations. Most migrants are recorded between the second week of Apr and the second week of May. Migration usually ends by the fourth week of May.

Breeding: Tree Swallows are uncommon to locally common summer residents. They are most common in the northern counties, particularly around the numerous wetlands of the Glaciated Northeast, and they are least common in the state's southern third, particularly west of the Ridge and Valley (see Map 15). Urban areas are still avoided except at Tinicum, but they readily nest around rural buildings. Populations have expanded dramatically in the past 30 years, at least in part because of the availability of nest boxes. The BBS shows a long-term increase of 3.8% per year statewide. Most eggs are reported between 8 May and 21 Jun, with second broods considered rare.[3]

Fall: Tree Swallows are more common in fall, especially at staging sites, than at any other time of year. Migration begins the last week of Jul or the first week of Aug and ends in late Oct. The greatest number of birds is seen from the third week of Aug to the second week of Sep, when thousands of staging birds can been seen lined up on utility lines or in dense flocks in trees along or near

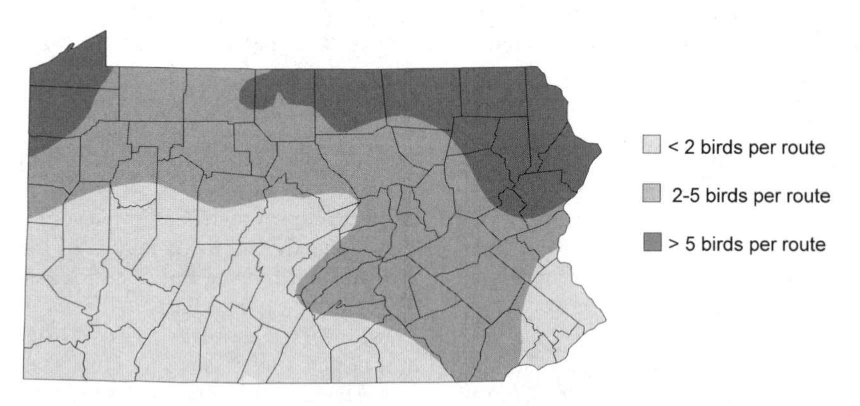

15. Tree Swallow relative abundance, based on BBS routes, 1985–1994

bodies of water. Stragglers occasionally remain to the first week of Dec. A very late bird was recorded on the Lititz CBC in Lancaster Co. on 29 Dec 1974.[4]

History: Tree Swallows were not widespread as breeding birds before the early 1900s. They were conspicuously absent from many summer bird lists (e.g., Baird 1845; Stone 1891), although Warren (1890) stated that they nested "more or less regularly and rather sparingly in nearly every county in the state." In stark contrast to Warren, Harlow (1913) considered Tree Swallows to be almost entirely limited to the "Canadian zone" fauna, found in the northern mountains. Competition with House Sparrows was mentioned by Warren (1890) as a limiting factor and may have affected the swallow's population at the end of the nineteenth century, when sparrow populations were rapidly expanding. By the early 1960s, Poole (1964) described the nesting status as "locally throughout the state, most abundantly about the lakes and beaver dams in the glaciated portions of northern Pennsylvania."

Comments: Tree Swallows generally nest at least 30 feet from each other, but sometimes within 3–10 feet. Competition for nest sites among other Tree Swallows and with Eastern Bluebirds may be reduced by placing boxes in pairs, approximately 10–20 feet apart.

[1] Scott, F.R., and D.A. Cutler. 1972. Middle Atlantic Coast region. AB 26:587.

[2] E. Fingerhood, pers. comm.

[3] Robertson, R.J., B.J. Stutchbury, and R.R. Cohen. 1992. Tree Swallow *(Tachycineta bicolor)*. BNA (A. Poole and Gill 1992), no. 11.

[4] Regennas, C. 1975. Lititz CBC. AB 29:283.

Violet-green Swallow *Tachycineta thalassina*

General status: Violet-green Swallows breed across western North America from Alaska south through Mexico and east to the eastern slopes of the Rocky Mountains. They winter from coastal and southern California south through Mexico and Central America. Vagrants have been reported as far east as Ohio, Nova Scotia, New Hampshire, and New Jersey (AOU 1998) and Pennsylvania. There is one accepted record in Pennsylvania.

Seasonal status and distribution

Fall: A Violet-green Swallow was discovered by J. Horn at Penn Forest Reservoir in Carbon Co. on 25 Aug 1995 and was well described and sketched. It was observed flying with a mixed flock of swallows.[1]

[1] Morris, B.L. 1995. Local notes: Carbon County. PB 9:146.

Northern Rough-winged Swallow
Stelgidopteryx serripennis

General status: Rough-winged swallows breed from Alaska to British Columbia east across southern Canada and south throughout the U.S. They winter in southern California, in Arizona, from southern Texas along the coast to Alabama, and in southern Florida. Most winter in Mexico and Central America. This nondescript brown swallow is probably underreported because of its resemblance to Bank and immature Tree swallows. They are quite adaptable, using a variety of locations for nesting. Rough-winged swallows are fairly common regular migrants and are local breeders statewide.

Habitat: During migration Northern Rough-winged Swallows are observed over water and along mountain ridges. During the breeding season they are found along creeks, rivers, ponds, and lakes that are near exposed steep banks containing burrows made by Belted Kingfishers and Bank Swallows. Rough-winged swallows also nest in drainage holes of bridges and dams.

Seasonal status and distribution

Spring: Most migrants are observed during this season. The earliest spring arrivals occur the fourth week of Mar. Their typical migration time is from the second or third week of Apr to the fourth week of May. The greatest number of birds is usually recorded from the last week of Apr to the second week of May. An exceptionally large concentration of more than 350 rough-winged swallows was observed at Donegal in Westmoreland Co. on 30 Apr 1969 (Leberman 1976). Lingering birds, probably nonbreeding individuals, may remain to the first or second week of Jun.

Breeding: Northern Rough-winged Swallows are uncommon to locally fairly common over most of the state, except in the north-central portion of the High Plateau, where they are rare. They are most often found along streams that break the tree canopy and in open settings; they also occur along rivers even in urban centers. In extensively forested sections of the High Plateau they are restricted to larger stream and river courses. Rough-winged swallows do not excavate their own cavities. Most egg sets have been found from 19 May to 20 Jun. As do other swallow species, they disperse from nesting areas shortly after young are independent in mid-Jul, gathering in flocks before the southward migration. The BBS shows no discernible change in breeding populations.

Fall: Migration is inconspicuous in fall, after the breeding season, with fewer birds seen than in spring. Southbound movement is detected in early Jul. Birds trickle southward in variable numbers, often mixed with other species of swallows through Aug and Sep, with stragglers to the third week of Oct. A late individual was at Harrisburg in Cumberland Co. on 29 Nov 1990.[1]

History: Rough-winged swallows were once largely restricted to southern areas of Pennsylvania (Stone 1894), but they have expanded their range northward. Todd (1940) suggested that there was considerable confusion between Northern Rough-winged and Bank swallows, making nineteenth-century reports unreliable. Harlow (1918) asserted that there "is no reason" to confuse the species. He included Monroe and Warren cos. in the breeding range and called them "common" in Centre Co.

[1] Hoffman, D. 1991. County reports—October through December 1990: Cumberland County. PB 4:154.

Bank Swallow ***Riparia riparia***

General status: In North America, Bank Swallows breed from Alaska south across Canada and the northern two-thirds of the U.S. They winter primarily in South America. Bank Swallows are encountered locally in large concentrations in migration and around breeding colonies. They are generally common to abundant regular migrants, often seen with other species of swallows. The breeding distribution is statewide, but colonies are widely scattered.

Habitat: During migration, Bank Swallows are found along mountain ridges, along Lake Erie, and generally over water. During the breeding season, they excavate nesting tunnels in exposed dirt, limestone, or sand banks, often along creeks, rivers, ponds, and lakes. They also use quarries and piles of silt. Bank Swallows are less associated with water during the breeding season than are Northern Rough-winged Swallows.

Seasonal status and distribution

Spring: They arrive in Pennsylvania as early as the fourth week of Mar. Their usual migration time is from the second or third week of Apr to the fourth week of May. The greatest number of migrants is seen from the fourth week of Apr to the second week of May.

Breeding: Bank Swallows are restricted in their distribution primarily by suitable nest sites. Around such nesting colonies the species may be abundant, but colonies are widely scattered. Colonies have been reported in almost every county, but most are in the Ridge and Valley and across the northern tier of counties. Colonies containing as many as 500 nesting holes have been reported along the Delaware and Susquehanna rivers, near Conneaut Lake, and along the Lake Erie Shore. An unusual density of Bank Swallow colonies is found along the escarpment of the Erie Lake Shore, where they are found along the bluffs and stream cuts at the edge of Lake Erie. Populations are poorly documented by the BBS because of their spotty distribution. Eggs were reported from mid-May to early Jul. A grass nest is constructed at the end of a 2-foot burrow excavated by the swallows and is sometimes reused in following years. Harlow's (1918) average date for 45 nests was 20 May.

Fall: Bank Swallows move through quickly during this season. Their typical migration period is from the third week of Jul to the second week of Oct, with the greatest number of birds seen before the third week of Aug. Hundreds of staging Bank Swallows can often be seen lining utility wires along lakes during peak migration. On 22 Jul 1965 a flock estimated at 8000–10,000 birds was seen at Tamarack Lake, Crawford Co.[1]

History: Early authors reported scattered colonies nearly statewide, but there was considerable confusion on species identification. Stone (1894) wrote, "in every case where the present species was reported nesting in southern Pennsylvania, investigation proved that the bird was the Rough-winged Swallow." Poole (unpbl. ms.), however, suggested that large rivers in the southeastern counties were the "main strongholds of the species." Warren's (1890) account is among the shortest for breeding birds; he stated "Common summer resident at many points along Delaware, Susquehanna and other large streams." Perhaps as many as 500 pairs bred in the sandy banks of one deep ravine near the Lake Erie Shore (Todd 1904). Bank Swallows reportedly declined severely during the early years of the twentieth century (Todd 1940), at least in western Pennsylvania. Todd (1940) also stated that manmade sand banks provide new, if sometimes hazardous, nest locations.

Comments: The Bank Swallow is the only species of passerine bird known to range unmodified (without subspecies) through Eurasia and North America.

[1] Leberman, R.C. 1965. Field notes. Sandpiper 8:12.

Cliff Swallow ***Petrochelidon pyrrhonota***

General status: Cliff Swallows breed from Alaska south across Canada and throughout most of the U.S., and they winter in South America. Populations have increased in recent years in Pennsylvania largely because of recent construction of concrete dams and bridges, upon which they have become increasingly dependent for nesting sites. They are rare to fairly common regular migrants, often seen with other species of swallows. Cliff Swallows nest in colonies over most of the state, especially in the Ridge and Valley and Glaciated Northeast regions. They frequently associate with Barn Swallows, with which they share a preference for human structures for nesting.

Habitat: During migration Cliff Swallows frequently are found near water or moving along mountain ridges. In Pennsylvania, nest sites are exclusively on human-built structures, including bridges and rural buildings with an overhang, such as barns, sheds, and pavilions.

Seasonal status and distribution

Spring: Their typical migration period is from the second or third week of Apr to the fourth week of May. The greatest number of birds is recorded between the first and third week of May. Nonbreeding stragglers remain to the first week of Jun.

Breeding: Nesting in colonies statewide in open areas or near water, Cliff Swallows are common in the Glaciated Northeast, more local and fairly common in the northern High Plateau, and uncommon and spotty in the Piedmont and west of the Allegheny Front. Cliff Swallows have colonized new areas as concrete bridges replace steel structures, for example in Clinton and Lycoming cos.[1] The concrete structures create a more secure base than steel structures for the swallows to attach their mud nests. The globular nest is constructed with a cavity and entrance hole (like a miniature cave) and is secured to a structure. The white eggs are spotted, as are Barn Swallow eggs. No trend was detected on the BBS, but Cliff Swallows have expanded their range south, nesting in Delaware Co. for the first time in 1997.[2] Nests with eggs are found from the end of May to late Jun, with some as late as 23 Jul (Poole, unpbl. ms.).

Fall: Southward movement begins the third or fourth week of Jul and continues to the second or third week of Sep. The majority passes through the state between the first and third week of Aug. There have been some notable concentrations of staging birds. Leberman (1976) cited three gatherings at different sites, ranging from 50 to 100 or more Cliff Swallows in 1971 and 1972. A very large concentration of 350 birds was at Mill Village on 14 Aug 1981.[3] Stragglers have been recorded until the third week of Oct.

History: Regional and statewide accounts of birds suggest a varied history. They were thought to have expanded east from the Midwest more than 150 years ago with the construction of buildings and bridges in rural areas.[4] The earliest ornithologists, such as Bartram, Wilson, and Audubon, failed to list Cliff Swallows as Pennsylvania birds, but Baird (1845) considered them abundant in Cumberland Co. Other mid-1800 authors stated that this species was rapidly expanding (e.g., Turnbull 1869). Todd (1940) stated that Cliff Swallows formerly bred as abundantly as Barn Swallows throughout western Pennsylvania but that they had declined severely. Explosive growth of House Sparrow populations during the late 1800s may have adversely affected Cliff Swallow populations, potentially eliminating this swallow, for example, from northwestern Chester Co. in the mid-1900s (Conway 1940).

[1] P. Schwalbe and G. Schwalbe, pers. comm.
[2] Pulcinella, N. 1997. Local notes: Delaware County. PB 11:95.
[3] Hall, G.A. 1982. Appalachian region. AB 36:178.
[4] Brown, C.R., and M.B. Brown. 1995. Cliff Swallow *(Petrochelidon pyrrhonota)*. BNA (A. Poole and Gill), no. 149.

Barn Swallow *Hirundo rustica*

General status: Barn Swallows breed in North America from Alaska across the southern half of Canada and throughout most of the U.S. They winter in Mexico south through Central America. Barn Swallows are common to abundant regular migrants and breeders throughout most of Pennsylvania. They are among the most widespread breeding birds in the state, found almost anywhere humans have settled.

Habitat: During migration Barn Swallows are observed flying overhead in al-

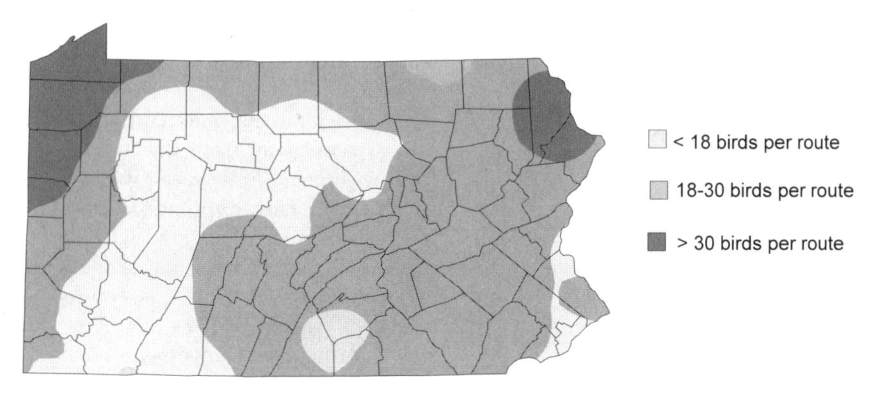

16. Barn Swallow relative abundance, based on BBS routes, 1985–1994

most any open areas, often near or over water and along mountain ridges. During the breeding season they are found over open fields in agricultural areas, around or in buildings (especially in barns that have a platform or a ledge protected from the weather), and around dams and bridges.

Seasonal status and distribution

Spring: Barn Swallows have returned as early as the third week of Mar, but their typical arrival time is the first week of Apr. An exceptionally early individual was discovered on 21 Feb 1984 at the Union City Dam in Erie Co.,[1] and one was found on 24–25 Feb 1992 at YCSP in Indiana Co.[2] Peak migration ranges from the third week of Apr to the second week of May. Migration usually ends by the fourth week of May.

Breeding: Barn Swallows are common nesters in open habitats statewide, except in uniformly forested regions, along ridgetops, or in urban centers. They will build a mud nest wherever there is a sheltered ledge. They are adaptable; Barn Swallows with young once were found in the hole in a bank on a heavily wooded undeveloped island in the Susquehanna River at Tunkhannock in Wyoming Co.[3] The BBS shows Barn Swallows least common in the central mountains, south through the Allegheny Mountains, and in the extreme southeastern counties (see Map 16), but they nest even within the limits of Philadelphia. Counts are highest in the Glaciated Northeast and Glaciated Northwest. A population decline of 1.2% per year is apparent on Pennsylvania BBS routes, with most of the decline since the mid-1980s. Changing agricultural practices, such as a decline in outbuildings, may be affecting numbers by reducing the availability of nest sites. Most egg sets were reported from 18 May to 9 Jul, with the major peak around 29 May and a small peak (second clutches) at the end of Jun. Late clutches have been found until 22 Jul.

Fall: Their typical migration period is from the third week of Jul to the second week of Oct. The greatest number of birds is recorded from the fourth week of Jul to the second week of Aug. It is not uncommon for some staging sites to contain thousands of birds. In the marshes around PISP at least 10,000 birds were estimated in mid-Aug 1996.[4] In the Conneaut Marsh at Geneva, Crawford Co., over 6000 were estimated roosting in the cattails on

26 Jul 1990.[5] Stragglers have remained as late as the second week of Nov. There are two very late fall records of Barn Swallows in Pennsylvania: 26 Dec 1965 at Carmichaels in Greene Co.[6] and 20 Dec 1980 at Chambersburg in Franklin Co.[7]

History: Before buildings were available these swallows undoubtedly were restricted to cliff walls, natural banks, and cave entrances, but few records exist of these natural nest sites in the East. Expansion of agriculture and the construction of farm buildings during the 1800s provided greatly expanded habitat for this species, but Todd (1940) thought that the decreased use of horses, and thus a decreased number of barns, reduced the Barn Swallow's population in the early 1900s.

[1] Hall, G.A. 1984. Appalachian region. AB 38:317.
[2] Higbee, M., and R. Higbee. 1992. County reports—January through March 1992: Indiana County. PB 6:37.
[3] W. Reid, pers. comm.
[4] G.M. McWilliams, pers. obs.
[5] Hall, G.A. 1990. Appalachian region. AB 44: 1133.
[6] Hall, G.A. 1966. Appalachian region. AFN 29: 423.
[7] Bender, M. 1981. Chambersburg CBC. AB 35:466.

Family Paridae: Chickadees and Titmice

Titmice have a short black bill with nostrils covered by tufts of stiff feathers. They are sociable birds, often seen in small groups or with mixed flocks of nuthatches, creepers, or kinglets. Of the four species found in Pennsylvania, three are permanent residents. They nest in tree cavities that they or woodpeckers have excavated, usually about 4–10 feet above the ground; Tufted Titmice sometimes nest as high as 80 or more feet above the ground (H. H. Harrison 1975). All species occasionally use artificial nest boxes. From five to ten white to creamy-colored eggs, evenly speckled or spotted, are laid in a nest lined with grasses, moss, feathers, fur, bark strips, snake skins, or hair. Male chickadees feed the females while they incubate; male Tufted Titmice call females away from the nest before feeding them. All species within this family are easily approached and are usually not easily disrupted when feeding or incubating. Boreal Chickadees are occasional visitors from the coniferous forests of Canada and the New England states during major movements of Black-capped Chickadees.

Carolina Chickadee *Poecile carolinensis*

General status: Carolina Chickadees are resident from New Jersey and Pennsylvania west to Kansas and Texas and east to Florida and the Atlantic coast. They are common permanent residents in the southeastern and southwestern corners of Pennsylvania. They rarely wander far from their known breeding range. A few occasionally associate with flocks of migrant Black-capped Chickadees and stray slightly outside their normal range. Such birds have been found at PNR.[1]

Habitat: Carolina Chickadees inhabit woodlands of all sizes, parks, and residential areas of towns and cities in hedgerows or brushy areas where at least a few trees are present. They are usually found at lower elevations than Black-capped Chickadees where the species overlap.

Seasonal status and distribution

Breeding: Carolina Chickadees are common residents in central Bucks Co., west across northern Montgomery and southern Berks cos., south of Blue Mountain near Harrisburg, and south and east of the Ridge and Valley region in Franklin Co. They also occur in valleys of the Potomac River watershed in Fulton Co. and are accidental in Bedford Co.[2] along the Maryland border. In the Southwest region, they are the common resident chickadees in Greene and Washington cos., extending just north to the southwestern corner of Westmoreland Co., across Allegheny Co. to southern Beaver Co., and just reaching into southeastern Butler Co.[3] They are easily confused with Black-capped Chickadees in the area of overlap, a band up to 12 miles wide in which hybrids sing both songs or birds learn the song of the other species.[2] The modest population increase reflected by the BBS may correspond to a range expansion, particularly in the east. In the 1960s, they occurred only accidentally in southern Berks Co. (Uhrich 1997) and probably not in Franklin Co. (Poole, unpbl. ms.). The nesting season and settings are very similar to those of Black-capped Chickadees, with eggs found from 28 Apr to 29 May. Nests have been reported in Jun. Numerous nests were once found in wooden fence posts. A nest box, with an entrance hole diameter of 1½ inches, will sometimes be used in a wooded setting.

Winter: During this season there is little evidence of movement. However, in the Ridge and Valley and in the Allegheny Mountain section, where they are local, they are rarely reported in winter. When not singing they may go undetected, especially when they mix with wintering flocks of Black-capped Chickadees. Carolina Chickadees are frequently observed in the winter at bird feeding stations.

History: The distribution of Carolina Chickadees was gradually being worked out during the later years of the 1800s. Baird (1845) failed to list them in Cumberland Co., although Audubon had distinguished the species in South Carolina in 1834 (AOU 1983). Todd (1940) indicated surprise to find Carolina Chickadees with Black-capped Chickadees along the Ohio River in Beaver Co. during the 1890s. Warren (1890) called Carolina Chickadees an "occasional summer resident," with a range south of central Lancaster and Chester cos. Frey (1943) included them in his list of Cumberland Co., where Baird failed to list them. A gradual northward range extension apparently has been ongoing for at least 100 years.

Comments: Carolina Chickadees are very similar to Black-capped Chickadees, and the two species are best separated by using a combination of factors including breeding range, plumage, and vocalizations.

[1] R.C. Leberman, pers. comm.

[2] Gill, F.B. 1992. Carolina Chickadee *(Parus carolinensis)*. BBA (Brauning 1992): 240–241.

[3] P. Hess, pers. comm.

Black-capped Chickadee ***Poecile atricapillus***

General status: Black-capped Chickadees are resident from Alaska across the southern half of Canada south in the U.S. to New Mexico in the West and through the Appalachian Mountains in the East. Among the most common and familiar birds in Pennsylvania, Black-capped Chickadees are regular breeding residents in all areas of the state except the southern corners. Some seasonal movement is detected, especially during invasion years. Banding recoveries suggest that Black-capped Chickadees tend to migrate south in the fall along the Appalachian Mountain system with a return flight in spring after a heavy fall movement.[1] They are regular visitors to bird feeders year-round.

Habitat: Black-capped Chickadees are found in woodlands of all sizes, parks, residential areas, and any hedgerows or brushy areas where at least a few trees are found.

Seasonal status and distribution

Spring: In spring, after winter invasions of Black-capped Chickadees, loose flocks of 30 or more have been seen leaving PISP and flying over Lake Erie in a northeasterly direction.[2] Birds that winter outside of their breeding range usually move north by the third week of Apr; stragglers leave by the third week of May.

Breeding: The southeastern extent of the Black-capped Chickadee's breeding distribution occurs from central Bucks west across northern Lancaster Co., the Conewago Mountains of northern York Co., and south through the Blue Ridge section. In the Southwest, they are found in the higher elevations from the Laurel Highlands in western Fayette Co., across to south of Pittsburgh, and to northern Washington Co.[3] Black-capped Chickadees overlap with Carolina Chickadees in a hybrid zone—a variable band about 12 miles wide in southern Berks Co.,[4] not defined by obvious ecological features. They may excavate a 1⅜-inch diameter hole in dead wood of a birch, aspen, or similar soft tree, often not more than 8 feet from the ground. Chickadees are increasing at 1.3% per year on BBS routes. Most eggs are collected between 29 Apr and 10 Jun, with a peak in mid- to late May.

Fall: Leberman (1976) wrote that in the Ligonier Valley they are occasionally abundant transients in fall, when hundreds can sometimes be seen in a single day. During invasion years birds move into areas where they do not breed about the first or second week of Oct. In most years, only a few birds are found outside their normal breeding range by the second or third week of Dec.

Winter: Black-capped Chickadees winter in every county, although population sizes vary from year to year. They are more frequently observed in winter than at any other season. Black-capped Chickadees can be found even in the coldest winter in deep forests, when few other birds are present.

History: Historically, Black-capped Chickadees have been considered ubiquitous and abundant. They may have expanded their range southward during the twentieth century. Beck (1924) did not list them nesting in northern Lancaster Co., where they have recently been found in the higher woodlands. This expansion apparently has not been at the expense of the Carolina Chickadee.

[1] Leberman, R.C. 1965. Bird banding at Powdermill, 1964. Research report no. 12. Powdermill Nature Reserve, Carnegie Museum of Natural History, Pittsburgh.

[2] G.M. McWilliams, pers. obs.

[3] P. Hess, pers. comm.

[4] Gill, F.B. 1992. Black-capped Chickadee *(Parus atricapillus)*. BBA (Brauning 1992):238–239.

Boreal Chickadee *Poecile hudsonicus*

General status: Boreal Chickadees are resident from Alaska across Canada and locally in the New England states south to New York. They join southward movements of Black-capped Chickadees and wander into Pennsylvania at rare intervals. Most birds observed in the state have been found in coniferous forests, associated with Black-capped Chickadees, or at bird feeding stations. Boreal Chickadees are currently listed as accidental because they have not been recorded in the state since 1984.

Seasonal status and distribution

Fall through spring: They were recorded in Pennsylvania during 19 of the 24 winters between 1960 and 1984. Since 1960 Boreal Chickadees have been recorded in Pennsylvania during the winters of 1960–1964, 1968–1971, and 1974–1984. Significant invasions occurred during 1961–1962, 1969–1970, 1975–1976, and 1981–1982. A state total of 12–24 or more birds was counted from each of those sets of years, including one-day counts of up to 10 birds at a site. As many as seven Boreal Chickadees wintered in 1962 at Media[1] in Delaware Co. Ten birds were at French Creek State Park in southern Berks Co. on 16 Nov 1969.[2] Most observations at a single site are of one or two birds. They have been recorded in at least 27 counties in every region in the state. Dates ranged from 21 Oct to 1 May with most sightings from Nov through Feb.

History: Conway (1940) stated that "specimens were secured in Lycoming, Pike, and Monroe counties" during the invasion of 1916–1917. In Jun 1917 two were discovered at Pocono Lake and two were near La Anna in Monroe Co. (Poole, unpbl. ms.), providing the only suggestion of breeding. Todd (1940) does not mention the occurrence of this species in western Pennsylvania. Poole (1964) listed Boreal Chickadees as rare and irregular winter visitants. He noted invasions during the winter of 1954–1955 and wrote that several specimens were taken. Apparently none of those specimens exists today.

Comments: Calls of one Boreal Chickadee were tape recorded at Wild Creek Reservoir in Carbon Co. on 3 Jan 1982,[3] but despite numerous sightings of Boreal Chickadees, very few have been documented with written descriptions or photographs.

[1] Schwalbe, P. 1962. The Boreal Chickadee in Delaware County. Cassinia 46:24–25.
[2] Scott, F.R., and D.A. Cutler. 1970. Middle Atlantic Coast region. AFN 24:29.
[3] R. Wiltraut, pers. comm.

Tufted Titmouse ***Baeolophus bicolor***

General status: Tufted Titmice are usually more conspicuous and vocal than their close relatives the chickadees, with which they frequently associate. They are resident from Maine and southern Ontario south throughout the eastern U.S. west to Texas. Titmice are permanent residents statewide. They are not as common in the Glaciated Northeast, High Plateau, and the Glaciated Northwest as in the remainder of the state. Tufted Titmice have been steadily spreading and increasing, especially across the northern half of the state.

Habitat: Titmice are found in woodlands, parks, and residential areas of towns and cities in hedgerows or brushy areas where at least a few mature trees are found. In mountainous regions of the state, most titmice are found near streams and lakes.

Seasonal status and distribution

Breeding: Titmice are common through the Ridge and Valley and the Southwest regions and uncommon across the northern tier of Pennsylvania (see Map 17). They have colonized deciduous forests statewide. Adaptable to human activities, titmice nest even in Pennsylvania's biggest cities as well as in the most extensive forests. The species' intermediate density in the Piedmont (see Map 17) reflects the intensive agriculture there and this species' dependence on woodlands. Titmice nest in abandoned woodpecker holes or natural cavities but do not excavate their own cavity. They occasionally will use a nest box. Nest placement varies widely, from low in a stump to high in a Downy Woodpecker's abandoned cavity. Egg sets are reported from mid-May to early Jun. The BBS shows a steady, long-term increase (3.0% per year) statewide, with the greatest increase since 1980. This corresponds with a significant range expansion throughout the state.

Winter: Tufted Titmice become more evident in winter, when they are observed at bird feeding stations.

History: Titmice were considered common in southern Pennsylvania during the mid-nineteenth century (Baird 1845). Warren's (1890) assertion that

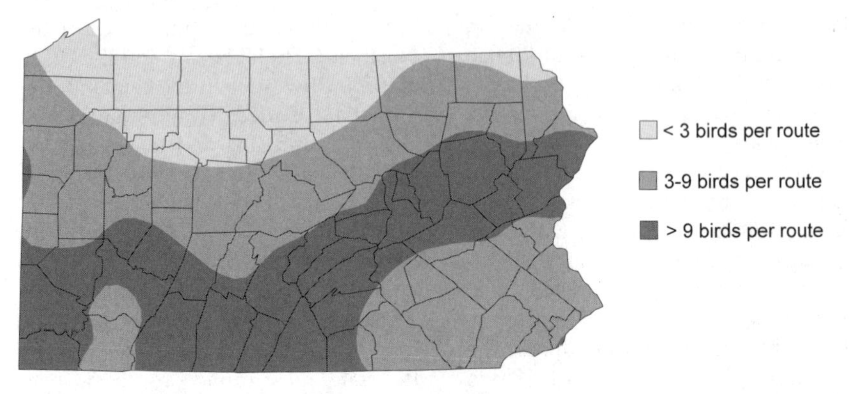

17. Tufted Titmouse relative abundance, based on BBS routes, 1985–1994

they were only rare winter visitors in the northern counties is corroborated by their conspicuous absence from the bird lists of his contemporaries (e.g., Cope 1898 in Susquehanna Co.; Stone 1891 in Luzerne Co.). Harlow (1918) considered titmice to be confined to the "Carolinian Faunal Region" of southern Pennsylvania (the Piedmont and Southwest regions); they were considered rare even in Berks Co. until the 1930s (Poole 1947). Titmice are among more than a dozen "southern" species (e.g., Red-bellied Woodpecker, Northern Cardinal) to experience a dramatic northward expansion while the state's woodlands were recovering from intensive cutting at the beginning of the twentieth century. They expanded north into Crawford Co. during the early decades of the twentieth century (Todd 1940). By the 1960s Poole (1964) stated that titmice occurred in practically every county.

Comments: They are reported to occasionally hybridize with chickadees. A bird photographed in Mar 1955 in Lititz, Lancaster Co., appeared to be a titmouse × chickadee (Poole, unpbl. ms.) as did a bird that visited a feeder in Audubon in Montgomery Co. from Sep to Dec 1956.[1]

[1] Northwood, J. d'A. 1957. A hybrid Chickadee × Tufted Titmouse. Cassinia 42:23.

Family Sittidae: Nuthatches

Nuthatches are small birds with a stubby tail, short neck, and slender bill. Nuthatches are easily recognized by their behavior of working their way head first down the trunk or branch of a tree in search of food. Three species have been found in Pennsylvania. A curious behavior noted only in the breeding season is bill sweeping; the pair swings their bills, often holding an insect, in a wide arc in or outside the nest cavity (H. H. Harrison 1975). The purpose of this behavior is apparently unknown.

Red-breasted Nuthatch *Sitta canadensis*

General status: Red-breasted Nuthatches breed from southern Alaska and across the southern half of Canada south through much of the West and from the New England states south through the Appalachian Mountains in the East. They winter primarily in the southern portion of their breeding range. Populations of Red-breasted Nuthatches are variable in all seasons in Pennsylvania. They are fairly common to common regular migrants during most years. They are irregular, local breeders statewide, but primarily across the northern half of the state. In winter Red-breasted Nuthatches are regular winter residents statewide, with highest numbers following strong fall migrations.

Habitat: Red-breasted Nuthatches are found primarily in coniferous forests, including pine plantings and mixed forests in all seasons.

Seasonal status and distribution

Spring: The earliest northbound birds are usually detected by the first week of Apr. Their typical migration period is from the second or third week of Apr to the third week of May. The greatest number of birds is recorded between the last week of Apr and the second week of May. Nonbreeding stragglers may remain to the first week of Jun.

Breeding: Red-breasted Nuthatches are rare or locally uncommon breeders. Populations are variable, inexplicably present in some years and absent in others. The Poconos probably support the most-regular breeding populations. Maturation of conifer plantations, many of which were planted by the CCC during the Great Depression, provided new habitat that began to be utilized in the early 1960s. The BBA project demonstrated a much wider distribution than had previously been reported, south to spruce plantations in York Co. on the Maryland line and west to Beaver Co. near Ohio. New nesting records continued to be found in the 1990s, with first county nesting records in Philadelphia Co. in 1995[1] and Allegheny Co. in 1996.[2] Over much of the state, Red-breasted Nuthatch territories occur, as Santner[3] stated, "as islands." Nest holes are excavated in dead limbs and trunks of conifers or old woodpecker holes up to 40 feet above the ground. Both sexes smear the outside entrance of the nest hole with resin throughout the nesting period, perhaps to keep insects or predators from entering. To avoid getting stuck to the sticky resin, female nuthatches enter the cavity by flying directly in. The nest is lined with fine grass, bark, moss, and usually five or six white eggs spotted with reddish brown are laid. Nest records are scarce; eggs have been found in May and young in Jun.

Fall: The earliest southbound birds may be detected by the first week of Aug. The majority begins moving south the second or third week of Sep. Peak migration may occur any time between the third week of Sep and the third week of Oct. Departure times are difficult to discern. When there is a fall irruption of nuthatches, numbers of birds may remain high into winter. On 19 Dec 1981 the Glenolden CBC in Delaware Co. recorded a total of 325 birds.[4]

Winter: After a fall irruption they may become locally abundant. Large numbers occurred during the fall-winter season of 1981–1982. On 19 Jan 1982 at Octoraro Lake in Lancaster Co., 300 were counted (Morrin et al. 1991). They have been recorded in every county in winter, and during most years they are rather rare, with only one or two birds recorded at each site. They are most frequently observed at bird feeding stations where pines or spruces are nearby.

History: Poole (unpbl. ms.) stated that "actual nesting records appear to be comparatively few in numbers." Those listed were all in northern forested counties from the Poconos to Pymatuning. The distribution described by Warren (1890) differed little from Poole's assessment.

[1] Fingerhood, E. 1995. Local notes: Philadelphia County. PB 9:95.
[2] Fialkovich, M. 1996. Local notes: Allegheny County. PB 10:96.
[3] Santner, S. 1992. Red-breasted Nuthatch *(Sitta canadensis)*. BBA (Brauning 1992):244–245.
[4] Haas, B.M., and F.C. Haas. 1982. Glenolden CBC. AB 36:495.

White-breasted Nuthatch ***Sitta carolinensis***

General status: White-breasted Nuthatches are residents in southern portions of Canada and across most of the U.S. They are residents throughout Pennsylvania. Population sizes may vary from year to year. Though they are sedentary by nature, fall movements have been detected at some sites.

Habitat: White-breasted Nuthatches prefer deciduous woodlands with mature trees. They are found in parks and in residential areas of towns and cities where at least a few mature trees are found.

Seasonal status and distribution

Breeding: White-breasted Nuthatches are fairly common in Pennsylvania, but are least common (see Map 18) and least uniformly distributed in the Piedmont. They tend to be less vocal during May and Jun than earlier in the breeding season and during winter. They are generally thought to be permanent residents (e.g., Wood 1979), but some northern populations undergo "irruptive movements" at infrequent intervals.[1] The nest site is a natural cavity or old woodpecker hole at various heights, often high above the ground, but these nuthatches do not generally excavate their own hole.[1] The cavity is lined with sticks, bark, grass, rootlets, and hair. White-breasted Nuthatches will nest in boxes placed in wooded settings. Most egg sets have been collected between 20 Apr and 21 May. Usually eight white eggs, densely spotted with brown and lavender, are laid. No significant change in population is apparent on BBS routes.

Fall: In some areas there is a noticeable fall movement. Some years there is an increase in the number of birds observed from the second or third week of Sep to at least the first week of Oct. White-breasted Nuthatch movement has been detected at the eastern hawk watches, such as Bake Oven Knob, where 81 were counted on 4 Oct 1975 (Morris et al. 1984). Peterjohn (1989) stated that in Ohio they periodically undertake noticeable fall movements and that these influxes are most ap-

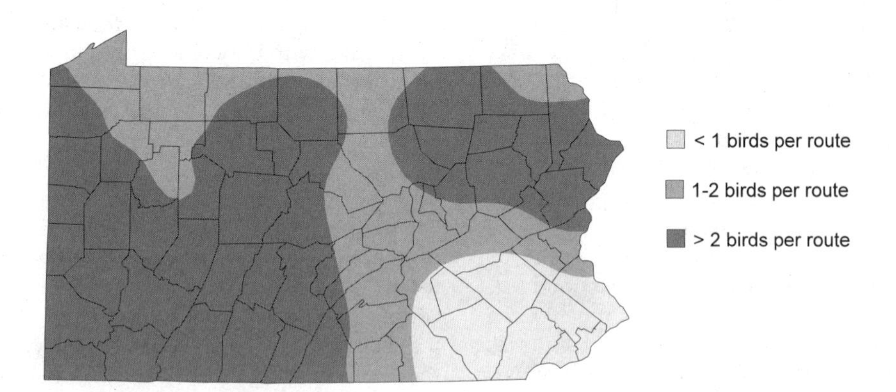

18. White-breasted Nuthatch relative abundance, based on BBS routes, 1985–1994

parent at lakefront migrant traps; this phenomenon has not been noted by authors in western Pennsylvania or at PISP.

Winter: White-breasted Nuthatches become more evident during the winter, when the trees are bare of leaves. They are frequently seen visiting bird feeding stations.

History: Warren (1890) considered this nuthatch a "common resident throughout the state," similar to the modern status described above. Even during the height of Pennsylvania's agricultural activity, sufficient large trees were available statewide to support this flexible species, even if at lower population levels.

Comments: Todd (1940) called this species a "true conservator of the forest, for it feeds largely on destructive insects."

[1] Pravosudov, V.V., and T.C. Grubb Jr. 1993. The White-breasted Nuthatch *(Sitta carolinensis)*. BNA (A. Poole and Gill 1993), no. 54.

Brown-headed Nuthatch *Sitta pusilla*

General status: Brown-headed Nuthatches are resident from Delaware south to Florida and west to eastern Texas. There are a few historical notes on the occurrence of these nuthatches in Pennsylvania, and at least one specimen was collected and preserved.

History: Turnbull (1869) listed this species as a "rare visitant to the lower counties." Warren (1890) mentioned a specimen that was taken near Philadelphia in the autumn about 1885 by C. D. Wood, but insufficient data were available to indicate that this specimen was taken within Pennsylvania. A Brown-headed Nuthatch was collected by L. W. Mengel at Boyertown in Berks Co. on 6 Sep 1894 (Poole 1964). The specimen is now in the Reading Museum. One was reported by R. Harlow from Franklin Co. at Monterey, on 12 Aug 1903, and still another was reported near Lewisburg by D. O. Karraker but with no date (Kunkle 1951).

Family Certhiidae: Creepers

The Brown Creeper is the only species in the family Certhiidae that occurs in North America. Creepers are small brown birds with a short, slender, decurved bill; stiff, pointed tail; and toenails that are rather long, curved, and sharp. The stiff, pointed tail and sharp, curved toenails help them creep along vertical branches and trunks of trees in search of insects. Creepers have a unique way of searching for food; they often start at the bottom of a tree and work their way up, and when they reach the top they fly to the bottom of another tree and repeat this behavior. Brown Creepers breed over most of Pennsylvania, primarily in extensively forested, mountainous areas.

Brown Creeper *Certhia americana*

General status: Though Brown Creepers are not always easy to find, their presence is usually given away by their high-pitched squeaky calls. Brown Creepers breed from Alaska and east across portions of southern Canada, in the Rocky Mountains west to the Pacific coast, and in the New England states south through the Appalachian Mountains. They winter over most of the U.S. Creepers are most easily found during migration, when they are fairly common and regular. They are regular breeders and winter residents across forested areas of the state.

Habitat: They prefer extensive mature deciduous or mixed woodlands during the breeding season but will inhabit second-growth woodlands during migration and in winter.

Seasonal status and distribution

Spring: Brown Creepers are most commonly seen during spring. Birds returning from southern wintering grounds appear in Pennsylvania the third or fourth week of Mar. The greatest number of migrants is observed during the second and third weeks of Apr. Migration has usually ended by the second week of May.

Breeding: These cryptic birds are uncommon to rare but widely distributed in the more heavily forested areas of the state. Brown Creepers are regular summer residents across the High Plateau but rare and irregular away from the mountains. They have been confirmed breeding in several southern-tier counties (e.g., Delaware and York), but not yet in Beaver, Chester, Greene, Montgomery, Philadelphia, or Washington. In all seasons they are associated with larger trees, occurring at much higher densities in old-growth forests than in the surrounding woodlands (Haney and Schaadt 1996). Nests are usually built less than 15 feet above the ground. They may take as long as four weeks to construct their nest behind a piece of loose bark on the trunk of a tree or, rarely, behind shutters of unoccupied cabins. Creepers have been known to use cavities in trees when no trees with loose bark are available. Availability of nest sites probably limits the population. Creepers are poorly documented by the BBS because of their low density and early nesting season, but they appear to be expanding their range south of traditional areas. They have benefited from the maturation of Pennsylvania's forests through the twentieth century. Not surprisingly, few nests with eggs have been found or described, but Poole (unpbl. ms.) listed full sets of eggs from 20 to 25 Apr. Usually four or five white eggs, spotted with reddish brown, are laid.

Fall: Their typical migration period is from the second or third week of Sep through Oct and Nov. Fewer birds are observed by the second or third week of Dec. Banding records suggest that peak migration occurs from the fourth week of Sep to the second or third week of Oct. A few lingering birds remain to winter.

Winter: They are uncommon to rare but widely distributed at this season, when they may move to lower elevations in mountainous regions. Creepers occasionally visit bird feeding stations where suet cakes are offered.

History: Early ornithological reports vary considerably. Before extensive forest clearing, creepers undoubtedly were widely distributed across Pennsylvania. Baird (1845) indicated the presence of the species in south-central Pennsylvania by stating "some resident," in contrast to the northerly distribution reported by many subsequent authors (e.g., Harlow 1918; Todd 1940). Since they are associated with larger timber, populations probably declined as forests were cleared. Stone (1891) considered the species "rather common in the forest region" of Luzerne Co. and he associated creepers with the "deep hemlock wood." In areas where more forests were cleared, creepers were often missing from bird lists (e.g., Cope 1898; Susquehanna Co.). The southward expansion described in recent times should best be considered a recovery of their historical range, even though Pennsylvania is close to the southern edge of their breeding range.

Family Troglodytidae: Wrens

Wrens are small, energetic, inquisitive, primarily insectivorous birds. Their bill is short to moderate in length and is thin, pointed, and slightly decurved. They often cock their tail upward. Five of the six species that occur in Pennsylvania also nest here. The sixth species, Bewick's Wren, formerly nested here. Nests of some species (House, Marsh, and Carolina wrens) are fairly conspicuous. Five to seven eggs form a normal clutch; they usually are white and speckled, spotted, or blotched with brown or red. Exceptions are the Marsh Wren, whose eggs are brownish red, and the Sedge Wren, whose eggs are unmarked.

Carolina Wren ***Thryothorus ludovicianus***

General status: The largest wrens found in the eastern U.S., Carolina Wrens are resident throughout the East from southern Ontario and the southern New England states west to Oklahoma and Texas. Carolinas are permanent residents across Pennsylvania. Severe winters with heavy snow cover drastically reduce populations.

Habitat: These wrens occupy dense underbrush and brush piles in residential areas, in parks, and around abandoned buildings and dense underbrush in wet woodlands. They are often found along streams, especially in mountainous regions.

Seasonal status and distribution

Breeding: Carolina Wren populations have undergone two complete cycles since 1960, expanding during years of mild weather and crashing during severe winters. At the height of their distribution, these wrens were fairly common across the southern half of the state and north along the major rivers, uncommon at lower elevations of the High Plateau along rivers, and uncommon near rural homes in the higher elevations and in heavily forested regions. The BBA project documented the Carolina Wren's distribution during a recovery, although the peak abundance and distribution occurred several years after BBA fieldwork. Populations crashed during the winter of 1977, grew to a peak in 1992, and again crashed in 1994. BBS and CBC data follow a similar pattern.[1] At low points, Carolina Wrens are uncommon in southern counties and rare and local along major rivers north of the Piedmont and Southwest regions. The hard winter of 1995 severely reduced populations, but far from eliminated, the species from Pennsylvania. In 1995, the Carolina Wren was still reported on 25 BBS routes widely scattered across the species' historical range but at less than half its abundance of just a few years earlier. The nest is a bulky structure of grass, leaves, and other debris, placed in a variety of settings such as on a shelf or in a notch in a building, in a natural cavity, or even in a hollow in the ground. Egg sets have been observed from 6 Apr through 7 Jul, reflecting sufficient time for at least two broods.

Winter: Harsh winters with deep snow or persistent ice on the vegetation can limit food availability for this species. Prolonged cold weather at the species' southern limits within the state may have the same results. If deep snow or extremely cold temperatures persist for long periods, Carolina Wrens may disperse from much of their range in northern Pennsylvania or suffer high rates of mortality. They are especially likely to suffer from harsh weather across the northern counties and south through the mountains of the High Plateau, Glaciated Northwest, and along the Lake Erie Shore. Several years may be required for populations to rebound to their previous numbers. Carolina Wrens visit feeders where suet is offered.

History: Although early ornithologists did not mention fluctuating populations, varying descriptions of the wren's abundance suggest their response to winter weather conditions. For example, observers in the 1860s (e.g., Libhart 1869; Turnbull 1869) described Carolina Wrens as rare birds, but by Warren's time (1890) they were common in the southern tier and were found "in nearly all parts of the state except in the highest mountainous regions." Carolina Wrens, true to their name, have long been associated with more southerly climates. Harlow (1918) stated that this species was "closely confined to the limits of the Carolinian fauna" (the Piedmont and Southwest regions). Poole (unpbl. ms.) described weather-induced declines in the Carolina Wren's population after the winters of 1918, 1932–1934, and 1958. Through the twentieth century, six major declines have been described in Carolina Wren populations; an average cycle of about 15 years.

[1] Hess, P. 1989. Carolina Wrens at Pittsburgh, 1970–1988: Persistence in a dangerous environment. PB 3:3–7.

Bewick's Wren ***Thryomanes bewickii***

General status: In the U.S., Bewick's Wrens breed from Washington State, south through California and east across the southern half of the U.S. to Pennsylvania (formerly). They winter in their western breeding range east to North Carolina and south to the Gulf coast and Florida. In Pennsylvania, they formerly bred primarily in the southwestern corner of the state, but they were also recorded breeding east across the southernmost counties and as far north as Centre Co. They are disappearing over most of their breeding range in the eastern U.S. and are now virtually gone from the state except as a rare vagrant. Bewick's Wrens are currently listed as accidental.

Seasonal status and distribution

Spring through fall: Until the early 1970s Bewick's Wrens were considered uncommon in the state and were recorded primarily in the Southwest region east to the Ridge and Valley, with scattered records north to Erie and Tioga cos. and east to the Coastal Plain. There apparently has been no record from the Piedmont since 1960. In western Pennsylvania there has been no reliable record since 1977 (Leberman 1988). After 1960 the only nesting records in Pennsylvania came from Lycoming Co. in the late 1970s (Schweinsberg 1988) and Waynesburg in Greene Co. on 17 May 1974.[1] Recently reviewed sight records from Cumberland, Fulton, and Perry cos. were not accepted. Former records indicate that the arrival time in Pennsylvania was around mid-Apr and the departure time in mid-Oct.

Winter: There has been no reliable winter record since 1960.

History: J. J. Audubon first described the Bewick's Wren in 1827 (AOU 1983). The first record in Pennsylvania was made in Cumberland Co. in Jun 1843, when a female was reported, and eggs were first collected in 1845 by Baird (1845). Turnbull (1869) reported that the species was a rare nester in eastern Pennsylvania, and Libhart (1869) noted nesting in Lancaster Co. Stone (1894) mentioned specimens taken in Montgomery and Cumberland cos. In southwestern Pennsylvania, Jacobs (1893) claimed that Bewick's Wren was a "common" breeder in Greene Co., and Harlow (1918) and Todd (1940) found them more numerous from Chestnut Ridge in Fayette Co. west to Greene Co. than in any other Pennsylvania counties. In the interior of the state nesting was documented as far north as Centre Co. (Todd 1940). Todd further stated that they once ranged north to Mercer Co. Barring the real possibility of confusion with Carolina Wrens, breeding records in the Ridge and Valley region, including Centre, Clinton, and Union cos. were noted until 1957.[2] A very late Bewick's Wren was found at Honey Brook on 14 Dec 1954, and possibly the same bird was there on 16 Jan 1955 (Poole, unpbl. ms.). One was at Elverson from 17 Dec 1954 to 15 Jan 1955 (Poole, unpbl. ms.). They were reported to occasionally winter in Greene Co. (Todd 1940).

Comments: The cause of the decline has been widely debated. Todd (1940) discussed the theory that expanding House Wren populations competed with the Bewick's; competition has in fact been observed in other areas in which the species overlap.[3] The decline has been across the eastern U.S. Bewick's Wrens are now listed as Extirpated in Pennsylvania by the OTC.

[1] Hall, G.A. 1974. Appalachian region. AB 28:803.
[2] Fingerhood, E. 1992. Bewick's Wren *(Thryomanes bewickii)*. BBA (Brauning 1992):435–436.
[3] Kennedy, E. D. 1998. The decline of Bewick's Wren: Interactions with House Wrens and changing land-use history. Abstracts of the North American Ornithological Conference 1998. St. Louis, Missouri.

House Wren *Troglodytes aedon*

General status: House Wrens breed across southern Canada and south through the western U.S. to Texas and east across the northern Gulf states to Georgia and South Carolina. They winter across the southern states and north casually to southern Ontario and New York. House Wrens are the most widespread and common of the wrens found in Pennsylvania. They have adapted well to human habitation and are most common in areas disturbed by humans, provided suitable foraging habitat is present. House Wrens are fairly common to common regular migrants and are regular breeders statewide. They have occasionally been recorded in winter in the Coastal Plain and Piedmont.

Habitat: House Wrens are usually found in thickets, especially bordering fence rows and woodlands, and are found in brush piles and brushy second-growth woodlands. They also inhabit overgrown areas in vacant lots, in parks, and around buildings in both rural and urban areas.

Seasonal status and distribution

Spring: In the southern portion of the state the earliest arrivals usually appear during the second or third week of Apr. At higher elevations and in the northern portion of the state they are usually not found until the fourth week of Apr. The greatest number of migrants is seen the first or second week of May. Very little movement is detected after the fourth week of May.

Breeding: House Wrens are common breeding birds statewide. They were absent from BBA blocks only in the most uniformly forested areas, but even there they occupy small clearings. Similarly, they are least common on BBS routes in the central, heavily forested High Plateau (see Map 19). The male builds several bulky stick nests that fill natural cavities or boxes placed for bluebirds. The nests are usually within 8 feet of the ground. The female then selects one of the sites and lines the nest with a soft cup, often using feathers. Pairs often raise at least two broods. Eggs have been found mainly from 19 May through 15 Jul, with nesting activity sometimes into Aug.

Fall: Because wrens become quite inconspicuous after singing stops, the start of migration is difficult to discern. The greatest number of birds is usually seen the last two weeks of Sep. Banding records also suggest that most birds pass through the state during the second half of Sep. By the second or third

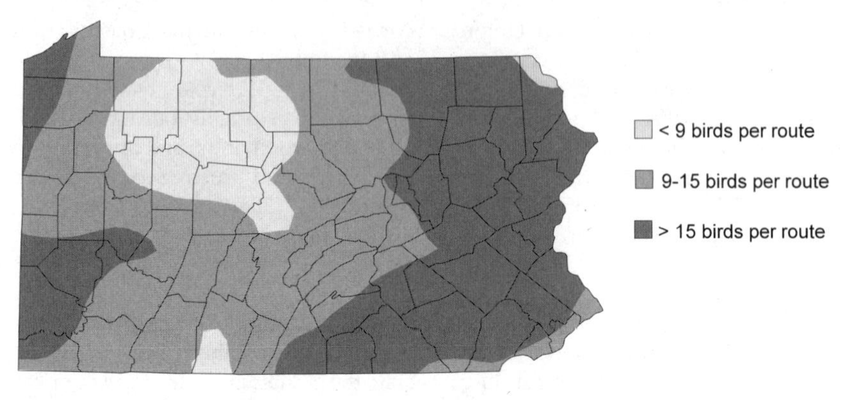

19. House Wren relative abundance, based on BBS routes, 1985–1994

week of Oct most House Wrens have left the state, but some linger to the first week of Jan or later.

Winter: House Wrens are accidental in winter in the Coastal Plain and Piedmont and have wintered once in the Southwest. In the Coastal Plain they have been observed in winter at Tinicum in Delaware[1] and Philadelphia cos.[2] Winter records in the Piedmont have come from Lake Ontelaunee in Berks Co. on 30 Jan 1972[3] and in Montgomery Co.[4] From the Southwest there was one in Allegheny Co. on 17 Jan 1982,[5] and one was at Bradys Run Park in Beaver Co. on 3 Feb 1998.[6]

History: House Wrens were generally described as common or abundant birds of open habitats and edges in the eastern half of the state but were absent in extensive forested regions such as North Mountain in Sullivan Co. (Dwight 1892) and Cambria Co. (Todd 1940). Audubon noted that this species was not frequently seen west of the Alleghenies (fide Todd 1940). Late nineteenth-century descriptions from Greene Co. (Jacobs 1893) and Westmoreland Co. (Townsend 1883) identified House Wrens as uncommon. Although habitat was suitable in Greene Co., House Wrens were apparently less common there than were Carolina and Bewick's wrens. However, in the Westmoreland Co. account (Townsend 1883) Bewick's and Carolina wrens were not listed, whereas House Wrens were "Apparently not common." Todd (1940) stated that "beyond question . . . there has been a tremendous increase in the numbers of this species in the past hundred years—particularly notable since the beginning of this century." This increase was partly attributed to the opening of the great forests but may also have been enhanced by placement of bird boxes for them and other cavity-nesting species.

Comments: House Wrens are known for their habit of puncturing eggs in nests near their own.

1 J.C. Miller, pers. comm.
2 J.C. Miller and E. Fingerhood, pers. comm.
3 R. Keller, pers. comm.
4 G. Freed, pers. comm.
5 Hall, G.A. 1982. Appalachian region. AB 36:294.
6 Haas, F.C., and B.M. Haas, eds. 1998. Birds of note—January through March 1998. PB 12:14.

Winter Wren ***Troglodytes troglodytes***

General status: Winter Wrens are usually found in constant motion on or near the ground in dense thickets or tangles. In North America they breed from coastal Alaska, south along the Pacific coast, east across central Canada to Newfoundland, and south through the New England states and the Appalachian Mountains. They winter across the southwestern U.S. and most of the eastern U.S. north to Massachusetts. Winter Wrens are fairly common regular migrants in Pennsylvania. They are most easily found during peak migration. During the breeding season they are best located by their beautiful melodic songs. Winter Wrens are local breeders primarily in the Glaciated Northeast and High Plateau. In winter most Winter Wrens are observed east of the Allegheny Front and in the Southwest.

Habitat: During migration they are found in thickets, brush piles, and dense undergrowth in woodlands. During the breeding season and in winter they are usually more confined to rock outcrops, dense undergrowth along mountain streams, steep slopes, bogs, or coniferous swamps, but they almost always can be found in extensive forests, especially those with hemlocks.

Seasonal status and distribution

Spring: Northbound birds are detected the third or fourth week of Mar. Most birds have returned to their breeding site by the first week of May. Peak migration occurs during the second and third weeks of Apr. Stragglers remain to the third week of May.

Breeding: Winter Wrens are uncommon breeders in the more heavily forested and mountainous sections of the Glaciated Northeast, High Plateau, and Allegheny Mountain section in Somerset, Westmoreland, and Fayette cos.; they are rare and local in the mountains of the Ridge and Valley (notably present along Blue Mountain). They are most frequently found above 2000 feet elevation.[1] Before 1980 the Winter Wren was reported on state BBS routes in only 5 of 14 years, but between 1980 and 1996 they were detected annually, increasing 13% per year on average despite a major decline in 1996. Despite that setback, territorial males were found outside of their historical range in 1997 for the first time, in Butler Co. at MSP,[2] in the Unami Creek valley of Montgomery Co.,[3] and in the Blue Ridge section in Adams Co.[4] Nest records are few, with sets of eggs from 20 May to 4 Jun (Poole, unpbl. ms.). The nest is placed on or close to the ground and is extremely well hidden in upturned roots or in a rocky crevice. The cavity is filled with moss and twigs and lined with hair and feathers.

Fall: Their usual migration time is from about the third week of Sep to the fourth week of Nov. Peak migration occurs from the first to the third week of Oct. Most migrants are gone from the state by the first or second week of Jan. A few birds remain to winter.

Winter: Winter residents are uncommon in the Coastal Plain and Piedmont. In the remainder of the state most winter records are from lower elevations along major river valleys and their tributaries, where they are rare. In the High Plateau they have been recorded in winter only in the Allegheny Mountain section, where they are rare but regular.

History: Winter Wrens were described as "common throughout the hemlock forest" (Stone 1891) and "abundant in the damper portions" (Dwight 1892), but always decidedly northerly as a breeding species. Notably, where the original forests had been cleared, this species was considered uncommon (e.g., Cope 1898; Susquehanna Co.). They were common in winter in the southern counties (Warren 1890), hence the name of this species. Poole (1964) thought that they had "greatly diminished as a breeding species since the pine and hemlock forests were cut," although his distribution closely resembles the BBA map. Winter Wrens were among the species considered by R. Mellon to have "rebounded in recent years" after being reduced in numbers.[5]

[1] Brauning, D.W. 1992. Winter Wren *(Troglodytes troglodytes)*. BBA (Brauning 1992):254–255.
[2] Hess, P. 1997. Local notes: Butler County. PB 11:92.
[3] Freed, G.L. 1997. Local notes: Montgomery County. PB 11:98.
[4] Kennel, E., and A. Kennel. 1997. Local notes: Adams County. PB 11:91.
[5] Mellon, R. 1990. An ornithological history of the Delaware Valley region. Cassinia 63:36–56.

Sedge Wren ***Cistothorus platensis***

General status: Sedge Wrens breed in North America from Alberta, across southern Canada to the New England states, and south to Arkansas, Kentucky, and Virginia. They winter along the Atlantic coast from Maryland through Florida and east across the Gulf states to Texas. The seasonal movements of Sedge Wrens are highly erratic and unpredictable. They may appear and (presumably) breed at almost any time from late spring to early fall. They seem to be absent from much of their range in Pennsylvania even where there is suitable habitat. Their absence may be in part because of the difficulty in observing them, as they usually remain hidden in dense grass. Sedge Wrens are rare, irregular migrants and breeders.

Habitat: Sedge Wrens prefer wet upland sedge meadows, hayfields, marshes bordered by sedges, and, occasionally, thickets and cattail marshes.

Seasonal status and distribution

Spring: Migration is difficult to discern because they are secretive and difficult to locate, especially when they are not singing. The first arrivals in spring may appear as early as the last week of Apr. They usually appear the first or second week of May. Birds that seem to be migrants continue to be seen until at least the first week of Jun. Most recent sightings of Sedge Wrens have been of single birds.

Breeding: Sedge Wrens are rare and irregular breeders, with records widely scattered across the state. Evidence of nesting in Pennsylvania includes one of a bird singing until 15 Aug 1978 at Struble Lake in Chester Co.[1] and several reports of Sedge Wrens at Hunlock's Creek in the Glaciated Northeast, including one bird reported on 27 Jul 1975,[2] three on 30 Aug 1974,[3] and one on 27 Jul 1980.[4] Two were at MCWMA 26 Jun 1974 (Morrin et al. 1991). The last confirmed nesting at PISP in Erie Co. was on Jul 1976 (Stull et al. 1985). They were found in late Jun east of Parker Dam State Park in 1980 in Clearfield Co.[5] Likely and confirmed breeding records since 1983 include reports along Stumpstown Road, Cumberland Co.[6] and near Uniontown in Fayette Co.[7] The most regular recent sites include those along Sugar Lake in the Erie National Wildlife Refuge (Crawford Co.) and the propagation area at MSP (Butler Co.) during most of the 1980s.[8] Records since 1990 include one record of two birds near Jacobsburg State Park, Northampton Co., from 21 May through Aug 1996;[9] confirmed nesting at Dunnings Creek, Bedford Co., in 1992;[10] a nest in a restored wetland in Mercer Co. in 1995 (Cashen and Brittingham 1998), and an unoccupied nest near Latrobe, Westmoreland Co., in 1997.[11] At most sites they were not observed in subsequent years. The nesting season is extended; nests with eggs have been found in Pennsylvania as early as 30 May (Beck 1955) and at various dates in Jun and Jul. Occasionally birds set up territories in Aug.[12] They nest colonially and frequently build dummy nests. Sedge Wren populations definitely have been hurt by the loss of wetlands and intensive grass cutting typical of modern agriculture.

Fall: Like spring migration, little is known about their southbound movements. Some observations in Jul and Aug may be of migrants, but sightings through Sep to the first week of Oct are most certainly migrants. There are two very late sightings of Sedge Wrens: one was seen at Frances Slocum State Park in Luzerne Co. on 25 Nov 1979,[13] and one was at Washington Boro in Lancaster Co. on 13 Nov 1993.[14]

History: Early references are enigmatic, suggesting that Sedge Wrens were rare birds locally found in various wetland habitats. The breeding status, and in some cases the very identity, of summer birds was questioned. Harlow (1913) identified only one definite breeding record as "authentic" (a set of eggs taken in Philadelphia), apparently dismissing Warren's (1890) reports out of hand. From 1930 to 1941 Sedge Wrens experienced a major population expansion. Singing males occurred in widely scattered sites around the state, including sites in Lebanon, Monroe, Somerset, and Union cos., and nests were found at Quakertown Marsh, Bucks Co.; marshes in Crawford and Centre cos.; near Lake Ontelaunee, Berks Co.; Long Pond, Monroe Co.; near White Oak Pond, Lancaster Co.; Coatesville, Chester Co.; Tinicum marshes, Delaware Co.; and "about fifteen nests" at Niagara Pond, Presque Isle, Erie Co., in Jun 1932 (Todd 1940). At some sites birds did not appear until Jul. Sedge Wrens were considered a "fairly common to uncommon transient and summer resident" near Lewisburg, Union Co., probably in the 1930s (Kunkle 1951). Most of these sites were not occupied again after the early 1940s.

Comments: The BBA project has shown that Sedge Wrens are probably declining in Pennsylvania, and the BBS data indicate a decline in the populations in the northeastern U.S. as well (Robbins et al. 1986). The decline may be attributed to habitat loss. The PGC has listed the Sedge Wren as Threatened. Some Sedge Wren sightings may be misidentified Marsh Wrens, especially during migration, when Sedge Wrens are occasionally found in shared habitats such as cattail marshes.

[1] Paxton, R.O., P.W. Smith, and D.A. Cutler. 1979. Hudson-Delaware region. AB 33:162.
[2] Buckley, P.A., R.O. Paxton, and D.A. Cutler. 1975. Hudson-Delaware region. AB 29:953.
[3] Buckley, P.A., R.O. Paxton, and D.A. Cutler. 1975. Hudson-Delaware region. AB 29:33.
[4] Boyle, W.J., Jr., R.O. Paxton, and D.A. Cutler. 1980. Hudson-Delaware region. AB 34:881.
[5] P. Schwalbe and G. Schwalbe, pers. comm.
[6] Hoffman, D. 1988. County reports—July through September 1988: Cumberland County. PB 2:103.
[7] Hall, G.A., 1985. Appalachian region. AB 39:91.
[8] Wilhelm, G. 1989. County reports—July through September 1989: Butler County. PB 3:102.
[9] R. Wiltraut, pers. comm.
[10] Shafer, J. County reports—April through June 1992. Bedford Co. PB 6:70.
[11] Leberman, R.C. 1997. Local notes: Westmoreland County. PB 11:101.
[12] Burns, J.T. 1982. Nests, territories, and reproduction of Sedge Wrens *(Cistothorus platensis)*. WB 94:338–349.
[13] W. Reid, pers. comm.
[14] Haas, F.C., and B.M. Haas, eds. 1994. Rare and unusual bird reports. PB 7:162.

Marsh Wren *Cistothorus palustris*

General status: Marsh Wrens are readily identified by the bubbly rolling songs they give both day and night. The loss of wetlands with emergent vegetation has diminished many populations of this highly habitat-specific species. Marsh Wrens breed across southern Canada and locally across the interior of the U.S. and along the Pacific, Atlantic, and Gulf coasts. They winter primarily along the coasts and across the southern U.S. Their migrant status is not well known in Pennsylvania, especially in spring. They are regular breeders mainly in the Coastal Plain, Glaciated Northwest, and Lake Erie Shore. They occasionally winter in the Coastal Plain.

Habitat: Marsh Wrens are restricted to large marshes with emergent vegetation, especially cattails.

Seasonal status and distribution

Spring: Birds generally are found only in areas where they breed. In the Coastal Plain the first appearance of Marsh Wrens may be as early as the first week of Apr. In all other areas of the state they usually arrive the third or fourth week of Apr. An exceptionally early Marsh Wren was found on 14 Mar 1982 at State College in Centre Co.,[1] and one at the "Muck" in Tioga Co. on 29 Mar 1998 was very early for this northern location.[2]

Breeding: Locally abundant, Marsh Wrens are colonial-breeding residents in marshes, usually larger than 20 acres (Cashen and Brittingham 1998) in the Glaciated Northwest and near Tinicum Refuge, Philadelphia and Delaware cos. The state's largest population occurs in the Conneaut Marsh in Crawford Co., where surveys near Geneva indicate a population of more than 500 territorial males in the 3000-acre marsh.[3] Colonies also are found in the Pymatuning Swamp north of Hartstown and in scattered wetlands in Crawford, Mercer, and Erie (including PISP) cos. More than 100 singing males were estimated in the Tinicum region of the Coastal Plain in 1997, where numbers have been declining.[4] Up to 15 singing males may be found in widely scattered cattail and sedge wetlands including Dunning's Creek in Bedford Co., in the "Muck" north of Wellsboro in Tioga Co., and formerly near the Cumberland Depot in Cumberland Co. Marsh Wrens are rare and irregular in scattered wetlands of southeastern and northeastern counties, and they are accidental over the remainder of the state. The nest is a globelike ball with a small entrance hole on the side, placed 1–3 feet over standing water. Most nests with eggs have been found between 30 May and 18 Jul; numerous dummy nests are constructed. Harlow's (1918) average date of full clutches was 15 Jun, but the nesting season is sometimes delayed because of high water or delayed vegetation growth.

Fall: Fall migration is difficult to discern, especially after they have stopped singing. On the Conejohela Flats in Lancaster Co., fall migrants were regularly seen from 1986 to 1990 (Morrin et al. 1991). The greatest number of Marsh Wrens in the Coastal Plain is found in Sep, but during mild falls, many birds may remain through the fall into winter. The Glenolden CBC recorded as many as 14 on 31 Dec 1966.[5] Most years they have departed from the state by the second week of Jan. Most reports of Marsh Wrens from the rest of the state are in Sep, and most birds depart before the third week of Oct. Stragglers may remain to at least the third week of Nov.

Winter: Tinicum is the only area in the state where Marsh Wrens have been recorded in winter. Most winter records are of birds surviving until the first severe weather has arrived, usually no later than the second or third week of Jan. However during the mild winter of 1967 about 20 birds successfully wintered there.[6]

History: The significant loss of wetland habitats resulted in declines in Marsh Wrens and other emergent-wetland obligates. Poole (unpbl. ms.) commented on the loss of habitat by saying "many of these smaller marshes have been or are still being drained, so that suitable Marsh Wren habitat is becoming scarcer from year to year." Many historical sites no longer support wrens, such as along the Delaware River in lower Bucks Co. and near New Galilee,

Beaver Co. (Todd 1940). As many as 13 nests were found in one day (30 May 1935) at Quakertown Marsh, Bucks Co. (Poole, unpbl. ms.).

Comments: The OTC lists the Marsh Wren as a Species of Special Concern, Candidate–At Risk.

[1] Hall, G.A. 1982. Appalachian region. AB 36:852.
[2] Haas, F.C., and B.M. Haas, eds. 1998. Birds of note—January through March 1998. PB 12:14.
[3] Brauning, D.W. 1993. Wetland nesting bird survey. Annual report. Pennsylvania Game Commission, Harrisburg.
[4] E. Fingerhood, pers. comm.
[5] Rigby, E.H. 1967. Glenolden CBC. AFN 21:152.
[6] Scott, F.R., and D.A. Cutler. 1967. Middle Atlantic Coast region. AFN 21:403.

Family Regulidae: Kinglets

Both of Pennsylvania's two species of kinglets are most easily observed during migration. They are easily recognized by their very small, plump, grayish olive body and short tail, and by their feeding behavior of hovering while picking food from a branch. Golden-crowned Kinglets breed in the state, but both species may be found here in winter.

Golden-crowned Kinglet *Regulus satrapa*

General status: Golden-crowned Kinglets are constantly on the move, flicking their wings, hovering, and flitting from branch to branch, giving their almost inaudible high-pitched calls. Golden-crowned Kinglets breed from Alaska across the forested regions of southern Canada and south through the mountains of the western U.S. and the New England states through the Appalachian Mountains. They winter throughout most of their breeding range. They are fairly common to abundant regular migrants, but numbers may vary from year to year, especially in spring. They are regular local breeders, primarily in the Ridge and Valley, Glaciated Northeast, and High Plateau. Breeding has increased in the state in recent years with the maturing of introduced pine and spruce plantings. Golden-crowned Kinglets winter throughout the state but are prone to yearly population fluctuations. They frequently flock with Ruby-crowned Kinglets when their migration periods overlap.

Habitat: During migration Golden-crowned Kinglets are found in any woodland, thicket, and hedgerow. In the breeding season they prefer pine and spruce stands and mature hemlock groves. In winter they are associated with conifers but occasionally are found in mixed brushy woodlands.

Seasonal status and distribution

Spring: Their typical migration period is from the second or third week of Mar to the first week of May. The greatest number is seen during the second and third weeks of Apr. Stragglers remain to the third week of May.

Breeding: Golden-crowned Kinglets are rare to uncommon, widely scattered summer residents across Pennsylvania's mountains (including the Laurel Highlands) and are rare away from the mountains.[1] They are locally fairly common in old-growth forests of mixed northern hardwoods (Haney and Schaadt 1996). The BBA project documented a much more extensive distribution than was reported as recently as the 1970s by Wood (1979), who identified only Monroe, Pike, and Wyoming cos. as the breeding range. The breeding range expanded in response to the growth of conifer stands (particularly spruce), many planted by the CCC in the 1930s. Confirmed breeding records now include Berks and York cos. in the Piedmont, Westmoreland Co. in the Allegheny Mountains, and Indiana Co. in the Southwest. Maturing groves of native red and black spruce in the Glaciated Northeast also contributed to the growing population. Kinglets are most likely to be found in conifers at least 30 feet tall and thickly spaced to provide a closed canopy in stands as small as 2 acres.[2] Golden-crowned Kinglets are not reported frequently enough to provide a meaningful population trend on Pennsylvania BBS routes. The nest is a small hanging cup, woven into high branches of a conifer. The few nests found in the state show that birds are incubating in Jun and Jul. From 5 to 10 spotted or blotched white or cream-colored eggs are laid.

Fall: The arrival time of this species during the fall is fairly consistent from year to year. Their usual migration period is from the third or fourth week of Sep through Dec. Peak migration ranges from the second week of Oct to the second week of Nov.

Winter: Golden-crowned Kinglets are known to winter in every county in the state, but numbers fluctuate from year to year depending on the severity of the weather. Even during mild winters, however, fewer than 10 or 15 birds usually are observed at each site. As winter progresses their numbers decline, and by Feb they may become difficult to locate. Golden-crowned Kinglets are frequently found associated with chickadees or titmice and may wander away from conifers in search of food.

History: The earliest confirmation of breeding Golden-crowned Kinglets in Pennsylvania came from Sullivan and Wyoming cos. near Lopez in 1890 (Stone 1900). Observations at the end of the 1800s and in the early 1900s also documented summering, and presumably breeding populations, in the Poconos and west through Lycoming Co. and possibly to McKean Co. (Harlow 1913; Todd 1940). Although Golden-crowned Kinglets may have been widespread before forests were cleared, there is little documentation. They apparently retreated during the twentieth century from much of this range, except around the Poconos, where they were consistently reported at Pocono Lake (Poole 1964; P. B. Street 1954).

[1] Mulvihill, R.S. 1992. Golden-crowned Kinglet *(Regulus satrapa)*. BBA (Brauning 1992):260–261.
[2] Andrle, R.F. 1971. Range extension of the Golden-crowned Kinglet in New York. WB 83: 313–316.

Ruby-crowned Kinglet ***Regulus calendula***

General status: Ruby-crowned Kinglets breed from Alaska across Canada south in the Rocky Mountain region in the West and to New York in the East. They winter across the southern U.S. Ruby-crowned Kinglets are fairly common to abundant regular migrants in Pennsylvania, but populations fluctuate from year to year. They are accidental summer visitors, and in winter they are regular but local. They frequently flock with Golden-crowned Kinglets when their migration periods overlap.

Habitat: During migration Ruby-crowned Kinglets are found in woodlands, thickets, and hedgerows. In winter they are more confined to conifers, dense thickets, and hedgerows.

Seasonal status and distribution

Spring: Some migrants have been recorded as early as the third or fourth week of Mar, but their typical migration period is from the first or second week of Apr to the third week of May. Peak migration usually occurs from the second week of Apr to the first week of May. Occasionally their peak migration overlaps that of Golden-crowned Kinglets and large numbers of both can be found together for a short time. Spring flights may be affected by previous fall flights as suggested by banding records at PNR. After a light migration during the fall of 1977 only half the usual number of birds were banded at PNR in the spring of 1978.[1] Stragglers may remain to the second week of Jun.

Fall: Earliest migrants may appear by the third week of Aug but generally not until the third or fourth week of Sep, when they usually accompany major flights of warblers. The greatest number of birds is recorded in a short period during the second or third week of Oct, when they are abundant. Numbers rapidly decrease through Nov and Dec, and by the first week of Jan most have departed. Banding records suggest a population crash in the severe winter of 1977; they were 56% below average at PNR[2] and were down 84% on Pennsylvania CBCs the next year.[3]

Winter: Most wintering birds are recorded in the southern half of the state, where they are uncommon to rare. In areas north and west of the Piedmont, most winter sightings are during Jan. In the Ridge and Valley and High Plateau they are found primarily along the major stream valleys in winter. Across the northern-tier counties of the High Plateau, they have been recorded in Cameron and Warren. In the Glaciated Northeast they have been recorded in winter in Bradford, Tioga, and Sullivan cos. In the Glaciated Northwest winter records have come from Crawford and Erie cos. Most sightings are of one or two birds at each site.

History: Little has changed in the history of migration of this species. Both Poole (1964) and Todd (1940) listed them as very common or abundant in migration. A rather late male was observed at Black Moshannon State Park in Centre Co. on 3 Jun 1957.[4] Apparently Ruby-crowned Kinglets were much rarer in winter than in recent times. Poole (1947) listed them as occasional in winter in Berks Co., and later Poole (unpbl. ms.) stated that there had been 15 observations in Jan and Feb. In other areas east of the Allegheny Front, Poole (unpbl. ms.) mentioned winter records (mostly only in early Jan) in Bucks, Delaware, Lackawanna, Union, Philadelphia, and Tioga cos. Todd (1940) considered the winter occurrence of this species as extraordinary in western Pennsylvania. He listed only two counties, Allegheny and Centre, for winter records, all in the first half of Jan.

[1] Hall, G.A. 1978. Appalachian region. AB 32:1007.
[2] Hall, G.A. 1978. Appalachian region. AB 32:205.
[3] Paxton, R.O., P.A. Buckley, and D.A. Cutler. 1978. Hudson-Delaware region. AB 32:329.
[4] Brooks, M. 1957. Appalachian region. AFN 11:405.

Family Sylviidae: Gnatcatchers

Only one species in the family Sylviidae, which consists mostly of Old World birds, is found in Pennsylvania. Blue-gray Gnatcatchers are small active birds with a rather long narrow tail which they usually cock upward.

Blue-gray Gnatcatcher *Polioptila caerulea*

General status: The aptly named Blue-gray Gnatcatchers are active and conspicuous as they hop from branch to branch with tail cocked. They breed primarily across the eastern and southwestern U.S. and winter from along the Atlantic coast from Virginia south along the Gulf coast and west across the U.S. to California. In the East they are casual north along the Atlantic coast to New England. Gnatcatchers are fairly common regular migrants and they breed throughout Pennsylvania, although they are more common in the southern portion of their range. They have expanded northward in Pennsylvania since the 1950s.

Habitat: Gnatcatchers are found in a variety of woodland types, especially woodland edges and thickets during migration. They breed in various forested settings, most frequently in oaks,[1] but they generally avoid northern hardwood forest settings.[2] They also use both dry upland and moist lowland woodlands during nesting.

Seasonal status and distribution

Spring: Migrants may appear as early as the last week of Mar. Their typical migration period is from the second or third week of Apr to the third week of May. The greatest number of birds is observed from the fourth week of Apr to the second week of May. A few late migrants linger to the first week of Jun. Twenty or more birds have been observed in a day at migration concentration points, such as North Park in Pittsburgh and PISP, but migrants are usually not seen in these numbers.

Breeding: Gnatcatchers are fairly common breeders in the southern Ridge and Valley and Southwest regions. They are uncommon to rare (or at least local) in the Piedmont and over the northern half of the state, including the Allegheny Mountains in Clearfield and Cambria cos. They tend to be associated with major river valleys across the northern tier. Regrowth and maturation of oak forests in Pennsylvania have undoubtedly benefited this species. The BBA project documented the most extensive distribution known, but BBS data since 1966 do not reflect the expansion. The reason may be that Jun is not the best month to detect gnatcatchers' singing. Most nests with eggs have been found from 5 May to 3 Jun, with Harlow's (1918) average date of 14 May. The nest is a compact cup bound by spider's web, decorated with lichen, and saddled on a horizontal branch well out from the trunk at various heights above ground. Four or five bluish white eggs are laid.

Fall: The status of gnatcatchers in fall is poorly defined. Usually, fewer birds are recorded in fall, when singing stops, than in spring and summer. An increase in birds at banding stations suggests southward movement as early as the third week of Jul, with a decrease in the number after the first week of Sep. The greatest number of birds are banded during the third and fourth weeks of Aug. Observations of single gnatcatchers may continue through the fall to the first week of Jan. A number of sightings have been made during Dec, especially in the Coastal Plain and the Piedmont, but the species is not known to winter.

History: Historical accounts indicate a radically different pattern than that seen in recent times. Various southeastern bird lists published in the late nineteenth century considered gnatcatchers "frequent" in summer (e.g., Delaware Co., Cassin 1862; Lancaster Co., Libhart 1869). By the end of the 1800s, Warren (1890) considered them rare and Harlow (1913) "accidental" east of the Alleghenies, suggesting a major decline. Jacobs (1893) called gnatcatchers "numerous" and referred to an association with walnuts and oaks in Greene Co. Townsend (1883) suggested their rarity in Westmoreland Co., in contrast with other southwestern reports of that era. The northern extent of the distribution at about 1900 appeared to be Beaver and Perry cos. (Poole, unpbl. ms.). Later, Todd (1940) identified Crawford Co. as the northwestern extent of their range but considered them "virtually unknown" in the Ridge and Valley. Poole noted that in the 1940s the gnatcatcher populations were expanding north and regaining their range in the east and stated that the "first definite nesting records for the southeastern counties during the present century" in Berks and Delaware cos. were in 1946. Kunkle (1951) considered gnatcatchers to have become fairly common transients and breeders in Union Co. from 1946 to 1951, when presumably they were expanding north along the Susquehanna River. The expansion continued north, for example, to Monroe Co. by 1960 (Poole 1964). One early winter bird was at Conestoga Township in Lancaster Co. on 17 Jan 1949 (Morrin et al. 1991).

Comments: A major expansion of the gnatcatcher's breeding range was also noted in Ontario in the 1940s.[3] Breeding has been noted in southern Maine since 1979 and was documented for the first time in the Maritime Provinces in 1989 (Erskines 1992). A female banded at PNR on 9 May 1969 was killed by a boy with a slingshot in Llano Grande, Jalisco, Mexico, on 1 Nov 1970.[4] This was the first recovery of a banded Blue-gray Gnatcatcher south of the U.S. and led to the discovery of a new migration route and wintering range for the eastern North American population.[5]

[1] Ellison, W.G. 1992. Blue-gray Gnatcatcher *(Polioptila caerulea)*. BNA (A. Poole and Gill 1992), no. 23.

[2] Leberman, R.C. 1992. Blue-gray Gnatcatcher *(Polioptila caerulea)*. BBA (Brauning 1992): 262–263.

[3] Sutherland, D.A., and M.E. Gartshore. 1987. Blue-gray Gnatcatcher *(Polioptila caerulea)*. Pages 318–319 *in* Atlas of breeding birds of Ontario, ed. M.D. Cadman, P.F.J. Eagles, and F.M. Helleiner. Univ. Waterloo Press, Waterloo, Canada.

[4] Leberman, R.C., and M.H. Clench. 1972. Birdbanding at Powdermill, 1971. Research report no. 30. Powdermill Nature Reserve, Carnegie Museum of Natural History, Pittsburgh.

[5] Parkes, K.C., and M.H. Clench. 1972. Recovery of a Pennsylvania-banded Blue-gray Gnatcatcher in western Mexico. Condor 74:222.

Family Turdidae: Thrushes

Most thrushes are ground-dwelling birds of the woodlands. They are best known for their beautiful songs. They frequently are observed in their typical feeding behavior, hopping a few feet and then picking at the ground or under leaf litter. Bluebirds and robins are distinctive in this family for their use of man-made structures for nesting. The six species of spot-breasted thrush (Veery, Gray-cheeked, Bicknell's, Swainson's, Hermit, and Wood) nest on the ground or in shrubs or trees, usually less than 10 feet above the ground. The cup-shaped nests of these birds are usually constructed of grasses and mosses. All thrushes nesting in Pennsylvania (except Swainson's Thrush) lay from three to six unmarked blue or bluish green eggs. Large nocturnal flights occur during migration, when hundreds can be heard flying overhead just before dawn.

Northern Wheatear *Oenanthe oenanthe*

General status: In North America, Northern Wheatears breed in Alaska and Yukon in the West and in Greenland, Ellesmere, and Baffin islands south to Labrador and northern Quebec in the East. They winter in northern Africa and across southern Asia to China. During migration they are casual along the Atlantic coast and inland. There have been two accepted records in Pennsylvania.

Seasonal status and distribution

Fall: A Northern Wheatear was found by A. Brady at Newtown, Bucks Co., on 25 Sep 1995 and was photographed.[1] Another identified by E. Wengard and colleagues in Walker Township in Juniata Co. from 4 to 7 Oct 1997 also was photographed.[2]

Comments: A bird was collected near Lansdale in Montgomery Co. on 7 Oct 1919, but the specimen apparently no longer exists.[3] An immature reported at Memorial Lake State Park in Lebanon Co. on 19 Oct 1997 is currently under review.[4]

[1] Pulcinella, N. 1995. Rare bird reports. PB 9:137.

[2] Haas, F.C., and B.M. Haas, eds. 1998. Birds of note—October through December, 1997. PB 11:231.

[3] Gillen, J.R., 1919. General notes. Cassinia 23:39.

Northern Wheatear at Walker Township, Juniata County, October 1997. (Photo: Franklin C. Haas)

[4] Miller, R. 1998. Local notes: Lebanon County. PB 11:240.

Eastern Bluebird ***Sialia sialis***

General status: Eastern Bluebirds are breeding residents primarily across southern Canada from Saskatchewan east to Nova Scotia and south across the eastern half of the U.S. They winter primarily across the southern half of the U.S. Eastern Bluebirds are fairly common regular migrants and breed throughout Pennsylvania away from urbanized centers. They winter in their breeding range except at higher elevations in the High Plateau. The number of nesting birds has increased since the 1950s with the erection of bluebird boxes. Many have become dependent upon artificial nest boxes for nesting; few birds now use natural cavities.

Habitat: Bluebirds can be found in a wide variety of open habitats where tree cavities or nest boxes are available. They prefer abandoned orchards, hayfields and pastures with scattered trees, forest edges, and openings.

Seasonal status and distribution

Spring: The timing of northward movement may depend upon the severity of winter weather. After mild winters the first movement may begin as soon as a warm front passes during the third or fourth week of Feb. After severe winters migration may not begin until the first or second week of Mar or later, especially in the High Plateau and in the snowbelt areas of the Glaciated Northwest. Peak migration is not well defined, but loose flocks containing from 5 to more than 25 birds have been recorded on days of warm fronts throughout Mar and Apr. Very little movement is detected after the second week of May.

Breeding: Uncommon to fairly common statewide, Eastern Bluebirds now nest in every county. Bluebirds are most commonly reported on BBS routes in the Ridge and Valley region and along the western-tier counties (see Map 20), and they are least frequently reported in the heavily forested northern tier and in the urbanized southeast. Even in extensively forested areas, they may be found in openings and clearcuts. They compete for nest boxes with Tree Swallows, House Wrens, and House Sparrows. European Starlings generally usurp natural sites, including natural cavities and holes excavated by woodpeckers. The BBS documented a long-

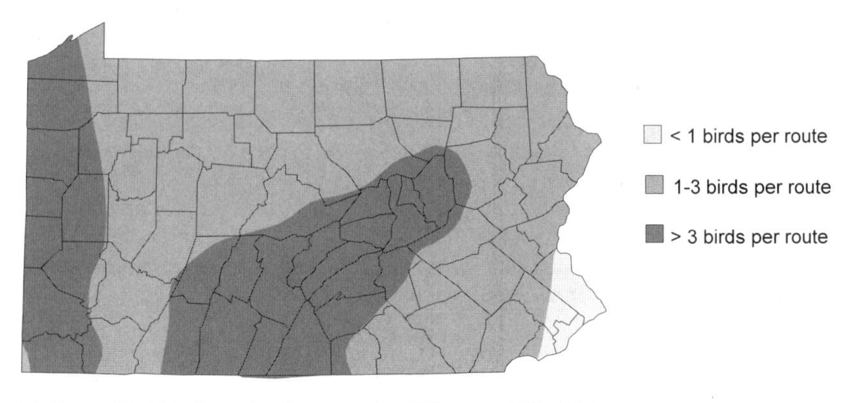

20. Eastern Bluebird relative abundance, based on BBS routes, 1985–1994

term expansion of 1.7% per year, with much of that occurring since 1980, but these data seem to underestimate the recovery of this species. Bluebirds have two or three broods between mid-Apr and early Aug.

Fall: As during the spring, it is difficult to determine their migration patterns during this season. When the last young have fledged, family groups may begin wandering with no indication of actual southbound movements. When colder weather arrives in late fall and food becomes scarce, some will withdraw to lower elevations or drift southward to seek more suitable conditions. Occasional notable flights have been observed in fall along the hawk ridges. For example, on 27 Oct 1963 several hundred were counted passing Mt. Bethel in Northampton Co.[1] In late fall they may become locally common, especially in the Piedmont, where CBC totals exceed 200 birds in some years.

Winter: Populations vary from year to year depending upon food availability and severity of weather. However, they have become more common winter residents while breeding populations have grown since the early 1970s. At most sites, wintering flocks consist of small family groups of four to six birds. During severe winters local bluebird flocks may be found with American Robins feeding on sumac or other fruit-bearing shrubs and trees. They are fairly widespread over the state. Fewer birds are recorded in mountainous areas and in the snowbelt of the Glaciated Northwest, where they may be absent during years of severe winters with cold temperatures and heavy snow cover. More birds winter in agricultural areas of the Piedmont than in any other areas of the state.

History: Bluebirds were considered abundant by many ornithologists in the nineteenth century (e.g., Cassin 1862; Libhart 1869; Turnbull 1869). A decline in numbers was noted at the end of the nineteenth century by Stone (1894) and Cope (1898) possibly in response to increasing House Sparrow populations or possibly as a result of heavy winter weather. They were still common, but were declining, in the early decades of the twentieth century (Harlow 1913; Sutton 1928a). Nest boxes were promoted by the PGC and local Audubon societies and were apparently distributed as early as the 1920s.[2] Populations nevertheless continued to decline, possibly as a result of increased use of pesticides during the 1950s. Poole (unpbl. ms.) considered 1960 to be "an all-time 'low'" point in the population of bluebirds and called them "rare" in southeastern counties (Poole 1964).

Comments: Nest boxes with a 1.5-inch diameter hole, placed in open habitats, have good prospects of attracting bluebirds anywhere in Pennsylvania except urban centers. Boxes placed in pairs, 10–20 feet apart, reduce swallow competition. Several useful books are available on this popular species (Zeleny 1976; Davis and Roca 1995).

[1] Scott, F.R., and D.A. Cutler. 1964. Middle Atlantic Coast region. AFN 18:22.
[2] Sutton, G.M. 1925. A year's program for bird protection. Pennsylvania Game Commission, Harrisburg.

Mountain Bluebird ***Sialia currucoides***

General status: These birds of the mountainous West breed from southern Alaska and southeast to Manitoba south in the Rocky Mountain region to New Mexico. Mountain Bluebirds winter from Montana south to the southwestern states east to Texas. They are casual as far east as Nova Scotia, Vermont, Massachusetts, New York, Pennsylvania, Maryland, Kentucky, and Mississippi (AOU 1988). In Pennsylvania they have been recorded at least five times, with all but one record in the Piedmont.

Seasonal status and distribution

Fall through spring: The first record in Pennsylvania was of a male discovered by B. Grimacy and colleagues at MCWMA in Lancaster Co. from 1 to 5 Mar 1974, and it was photographed.[1] A male observed and photographed by R. Wiltraut at Beltzville State Park in Carbon Co. on 16 Dec 1984 remained to 22 Mar 1985.[2] Possibly the same bird, found by R. Wiltraut and co-workers, returned to Beltzville State Park on 22 Dec 1985 and remained to 26 Jan 1986.[3] A female was found by D. Heathcote and co-workers at Long Arm Dam in York Co. on 23 Dec 1988 and remained to 19 Mar 1989.[4] It was also photographed. The most recent sighting was of one photographed near Shirleysville in Huntingdon Co. on 15–16 Apr 1995 by T. Fultz.[5]

[1] Scott, F.R., and D.A. Cutler. 1974. Middle Atlantic Coast region. AB 28:624.
[2] Boyle, W.J., Jr., R.O. Paxton, and D.A. Cutler. 1985. Hudson-Delaware region. AB 39:152–153.
[3] Boyle, W.J., Jr., R.O. Paxton, and D.A. Cutler. 1985. Hudson-Delaware region. AB 40:263.
[4] Boyle, W.J., Jr., R.O. Paxton, and D.A. Cutler. 1989. Hudson-Delaware region. AB 43:292.
[5] Grove, G. 1995. Local notes: Huntingdon County. PB 9:93.

Mountain Bluebird at Beltzville Lake, Carbon County, December 1984. (Photo: Rick Wiltraut)

Townsend's Solitaire ***Myadestes townsendi***

General status: Townsend's Solitaires breed from southern Alaska south in the Rocky Mountain region to New Mexico. They winter over their entire breeding range in the U.S. and east to North Dakota, Missouri, and Texas. Solitaires sometimes wander great distances and are casual as far east as Newfoundland, south through the New England states, New York, Pennsylvania, and New Jersey (AOU 1998). There are five accepted records in Pennsylvania, all since 1986.

Seasonal status and distribution

Spring: A Townsend's Solitaire was photographed by G. Dewaghe at Revere in Bucks Co. on 22–23 Mar 1993.[1]

Fall: One was well described by J. Bouton and H. Fink at HMS on 15 Oct 1989,[2] and one was identified in flight by G. Wilhelm on 18 Dec 1993 at MSP in Butler Co.[3] On 10 Oct 1995 one was found at SGL 106 in Schuylkill Co. by S. Weidensaul.[4,5]

Winter: The first accepted record in Pennsylvania was of one photographed by J. Pushcock along the Appalachian Trail near Lehigh River Gap in Lehigh/Carbon Co. on 18 Jan 1986.[6]

[1] Haas, F.C., and B.M. Haas, eds. 1993. Photographic highlights. PB 7:19.
[2] Paxton, R.O., W. J. Boyle Jr., and D.A. Cutler. 1990. Hudson-Delaware Region. AB 44:65.
[3] Pulcinella, N. 1995. Sixth report of the Pennsylvania Ornithological Records Committee (P.O.R.C.) June 1995. PB 9:72.
[4] Haas, F.C., and B.M. Haas, eds. 1996. Birds of note—October through December 1995. PB 9:209.
[5] Pulcinella, N. 1996. Seventh report of the Pennsylvania Ornithological Records Committee (P.O.R.C.) June 1996. PB 10:50.
[6] Boyle, W.J., Jr., R.O. Paxton, and D.A. Cutler. 1986. Hudson-Delaware region. AB 40:263.

Veery ***Catharus fuscescens***

General status: One of the six species of spot-breasted thrush, Veeries breed across southern Canada south in the Rocky Mountains to Utah and Colorado in the West and in the New England states through the Appalachian Mountains to Georgia in the East. They winter in South America. Only three winter records, in Dec and Jan, exist for this species in the Northern Hemisphere: one record each in Ontario, Connecticut, and Louisiana. Veeries are fairly common regular migrants in Pennsylvania, but they are most conspicuous in spring when in full song. They are regular breeders across much of the state but only locally in the southern half of the state.

Habitat: During migration Veeries are found in woodlands, thickets, parks, and even suburban yards surrounded by some cover. They prefer moist woodlands with dense shrub understory during the breeding season.

Seasonal status and distribution

Spring: Early migrants are occasionally recorded the first week of Apr during warm springs followed by mild winters. Spring migration usually begins in the last week of Apr or the first week of May and continues until the last week of May. Most migrants are recorded during the second or third week of May.

Breeding: Veeries are common in the northern half of the state and locally south into the Laurel Highlands; they are rare to uncommon and local in southern counties but do not regularly nest from Pittsburgh south. They are found across the range of elevations found in the state and are not particularly sensitive to forest fragmentation (Robbins et al. 1989), but their distribution tends to be patchy. On BBS routes they are most abundant in the Glaciated Northeast and in the High Plateau from Potter Co. west to Crawford Co. in the Glaciated Northwest. They are uncommon in the eastern Piedmont, Blue Ridge section, Allegheny Mountain section, and north of Pittsburgh in the Southwest region. They are rare and irregular in the southern Ridge and Valley and western Piedmont. Eggs are laid in a bulky nest on a base of leaves on the ground or in a low shrub. Eggs are usually found from 24 May to 20 Jun, reflecting only one brood per year, although late nests have been found in Jul. Harlow's average date for 95 sets in the Poconos was 1 Jun (P. B. Street 1954). Singing ceases in Jul, after which birds are difficult to detect. The state BBS shows no population change since 1966.

Fall: The beginning of fall migration is not well defined. Probable migrants have been recorded at PNR as early as the first week of Jul, but most migrants probably begin moving the second week of Aug. Banding records suggest that the greatest numbers of birds pass through Pennsylvania during the first week of Sep. Birds become scarce by the third week of Sep, and most have departed by the fourth week of Sep. Stragglers may remain to the second week of Oct. One found on 21 Dec 1979 at Warren was a very late record.[1]

History: There has been some confusion among early authors as to the breeding status and distribution of Veeries in the state. Contradictory historical references suggest that Veeries were spottily distributed, locally common, but absent from some areas. For example, Stone (1891) did not list them among breeding birds in Luzerne Co., but Cope (1898) considered Veeries abundant, even more plentiful than either Wood or Hermit thrushes, just 20 miles to the north in Susquehanna Co. Local experience, or loss of forest cover, may have biased some authors such as Warren (1890), who stated that Veeries bred sparingly in the northern mountains, and Harlow (1913), who considered them rare in the northeast. Scattered confirmed breeding observations were reported south even to Philadelphia Co. in the 1940s.[2] Poole (1964) suggested that Veeries were expanding, particularly in eastern Pennsylvania. Todd (1940) referred to the spotty and possibly irregular pattern of occurrence in summer but presented a distribution map in western Pennsylvania that corresponds almost precisely to the pattern shown by the BBA.

Comments: Veeries are intermediate in relative abundance between Hermit and Wood thrushes statewide, with an average of about three birds per BBS route statewide.

[1] Hall, G.A. 1980. Appalachian region. AB 34:273.
[2] Bond J., and R. Miller 1942. General notes. Cassinia 32:48.

Gray-cheeked Thrush ***Catharus minimus***

General status: Gray-cheeked Thrushes breed from Alaska east across Mackenzie, northern Ontario, and Quebec to Newfoundland. They winter primarily in South America at least to Brazil. Gray-cheeked Thrushes are rather secretive and inconspicuous. Banding records and the detection of nocturnal flight calls indicate that more birds pass through the state than field observations indicate. They are generally considered uncommon to rare regular migrants. One summer record exists.

Habitat: Gray-cheeked Thrushes are found in woodlands, often with a dense understory and in parks and on suburban yards surrounded by some cover.

Seasonal status and distribution

Spring: Gray-cheeked Thrushes are usually the last of the spot-breasted thrushes to appear in Pennsylvania. Their usual migration period is from the second to the fourth week of May. Birds pass through the state rather quickly, with most recorded during the third and fourth weeks of May. Stragglers have been recorded to the second week of Jun.

Summer: The only reliable summer record of Gray-cheeked Thrush in Pennsylvania was of a bird found singing in late Jun at State College in 1982.[1] No breeding evidence has been reported for this species in the state.

Fall: Most fall migrants are recorded at banding stations. Earliest migrants appear the last week of Aug. Their primary migration period is from the second week of Sep to the third week of Oct. Banding records suggest peak migration during the fourth week of Sep and the first week of Oct. At banding stations such as PNR, 20 or more birds may be banded in a day. On 25 Sep 1965 an extrordinary total of 52 was banded at PNR (Leberman 1976). Stragglers have been recorded until the third week of Nov.

History: It is difficult to determine the historical status of Gray-cheeked Thrush in Pennsylvania, as difficulty in separating this species from Swainson's Thrush was noted by many authors. Todd (1940) considered Gray-checked Thrushes to be more frequently observed in fall than in spring. They were reported to be fairly numerous on Presque Isle in the fall of 1900, though Todd (1940) saw no birds in the spring of that year. Poole (unpbl. ms.) considered the species to be "sometimes rather common in spring, but usually rare or often undetected in the fall." Timing of migration has changed little in recorded history.

[1] Hall, G.A. 1982. Appalachian region. AB 36:977.

Bicknell's Thrush ***Catharus bicknelli***

General status: Until 1995 this spot-breasted thrush was considered a subspecies *(C. m. bicknelli)* of the Gray-checked Thrush. Bicknell's current breeding range is from the eastern edge of Quebec along the St. Lawrence River, including New Brunswick and northern Nova Scotia, south through northern New England to the Catskills in New York.[1] They winter in Cuba, Hispaniola, and Puerto Rico (AOU 1998). Very little is known about their migrant status; away from their breeding range they are difficult to separate from Gray-cheeked Thrushes, except in the hand. As a result, there is little information about the occurrence of the species in Pennsylvania. The inclusion of Bicknell's Thrush in the state is based on specimens collected and banding records from before 1979. The specimens collected suggest that migration timing is very similar to that of Gray-cheeked Thrush. Most records are from the southeastern portion of the state. Apparently there has been no accepted spring record since 1953.

Seasonal status and distribution

Fall: Several Bicknell's Thrushes have been collected, and at least three have been banded between 20 Sep and 16 Oct: the most recent record was of a bird banded by R. F. Leberman at PISP on 25 Sep 1996.[2] A few have also been banded at PNR, including one collected on 30 Sep 1979.[3]

History: Two birds were collected in spring: one at Chestnut Hill by W. L. Abbott in Philadelphia Co. on 19 May 1877 and another at Kennett Square by C. J. Pennock in Chester Co. on 20 May 1909 (Poole, unpbl. ms.). Both specimens are at the ANSP. A Bicknell's Thrush was banded by M. Emlen at Germantown in Philadelphia on 2 Oct 1956.[4]

Comments: This species' recent split presents a new challenge to field observers. At present, the only reliable field distinction between Gray-cheeked and Bicknell's thrushes is their song. It is likely that at least a few Bicknell's Thrushes pass through Pennsylvania east of the Allegheny Front each year.

Until a safe method of field identification is discovered or banders take a closer look at the Gray-cheeked Thrushes they capture, Bicknell's Thrushes are likely to migrate through the state unidentified.

[1] Rimmer, C. 1996. A closer look: Bicknell's Thrush. Birding 28:118–123.
[2] Leberman, R.F. 1997. Atlantic Flyway review, region 3, ed. E. Brooks. North American Bird Bander 22:147.
[3] K.C. Parkes, pers. comm.
[4] Ulmer, F.A. 1956. General notes (other records). Cassinia 42:32.

Swainson's Thrush ***Catharus ustulatus***

General status: The breeding range of Swainson's Thrush is from Alaska across Canada south in the Rocky Mountain region to Arizona and New Mexico and the New England states through the Appalachian Mountains to West Virginia. They winter from Central America to South America. A species of northern forests, Swainson's Thrush is the most common migrant spot-breasted thrush in Pennsylvania. However, they appear to have declined in recent years as spring and fall migrants. They are fairly common to common regular migrants over most of the state. Migrant populations vary from year to year. In Pennsylvania, Swainson's Thrushes breed almost exclusively in mature forests in the northern-tier counties of the High Plateau.

Habitat: During migration Swainson's Thrushes inhabit woodlands, often with a dense understory, and parks and suburban yards surrounded by some cover. During the breeding season they are found primarily in mature hemlock forests above 2000 feet elevation.

Seasonal status and distribution

Spring: The earliest spring arrivals appear the fourth week of Apr. Their typical migration period is from the first or second week of May to the fourth week of May. Peak migration is usually during the third and fourth weeks of May. Stragglers away from breeding sites remain until the first week of Jun.

Breeding: Swainson's Thrushes are rare and extremely local breeders in northern counties. The largest population is in the 4000-acre old-growth forests of Tionesta Natural and Scenic Areas and Heart's Content (Allegheny National Forest) of McKean and Warren cos., where they are the most common thrush (Haney and Schaadt 1996). Swainson's Thrushes are regular but rare in scattered mature woods of Potter, northern Lycoming and Clinton, Sullivan, and formerly Monroe cos. Nesting records also have come from Elk Co. and Cook Forest State Park, Forest Co. They were recorded during about half of the years on one Sullivan Co. BBS route and are casually on 12 others. Pennsylvania is at the southern edge of Swainson's Thrushes' breeding distribution, except for a disjunct population in the high spruce forests of West Virginia (Buckelew and Hall 1994). The nest is a bulky cup of twigs and diverse vegetation, usually placed low (up to 10 feet) in a coniferous tree or shrub. They lay three or four pale blue eggs blotched or spotted with brown.

Fall: First migrants are detected the first or second week of Aug. One banded on 3 Jul 1983 at PNR[1] and others banded at this site later in Jul suggest that migration may begin much earlier, even before the start of the postbreeding molt. These birds are always adults that probably failed in their nesting attempts and wandered from breeding territories.[2] Migration continues to the fourth week of Oct. Migration peaks from the second to the fourth week of Sep. Stragglers sometimes remain as late as early Jan.

Winter: One apparently injured bird spent the winter of 1965 at a feeding station east of Jones Mills in Westmoreland Co. (Leberman 1976).

History: Warren (1890) was the first to identify Swainson's Thrushes as breeders in northern Pennsylvania. Earlier bird students probably overlooked them. Subsequent authors identified them as locally common in northern extensively forested regions, for example in Tamarack Swamp, Clinton Co. (Cope 1902) and at North Mountain, Sullivan Co. (Dwight 1892). Poole (unpbl. ms.) thought that Swainson's Thrushes "may be found nesting in suitable spots along the entire Allegheny Mountain range, south to the Maryland line." He was almost certainly correct in indicating that Baird's (1845) statement that this species was abundant in summer near Carlisle was obviously in error. Swainson's Thrushes apparently disappeared from much of their range when the northern forests were logged. Reimann (1947) failed to find them in Tamarack Swamp, and P. B. Street (1954) considered them a former breeder in the Poconos.

Comments: It is likely that at least some of the records, especially those in late fall, are misidentified Hermit Thrushes. Natural deterioration of Pennsylvania's few old-growth stands and harvest of other mature woodlands could significantly affect the population of this species. A Swainson's Thrush banded on 26 Sep 1964 at PNR was recovered at Vergara, Colombia, South America, on 25 Nov 1964, approximately 2500 air-miles away.[3] A Swainson's Thrush banded on 24 Sep 1966 at PNR was killed by an Indian blowgun near Sinchi-Yacu, Peru. This banding recovery was a total of 3150 air-miles away, PNR's long-distance recovery record.[4] Swainson's Thrushes are listed as Rare by the OTC.

[1] Hall, G.A. 1983. Appalachian region. AB 37: 989.

[2] R.C. Leberman, pers. comm.

[3] Leberman, R.C., and M. A. Heimerdinger, 1966. Bird-banding at Powdermill, 1965. Research report no. 15. Powdermill Nature Reserve, Carnegie Museum of Natural History, Pittsburgh.

[4] Leberman, R.C., and M.H. Clench. 1972. Birdbanding at Powdermill, 1971. Research report no. 30. Powdermill Nature Reserve, Carnegie Museum of Natural History, Pittsburgh.

Hermit Thrush *Catharus guttatus*

General status: Hermit Thushes breed from Alaska across Canada south in the Rocky Mountain region to Arizona and New Mexico and in the East through the New England states and Appalachian Mountains to North Carolina and Tennessee. In winter they are found over most of the U.S. except in the higher elevations of the Rocky and Appalachian mountains. Hermit Thrushes are uncommon to fairly common regular migrants in all regions of Pennsylvania. Of the six species of spot-breasted thrushes, they are the first to appear in the spring and the last to leave in the fall, and they are the only one that remains in the state through winter. They nest in the forested mountains across the state. Hermit Thrushes winter at low elevations, primarily in the southeast.

Habitat: During migration they are found in mixed woodlands and brushy edges, shaded suburban lawns, and parks with shrubby or wooded margins. During the nesting season they are found in a variety of forested habitats, usually those containing some conifer cover and generally above 1800 feet elevation. In winter they usually prefer thickets and brushy woodland edges in riparian areas with fruit-bearing trees or shrubs.

Seasonal status and distribution

Spring: Their typical migration period is from about the second week of Mar to the second or third week of May. Migration peaks during the second and third weeks in Apr.

Breeding: Hermit Thrushes are common widespread nesters in forested settings in the High Plateau and Glaciated Northeast, south locally through the

Allegheny Mountains to the Maryland border. They are uncommon and local in extensive forested mountains in the Ridge and Valley and rare in the Glaciated Northwest and the Blue Ridge Mountains. On BBS routes they are most common in extensive northern hardwood forests of the High Plateau (Cameron, McKean, and Potter cos.) and in the Pocono section. A population increase of 9% per year since 1980 on BBS routes is reflected in the expansion into the Blue Ridge section during the late 1980s,[1] and nesting near Hawk Mountain, Berks Co., since 1980 (Uhrich 1997). They are notably absent from the Pymatuning area, where Poole (1964) indicated isolated instances of nesting. Nests, hidden on the ground under low branches or ferns, are usually lined with conifer needles or various rootlets. Eggs are most often found between 12 May and 29 Jun, with nests as late as 10 Aug. This long nesting season suggests two broods.

Fall: Hermit Thrushes begin migrating through the state in the last week of Sep, when the other spot-breasted thrushes have already left. Migration usually peaks during Oct. By the second week of Nov most birds have moved south. Stragglers remain to the first or second week of Jan, and a few may remain to winter.

Winter: More wintering birds are observed in the Piedmont, where they are uncommon, than in any other region. They are rare and local in all other regions in winter. In areas west of the Allegheny Front they are found only during mild winters. During severe winters Hermit Thrushes usually are not found after the second or third week of Jan west of the Allegheny Front. In the upper and central High Plateau region they are absent in winter except for a record from Columbia in Bradford Co.[2] Most sightings are of single birds, often associated with American Robins, usually where fruit-bearing trees or shrubs are present.

History: Throughout history the abundance of Hermit Thrushes as breeders in the state has depended on the presence of mixed forests. During periods when forests were cleared, their numbers declined, only to rebound when the forests regenerated. Harlow (1913) called Hermit Thrushes "especially abundant in the northern tier where the conditions are more boreal." Some early authors called these thrushes rare in the northern mountains (Warren 1890; Sutton 1928a), possibly reflecting the loss of northern forested habitat at that time.[1] They were notable as breeding in the Black Moshannon area of Centre Co. (Burleigh 1931) and at Laurel Hill, Westmoreland Co. (Todd 1940), where they are now common.

Comments: Hermit Thrushes are not as common or as widespread in Pennsylvania as either Veeries or Wood Thrushes, even in the High Plateau region. It has been suggested that Wood Thrushes may be dominant over Veeries and Hermit Thrushes, possibly excluding them from breeding territories (Ehrlich et al. 1988). As a result, Hermit Thrushes tend to occur at higher elevations or in woodlands more dominated by conifers than do Wood Thrushes. There, they replace Wood Thrushes but may be matched in abundance by Veeries. With these three species overlapping in at least a third of the state, opportunities abound to study their interspecific interactions.

[1] Mulvihill, R.S. 1992. Hermit Thrush *(Catharus guttatus)*. BBA (Brauning 1992):270–271.
[2] T. Gerlach, pers. comm.

Wood Thrush *Hylocichla mustelina*

General status: Wood Thrushes breed from southeastern Canada across the eastern U.S. west to South Dakota, Kansas, and Texas. They winter from

Mexico south through Central America to Colombia. Wood Thrushes are the most widely distributed of the spot-breasted thrushes in Pennsylvania. They are fairly common regular migrants, and their breeding distribution is statewide.

Habitat: During migration Wood Thrushes are found in mixed woodlands and brushy edges, shaded suburban lawns, and parks with shrubby or wooded margins. In the breeding season they prefer moist woodlands but will also use a variety of habitats, including dry hillsides, parks, orchards, and woodlots in suburbs.

Seasonal status and distribution

Spring: Wood Thrushes have been recorded as early as the first week of Apr. Their regular migration period is from the third or fourth week of Apr to the fourth week of May. The greatest number of birds is recorded during the second and third weeks of May.

Breeding: Wood Thrushes are fairly common breeders. They are generally found at lower elevations than Hermit Thrushes or Veeries. Although found in 91% of BBA project blocks, their relative abundance on BBS routes varies markedly across the state and in complex ways. They are least common in highly agricultural areas (Chester to York cos.), urban centers, and in parts of the northern-tier counties (see Map 21). The BBS reports a decline of 2% per year in Pennsylvania since 1966, but this decline has not altered its status in the state. From a large collection of egg sets (more than 150), most were found from 19 May to 22 Jun. Average dates for nests with eggs vary from 26 May in southern Pennsylvania (Harlow 1918) to 5 Jun in the Poconos (P. B. Street 1954). Nests with young have been found as late as 22 Aug (Sutton 1928b). Wood Thrush nests are similar to the robin's, but smaller (H. H. Harrison 1975) and are placed on a base of leaves. Their nests are easier to find than those of many other migrant birds, often placed less than 10 feet above the ground in the crotch of a sapling or on a horizontal branch of a tree.

Fall: The beginning of migration is difficult to discern because of their wide breeding distribution in the state. Migration becomes evident by late Aug, when there is a noticeable increase in numbers. Migration usually peaks during the second and third weeks of Sep. Most birds are gone by the first week of Oct, but stragglers may remain to the fourth week of Dec. A very late Wood Thrush was seen in a flock of Eastern Bluebirds near PNR on 14 Jan 1985.[1]

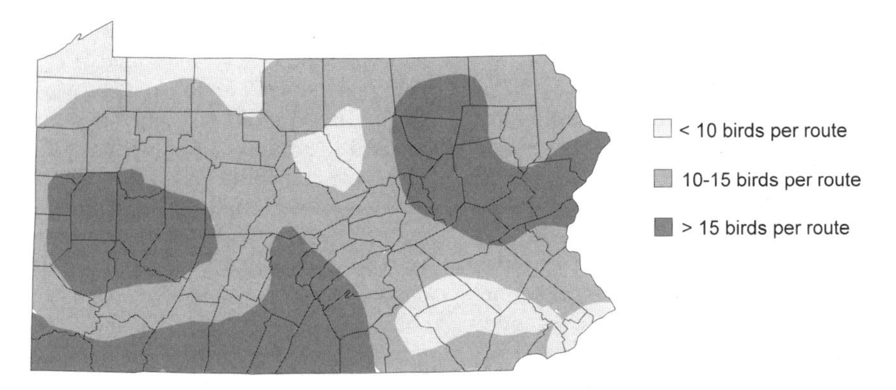

21. Wood Thrush relative abundance, based on BBS routes, 1985–1994

There have been no midwinter records in the state.

History: Harlow (1913) described the general southerly distribution of Wood Thrushes when stating that they were an "abundant breeding bird over the whole southern half of the state" but a rare summer resident in the northern counties. Other historical descriptions suggest that they were less common in the northern, higher sections of the state (e.g., Warren 1890; Stone 1891) than typically described today. However, there are notable exceptions, including Dwight's (1892) description; he listed Wood Thrushes as "an abundant species, found in the woods to the highest peaks."

Comments: Wood Thrushes became the focus of intensive breeding studies across the continent during the late 1980s that added tremendously to our knowledge of this species. Nesting success is much lower in smaller woodlots than in more extensive forest settings in Pennsylvania, suggesting that many territorial birds in predominantly agricultural or suburban areas are not successful and are replenished by immigrants from more extensively forested regions.[2]

[1] Hall, G.A. 1985. Appalachian region. AB 39:167.
[2] Hoover, J.P., M.C. Brittingham, and L.J. Goodrich. 1995. Effects of forest patch size on nesting success of Wood Thrushes. Auk 112:146–155.

American Robin ***Turdus migratorius***

General status: American Robins breed throughout North America and winter from southern Canada south across the U.S. They are among the most familiar and widespread birds in Pennsylvania. Robins have successfully adapted to human habitat alterations and activities and are abundant residents in suburban areas. They are common to abundant regular migrants, and they breed wherever suitable ledges or trees for nesting exist. The winter range of the American Robin is statewide, but numbers of birds vary in different regions of the state, depending upon severity of winter and availability of food.

Habitat: During migration and the breeding season, robins use a wide variety of habitats from woodlands to suburban lawns. In winter they prefer thickets, brushy woodland edges, and city parks, especially where fruit-bearing trees and shrubs or seeps are found.

Seasonal status and distribution

Spring: Flocks of robins frequently move around throughout the winter searching for food; usually by the third week of Feb flocks begin a northerly movement. By the second or third week of Mar robins are widespread and abundant. Numbers remain high until the second or third week of May. Spring flocks during days with warm southerly winds may number in the hundreds or thousands. By the third or fourth week of May, migration has ended.

Breeding: Common breeders in open habitats statewide, American Robins were the most widespread birds in the BBA[1] and second most common on recent BBS routes in Pennsylvania. They are least common in the extensively forested north-central mountains and are dramatically more common in the Piedmont and western counties (see Map 22) than elsewhere. The long-term population trend is slightly negative (1% per year) statewide since 1966. Most nests with eggs have been found between 18 Apr and 28 Jun, with a range of 10 Apr and 15 Jul and casually later. Harlow's (1918) average of first clutches was 20 Apr; second clutches, 25 May; and third clutches, 4 Jul. Todd (1940) cited an active nest on 9 Dec 1889. The robin's nest is probably the most frequently seen of wild birds' in the state. It is a mud-and-grass structure placed

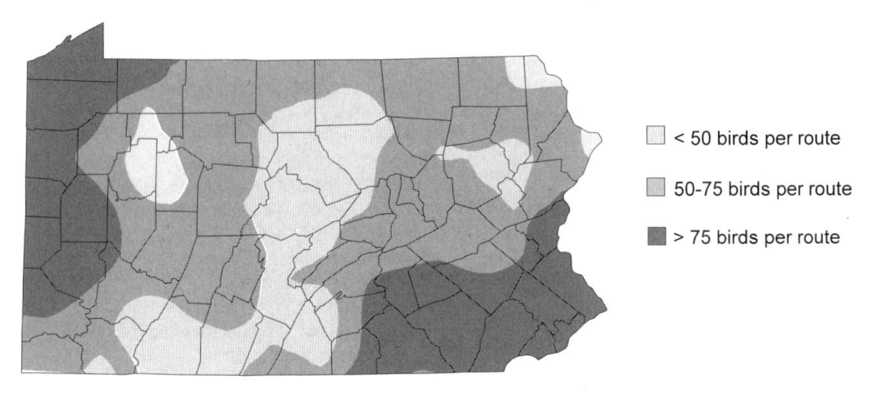

22. American Robin relative abundance, based on BBS routes, 1985–1994

in a range of settings, most often at the fork of a branch or in the crotch of a tree, but also on building ledges. It is frequently found less than 20 feet from the ground. During dry summers, robins are less frequently observed on lawns, having moved to damp woodlands where food is more plentiful.

Fall: The wide breeding distribution and abundance of robins make it difficult to discern the beginning of migration. Birds begin to gather in flocks by mid-Aug, and southward movement is usually well under way by mid-Sep. Migrant flocks continue to pass through the state to at least mid-Nov. In some years, usually by late Oct, gathering flocks form large roosts that may number in the tens of thousands. Robins continue to use large communal roosts through Dec and into the winter. CBCs occasionally document these concentrations, such as the count of 12,642 on the Pittsburgh CBC in 1980.[2]

Winter: Robins are found throughout the state, but the number of birds varies from year to year. They are usually found in nomadic local flocks of fewer than 10 to several thousand birds. Wherever fruit-bearing trees and shrubs such as northern mountain ash or sumac are present, they remain until the food has been consumed and then move on in search of other food sources. They are frequently found in woodlands at seeps or springs that remain open, especially where there is heavy snow cover. Robins may be absent in some areas for weeks during the winter; then flocks may suddenly appear. A sudden appearance of robins in midwinter may be birds from a more-northern population. They are less frequent in winter in most of the High Plateau, and in the snowbelt area of the Glaciated Northwest.

Comments: For generations, people believed that when a robin was seen after the first snowmelt in late winter, spring was not far away. Several subspecies have been described in North America, but little is known of their status and distribution in Pennsylvania.

[1] Santner, S. 1992. American Robin *(Turdus migratorius)*. BBA (Brauning 1992): 274–275.
[2] Van Cleve, B. 1981. Pittsburgh CBC. AB 35:471.

Varied Thrush ***Ixoreus naevius***

General status: Varied Thrushes breed from Alaska south to California and winter from southern Alaska south through Washington, Oregon, and California to Arizona. They are casual over most of eastern North America. Vagrants are generally found with flocks of American Robins. Apparently they were first reported in Pennsylvania around 1966. Since then Varied Thrushes have been recorded in at least 16 out of the last 30 years, with no more than a two-year gap between sightings. There are at least 21 records from the state, with most from the Piedmont. Several of those records have been documented with photographs.

Habitat: Varied Thrushes are observed primarily at bird feeding stations but also have been found in hedgerows, thickets, or woodland edges where fruit-bearing trees or shrubs are found.

Seasonal status and distribution

Fall through spring: The first record in Pennsylvania was of a bird photographed by D. Ostrander on 19 Jan 1966 near Tionesta in Forest Co.[1] Sightings range from Nov to Apr, with most occurring after the first week of Jan. In the Piedmont they have been reported from Berks, Bucks, Chester, Dauphin, Delaware, Franklin, Lancaster, Montgomery, and Northampton cos. There is one record from the Ridge and Valley: a bird found at Duboistown in Lycoming Co. in mid-Dec 1987 remained to 17 Apr 1988.[2,3] In the Glaciated Northeast one was recorded once at Pocono Lake in Monroe Co. from Nov 1971 to Apr 1972.[4] West of the Allegheny Front they have been recorded in the High Plateau twice: one was near Tionesta in Forest Co. from about Jan to Mar 1966;[1] one was present from 1 to 10 Jan 1977 at Tidioute in Warren Co.[5] In the Southwest they have been recorded four times, all in the Pittsburgh area of Allegheny Co. They were reported once each in 1973,[6] 1981,[6] 1981,[7] and 1998.[8] In the Glaciated Northwest one was in Greene Township in Erie Co. from 10 to 21 Mar 1993 and was photographed.[9] No accepted records have come from the opposite corners of the state (i.e., the Coastal Plain or the Lake Erie Shore).

[1] Leberman, R.C. 1966. Field notes. Sandpiper 10:11–12.
[2] Hall, G.A. 1988. Appalachian region. AB 42:265.
[3] Hall, G.A. 1988. Appalachian region. AB 42:435.
[4] Street, P.B. 1973. Varied Thrush wintering in the Pocono Mountains. Cassinia 54:25.
[5] T. Grisez, pers. comm.
[6] Hall, G.A. 1973. Appalachian region. AB 27:616.
[7] Hall, G.A. 1981. Appalachian region. AB 35:300.
[8] J. Hoffmann, pers. comm.
[9] Hall, G.A. 1993. Appalachian region. AB 47:414.

Family Mimidae: Mockingbirds, Thrashers, and Allies

Birds of the family Mimidae are best known for their beautifully varied songs and for the ability of some species to mimic. They are structurally intermediate between wrens and thrushes, having a long tail, long decurved bill, and rictal bristles. Mockingbirds and thrashers spend much of their time on or near the ground, often hidden in heavy undergrowth, but when singing, they often perch exposed high in a tree, or on a rooftop or television antenna. They feed on a variety of seeds, fruits, and berries. Three species are found in Pennsylvania: Gray Catbird, Northern Mockingbird, and Brown Thrasher.

Gray Catbird *Dumetella carolinensis*

General status: Gray Catbirds breed across southern Canada and the U.S. except in Texas and across the Southwest north to Oregon. They winter along the Atlantic coastal lowlands from southern New England south and along the Gulf coast. In Pennsylvania, Gray Catbirds are fairly common to common regular migrants with a statewide breeding distribution. A few winter, primarily in the southeastern portion of the state, at low elevations and along major river valleys.

Habitat: Catbirds are found in a variety of habitats including thickets, hedgerows, woodland edges, and clearcuts. In winter they frequent the same habitat but usually where fruit-bearing trees and shrubs are present.

Seasonal status and distribution

Spring: The first migrants may be seen as early as the second week of Apr. Their usual migration period is from the third or fourth week of Apr to late May. The greatest number of birds is seen during the second and third weeks of May.

Breeding: Catbirds are common nesters in most of Pennsylvania. The lowest frequency of reports on BBS routes (see Map 23) comes from the Allegheny National Forest region; the highest is in the southeast. Catbirds are found in brushy clearcuts from Pennsylvania's highest mountains to urban yards at sea level, provided shrubs are available. No long-term population change is apparent on the BBS routes. The nest is a bulky cup lined with rootlets, most often 3–6 feet from the ground, placed toward the top of a bush or a thicket. Usually four glossy, deep blue-green eggs are laid. Most nests with eggs have been found from 16 May to 27 Jun. Late nests may be active through mid-Jul. This species is rarely parasitized by cowbirds, but when a parasitic egg is found the female catbird ejects it (H. H. Harrison 1975).

Fall: Migration probably begins some time in Aug, but few migrants are detected before the first week of Sep. The greatest number of birds passes through the state during the third and fourth weeks of Sep. Most birds have left the state by the fourth week of Oct. In a mild fall, a few birds linger until the first or second week of Jan or until the first severe winter weather arrives.

Winter: Gray Catbirds are rare but regular winter visitors in the Coastal Plain and perhaps in the southernmost areas of the Piedmont. No winter record is known from the Glaciated Northeast. In the High Plateau one was recorded in Jan and Feb 1987 at Saybrook in Warren Co.[1] They are casual to accidental elsewhere in the state and many do not survive the winter.

History: Catbirds were consistently considered common or abundant birds

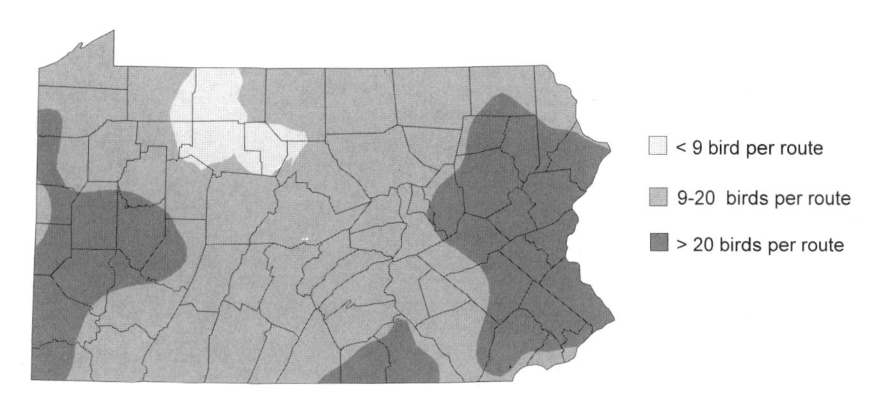

23. Gray Catbird relative abundance, based on BBS routes, 1985–1994

statewide. They undoubtedly became more common as original forests were cut during the nineteenth century.

[1] Hall, G.A. 1987. Appalachian region. AB 41:283.

Northern Mockingbird ***Mimus polyglottos***

General status: Northern Mockingbirds are aggressive and will readily drive away birds larger than themselves as well as cats and dogs. They breed from California east across the U.S. to New England, and they winter throughout most of their breeding range. Mockingbirds are residents primarily east of the Allegheny Front and in the southwestern corner of the state. They are rare residents in the High Plateau, Glaciated Northwest, and Lake Erie Shore, where some seasonal movement is noted. Mockingbirds have been extending their breeding range in Pennsylvania to the north and west.

Habitat: Mockingbirds inhabit thickets and hedgerows, especially those containing multiflora rose, and shaded parks and suburban areas of towns and cities, woodland edges, and rural roadsides. In winter they are found around fruit-bearing trees and shrubs and at bird feeding stations.

Seasonal status and distribution

Spring: Some movement is detected in spring. Away from breeding sites a few birds are observed from about the second week of Apr to the second week of May, with individuals remaining to the first week of Jun.

Breeding: Mockingbirds are fairly common residents in open residential and agricultural areas of the Piedmont and southern Ridge and Valley. They are uncommon to fairly common in the valleys of the northeastern Ridge and Valley and from Pittsburgh south, and generally are rare and irregular elsewhere. In the Glaciated Northeast, mockingbirds are uncommon but regular in the lowlands along the North Branch Susquehanna River north to the New York State line. The distribution during the BBA project reflected the broadest to date, and Mockingbirds continued to expand their range in and north of Pittsburgh in the 1990s.[1] They regularly nest in the southern sections of the Allegheny Mountains and Glaciated Northwest, but are irregular vagrants in all seasons in the northern parts of these regions. They sometimes retreat from their northern range after hard winters. Their bulky nest is usually 3–8 feet from the ground in a shrub, often close to a house (H. H. Harrison 1975). Nests with eggs are found after 10 May and at least until 24 Jun (Poole, unpbl. ms.). Usually four blue-green eggs, heavily blotched or spotted with brown, are laid.

Fall: Movement is occasionally reported from Sep through Nov in areas where they rarely breed.

Winter: Mockingbirds winter throughout their breeding range, but some may withdraw during severe winters from high elevations in the High Plateau, from the Ridge and Valley, and from the snowbelt area of the Glaciated Northwest. They vigorously defend fruit-laden shrubs from other species during the winter. Widespread planting of multiflora rose may contribute to overwinter survival and to their northward range expansion.

History: The earliest naturalists (e.g., Barton, Wilson, Bartram) considered mockingbirds partial migrants in the Philadelphia area shortly after the founding of the country. By the early 1800s, unregulated capture of mockingbirds as cage birds had "rendered this bird extremely scarce for an extent of several miles around the city [Philadelphia]" (Wilson 1808–1814). Mockingbirds were very rare or local in the nineteenth century (e.g., Baird 1845; Libhart 1869) in what is now the heart of their range in the state. Early in the twentieth century, Harlow (1913) said that in the southeast they were "so rare and irregular that it might almost be called accidental." A single breeding record in 1912 was noted in Greene Co. (Harlow 1913). Brushy regrowth that followed clearing of the original forests and farm abandonment enhanced the expansion of this species (Todd 1940). The Migratory Bird Act of 1913 prohibited capture of this and other native birds, thereby paving the way for the mockingbird's modern expansion. Sutton (1928a) noted the gradual northward expansion, which has continued through the century, but he still considered mockingbirds to be rare. The first nesting record in Berks Co. was in 1937 (Poole 1947). By the late 1950s, mockingbirds had become established as permanent residents north to Centre Co. (Wood 1958).

[1] P. Hess, pers. comm.

Brown Thrasher ***Toxostoma rufum***

General status: Brown Thrashers breed from Alberta east across southern Canada to New Brunswick and south to Texas and Florida. They winter primarily from Maryland south to Florida and southwest to Arizona. Brown Thrashers are fairly common regular migrants in Pennsylvania. They breed throughout the state, but numbers have declined. A few thrashers winter, primarily across the southern half of the state.

Habitat: Brown Thrashers prefer thickets, especially multiflora rose and hedgerows in suburban areas, overgrown fields, brushy pastures, and woodland edges. In some winters a few may be found where fruit-bearing trees and shrubs are found.

Seasonal status and distribution

Spring: The earliest migrants may appear the fourth week of Mar. Their typical migration period is from the second or third week of Apr to the last week of May. Peak migration usually occurs from the fourth week of Apr to the second week of May.

Breeding: The Brown Thrasher is fairly uniformly distributed across the state, with somewhat higher populations in southern and western counties, according to BBS data. Territorial thrashers sing noisily from high perches at the beginning of their nesting season in mid-Apr, but are less vocal by mid-May. Most eggs have been found from 3 May to 1 Jul; second or replacement clutches continue into Jul and, rarely, mid-Aug. The nest is similar to a catbird's large and loosely constructed home, usually placed in a shrub less than 6 feet from the ground, or sometimes on the ground. Up to five bluish white eggs, heavily marked with fine reddish brown spots, are laid. BBS routes showed a steady decline of 3.1% per year until the mid-1980s, and relatively stable populations since.

Fall: Their secretive behavior makes the beginning of migration difficult to discern, but birds become evident in Sep, with increased sightings during the third or fourth week. By the third week of Oct most have left the state, with stragglers remaining through the first or second week of Jan.

Winter: Brown Thrashers are casual to accidental in winter mainly across the lower elevations of the southern half of the state; they are more frequent in the Coastal Plain and in the southern portion of the Piedmont than elsewhere. Apparently they have not been recorded from the northern-tier counties or in the Glaciated Northwest.

History: Clearing the original forests opened habitat significantly for thrashers. Harlow (1913) and others suggested that they were expanding in the northern counties. Todd (1940) thought thrashers were "decidedly less common in the more northern and elevated parts" of the state and had recently expanded into the High Plateau region.

Family Sturnidae: Starlings

Only one species of the large, widespread Old World family Sturnidae has become an established resident over most of North America since it was introduced. European Starlings have successfully flourished wherever humans have altered the natural surroundings. They are quite noisy, often imitating the songs of other birds, and are gregarious, often forming huge flocks during migration and at winter roosts. Their diversified diet includes both animal and plant matter.

European Starling *Sturnus vulgaris*

General status: Introduced from Europe, European Starlings are resident from Alaska across Canada and south throughout the U.S. They first nested in Pennsylvania around the beginning of the twentieth century and are now among the most abundant and adaptable birds in Pennsylvania. They are permanent residents statewide, but immense flocks gather, especially in fall, and frequently join migrating flocks of blackbirds.

Habitat: Starlings are well adapted to human habitations. They can be found almost anywhere, particularly in agricultural areas, in parks, and in suburban and urban areas of towns and cities, but they are absent from extensive forested areas. Starlings are frequently found at garbage dumps and along sidewalks and city streets where they search for food scraps.

Seasonal status and distribution

Spring: The number of starlings seems to increase as they join northbound flocks of blackbirds as soon as the first warm front passes through the state in late Feb or early Mar. The largest flocks are seen between the second week of Mar and the second week of Apr. Birds disperse to breeding territories generally around the end of Apr or the first week of May. Flock sizes are usually smaller in spring than in fall.

Breeding: Starlings are most common in the Piedmont and Glaciated Northwest (see Map 24), where they may outnumber all other species. They are restricted to fields and towns in extensively forested areas of the central and northern counties. In part because they are so conspicuous, starlings are the most abundant birds on BBS routes, averaging 81 birds per route statewide. However, they were not among the top 10 most frequently reported in BBA blocks because of their absence from uniformly forested High Plateau and Pike Co. blocks.[1] The starling population has declined at 2.3% per year on BBS routes since 1966. A cavity nester, starlings occupy a variety of natural and manmade cavities, including holes in buildings, and they often displace other cavity nesters. The cavity entrance must be greater than 1½ inches. The nesting season is long, with some nests reported as early as Jan. Most nests with eggs were found from 17 Mar to 4 Jun, but some have been found as late as 23 Jun (P. B. Street 1954). From four to six pale unmarked bluish eggs are laid.

Fall: Juvenile starlings gather in flocks as early as mid- to late Jun, and numbers build with the addition of new broods. Flocks containing thousands of birds can be seen in fields in Aug, with the greatest number recorded in Sep

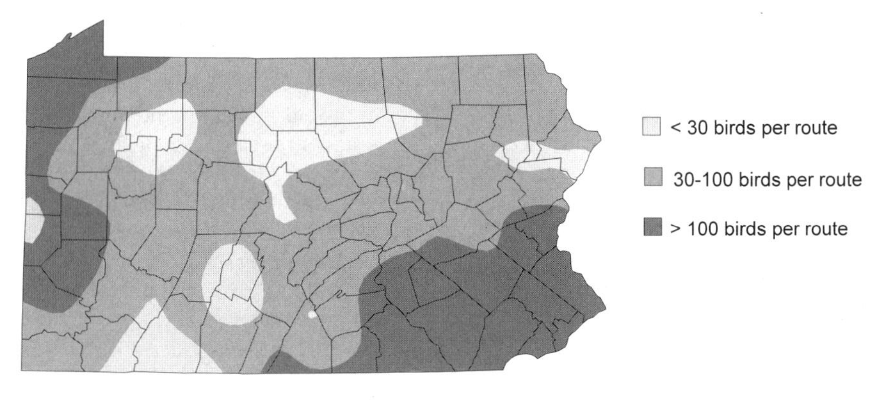

24. European Starling relative abundance, based on BBS routes, 1985–1994

and Oct. Roosting flocks may grow in size until Dec to over 100,000 at some sites.

Winter: Starlings disperse from their summer range as cold weather approaches, usually by the first or second week of Jan, and as food becomes more difficult to find. The largest flocks are usually found in agricultural areas, especially around freshly spread manure and at landfills. Their varied diet even during the most severe winters allows numbers to remain throughout the state around human habitation. Starlings often perch around the tops of chimneys to warm themselves during very cold periods. They are absent in unbroken forested areas where there are few people (e.g., in the northern and central portions of the High Plateau).

History: Since the release of 60 birds in 1890 in New York City, starlings have expanded rapidly to a population of approximately 200 million birds nationwide.[2,3] The first published record of starlings in Pennsylvania was of a pair that raised a brood of young in Trevose (Bucks Co.) in 1904.[4] By 1913 they had spread across the southeastern counties (Harlow 1913). Starlings reached Crawford Co. in 1922 and within a few years had become "menacingly abundant" around Pymatuning (Sutton 1928b). Harlow (1913) predicted, and was proved correct, that starlings would spread statewide during the first two and a half decades of the twentieth century. By the late 1920s, they were already "one of the most abundant birds of most of Pennsylvania" and were found in every county (Sutton 1928a).

[1] Master, T.L. 1992. European Starling *(Sturnus vulgaris)*. BBA (Brauning 1992):286–287.
[2] Kessel, B. 1957. A study of the breeding biology of the European Starling *(Sturnus vulgaris)* in North America. American Midland Naturalist 58:257–331.
[3] Cabe, P.R. 1993. European Starling *(Sturnus vulgaris)*. BNA (A. Poole and Gill 1993), no. 48.
[4] Stone, W. 1908. European Starling *(Sturnus vulgaris)* in Pennsylvania, New Jersey, and Delaware. Auk 25:221–222.

Family Motacillidae: Pipits

Most species in the family Motacillidae, which includes pipits and wagtails, are found in the Old World. Pipits are ground-inhabiting birds with a moderate to long, slender tail and a short, slender, pointed bill. They have long secondaries and long hind toes and toenails. They walk rather than hop and constantly wag their tails. Six species of pipit (genus *Anthus*) have been recorded in North America; only one has occurred in Pennsylvania, the American Pipit. It is strictly a migrant and winter visitor.

American Pipit ***Anthus rubescens***

General status: American Pipits breed in North America from Alaska across the Canadian Arctic and south in the East to Newfoundland and locally to New Hampshire. They winter in the southern half of the U.S. and north along the Pacific coast to British Columbia, and along the Atlantic coast to New York. In Pennsylvania American Pipits are regular migrants that frequent barren wide open spaces. The number of birds seen during migration is highly variable. At some sites, especially in the Piedmont, Glaciated Northwest, and the Lake Erie Shore, they may be common to abundant. Pipits are accidental in summer and are irregular winter visitors, with most records from east of the Allegheny Front.

Habitat: Pipits inhabit freshly plowed or short grassy fields in agricultural areas or in strip mines, airports, mudflats, and beaches. They are often seen along rural roads after a heavy spring snowfall.

Seasonal status and distribution

Spring: The migration period may vary considerably from year to year depending on weather conditions and the amount of snow cover. In some years, migration may begin as early as the third week of Feb, but generally first arrivals are not seen until at least the second week of Mar. At most sites the greatest number of migrants may be seen any time from about the fourth week of Mar to about the third week of Apr. On 20 Mar 1977 more than 1000 pipits were found between Milton in Northumberland Co. and Washingtonville in Montour Co. (Schweinsberg 1988). At PISP birds are rarely reported before the third week of Apr. The greatest number of birds at this site is observed during the second and third weeks of May. Migration is usually over at this time elsewhere in the state except for a few stragglers. Small flocks continue to be seen at PISP until the fourth week of May, with stragglers to the second week of Jun in some years.

Summer: A single bird present from 4 to 10 Jul 1990 at PISP is the only summer record in Pennsylvania.[1]

Fall: The earliest southbound birds may begin returning the first week of Sep, but most migrants appear after the third or fourth week of Sep. The greatest numbers accompany cold fronts in Oct and Nov. Flock sizes diminish through Dec, but large numbers have been recorded, such as the 280 counted on the Southern Lancaster Co. CBC on 16 Dec 1990 (Morrin et al. 1991). Lingering birds may remain to the first or second week of Jan.

Winter: A few pipits may remain during mild winters when there is little snow cover. They may suddenly appear at any time in Jan or Feb. Birds arriving during the first two weeks in Feb may be early migrants. Most winter records come from the Coastal Plain and southern portion of the Piedmont. In the Ridge and Valley they have been reported in winter from Columbia, Centre, and Huntingdon cos. West of the Allegheny Front they have been recorded in winter in Cambria, Erie, Lawrence, and Westmoreland cos.

History: Todd (1940) reported that in western Pennsylvania American Pipits were common and regular only along the shores of Lake Erie. Most were found along beaches, but they were also reported along the shores of lakes and ponds and on freshly plowed fields in other parts of the state. In eastern Pennsylvania, Poole (unpbl. ms.) reported them as irregular. S. H. Dyke reported that large flocks occur in fall along the lower Susquehanna.[2] In winter all but one record were from east of the Allegheny Front, primarily in the southeastern counties. The only record outside this area was a report of seven or eight birds near Wellsboro, in Tioga Co., on 7 Feb 1949 (Poole, unpbl. ms.). Todd reported them as rather rare and irregular in Allegheny, Beaver, Butler, Centre, Huntingdon, Somerset, and Warren cos. Sutton (1928b) found them to be fairly regular in Crawford Co. in the Pymatuning and Conneaut area.

[1] Hall, G.A. 1990. Appalachian region. AB 44: 1134.
[2] Dyke, S.H. 1955. Shorebirds on the Conejohela Flats. Atlantic Naturalist 10:260–268.

Family Bombycillidae: Waxwings

Named for the waxy red appendages at the tips of their secondaries, waxwings are small, short-tailed birds with soft, sleek, grayish or brownish plumage and a sharply pointed crest. The bill is short, flat, and broad at the base with a slight hook to the tip. Even during the breeding season, they are usually seen in flocks. Their diet consists mainly of fruits and berries, but, like flycatchers, in the summer they eat insects that they catch on the wing. Two of the three species in this family are recorded in Pennsylvania, Bohemians and Cedars. Bohemian Waxwings are rare vagrants and usually have been seen with feeding flocks of Cedar Waxwings. Cedar Waxwings are widespread breeders in the state.

Bohemian Waxwing *Bombycilla garrulus*

General status: In North America, Bohemian Waxwings breed from Alaska across Canada to extreme west-central Quebec and south to Washington, Oregon, and Idaho. They winter primarily across southern Canada to Ontario and south through most of the western U.S. Bohemian Waxwings sporadically wander east to Newfoundland and south to Ohio, Pennsylvania, Virginia, and New Jersey. Specimens were collected in Pennsylvania near the end of the nineteenth century, but apparently they have been lost. There have been many sightings of this species in Pennsylvania, but few are ever documented. They have been reported throughout the state except in the Coastal Plain and the Glaciated Northeast. Most sightings of Bohemian Waxwings in Pennsylvania have been with flocks of Cedar Waxwings.

Habitat: They are found where fruit-bearing shrubs or trees are present. Bohemian Waxwings are particularly attracted to northern mountain ash trees. Most observations have been in or near towns, cities, and parks.

Seasonal status and distribution

Fall through spring: Bohemian Waxwings have been seen from 19 Oct to 9 May, with most sightings, up to six birds, from Nov to Feb. If there is an abundant supply of fruit, they may remain for several weeks. Most reports have come during years of heavy invasions into the northeastern U.S. Most sightings have been from the Lake Erie Shore, but they have also been reported from Allegheny, Berks, Bucks, Butler, Crawford, Lancaster, Luzerne, Mercer, Mifflin, Montgomery, Union, and Warren cos.

History: The earliest reported sighting of a Bohemian Waxwing appears to have been of one captured in 1860 in Chester Co. (Burns 1919). Cassin (1862) reported this species to be rare in Delaware Co., and only in winter. Warren (1890) reported that birds were taken at irregular intervals during a period of 25 years and cited a record of one captured from a flock of about 20 in winter from a pine forest in northern Elk Co. He also mentioned reports of Bohemian Waxwings from Clinton, Lackawanna, Lancaster, and Northampton cos. Todd (1940) reported a bird shot from a flock of 10 or 12 birds near Natrona in Allegheny Co. on 24 Dec 1897, and a flock of 12 was near Renovo in Clinton Co. on 5 Mar 1912 (Todd 1940). They also were reported from Butler, Bradford, McKean, and Mercer cos. (Todd 1940).

Comments: Probably more Bohemian Waxwings invade Pennsylvania than records indicate, especially along the mountain ridges and the High Plateau where there are few observers. They may also be easily overlooked, especially when one or two Bohemian Waxwings are mixed in flocks of several hundred Cedar Waxwings.

Cedar Waxwing ***Bombycilla cedrorum***

General status: Cedar Waxwings breed from Alaska south across Canada and the northern half of the U.S. They winter across southern Canada and throughout the U.S. Cedar Waxwings are almost always found in tight flocks. Their movements are highly erratic throughout Pennsylvania in all seasons, with numbers varying from year to year. During some years they may be rather uncommon and local; in other years they may be abundant and widespread, especially as migrants.

Habitat: Waxwings use a wide variety of habitats where fruit-bearing shrubs or trees are found, especially near water. These areas include open secondary woodlands, woodland edges, thickets and hedgerows, and parks and suburban areas. They are particularly attracted to northern mountain ash trees in winter but are often seen flying overhead in any habitat.

Seasonal status and distribution

Spring: These nomadic birds may appear at almost any time. There is usually one influx in Feb and Mar and then another during the second or third week of May. Usually the earlier movement contains smaller flocks than the later. Flock sizes may vary from a few to several hundred birds. By the first week of Jun migration has ceased and nesting has started.

Breeding: Although they are most common (on the BBS) across the northern forested mountains, waxwings are birds of forest openings, often heard flying over woodlands but not found under closed canopy. They are least common, but occur widely, in the Piedmont in summer and are rare and local in urbanized Philadelphia and Delaware cos. (see Map 25). They are one of only a few species that flock through the breeding season. Groups range widely in search of fruit and concentrations of flying insects. They also nest in groups but are not strictly colonial nesters. Their nest is built of woven grasses, twigs, plant stems, and cottony fibers (especially cottonwood) and lined with fine grasses, rootlets, and plant down. The nest is constructed on a horizontal limb less than 50 feet above the ground. While many nests are begun in late May, nesting regularly occurs through the middle of Aug and occasionally into Sep.[1] Most egg sets have been collected between 29 May and 13 Aug. Harlow (1918) gave an average date of 25 Jun for complete clutches, later than the average date for most songbirds. Four or five gray or bluish gray eggs, lightly and irregularly spotted with brown and lightly blotched with darker gray, are laid. No long-term population trend is apparent on BBS routes.

Fall: As fall approaches, flock sizes increase, and by late Aug flocks may contain several hundred birds. Numbers remain high through the fall and are es-

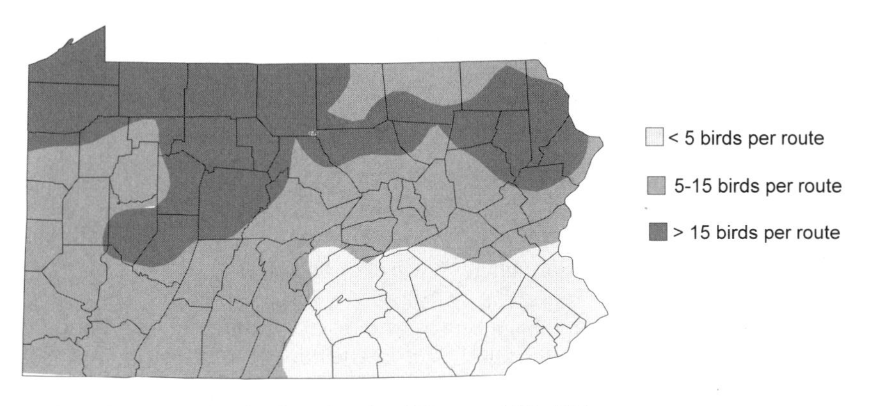

25. Cedar Waxwing relative abundance, based on BBS routes, 1985–1994

pecially notable at hawk watches along the mountain ridges, as long as fruit-laden trees are available. By mid-Jan flocks start to wander farther, and numbers begin to decrease as their food sources become depleted.

Winter: They depend upon the productivity of fruit-bearing trees, shrubs, and vines for their survival more so in winter than at any other time of year. When they find these food sources, they may strip every bush and tree of fruit in a few hours or days and then move on. They may be absent from many parts of the state during some winters, while in other winters they may be rather common and widespread.

History: Cedar Waxwings were considered common or abundant at the end of the nineteenth century, even in southeastern counties where they are recently least common.

Comments: There have been several recoveries of Cedar Waxwings south of the U.S. border that were banded at PNR: an immature Cedar Waxwing banded on Aug 15 Aug 1962, was found dead at Tlalnepantla, Mexico in Feb 1963;[2] one banded on 12 Sep 1968 was found injured on 17 Jan 1972 at La Concordia, Chiapas, Mexico;[3] a Cedar Waxwing banded on 26 May 1972 was recovered on 25 Feb 1973 at Santa Eulalia at about 10,000 feet in the Cuchumatanes Mountains, Huehuetenango, Guatemala;[4] one banded 23 Oct 1982 was recovered at Acatzingo, Mexico in Feb 1986;[5] and another banded on 11 Aug 1989 was found dead near Ejutla, Oaxacam, Mexico, three months later. This was the seventh record south of the border (6 in Mexico and 1 in Guatemala).[6]

[1] Gross, D.A. 1992. Cedar Waxwing *(Bombycilla cedrorum)*. BBA (Brauning 1992):282–283.

[2] Leberman, R.C., and M.A. Heimerdinger. 1966. Bird-banding at Powdermill, 1965. Research report no. 15. Powdermill Nature Reserve, Carnegie Museum of Natural History, Pittsburgh.

[3] Leberman, R.C., and M.H. Clench. 1973. Bird-banding at Powdermill, 1972. Research report no. 31. Powdermill Nature Reserve, Carnegie Museum of Natural History, Pittsburgh.

[4] Leberman, R.C., and M.A. Clench. 1975. Bird-banding at Powdermill, 1974. Research report no. 35. Powdermill Nature Reserve, Carnegie Museum of Natural History, Pittsburgh.

[5] Leberman, R.C., and R.S. Mulvihill. 1987. Bird-banding at Powdermill, 1986. Research report no. 47. Powdermill Nature Reserve, Carnegie Museum of Natural History, Pittsburgh.

[6] Mulvihill, R.S., and R.C. Leberman. 1990. Bird-banding at Powdermill, 1989. Research report no. 50. Powdermill Nature Reserve, Carnegie Museum of Natural History, Pittsburgh.

Family Parulidae: Wood-Warblers

The family Parulidae, wood-warblers, are small, active, often colorful birds averaging about 5 inches in length. They are confined to the western hemisphere. Most spend the winter in Central and South America and the Caribbean. Wood-warblers are most easily observed during spring migration when the brightly colored males appear, actively singing, often before the foliage is fully errupted. In the fall many males lose their brilliant colors and pass through the state nearly unnoticed in the company of drab immatures and females. Forty species of wood-warblers, about half of all the species in this family, have been recorded in the state. Of this number, 30 species are known to breed in the state and 10 are strictly migrants or rare visitors. Most build a cup-shaped nest made of grasses, various plant fibers, mosses, pine needles, strips of bark or leaves, hair, and rootlets. Some species nest on the ground; others build their nest in a shrub or a tree at various heights. Prothonotary Warblers are the only cavity-nesting wood-warblers in the eastern U.S. Most wood-warblers lay from three to six white or cream-colored eggs, usually spotted or blotched on the large end. Wood-warblers' diet consists mostly of insects, but those few species that attempt to winter will eat fruit or seeds or will come to suet at feeders.

Blue-winged Warbler *Vermivora pinus*

General status: Blue-winged Warblers breed from Maine south through the New England states and the Appalachian Mountains west to Minnesota and Arkansas. They winter from southern Mexico south through Central America. Blue-winged Warblers are uncommon to fairly common regular migrants in Pennsylvania. They have been expanding their breeding range and are now found nearly throughout the state. They frequently hybridize with and replace local populations of Golden-winged Warblers.

Habitat: Blue-winged Warblers are found in overgrown fields, power-line cuts, open brushy second-growth woodlands or woodland edges, thickets, and shrubby swamps.

Seasonal status and distribution

Spring: Their typical migration period is from the third or fourth week of Apr to the fourth week of May. The greatest number is seen during the first or second week of May.

Breeding: Blue-winged Warblers are locally fairly common in early successional habitats in the Southwest and Glaciated Northwest and locally uncommon to fairly common in the Piedmont and Glaciated Northeast. In the Ridge and Valley and High Plateau they are rare but expanding into the periphery of these regions. Blue-winged Warbler populations are increasing at a time when many early successional species are declining; the BBS reports a healthy 6.5% per year increase since 1966. The current distribution is the widest recorded, but from a few central counties, notably Clearfield and Clinton, reports come only rarely. The nest is made of dried leaves, grass, or vines, placed on or just above the ground. It is usually well concealed at the base of a shrub or small tree. The nesting season is brief; singing drops off by late Jun, and most eggs have been found from 23 May to 14 Jun.

Fall: Fewer birds are observed in fall than in spring in most localities. The beginning of migration is difficult to detect, but birds begin to be seen at migrant traps during the third or fourth week of Aug. Banding records suggest that the greatest number of birds pass through the state during the first and second weeks of Sep. Migration usually ends by the fourth week of Sep. Stragglers remain to the first week of Oct.

History: The earliest accounts typically listed Blue-winged Warblers as frequent during the breeding season in the southeast. Warren (1890) considered them common there and rare elsewhere. All of the early-twentieth-century eggs sets were from southeastern counties. Blue-winged Warblers expanded dramatically into the northwest and southeast during the early decades of the 1900s. Harlow (1913) considered them "abundant" in Pennsylvania's southeastern corner, locally as far north as Berks and Lehigh cos. and west to York and Cumberland cos. Although Todd did not record them in the Pymatuning region late in the 1890s, Sutton (1928b) stated that they were among "the most abundant, widely distributed, and characteristic members of its family in the region" (the old Pymatuning swamp). Jacobs (1893) did not list Blue-winged Warblers in Greene Co. Until the 1930s they remained "one of the rarest of the warblers" in southwestern Pennsylvania (Todd 1940). Blue-winged Warblers rapidly expanded through the Ohio River drainage[1] and had become well established across western Pennsylvania, where they remain locally common, by the mid-twentieth century.

Comments: Hybridization between Blue-winged and Golden-winged warblers produces two distinct phenotypes: Brewster's and Lawrence's (see Golden-winged Warbler). The form most resembling the Blue-winged Warbler, the Brewster's Warbler, is the more frequently seen and has been recorded throughout the state. Mated pairs of Brewster's as well as Brewster's mated with Blue-winged and Golden-winged Warblers have been found in Pennsylvania. A pair of Brewster's Warblers successfully nested in Allegheny Co. in 1986.[2] Scattered sightings of Brewster's date from the late 1800s (Poole, unpbl. ms.), indicating the early intergradation of these two species. Caution should be used when identifying Blue-winged and Golden-winged warblers by song, as either species may give the other's song or variations of the two.

[1] Gill, F.B. 1980. Historical aspects of hybridization between Blue-winged and Golden-winged warblers. Auk. 97:1–18.

[2] Hall, G.A. 1986. Appalachian region. AB 40: 1204.

Golden-winged Warbler *Vermivora chrysoptera*

General status: Golden-winged Warblers breed from New Hampshire south to Georgia in the Appalachian Mountains west across Ohio (formerly) and southern Canada to Minnesota and Missouri. They winter from Mexico through Central America. In Pennsylvania, Golden-winged Warblers are rarely observed during migration away from breeding sites. They breed primarily in the mountainous areas of the Ridge and Valley and Allegheny Mountain section. Both hybridization and competition with Blue-winged Warblers are contributing to the national decline of this species.[1,2] Birds in migration are more often seen in spring than in fall.

Habitat: Golden-winged Warblers are locally distributed in overgrown fields, power cuts, shrubby early-successional habitats such as oak barrens (notably in the Poconos), three- to eight-year-old clearcuts, and open swampy forests. They may occasionally be found in open woods that have a dense understory, such as partial timber cuts and forests damaged by gypsy moths.

Seasonal status and distribution

Spring: They may appear during the fourth week of Apr or the first week of May. Most are found during the second or third week of May, and migration usually ends by the fourth week of May.

Stragglers occasionally remain until the first week of Jun in areas where they are not known to nest.

Breeding: Golden-winged Warblers are rare to locally fairly common breeders throughout the Ridge and Valley and the Poconos, rare in the Southwest and Allegheny Mountain section, and irregular across most of the remaining High Plateau. They are generally absent in summer from the Piedmont. Golden-winged Warblers have declined severely in Pennsylvania (6.9% per year on BBS routes since 1966), in part as a result of competition with (or genetic swamping by) Blue-winged Warblers. They have retracted their range in western counties as Blue-winged Warblers have expanded theirs. Nesting characteristics of the Golden-winged are similar to those of the Blue-winged Warbler. Most eggs have been found from 18 May to 7 Jun.

Fall: Leberman (1976) reported that "by mid-July considerable movement is obvious" in the Ligonier Valley. Banding records suggest a movement of birds during the fourth week of Jul and the first week of Aug. Records also suggest another movement during the fourth week of Aug and the first week of Sep. By the third week of Sep most birds have left the state. Stragglers remain to the third week of Oct.

History: Golden-winged Warblers appeared on few state bird lists of the late nineteenth century. They were considered rare in eastern counties (Turnbull 1869) but more common and widely distributed in the west. Although they were widely distributed, Harlow (1913) never knew them as abundant. Golden-winged Warblers were considered more common than Blue-winged Warblers in the southwestern counties, although still uncommon (Todd 1940), and in the Glaciated Northeast Golden-winged Warblers outnumbered Blue-winged Warblers in the 1950s,[3] a pattern that had reversed by the 1980s. Poole (unpbl. ms.) called the Golden-winged Warbler "a summer resident of peculiarly local and interrupted distribution" over the state, except generally absent from the southeast.

Comments: The recessive hybrid of the Blue-winged and the Golden-winged warbler is called the Lawrence's Warbler, the rarer of the two hybrid forms (see Blue-winged Warbler). Scattered reports of this hybrid come from throughout the state, with most sightings in spring. Except at the PNR banding station, Lawrence's Warblers have not been reported more than two or three times at most sites. Golden-winged Warblers are listed among 16 species on a "Watch List" as "the highest under-recognized avian conservation priorities in the continental U.S. and Canada."[4] A notable portion of this species' range falls in Pennsylvania, ranking the state high in the responsibility to sustain the species (Rosenberg and Wells 1996).

[1] Gill, F.B. 1980. Historical aspects of hybridization between Blue-winged and Golden-winged warblers. Auk 97:1–18.

[2] Confer, J.L., and K. Knapp. 1981. Golden-winged and Blue-winged warblers: The relative success of a habitat specialist and a generalist. Auk 98:108–114.

[3] W. Reid, pers. comm.

[4] Pashley, D. 1996. Watch list. Field notes 50:129–134.

Tennessee Warbler ***Vermivora peregrina***

General status: Tennessee Warblers breed from Alaska across Canada and south to Montana in the West and to New York in the East. They winter primarily from Mexico south to South America. Tennessee Warblers are fairly common to common migrants across Pennsylvania. During some years they may be only uncommon to rare migrants in the Coastal Plain and the southeastern-most counties of the Piedmont. The number of birds recorded from year to year varies west of the Allegheny Front. Tennessee Warblers are accidental in summer.

Habitat: They are found in woodlands and woodland edges, usually from the mid to the upper level of the canopy, occasionally in the brushy understory of woodlands or in thickets.

Seasonal status and distribution

Spring: The earliest records are in the third or fourth week of Apr, but Tennessee Warblers are rarely reported before the first week of May. The greatest number of birds is recorded during the second and third weeks of May. Migration has usually ended by the fourth week of May, with stragglers to the third week of Jun.

Summer: There have been several reports of Tennessee Warblers during this period. Included among these are a report of one seen and heard in full song on 24 Jun 1968 near Ramsey in Lycoming Co.,[1] one at Ligonier in Westmoreland Co. on 27 Jun 1981,[2] and one banded at PNR on 14 Jul 1981.[2] A bird banded at PNR on 29 Jun 1987 was molting some primaries.[3] There has been no evidence of breeding in Pennsylvania.

Fall: Probable migrants have appeared exceptionally early, especially at the banding station, PNR, where they have been netted from 22 Jul through the first week of Aug. These birds, which had not completed their fall molt (Leberman 1976), were unusual because warblers do not usually migrate before molting, and Tennessee Warblers are not known to breed in the state. Their typical migration period is from the third or fourth week of Aug to the second week of Oct. The greatest number of birds is recorded during the second and third weeks of Sep. Many Tennessee Warblers go undetected except at banding stations, such as PNR, where as many as a hundred birds have been banded in a single day (Leberman 1976). Stragglers have been recorded to the second week of Nov.

History: Poole (unpbl. ms.) also noted that in eastern Pennsylvania the number of migrants observed varies from year to year: in some years they were common to abundant in fall; in other years they were rare or absent. Very little difference in abundance was noted by authors west of the mountains. However, there was some disagreement between Sutton (1928b), who found them to be rare in spring in the Pymatuning area, and Grimm (1952), who found them to be abundant in the same area. Unlike records of recent years, there was no summer record of Tennessee Warbler before the middle of Aug. Todd (1940) wrote that they were rare and irregular transients in the High Plateau.

Comments: They are frequently misidentified in fall, when many immature birds are very yellow. They are sometimes confused with Warbling and Philadelphia vireos, as well as with Orange-crowned Warblers. Unique was a hybrid Tennessee Warbler × Nashville Warbler netted at PNR in 1979.[4]

[1] Hall, G.A. 1968. Appalachian region. AFN 22: 608.

[2] Hall, G.A. 1981. Appalachian region. AB 35:938.

[3] Hall, G.A. 1987. Appalachian region. AB 41: 1437.

[4] Leberman, R.C. 1981. Bird-banding at Powdermill, 1980. Research report no. 41. Powdermill Nature Reserve, Carnegie Museum of Natural History, Pittsburgh.

Orange-crowned Warbler *Vermivora celata*

General status: Orange-crowned Warblers breed from Alaska south through the Rocky Mountain region and east across Canada to Labrador. They winter primarily along the Pacific coast from Washington south to Arizona, east across the southern states, and north along the Atlantic coast to Virginia. They are far more common in western North America than in the East. In Pennsylvania they are rare regular migrants. Orange-crowned Warblers are not known to be regular at any site, except at PNR (where banding records suggest that they are regular only in the fall) and at PISP, where they are regular in both spring and fall. These nondescript warblers are easily overlooked and are undoubtedly more widespread than records indicate. Apparently they have not been recorded in the High Plateau, except in Warren Co. and in the Allegheny Mountain section. Orange-crowned Warblers are casual in winter.

Habitat: They are usually observed within a few feet of the ground in the brushy understory of woodlands and woodland edges, thickets, tangles, and weedy fields. In early spring, they may be seen in mixed warbler flocks high in the tree canopy.[1]

Seasonal status and distribution

Spring: Migrants may appear as early as the second week of Apr, but most are not reported until the fourth week of Apr or the first week of May. Birds seen in Mar are likely to be overwintering birds rather than early migrants. Most spring birds are reported during the second and third weeks of May, but they are usually silent on their passage through Pennsylvania.

Fall: Orange-crowned Warblers are among the last of the warblers to migrate through the state. They have been recorded as early as the first week of Sep, and their typical migration period is from the third or fourth week of Sep to the fourth week of Oct. The majority are recorded during the first and second weeks of Oct. Larger numbers are recorded at PNR than at other sites in the state; more are banded there than are observed in the field. Seasonal banding totals have reached 24 birds at PNR.[2] Stragglers remain into Dec.

Winter: All but one of the winter records is from the Coastal Plain or Piedmont. In the Coastal Plain most records have been at Tinicum or in the vicinity of Philadelphia.[3] In the Piedmont they have been reported from Montgomery Co. in Flourtown from 24 Jan to 21 Feb 1960[4] and in Towamencin Township in Feb 1998;[5] at Blue Marsh Reservoir in Berks Co. on 9 Jan 1986;[6] at Struble Lake in Chester Co. to Feb 1993;[7] and in Solesbury Township in Bucks Co. on 4 Feb 1995.[8] The only winter record away from the southeast was in Luzerne Co. on 17 Feb 1975.[9]

History: In western Pennsylvania Todd (1940) had little data for this species. He collected an Orange-crowned Warbler in unusual circumstances: one was singing in the top of a tree. Poole (1964) listed this species as a rare transient in spring and fall and stated in his unpublished manuscript that numerous observations had been made from all sections of the state. He also cited records in southeastern Pennsylvania from mid- to late Jan in Berks and Montgomery cos. and in early Jan from Allegheny and Centre cos.

Comments: Tennessee Warbler fall plumage may be tinged with yellow that closely resembles the Orange-crowned Warbler's. Misidentifications may occur often from late Aug to mid-Sep, when many Tennessee Warblers are passing through the state.

[1] R.C. Leberman and R.S. Mulvihill, pers. comm.
[2] Hall, G.A. 1968. Appalachian region. AFN 22:40.
[3] J.C. Miller and E. Fingerhood, pers. comm.
[4] Schwalbe, P. 1960. Field notes. Cassinia 45:18.
[5] Crilley, K. 1998. Local notes: Montgomery County. PB 12:25.

[6] R. Keller, pers. comm.
[7] Haas, F.C., and B.M. Haas, eds. 1993. Rare and unusual bird reports. PB 7:18.
[8] Kitson, K. 1995. Local notes: Bucks County. PB 9:31.
[9] W. Reid, pers. comm.

Nashville Warbler ***Vermivora ruficapilla***

General status: Nashville Warblers breed from British Columbia south to California in the West and from Manitoba east to Nova Scotia and south to Wisconsin, Michigan, West Virginia, and Maryland in the East. They winter from Mexico to Central America and casually in southern Texas, along the Gulf coast, and Florida. Nashville Warblers are fairly common regular migrants in Pennsylvania except in the Coastal Plain and Piedmont, where they are uncommon to rare. They breed widely, but spottily, in the state except in the southern corners. Nashville Warblers are accidental in winter.

Habitat: During migration Nashvilles prefer woodlands, woodland edges, and thickets. During the breeding season they are almost always found above 1000 feet elevation in overgrown clearcuts and strip mines; in dry barrens with oak, birch, and pitch pine; or in bogs.

Seasonal status and distribution

Spring: Spring arrivals may appear as early as the third week of Apr. Their typical migration period is from the fourth week of Apr to the third week of May. The greatest number of birds is observed during the first and second weeks of May. Stragglers remain to the fourth week of May.

Breeding: Generally rare but widely distributed in the northern and mountainous regions, Nashville Warblers are breeding residents in scattered clusters outside the Piedmont and southern Southwest regions. In some areas, notably the Pocono section where they were reported in 32% of BBA blocks,[1] they may be locally fairly common. There, they are most common in scrub oak barrens, such as that found around Long Pond, Monroe Co. No population trend is detectable on BBS routes because of their patchy distribution and generally low numbers. The nest is placed on the ground, concealed by dense foliage. The few nests found in Pennsylvania have had eggs from late May through mid-Jun. Unusual in the Piedmont, well below 1000 feet, was one seen carrying food, suggesting local breeding from 24 Jun to 1 Jul 1973, at Fort Washington State Park in Montgomery Co.[2]

Fall: The first southerly movement may be detected as early as the second week of Aug. Their typical migration period is from the third week of Aug to the fourth week of Oct. Peak migration may occur any time during the third week of Sep to the first week of Oct. Stragglers have been recorded to the fourth week of Dec.

Winter: Nashville Warblers have been recorded in winter four times. One came to a feeder at Kennett Square in Chester Co. from 22 Dec 1979 to 24 Jan 1980,[3] and an immature was visiting a suet feeder near Landisville in Lancaster Co. from 6 Feb to 8 Apr 1991 (Morrin et al. 1991). One survived the winter of 1998 in Huntingdon Valley in Montgomery Co.[4] One was present at Wyoming in Luzerne Co. on 28–29 Jan 1967.[5]

History: Some authorities believed that Nashville Warblers were more common and regular migrants west of the mountains than east of them. The spotty breeding distribution is reflected in historical records that include North Mountain in Sullivan and Wyoming cos. (Stone 1900); Hawk Mountain area, Berks Co. (Poole, unpbl. ms.); and Chestnut Ridge, Westmoreland Co. (Todd 1940). During the BBA project, they were notably absent from historical locations, such as the Pymatuning area (Sutton 1928b; Grimm 1952) and the Blue Ridge (Baird 1845). The first set of eggs in Pennsylvania was taken in 1907 at Pocono Lake, Monroe Co. (J. F. Street 1915).

[1] Santner, S. 1992. Nashville Warbler *(Vermivora ruficapilla)*. BBA (Brauning 1992):304–305.
[2] Scott, F.R., and D.A. Cutler. 1973. Middle Atlantic Coast region. AB 27:856.
[3] Richards, K.E., R.O. Paxton, and D.A. Cutler. 1980. Hudson-Delaware region. AB 34:259.
[4] Crilley, K. 1998. Local notes: Montgomery County. PB 12:25.
[5] Carleton, G. 1967. Hudson St. Lawrence region. AFN 21:400.

Northern Parula ***Parula americana***

General status: Northern Parulas breed across southern Canada from Manitoba to Nova Scotia and south across the eastern U.S., except in the Midwest-Great Lakes region and southern New England. They winter in the West Indies and Mexico south to South America. Parulas are uncommon to fairly common regular migrants in Pennsylvania. They breed at widely scattered sites throughout the state.

Habitat: A range of habitats is used, including mature woodlands with hemlocks, scattered tall ornamental spruces in open wooded settings or rural yards, and large trees (often sycamores) along streams. Tall trees in moist woods provide a common denominator to this diverse breeding habitat. Parulas may also be found in brushy woodland understory or edges during migration.

Seasonal status and distribution

Spring: A few migrants have been found as early as the first week of Apr, but their typical migration period is from the third or fourth week of Apr to the fourth week of May. At northern sites, birds do not usually appear before the first week of May. Most parulas are seen from the first to the third week of May.

Breeding: Northern Parulas are rare to uncommon summer residents statewide, except in the southern Allegheny Mountains of Fayette, Somerset, and Westmoreland cos., where they may be fairly common. During the BBA project they were absent, or nearly so, from the Coastal Plain, Glaciated Northwest, and Erie Lake Shore. In the Piedmont, Ridge and Valley, and High Plateau they may be regular at scattered sites but are absent from other superficially similar areas. In Pennsylvania, the nest is loosely constructed, hanging from a high tree branch. The few nests seen with eggs have been found from late May through Jun. The population has been increasing since the 1960s on BBS routes in the state.

Fall: Their usual migration season is from the fourth week of Aug to the second week of Oct. Most birds are reported between the second and the fourth week of Sep. Stragglers remain to the first week of Nov. One was found on the late date of 7 Dec 1978 at State College.[1]

History: Parulas historically were more common in the northern-tier counties and much less common in the south than described above. Warren (1890) considered them "generally throughout the state in damp forest and swampy wooded thickets, where the long tufts of gray lichens (*Usnea barbata* and its varieties) are abundant." He also stated that they were more common in the northern and eastern portions of the state than elsewhere. In sharp contrast to recent patterns, were those described by Todd (1940), who stated that parulas were "common throughout the high country of the Allegheny Plateau and also in the valleys and ravines of the ridge and valley section where hemlocks remain." They were "rather common" in "certain localities" on North Mountain, Wyoming Co. (Dwight 1892). A general decline in this species apparently occurred during the early years of the 1900s.

Comments: Although not dependent on *Usnea*, they use the lichens in their nests when available.

[1] Hall, G.A. 1979. Appalachian region. AB 33:282.

Yellow Warbler ***Dendroica petechia***

General status: These brilliant yellow birds breed from Alaska across Canada and throughout the U.S. except from Texas east across the Gulf states and in Florida. Yellow Warblers winter from southern California and Arizona south through South America. In Pennsylvania they are common to abundant regular migrants and breed over the entire state except in extensively forested areas.

Habitat: Yellow Warblers inhabit thickets, shrubs, and brushy areas; frequently along streams, and lakes. They are also found in shrubby patches of brush near agricultural fields, in parks, suburban and urban areas.

Seasonal status and distribution

Spring: Yellow Warblers have arrived as early as the second week of Apr. A very early one was reported on 27 Mar 1998 at Harrisburg in Dauphin Co.[1] Their typical migration period is from the third or fourth week of Apr to the fourth week of May. Peak migration occurs during the second and third weeks of May.

Breeding: Fairly common to common statewide, Yellow Warblers are most abundant (more than 20 birds per route) in northern and western counties and somewhat less common in the central Susquehanna Valley (see Map 26). They occupy a small territory, may occur in high densities, and are locally common. They are found at woodland borders, but uncommonly in small, uniform clearcuts (Yohn and Yahner 1995). Yellow Warblers arrive on their summer territories earlier than many *Dendroica* warblers in Pennsylvania, but most nests with eggs have been found during a period typical of this group: from 17 May to 19 Jun. A few egg sets have been found as early as 9 May (Poole, unpbl. ms.). Nests are neat and compact, distinctively covered with milkweed fibers or other types of plant down that give them a cottony appearance. There is no evidence of second broods; Yellow Warblers become quiet and inconspicuous after young fledge in late Jun and thus escape observation until the fall migration (Todd 1940). The BBS population trend has been variable, but generally positive (1.9% per year) since 1966.

Fall: The fall movement of Yellow Warblers is difficult to discern over most of the state as birds become increasingly

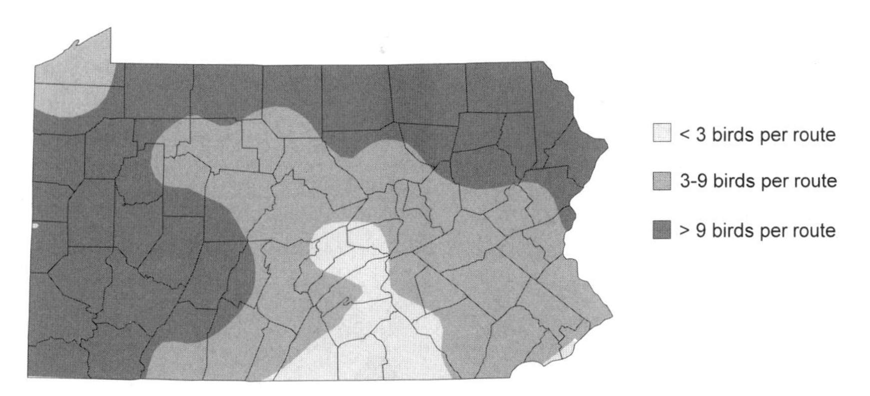

26. Yellow Warbler relative abundance, based on BBS routes, 1985–1994

hard to find after singing has stopped. However, migration is quite evident at the Lake Erie Shore, especially at PISP. At this site they are typically observed from the third week of Jul to the second week of Sep. Peak migration is during the fourth week of Jul and the first week of Aug, when over 100 birds have been banded at PISP in a single day. Stragglers have been recorded to the third week of Nov. Two very late birds were found on 10 Dec 1990 at Tinicum.[2] An exceptionally late Yellow Warbler remained to 11 Jan 1998 at the Philadelphia Sewage Treatment Plant in Philadelphia Co.[3]

History: Away from extensively forested regions, Yellow Warblers always were considered common or abundant. The clearing of forests for farmland in the nineteenth century opened extensive habitat for this adaptable species, and abandonment of many of those farms in the early twentieth century further enhanced habitat, for a time.

[1] Williams, R. 1998. Local notes: Dauphin County. PB 12:17.

[2] Guarente, A. 1990. County reports—October through December 1990: Delaware County. PB 4:155.

[3] Floyd, T. 1998. Local notes: Philadelphia County. PB 12:25.

Chestnut-sided Warbler *Dendroica pensylvanica*

General status: Chestnut-sided Warblers breed primarily in Canada from British Columbia east to Nova Scotia south to Arkansas, Indiana, and Pennsylvania, and through the Appalachian Mountains to Georgia. They winter primarily in Mexico and Central America. Chestnut-sided Warblers are uncommon to fairly common regular migrants in Pennsylvania. They breed widely across the northern half of the state, south through the Allegheny Mountains, and locally elsewhere in the state.

Habitat: During migration they occupy woodlands, woodland edges, brushy areas, and thickets. During the breeding season they prefer open woodlands with brushy understory and tangles, as well as overgrown pastures and regenerating clearcuts, primarily in extensively forested areas.

Seasonal status and distribution

Spring: Chestnut-sided Warblers are more frequently observed at this season than in fall. Early migrants arrive as early as the fourth week of Apr. Their usual migration period is from the first to the fourth week of May, and peak migration usually occurs during the second and third weeks of May. Stragglers away from breeding areas are observed until the first week of Jun.

Breeding: Chestnut-sided Warblers are uncommon to locally common statewide. They reach their highest abundance in the High Plateau region in large clearcuts. They are regular, but rare, in scattered wooded sites in the eastern Piedmont, including Ridley Creek State Park, Delaware Co., and they are distributed widely in the hills of southern Berks and northern Chester cos. (Uhrich 1997). They are rare, local breeders throughout most of the Ridge and Valley and are generally absent in summer from the Coastal Plain, the western Piedmont, and west and south of the Monongahela and Ohio rivers in the Southwest region. The nesting season is typical of wood-warblers; most nests with eggs have been found from 23 May through 21 Jun. Their nest is generally in a small deciduous shrub or a blackberry clump, 1–3 feet from the ground. A significant increase (4.5% per year) in numbers on BBS routes has been seen since 1980—an increase based primarily on populations in the northern mountains—but they have declined in some areas as forests have matured.

Fall: Southbound movement may be detected as early as the second week of Aug. The greatest number of birds is noted during the first three weeks of Sep. Migration continues to the first week of Oct, and stragglers have been recorded until the third week of Nov.

History: Chestnut-sided Warblers have benefited significantly from human alterations in the eastern deciduous forest. Ornithologists after the Colonial era (e.g., Audubon, Wilson), but before extensive lumbering in the Appalachian mountain forests, scarcely ever observed this species.[1] After the advent of extensive timber harvesting in the early 1800s, Chestnut-sided Warblers rapidly became widespread and common. By the time Baird (1845) provided a list of the birds of the Carlisle area (Cumberland Co.), they were "Very abundant some seasons, Summer." Historical accounts since the mid-nineteenth century closely match the current distribution. Todd's (1940) map of breeding distribution for this species matches the BBA project results[2] with remarkable detail, such as records in northern Allegheny Co. and along Chestnut Ridge in southern Westmoreland Co. Breeding records in Fulton and Franklin cos. reported by Todd (1940) are exceptions; the species was not found there during the BBA project.

[1] Richardson, M., and D.W. Brauning. 1995. Chestnut-sided Warbler *(Dendroica pensylvanica)*. BNA (A. Poole and Gill 1995), no. 190.
[2] Schwalbe, P.W. 1992. Chestnut-sided Warbler *(Dendroica pensylvanica)*. BBA (Brauning 1992):310–311.

Magnolia Warbler ***Dendroica magnolia***

General status: Magnolia Warblers breed across central and southern Canada south to Minnesota east through the Great Lakes region to Pennsylvania and south through the Appalachian Mountains to North Carolina and Tennessee. Their winter range is from Mexico through Central America. Magnolia Warblers are fairly common to common regular migrants in Pennsylvania. They breed mainly in the High Plateau and in the Glaciated Northeast and are far less common as breeders elsewhere in mountainous portions of the state.

Habitat: Magnolia Warblers are found in woodlands, woodland edges, and brushy areas and thickets during migration. During the breeding season they are always associated with conifers, inhabiting mixed and coniferous forests containing cool damp hemlock ravines, swamps, or red spruces, primarily above 1500 feet elevation. They often are found in small groups of hemlocks within extensive deciduous woodlands. They may be locally fairly common in dense stands of young conifers or in old-growth northern hardwood forest.

Seasonal status and distribution

Spring: Magnolia Warblers may arrive as early as the fourth week of Apr. Their typical migration period is from the first to the fourth week of May. Peak migration usually occurs during the second and third weeks of May. Nonbreeding stragglers remain to the second week of Jun.

Breeding: Magnolia Warblers are fairly common in the High Plateau and Pocono section and are uncommon in the Allegheny Mountains south to the Maryland line.[1] Scattered breeding records occur regularly in hemlock-lined ravines, bogs, and swamps in the Glaciated Northeast and Glaciated Northwest regions, such as at Slippery Rock Creek, Butler Co.; in conifer groves in Centre and Union cos.; and rarely in hemlocks elsewhere in the Ridge and Valley. During the breeding season Magnolia Warblers are always associated with conifers (typically hemlock or spruce). The nest, made of fine twigs and grass, is lined with black rootlets and usually is placed 3–20 feet from the ground; it is well-concealed either in the top of a small conifer or well out on a side branch (H. H. Harrison 1975). Their numbers since 1980 have increased sharply on BBS routes, but they are poorly represented because of their specialized habitat requirement. The nesting season is short, with no evidence of second broods. Most nests with eggs have been found from 27 May to 13 Jun.

Fall: The first migrants may be found beginning the second week of Aug. Their typical migration period is from the fourth week of Aug to the fourth week of Oct. Peak migration occurs from the first to the third week of Sep. Stragglers have remained to the fourth week of Nov.

History: Historical descriptions of distribution and abundance of Magnolia Warblers differ little from recent accounts, but the birds appear to have retreated from some peripheral areas away from the mountains. Todd (1940) listed many locations in which he encountered Magnolia Warblers near the edge of their current distribution, including some in which they were not reported during the BBA project (e.g., Bear Meadows, Centre Co.; Watson Run, southeast Butler Co.; northern Erie Co.). Other peripheral areas noted by Poole (unpbl. ms.), notably sites along Blue Mountain such as near Hawk Mountain, Berks Co., were again documented during the BBA project.

[1] Mulvihill, R.S. 1992. Magnolia Warbler *(Dendroica magnolia)*. BBA (Brauning 1992):312–313.

Cape May Warbler *Dendroica tigrina*

General status: Cape May Warblers breed across southern Canada from British Columbia and southern Mackenzie east to Newfoundland south to Minnesota, Wisconsin, Michigan, and New York. They winter from coastal southern Florida south through the West Indies, Mexico, and Central America and are casual north to Canada. In Pennsylvania, Cape May Warblers are fairly common regular migrants especially in woodlands dominated by evergreens, but the number of migrants fluctuates from year to year. Cape May Warblers are accidental in winter.

Habitat: Woodlands, primarily those with larch, spruce or pine, are the preferred habitats of Cape May Warblers. Cape May Warblers are frequently found foraging among the higher branches.

Seasonal status and distribution

Spring: Over most of the state they are uncommon to rare in spring. Cape May Warblers occasionally appear as early as the third week of Apr. One present from 3 Mar to 9 Apr 1983 at Shelocta in Indiana Co. was almost certainly an overwintering individual rather than an early spring migrant.[1] An early spring record was of a bird reported on 2 Apr 1997 in Lebanon Co.[2] Their usual migration period is from the first to the fourth week of May, with the majority seen during the second or third week. Two birds found on 14 Jun 1981 at Wild Creek Reservoir in Carbon Co. were probably lingering migrants.[3]

Fall: The greatest number of birds is found during fall, when they may be locally common. They have been recorded as early as the second week of Jul. An early fall record includes a bird found on 30 Jul 1996 near Tunkhannock in Wyoming Co.[4] Their typical migration period is from the fourth week of Aug to the first week of Oct. Peak migration may occur any time between the first and third week of Sep. An unusually high count of 219 was made at Hooversville in Somerset Co. on 15 Sep 1981.[5] Stragglers have been recorded until the fourth week of Dec, and a few have remained longer during mild weather and have survived through the winter.

Winter: There have been four reports of Cape May Warblers in winter at bird feeding stations. One survived until at least 1 Feb 1988 at Philadelphia.[6] In 1995 at Reading in Berks Co. one was first discovered on 12 Feb and was last seen on 5 Mar 1995;[7] a lingering bird remained at Milton in Northumberland Co. until 11 Jan 1982;[8] one was at North Park in Pittsburgh, Allegheny Co., to at least 29 Feb 1980.[9]

History: Cape May Warbler populations through history have been dramatically affected by spruce budworm outbreaks in Canadian forests. This effect may account for their irregularity during migration in Pennsylvania. Turnbull (1869) and Pennock (1886) considered Cape May Warblers to be very rare in Pennsylvania. Poole (unpbl. ms.) listed them as irregular and often rare in spring but reported that he had counted at least 40 along a spruce-bordered road on 13 May 1961 at Lake Ontelaunee in Berks Co. He considered them as common in fall, with a few remaining into winter. They were recorded in Chester Co. at Kimberton until 29 Jan 1950[10] and at Exton until 31 Jan 1953;[11] at Reading in Berks Co. in 1958 (Poole unpbl. ms.); and three times in the State College area during the winters of 1947, 1953, and 1956 (Wood 1983).

[1] M. Higbee and R. Higbee, pers. comm.
[2] Miller, R. 1997. Local notes: Lebanon County. PB 11:97.
[3] Boyle, W.J., Jr., R.O. Paxton, and D.A. Cutler. 1981. Hudson-Delaware region. AB 35:925.
[4] W. Reid, pers. comm.
[5] Hall, G.A. 1982. Appalachian region. AB 36:178.
[6] Boyle, W.J., Jr., R.O. Paxton, and D.A. Cutler. 1988. Hudson-Delaware region. AB 42:242.
[7] Keller, R. 1995. Local notes: Berks County. PB 9:31.
[8] Hall, G.A. 1982. Appalachian region. AB 36:295.
[9] Hall, G.A. 1980. Appalachian region. AB 34:273.
[10] Gillespie, J. 1950. General notes (other unusual records). Cassinia 38:38.
[11] Street, P.B. 1953. General notes (other unusual records). Cassinia 40:38.

Black-throated Blue Warbler *Dendroica caerulescens*

General status: Black-throated Blue Warblers breed across southern Canada from Saskatchewan to Nova Scotia and south to Minnesota, Wisconsin, Michigan, Pennsylvania, and through the Appalachian Mountains to Georgia. They winter in the West Indies and Mexico south to South America. Black-throated Blue Warblers are fairly common regular migrants throughout Pennsylvania. They breed primarily in the High Plateau and in the Pocono section, but are less common and local in the mountains elsewhere.

Habitat: During migration Black-throated Blue Warblers are found in woodlands, woodland edges, and brushy areas and thickets. During the breeding season they are found primarily above 1000 feet elevation in large tracts of northern hardwood forests with a dense understory of shrubs such as mountain laurel or rhododendron.

Seasonal status and distribution

Spring: Birds may appear as early as the third week of Apr. Their typical migration period is from the first to the fourth week of May. Peak migration occurs during the second and third weeks of May. Nonbreeding stragglers have been recorded to the third week of Jun.

Breeding: Black-throated Blue Warblers are locally fairly common along slopes of the High Plateau and in the Poconos. They are uncommon in the glaciated regions and the northern Ridge and Valley (Lycoming to Huntingdon cos.) but are rare elsewhere on the slopes of the Ridge and Valley. Black-throated Blue

Warblers were found breeding for the first time in the twentieth century in the Blue Ridge section of Adams and Cumberland cos. in 1996.[1,2] Their highest density was found along the Allegheny Front in a narrow band between 2200 and 2450 feet elevation.[3] Nests are built of bark, dead wood, and rootlets, close (less than 3 feet) to the ground, woven in the forked branch of a rhododendron, mountain laurel, shrub, or sapling (H. H. Harrison 1975). Most nests with eggs have been found from 25 May to 21 Jun. BBS trends since 1980 show a population increase.

Fall: First fall migrants may be detected by the third week of Aug. Their typical migration time is from the fourth week of Aug to the third week of Oct. The greatest number of birds may occur any time from the third week of Sep to the first week of Oct. Stragglers have remained to the fourth week of Nov. A later than usual record was one found on 24 Dec 1984 at Falls in Wyoming Co.[4]

History: Harlow (1913) considered this species to be common across the northern mountainous counties, as they are today. In the mid-1800s they were noted to be common in summer (Baird 1845) in Cumberland Co., where Black-throated Blue Warblers were observed again in 1996. Poole (unpbl. ms.) thought that "since 1958, an alarming reduction in the number of this species and of various other insectivorous birds has been noticed, and is generally attributed, at least in part, to the widespread use of powerful insecticides." The current status of this species indicates that decline did not continue. A rather early male Black-throated Blue Warbler was shot on 6 Apr 1893 in Chestnut Hill, Philadelphia Co. (Poole, unpbl. ms.).

[1] Kennel, E., and A. Kennel 1996. Local notes: Adams County. PB 10:96.

[2] Hoffman, D. 1996. Local notes: Cumberland County. PB 10:99.

[3] Graves, G.R. 1997. Geographic clines of age rations of Black-throated Blue Warblers *(Dendroica caerulescens)*. Ecology 78:2524–2531.

[4] Boyle, W.J., Jr., R.O. Paxton, and D.A. Cutler. 1985. Hudson-Delaware region. AB 39:153.

Yellow-rumped Warbler ***Dendroica coronata***

General status: Yellow-rumped Warblers breed from Alaska across Canada south through the western U.S. and in the east to Virginia and Maryland. They winter from Washington south through California and Mexico in the West and from southern Ontario and New England south across the eastern U.S. Yellow-rumped Warblers are probably the most widespread and common of the warblers during migration in Pennsylvania. They are the first warblers to arrive in spring and the last to leave in fall. Yellow-rumped Warblers are very hardy; many regularly spend the coldest months of the year in the state. They are common to abundant regular migrants. They breed primarily in the Glaciated Northeast and in the north-central portion of the High Plateau, but they have been rapidly expanding their range during the last few decades. More wintering Yellow-rumped Warblers are found at PISP than at any other site in the state. A few may remain to winter at scattered locations statewide, mostly across the southern half of the state.

Habitat: During migration Yellow-rumped Warblers inhabit woodlands, woodland edges, brushy areas, and thickets. During the breeding season they inhabit primarily northern hardwood forests above 1000 feet elevation, especially near swamps or bogs in hemlock, pines, or plantings of spruce or larch. In winter, they can be found at lower elevations in woodlands with dense understory, brushy areas, thickets, and shrubbery or hedgerows, especially those bearing fruit.

Seasonal status and distribution

Spring: Migrants usually first appear the

second week of Apr. The number of birds increases through Apr and usually peaks from the fourth week of Apr to the second week of May. Yellow-rumped Warblers usually are abundant, outnumbering other warblers during their peak. By the fourth week of May migration has ended, with nonbreeding stragglers to the first week of Jun.

Breeding: Yellow-rumped Warblers are fairly common breeding residents in the northeastern mountains, including the central Poconos, Pike Co., and North Mountain (Sullivan and Wyoming cos.). They are uncommon to rare across the High Plateau west to Crawford Co.[1] and are rare, local, and irregular south to the Westmoreland-Somerset co. line[1] and at scattered points in the Ridge and Valley. The BBA project presented a dramatic expansion in range that continued into the 1990s west and south from the Poconos.[2] They had been detected on 15 BBS routes by 1996 and have expanded at the rate of 13% per year since 1980. New summer locations in 1996 and 1997 were reported south into northern Huntingdon Co.,[3] the Blue Ridge in Cumberland Co.,[4] and YCSP in Indiana Co.[5] Yellow-rumped Warblers arrive on breeding territory by late Apr. The few documented nests have been found over a long season, beginning in May. Nests are placed in a variety of settings, often in conifers but also in deciduous trees. A characteristic feature is feathers woven into the nest, with loose ends partly covering the eggs (H. H. Harrison 1975).

Fall: Yellow-rumped Warblers are even more abundant in fall than in spring. The earliest migrants may appear the third week of Aug. Their typical migration period is from the second week of Sep to the first week of Nov. Numbers usually peak during the second or third week of Oct. If the weather remains mild, a few may remain until the first severe winter weather arrives, usually during the first or second week of Jan.

Winter: The largest concentrations of wintering birds in the state are found at PISP, where flock sizes may contain over 50 birds in some winters.[6] Their abundance at this northerly site is attributable to their food source, bayberry. In the southeastern portion of the state, where they are uncommon to rare winter residents, they have been observed eating mainly poison ivy berries.[7] During mild winters they may be fairly common locally, especially along river valleys in the Ridge and Valley and at low elevations elsewhere in the southern half of the state. Yellow-rumped Warblers are casual or accidental in winter across the northern half of the state, where they occasionally visit bird feeding stations.

History: Warren (1890) observed summering Yellow-rumped Warblers (called Myrtle Warblers) in Cameron and McKean cos. and predicted that they would nest "regularly, but sparingly, in some of our secluded and higher mountainous districts." Although they had been well known as an abundant migrant, nesting was not confirmed in the state until 1949 (P. B. Street 1954). Until the current period, nesting was firmly documented only near Pocono Lake, Monroe Co., although they had been noticed during the nesting period at other localities in the Pocono section and even elsewhere in the state (Poole 1964). It was not until recent times that Warren's prediction was realized.

Comments: The subspecies *D. c. auduboni*, known as Audubon's Warbler, is easily distinguished from *D. c. coronata* in the field. Audubon's Warbler has been recorded in Pennsylvania at least a dozen times. Three specimens were collected in the state before 1960: one was found near Reading on 14 Oct 1888 (Poole, unpbl. ms.); a female was captured on 8 Nov 1889 in Chester Co. (Poole, unpbl. ms.); and one was found dead near Lititz in Lancaster Co. on 9 Jan 1954 (Poole, unpbl. ms.). More recently, all but one record have been during spring or fall migration from the counties of Bedford, Berks, Bucks (one banded),[8] Butler, Erie, Lancaster, and York. The only winter record is of one at Tinicum on 14 Feb 1980 that may have been a hybrid between the two subspecies.[9]

[1] R.C. Leberman, pers. comm.
[2] Gross, D.A. 1992. Yellow-rumped Warbler *(Dendroica coronata)*. BBA (Brauning 1992):316–317.
[3] Grove, G. 1997. Local notes: Huntingdon County. PB 11:96.
[4] Hoffman, D. 1997. Local notes: Cumberland County. PB 11:94.
[5] Higbee, M., and R. Higbee. 1997. Local notes: Indiana County. PB 11:96.
[6] G.M. McWilliams, pers. obs.
[7] R. Wiltraut, pers. comm.
[8] Scott, F.R., and D.A. Cutler. 1967. Middle Atlantic Coast region. AFN 21:18.
[9] Richards, K.E., R.O. Paxton, and D.A. Cutler. 1980. Hudson-Delaware region. AB 34:259.

Black-throated Gray Warbler *Dendroica nigrescens*

General status: Black-throated Gray Warblers breed from southern British Columbia south through California and east to New Mexico, Colorado, and Wyoming. They winter in coastal central and southern California, southern Arizona, and south into Mexico. There are eight recent records from Pennsylvania, five of which were recorded between 1972 and 1974. All records have been from east of the Allegheny Front and in the fall, except the record of one bird that remained into the winter.

Seasonal status and distribution

Fall to winter: In 1966 a male Black-throated Gray Warbler in fall plumage was was discovered by W. Reid at Wyoming in Luzerne Co. on 9 Sep.[1] In 1972 three different birds were reported in the state: a female was photographed by R. Wiltraut on 20 Sep at Whitehall in Lehigh Co.;[2] one was seen by H. Alexander at Swarthmore in Delaware Co. from 27 Nov to 7 Dec;[3] and one reported by M. Wood was accidentally killed in a banding trap at State College on 28 Sep.[4] A Black-throated Gray Warbler was picked up dead by C. Wonderly at Gladwyne in Montgomery Co. on 15 Jan 1973.[3] A bird was discovered by E. Budd at State College in Centre Co. on 10 Aug 1974,[5] and another was reported by K. Hiller near Berwick in Columbia Co. on 18 Sep 1988.[6] The most recent record was of a female reported by J. Silagy. It apparently appeared at a feeder in Bern Township, Berks Co., in late Nov or early Dec 1996 but was not reported until 19 Jan 1997. It remained until at least 13 Feb 1997.[7]

History: The only historical record of this species in Pennsylvania was of a male collected by B. H. Warren in Wyoming Co. on 10 May 1912; it was placed in the ANSP. The bird was originally mislabeled Blackpoll (Poole 1964).

[1] Scott, F.R., and D.A. Cutler. 1966. Middle Atlantic Coast region. AFN 21:410.
[2] Scott, F.R., and D.A. Cutler. 1973. Middle Atlantic Coast region. AB 27:39.
[3] Scott, F.R., and D.A. Cutler. 1973. Middle Atlantic Coast region. AB 27:599.
[4] Hall, G.A. 1973. Appalachian region. AB 27:62.
[5] Hall, G.A. 1975. Appalachian region. AB 29:61.
[6] Gross, D.A. 1988. County reports—July through September 1988: Columbia County. PB 2:102.
[7] Keller, R. 1997. Local notes: Berks County. PB 11:27.

Black-throated Green Warbler *Dendroica virens*

General status: Black-throated Green Warblers breed across Canada from British Columbia to Newfoundland south to Illinois, Ohio, New Jersey, and through the Appalachian region to Alabama and Georgia. They winter in southern Texas, Louisiana, and Florida south to Mexico and Central America. Black-throated Green Warblers are fairly common regular migrants across Pennsylvania. They breed in the northern two-thirds of the state, south through the High Plateau, and locally at higher elevations of the Ridge and Valley.

Habitat: They can be found in woodlands, woodland edges, brushy areas, and thickets during migration. During the breeding season Black-throated Green Warblers prefer northern hardwood forests, especially those containing mature hemlock and white pine, but they also occur in deciduous woodlands, at least in the northern mountains. In Indiana Co. they frequently inhabit red pine plantations.[1]

Seasonal status and distribution

Spring: The earliest migrants may appear the second week of Apr. One on 8 Apr 1978 in Westmoreland Co. was an exceptionally early record.[2] Their typical migration period is from the third or fourth week of Apr to the fourth week of May or the first week of Jun. Peak migration occurs during the first and second weeks of May.

Breeding: Black-throated Green Warblers are fairly common to common in a variety of forested settings over the northern two-thirds of Pennsylvania and south in the Allegheny Mountains to Maryland and West Virginia. They are local and uncommon in the mountains of the southern Ridge and Valley and in the Blue Ridge. Black-throated Green Warblers are the most widely distributed and common of the northern warblers and are absent in the breeding season only from the Coastal Plain, Piedmont, and southwestern-tier counties (Lawrence Co. south to Greene Co.). On BBS routes they are most common in the High Plateau, from Sullivan Co. west to Warren Co. Black-throated Green Warblers are less area-sensitive than are some forest species; they regularly occur in small woodlots in the north. They occur in suitable mixed forests down to 600 or 800 feet in elevation, even at the southern edge of their range at Hawk Mountain, Berks Co.[3] Nests are compact and usually are placed high in a conifer. Most egg sets have been found from 24 May to 21 Jun, with extremes to 6 Jul. An early date for a completed nest without eggs was 12 May (Grimm 1952).

Fall: Their usual migration time is from the fourth week of Jul or the first week of Aug to the second or third week of Oct. Peak migration occurs from the second to the fourth week of Sep. An unusually high count of 100 birds occurred during a heavy migration on 21 Sep 1996 at HMS.[4] Stragglers remain to the fourth week of Nov.

History: Strangely, this warbler was not listed as a summer resident by Baird (1845) in Cumberland Co., where Frey (1943) found them to be fairly common. No significant difference in distribution is otherwise noted from historical accounts, although Warren's (1890) reference to adults in Jun and Jul in Chester and Delaware cos. is ambiguous.

[1] M. Higbee and R. Higbee, pers. comm.

[2] Hess, P. 1978. Area bird summaries for April. ASWPB 42:10.

[3] L. Goodrich, pers. comm.

[4] Keller, R. 1996. Local notes: Berks County. PB 10:161.

Townsend's Warbler *Dendroica townsendi*

General status: Townsend's Warblers breed from southern Alaska south through British Columbia in Canada to Oregon and east to Montana. They winter along the Pacific coast from Washington to California and in Mexico through Central America. Pennsylvania has two valid records, a recent record and one based on a specimen collected in the nineteenth century.

Seasonal status and distribution

Fall to spring: A male Townsend's Warbler was reported coming to a feeder at Drums in Luzerne Co. on 15 Dec 1997.

Townsend's Warbler at Conyngham, Luzerne County, February 1998. (Photo: Rick Wiltraut)

It was confirmed by A. Gregory and J. Heuges on 17 Dec 1997 and was later photographed by many observers[1] and remained to late Mar 1998.[2]

History: A male was collected by C. D. Wood near Brandywine in Chester Co. on 12 May 1868 (Turnbull 1869). The specimen was eventually placed in the ANSP.

[1] Pulcinella, N. 1998. Rare bird reports. PB 11:226–227.
[2] Koval, R. 1998. Local notes: Luzerne County. PB 12:23.

Blackburnian Warbler ***Dendroica fusca***

General status: Blackburnian Warblers breed in Canada from Alberta east to Newfoundland and south in the U.S. to Minnesota, Wisconsin, Michigan, and Pennsylvania and in the Appalachians to Georgia. They winter in Central and South America. Blackburnian Warblers are uncommon to fairly common regular migrants throughout Pennsylvania. They breed primarily in the northern mountains and in heavily forested parts of the northern Ridge and Valley.

Habitat: During migration Blackburnian Warblers inhabit woodlands and woodland edges, especially in mature beech-hemlock forests. During the breeding season they are found primarily in large tracts of mature mixed northern hardwood forests, especially those with hemlocks or pines. Nesting birds also are found in large sycamores along rivers and in pine and spruce plantations.

Seasonal status and distribution

Spring: More birds are observed during spring than fall migration. Migrants appear as early as the third week of Apr, but two extraordinarily early records have been reported. A window-killed bird was found on 16 Mar 1982 at Pittsburgh in Allegheny Co.,[1] and another was seen during an unusually warm spring at Mt. Davis, Somerset Co., on 31 Mar 1998.[2] Their typical migration period is from the first to the fourth week of May. Peak migration usually occurs during the second or third week of May. Stragglers remain to the second week of Jun.

Breeding: Blackburnian Warblers are uncommon to fairly common summer residents in the High Plateau, Poconos, and northern portions of the Ridge and Valley. They are uncommon and local at higher elevations in the Glaciated Northeast and Glaciated Northwest regions, south through the Alleghenies, and in the southern Ridge and Valley. They are most common on BBS routes in the High Plateau. Extralimital breeding territories occur, such as along Slippery Rock Creek, Butler Co., and in the Blue Ridge section. Although not restricted to hemlocks or conifers, Blackburnian Warblers become more common with increasing coniferous cover (Beals 1960). Territorial densities averaged 46 times as high within old-growth hemlock forests in the Allegheny National Forest as in younger forests having a history of human disturbance (Haney and Schaadt 1996). No population trend is evident on BBS routes. Most nests are inaccessible, usually far out on a high branch of a tall conifer. Of 21 egg sets, most were found from 24 May to 19 Jun.

Fall: Migrants first are detected the third week of Aug, with the peak number occurring from the fourth week of Aug to the third week of Sep. Daily highs are usually fewer than five birds, but at HMS 25 birds were recorded on 21 Sep 1996.[3] Migration usually ends by the fourth week of Sep, with stragglers recorded to the second week of Oct.

History: Early ornithologists (e.g., Dwight 1892; Cope 1902) considered Blackburnian Warblers to be locally abundant in remnant old forests in Pennsylvania's mountains, including thick hemlock growth in Susquehanna Co., where they subsequently have been scarce. They were observed by J. F. Street (1921) on the ridges of Franklin Co., but Harlow's (1913) nesting season reference to Bedford and Fulton cos. is ambiguous, because he later stated that Blackburnian Warblers nest "in the mountainous regions of Pennsylvania from Huntingdon County north" (Harlow 1918). Sutton (1928b) considered them rare and local summer residents around the Pymatuning Swamp, where they were not confirmed during the BBA project.[4] They appear to have retreated from some peripheral parts of their range.

[1] Hall, G.A. 1982. Appalachian region. AB 36:853.

[2] J. Tilley, pers. comm.

[3] Keller, R. 1996. Local notes: Berks County. PB 10:161.

[4] Mulvihill, R.S. 1992. Blackburnian Warbler *(Dendroica fusca)*. BBA (Brauning 1992):320–321.

Yellow-throated Warbler ***Dendroica dominica***

General status: Yellow-throated Warblers breed from Texas across Florida and north to Wisconsin, Ohio, Pennsylvania, and Connecticut. In the U.S. they winter from southern Texas along the Gulf coast through Florida and north along the Atlantic coast to South Carolina. Yellow-throated Warblers have expanded northward along sycamore-lined streams into many portions of Pennsylvania, especially since the early 1970s, and now are fairly common in the Southwest region, but they are still uncommon to rare across the rest of the state.

Habitat: Away from breeding sites, Yellow-throated Warblers are observed during migration in woodlands or woodland edges, especially those with pines. They prefer mature sycamore-lined streams and sometimes occur in white pines during the breeding season.

Seasonal status and distribution

Spring: The earliest birds may arrive the first week of Apr, but most arrive the third or fourth week of Apr. An early migrant was found on 26 Mar 1977 at Cook Forest, Forest Co.[1] It is not clear when migration ends, but birds that have overflown their breeding range have been recorded to 12 May at PISP in Erie Co.[2]

Breeding: Yellow-throated Warblers are fairly common in forested settings of Pennsylvania's Southwest (Greene, Fayette, and Westmoreland cos.) and north to Armstrong, Indiana, and Lawrence cos.; they are rare and local at scattered sites north along the Allegheny River to points in Warren Co.[3] They are local and uncommon along the lower Susquehanna River in Lancaster Co. and rare at a few widely scattered locations along major tributaries, such as the Juniata River in Mifflin and Juniata cos. Isolated populations of a few pairs are regular in sycamores along Spruce Creek in Huntingdon Co. and Little Pine and Lycoming creeks in Lycoming Co. Similarly, they occur along wooded shores of the Delaware River and major tributaries north to Easton, Northampton Co. Additional scattered sites through the central counties are to be expected. They have dramatically expanded their range in Pennsylvania since the mid-1970s. For example, they were "unknown in the Ligonier Valley [Westmoreland Co.] before the spring of 1974" (Leberman 1976) and were first noted in Greene Co. in 1971 (Bell 1994), where they subsequently became regular and fairly common. Yellow-throated Warblers were first confirmed breeding in Westmoreland Co. at PNR in 1978, in Indiana Co. in 1987, and in Lycoming Co. in 1988, and they were on territory in Northampton Co. in 1991. Few nests have been found in Pennsylvania, but they are likely to be high in a tall white pine or sycamore. Yellow-throated Warblers nest early; a female was seen feeding nestlings as early as 15 Apr.[4] Nests found in 1995 and 1997 along the Lehigh River in Northampton Co. were in an upright crotch high in a sycamore tree.[5]

Fall: Fall migration is difficult to discern because Yellow-throated Warblers are usually found only where they also nest. Banding records suggest that migrants may be found from about the third week of Aug to the fourth week of Sep. The few fall sight records of Yellow-throated Warblers are also during this period in various parts of southern Pennsylvania. Stragglers have been recorded to the first week of Oct. There are two unusually late fall records. One bird was found on 2 Nov 1997 at Rohrsburg in Columbia Co.[6] and one even later on 31 Dec 1994 at Newville in Cumberland Co.[7]

Winter: There is one winter record: a bird visiting a feeder at Pittsburgh in Allegheny Co. remained until 19 Jan 1997 and was then captured in a weakened condition and was taken to the National Aviary in Pittsburgh.[8]

History: The first reference to Yellow-throated Warbler in Pennsylvania appears to have been made by Turnbull (1869), who considered it "a rare straggler to the lower counties of Pennsylvania and New Jersey." Warren (1890) added personal observations that suggested it was a "rare, local breeder" and cited summer records from Chester and Delaware cos. Other authors cited records from spring through fall from various parts of the Coastal Plain and Piedmont after the turn of the twentieth century. There was not a noticeable increase in reports from these regions until the 1950s. Yellow-throated Warblers were notably absent from the list of breeding birds in Greene Co. (Jacobs 1893). They were not recorded in the western part of the state until 1953, when Bent (1953) recorded one at Frankford Spring in Beaver Co. and Parkes (1956) reported on one in Pittsburgh in Allegheny Co. on 4 May 1953. Only since the 1970s could Yellow-throated Warblers be considered regular nesting birds in Pennsylvania.

Comments: Two recognized subspecies have occurred in the state: the yellow-lored (area between a bird's eye and bill) *D. d. dominica* and the white-lored *D. d. albinora. Dominica* is documented by a specimen (Poole, unpbl. ms.), but *albinora,* sometimes called the "Sycamore" warbler, appears to be the breeding resident in most of Pennsylvania and west of the Appalachians in West Virginia (Hall 1983).

[1] Hall, G.A. 1978. Appalachian region. AB 31: 1002.

[2] G.M. McWilliams, pers. recs.
[3] Grisez, T. 1997. Local notes: Warren County. PB 11:101.
[4] R.C. Leberman, pers. comm.
[5] R. Wiltraut, pers. comm.
[6] Haas, F.C., and B.M. Haas, eds. 1998. Birds of note—October through December 1997. PB 11:231.
[7] Haas, F.C., and B.M. Haas, eds. 1995. Rare and unusual bird reports. PB 8:230.
[8] Fialkovich, M. 1997. Local notes: Allegheny County. PB 11:26.

Pine Warbler ***Dendroica pinus***

General status: Pine Warblers breed across southern Canada from Manitoba east to New Brunswick south in the eastern U.S. to Texas and Florida, except in the Midwest. They winter from Maryland south along the Atlantic and Gulf coasts to Texas. In Pennsylvania Pine Warblers are uncommon to rare regular migrants. They breed at widely scattered sites throughout most of the state. Pine Warblers occasionally winter in the state, especially in the southeastern part.

Habitat: During migration Pine Warblers prefer pines and larches, but they may also occur in deciduous woodlands and edges. During the breeding season they are often found in pine barrens or plantations and, rarely, in deciduous woodlands that have scattered pines.

Seasonal status and distribution

Spring: Outside of wintering sites, Pine Warblers may arrive as early as the first week of Mar. Their typical migration period is from the first or second week of Apr to the third week of May. Peak migration occurs from the third week of Apr to the first week of May. Stragglers remain to the fourth week of May.

Breeding: Pine Warblers are uncommon to rare during the breeding season at most sites. They are locally regular and fairly common at some locations in eastern counties but generally are rare west of the Allegheny Front and are largely absent from the western-tier counties. Strongholds appear to be the Blue Ridge, Pike Co., the Poconos, and locally in the Ridge and Valley.[1] Although reported on 23 BBS routes in the state, they are among the most irregularly reported warblers, detected only once or twice on over half those routes and regularly on only a few in the Ridge and Valley. Pine Warblers may be overlooked in some locations because their song could be confused with that of Chipping Sparrows, juncos, or Yellow-rumped Warblers, and one or more of those species may overlap the Pine Warblers' range across Pennsylvania. They appear on territory in early Apr, which is very early for a warbler. The nest is a neat cup, placed high on the horizontal branch of a pine, but few have been found in Pennsylvania. There is little information on population trends.

Fall: Fewer birds are recorded in fall than in spring, in part because of the difficulty in separating them from similar-looking species such as Cape May and Blackpoll warblers. The earliest migrants away from breeding sites may begin to appear the first or second week of Aug; most are observed between the third week of Sep and the second week of Oct. A few birds may remain through the fall and into winter if the weather remains mild.

Winter: Winter records are mainly from the Piedmont, where they have been recorded in Bucks, Chester, Delaware, Lancaster, and York cos. The only other sites east of the Allegheny Front where Pine Warblers have been recorded in winter are at Pennypack Park in Philadelphia Co.,[2] at Wyoming in Luzerne Co.,[3] and in Wyoming Co.[4] West of the Allegheny Front they have been recorded in winter at West Newton in Westmoreland Co.[5] and at the Lake Erie Shore at three sites.[6]

History: Historically Pine Warblers have been poorly documented; their range in Pennsylvania is not well defined. Early descriptions are not substantially different from the recent, uncertain, descriptions. They were generally considered uncommon breeders over a large section of Pennsylvania, but predominantly restricted to the mountainous regions (Harlow 1913). Dwight (1892) considered the stands of yellow pines near Towanda (Bradford Co.) to be "full of Pine Warblers," where the BBA project did not report them.[1] Todd (1940) considered this species to be a rare migrant and exceedingly rare west of the Allegheny River. Poole (1964) listed it as an uncommon transient in spring and rare in fall over most of the state. Historical winter records have come from the counties of Chester, Delaware, and Lancaster and from near Philadelphia.

[1] Santner, S. 1992. Pine Warbler *(Dendroica pinus)*. BBA (Brauning 1992):324–325.
[2] E. Fingerhood, pers. comm.
[3] Hall, G.A. 1966. Appalachian region. AFN 20:405.
[4] Haas, F.C., and B.M. Haas, eds. 1995. Seasonal occurrence tables—January through March 1995. PB 9:45.
[5] Leberman, R.C., and R.S. Mulvihill. 1990. County reports—January through March 1990: Westmoreland County. PB 4:40.
[6] G.M. McWilliams, pers. recs.

Kirtland's Warbler ***Dendroica kirtlandii***

General status: This federally Endangered species has a limited breeding range in jack pine forests of north-central Michigan. The Kirtland's Warbler's only known wintering site is in the Bahamas, with most specimens from New Providence and Eleuthera, south to Caicos (Dunn and Garrett 1997). They are very rarely found anywhere in North America during their migration period. There are four accepted records of this species in Pennsylvania, two in spring and two in fall.

Seasonal status and distribution

Spring: One was carefully studied and heard by K. Gabler and other birders on 14 May 1994 in an area known as Little Cove in Warren Township, Franklin Co.[1] Another Kirtland's Warbler was observed by N. Hall and B. McChesney at PISP in Erie Co. on 14 May 1997. It was a male in full song observed from only a few feet away.[2]

Fall: Late in the morning of 21 Sep 1971 a hatching-year Kirtland's Warbler was captured and banded by R. C. Leberman at PNR. The bird remained around the reserve for 11 days and was seen and photographed by many people. It was recaptured on 26 Sep and again 2 Oct 1971. At the time, this was only the second Kirtland's Warbler ever banded on migration outside of Michigan.[3] One was observed by K. Knight at Egleman's Park in Berks Co. on 5 Sep 1996.[4]

[1] Pulcinella, N. 1994. Rare bird reports. PB 8:85.
[2] Haas, F.C., and B.M. Haas, eds. 1997. Birds of note—April through June 1997. PB 11:90.
[3] Parkes, K.C., and R.C. Leberman. 1971. A Kirtland's Warbler banded at Powdermill. Educational release no. 89. Powdermill Nature Reserve, Carnegie Museum of Natural History, Pittsburgh.
[4] Keller, R. 1996. Local notes: Berks County. PB 10:161.

Prairie Warbler ***Dendroica discolor***

General status: Prairie Warblers breed primarily from New Hampshire south to Florida and west across southern Ontario to Missouri and Texas. They winter from central Florida south through the Bahama Islands, West Indies, and islands off the coast of Central America. Prairie Warblers are uncommon to rare regular migrants. They breed primarily in the Ridge and Valley and Southwest regions. Elsewhere they are rare and widely scattered. Their breeding range has been expanding northward in recent years.

Habitat: During migration Prairie Warblers are found in overgrown fields with small trees or saplings, brushy areas, and scrubby pine forests. During the breeding season they are more typically associated with scrubby pine stands, overgrown fields dispersed with small pines or cedars, and Christmas tree plantings. They will also use power-line corridors and overgrown fields with crab apple, hawthorn, or other deciduous shrubs.

Seasonal status and distribution

Spring: Prairie Warblers are more frequently reported in spring than in fall. Birds have been recorded as early as the second week of Apr, but their typical migration period is from the third or fourth week of Apr to the third week of May. Peak migration occurs during the first or second week of May. They are rarely observed after the third week of May away from breeding sites.

Breeding: Prairie Warblers are fairly common locally in the Southwest region, uncommon in the Ridge and Valley, and locally north along the broad valley of the North Branch of the Susquehanna River to New York in Bradford Co. They are rare in the Piedmont, through the extensively forested High Plateau, and in the Glaciated Northwest. Prairie Warblers have expanded their range in recent years, although the BBS suggests a population decline in the Ridge and Valley and Southwest regions. They have become most common in the Southwest region and have expanded north, at least irregularly, into every county except possibly Warren. The nesting season is within the period typical of long-distance migrants, but few nest dates are available.

Fall: Migration is difficult to discern; they are rarely encountered away from breeding sites during this season. Records away from breeding sites suggest that migrants may be southbound by the third week of Aug. There does not seem to be a peak period in migration. Birds may still be present through the first week of Oct. Stragglers have been recorded to the second week of Nov. A Prairie Warbler on 18 Dec 1994 at Mehoopany in Wyoming Co. was very late.[1]

History: Prairie Warblers were rare and restricted to southeastern counties before 1900 (Harlow 1913). The scarcity of egg sets in collections reflects this history. They were first documented nesting in the southwestern portion of the state in 1939[2] and rapidly expanded from there during the 1940s and 1950s.[3] The expansion paralleled that in West Virginia during the same period (Buckelew and Hall 1994). Prairie Warblers were found north to at least Centre and Union cos. by the late 1950s (Wood 1958) but were not found in the Glaciated Northeast until 1966.[4]

Comments: The name *prairie* was given by Wilson to describe the shrubby barren habitat typical of this warbler. Since modern usage of the word refers to grassy habitat, perhaps a better name for this warbler would be scrub warbler.[5]

[1] Haas, F.C., and B.M. Haas, eds. 1995. Rare and unusual bird reports. PB 8:230.

[2] Dickey, S.S. 1941. The Prairie Warbler in Greene Co., Pennsylvania. Cardinal 5:125–129.

[3] Leberman, R.C. 1992. Prairie Warbler *(Dendroica discolor)*. BBA (Brauning 1992):326–327.

[4] W. Reid, pers. comm.

[5] Sprunt, A., Jr. 1957. Prairie Warbler *(Dendroica discolor)*. Pages 182–184 *in* The warblers of America, ed. L. Griscom and A. Sprunt Jr. Devin-Adair Company, New York.

Palm Warbler ***Dendroica palmarum***

General status: Palm Warblers breed in Canada from Yukon and southern Mackenzie east to Newfoundland and south to Minnesota, Wisconsin, Michigan, and Maine. They winter in the U.S. from Delaware south along the Atlantic coast states and the Gulf coast. Palm Warblers are rare to locally common regular migrants in Pennsylvania. Migrant populations may vary in number from year to year. They are accidental in winter, with most winter records from the southeastern portion of the state. Two distinct subspecies are found in the state.

Habitat: Palm Warblers are usually found on or near the ground in woodland edges, thickets, and brushy areas and at PISP on grassy sand dunes.

Seasonal status and distribution

Spring: Earliest arrivals may appear the fourth week of Mar. Their usual migration period is from the second or third week of Apr to the third week of May. The greatest number of migrants is reported during the third and fourth weeks of Apr east of the Allegheny Front, while to the west the largest number is reported during the first and second weeks of May. Stragglers remain until the fourth week of May.

Fall: Palm Warblers are among the last species of warblers to migrate south through Pennsylvania. Fall migrants have arrived as early as the fourth week of Aug. Peak migration may occur any time from the third week of Sep to the second week of Oct. Migration continues until the third or fourth week of Oct. If the weather remains mild, a few remain to the first week of Jan or later.

Winter: Most winter records are from east of the Allegheny Front, where they have been recorded in Philadelphia Co.,[1] Lancaster in Lancaster Co.,[2] twice in Berks Co. at Lake Ontelaunee,[3] Cumberland Co. at Big Spring Creek,[4] Centre Co. at State College,[5] and Union Co. at Lewisburg (Schweinsberg 1988). West of the Allegheny Front they have been recorded in winter in the Southwest at two sites: one bird was in Allegheny Co. at Pittsburgh on 26 Jan 1982,[6] and one successfully wintered in Greene Co. at Clarksville during the winter of 1961–1962.[7] Very little is mentioned in published literature as to which subspecies is here in winter.

History: Both Poole (1964) and Todd (1940) distinguished the two subspecies, *D. palmarum hypochrysea* and *D. p. palmarum*, and treated their status and distribution separately. Todd (1940) wrote that the western subspecies, but not the eastern subspecies, was found in western Pennsylvania. Poole (unpbl. ms.) stated that the western subspecies had been collected on several occasions in eastern Pennsylvania. He listed the eastern subspecies as more common in spring than in fall while the reverse was true of the western subspecies.

Comments: Apparently the greatest number of birds east of the Allegheny Front are of the eastern subspecies, *D. p. hypochrysea*, and the greatest number west of the Allegheny Front are of the western subspecies, *D. p. palmarum*. Although they are distinguishable in the field, few reports identify subspecies. Apparently birds of the western subspecies appear in greater numbers in the eastern part of the state than those of the eastern subspecies do in the western part of the state.

[1] Fingerhood, E. 1993. Notes from the field: Philadelphia County. PB 7:21.
[2] Haas, F.C., and B.M. Haas, eds. 1993. Rare and unusual bird reports. PB 7:18.
[3] R. Keller, pers. comm.
[4] J. Earle, pers. comm.
[5] Hall, G.A. 1983. Appalachian region. AB 37:302.
[6] Hall, G.A. 1982. Appalachian region. AB 36:295.
[7] Hall, G.A. 1962. Appalachian region. AFN 16:330.

Bay-breasted Warbler *Dendroica castanea*

General status: Bay-breasted Warblers breed in Canada from British Columbia and southern Mackenzie east to Newfoundland and south to Minnesota, Michigan, and New York. They winter from Central to South America. Bay-breasted Warblers are fairly common to common regular migrants over most of Pennsylvania. In the Coastal Plain and the southern portion of the Piedmont they are uncommon to rare regular migrants. The number of birds may vary from year to year. They are accidental in summer in Pennsylvania.

Habitat: Woodlands and woodland edges as well as pines or spruce plantings are used by this species in migration.

Seasonal status and distribution

Spring: Bay-breasted Warblers are among the last warblers to arrive in the state; the earliest may appear the first week of May. Their typical migration period is from the second week of May to the first week of Jun. The greatest number of birds is observed during the third and fourth weeks of May. At peak migration, 100 or more Bay-breasted Warblers are occasionally observed at PISP.[1]

Summer: Two recent records fall within this season. An adult male was discovered singing on territory for several weeks during Jun 1995 in an old-growth beech-hemlock stand near Ludlow in the Allegheny National Forest, McKean Co.[2] No evidence of nesting was observed. Another bird was reported in Sullivan Co. on 24 Jun 1997.[3]

Fall: Their usual migration period is from the third week of Aug to the second week of Oct. An early fall migrant was seen 30 Jul 1996 at Tunkhannock in Wyoming Co.[4] Peak migration occurs during the second and third weeks of Sep. Stragglers have been recorded to the second week of Nov.

History: Todd (1940) cited an early fall migrant from Pittsburgh on 30 Jul 1925, and Poole (unpbl. ms.) listed one at HMS on 7 Aug 1957.

[1] G.M. McWilliams, pers. obs.

[2] J. C. Haney, pers. comm.

[3] Haas, F.C., and B.M. Haas, eds. 1997. Seasonal occurrence tables—April through June 1997. PB 11:121.

[4] W. Reid, pers. comm.

Blackpoll Warbler ***Dendroica striata***

General status: Blackpoll Warblers breed from Alaska across central Canada to Newfoundland south mainly to New York and Massachusetts. They winter in South America. The arrival of Blackpoll Warblers heralds the end of the spring warbler migration; they usually are among the last warblers to arrive. In the fall, Blackpoll Warblers are noted for their long-distance trans-Atlantic flights to their wintering grounds, with some birds flying nonstop from the Atlantic coast to South America. They are fairly common regular migrants in Pennsylvania, but the number of birds may vary from year to year. Blackpoll Warblers were recently discovered breeding in the state.

Habitat: During migration, Blackpoll Warblers are usually found foraging in the tops of trees in woodlands and woodland edges. Where breeding has recently been confirmed, they were found in a large boreal conifer swamp at an elevation of 2200 feet.[1]

Seasonal status and distribution

Spring: Earliest arrivals have appeared the first week of May, but the typical migration period is from the second week of May to the first week of Jun. They migrate through Pennsylvania rather quickly, the peak occurring during the third or fourth week of May. Stragglers remain to the third week of Jun.

Breeding: Blackpoll Warblers were confirmed breeding in Pennsylvania for the first time in Wyoming Co. in 1994 and have been reported nesting annually since. A territorial male was discovered by T. Davis and J. Lundgren, biologists for the Nature Conservancy, on a survey of Coalbed Swamp in 1993. A nest discovered on 29 Jun 1994 in Coalbed Swamp by D. A. Gross and D. W. Brauning contained young.[1] The nest was typical of the species; it was in a spruce sapling within a few feet of the ground and was lined with feathers. Up to 10 territorial males were observed in that and nearby wetlands each year, indicating that this is an established although disjunct, population.[2] Two separate territorial males were reported in 1987 during the BBA project in Wayne Co., but no further evidence of nesting was provided. A singing male Blackpoll Warbler was at Jacobsburg State Park in Northampton Co. on 26 Jun 1997.[3]

Fall: The Blackpoll Warbler's typical migration period is from the second week of Aug to about the third week of Oct. Peak migration occurs from the third week of Sep to the first week of Oct. Leberman (1976) mentioned that hundreds of migrants may be seen moving through the treetops at Laurel Hill during the second half of Sep. Stragglers have remained to the third week of Nov.

History: Blackpoll Warblers were listed by early authors as numerous or abundant during migration. They were suspected of breeding in Pennsylvania in 1959, when three birds, including two males and a female, were discovered near Lake Ganoga in Sullivan Co. on 20–21 Jun at an elevation of 2200 feet elevation (Poole, unpbl. ms.). Notably, this location is just 25 miles from Coalbed, suggesting a long history of the species in the region. Other Jun records in southeastern Pennsylvania, compiled by E. Poole (unpbl. ms.), were of strays and late migrants that had no relation to breeding attempts. A summer record of a bird seen on 10 Jul 1946 at Pymatuning Lake in Crawford Co. was probably a nonbreeding stray (Poole, unpbl. ms.).

Comments: The recent documentation of Blackpolls nesting in Pennsylvania possibly reflects a lack of information from remote northern habitats rather than recent colonization. Most of Coalbed Swamp was not surveyed during the BBA project,[2] and it may never have been visited by an ornithologist before the Nature Conservancy's assessment in 1993. Blackpoll Warblers and nesting Yellow-bellied Flycatchers were documented at this site on a tip from the Nature Conservancy. The closest nesting population is in the Catskills of New York (Andrle and Carroll 1988). This species appears to be less common now as a migrant than in historical times. In some years, they may be rather scarce. In fall they are frequently confused with Bay-breasted Warblers and occasionally with other species, including Cape May and Pine warblers.

1 Gross, D.A. 1994. Discovery of a Blackpoll Warbler nest: A first for Pennsylvania, Wyoming County. PB 8:128–132.
2 D.A. Gross, pers. comm.
3 Wiltraut, R. 1997. Local notes: Northampton County. PB 11:99.

Cerulean Warbler *Dendroica cerulea*

General status: Cerulean Warblers breed from Minnesota east across southern Ontario to Rhode Island south to Arkansas, Mississippi, and the eastern edge of the Appalachians. They winter in South America. Cerulean Warblers are fairly common migrants in Pennsylvania. Cerulean Warblers breed locally across the state at widely scattered sites and are most common in the Southwest region.

Habitat: Cerulean Warblers usually are found in the open canopy of mature trees—primarily in oaks, sycamores, and slippery elms—along the edges of stream valleys, or in oak forests at higher elevations along the mountain ridges.

Seasonal status and distribution

Spring: During migration Cerulean Warblers are observed mainly during spring. Earliest arrivals appear the fourth week of Apr. Their usual migration time is from the first to the fourth week of May, and the peak of migration is during the second or third week of May. On 29 Apr 1981 there was an unusually heavy movement of warblers at the PNR banding station, when 20 species were recorded, including a remarkable count of 80 ceruleans.[1]

Breeding: Cerulean Warblers are fairly common breeders in the Southwest (particularly in Fayette, Greene, Washington, and Westmoreland cos.), where they occur in a variety of wooded settings. Cerulean Warblers are rare to fairly common locally across the southern half of state, on the slopes of ridges, or along larger rivers, and irregularly north to the New York line. Although confirmed breeding records are scant, full clutches have been observed from 20 May to 3 Jun. The nest is a compact cup of plant fibers, bark strips, and spider webs, usually placed high (more than 30 feet) in a deciduous tree (often an oak) far from the trunk. Cerulean populations have experienced a significant decline since 1966 in the BBS region that includes the Pennsylvania Southwest. This decline corresponds with a surveywide decline of 3.7% per year.

Fall: They are less frequently observed in fall than in spring. After singing has stopped they become difficult to find, and their habit of remaining high in the treetops compounds that difficulty. In Lycoming Co. migrant Cerulean Warblers have been observed in late Jul and early Aug.[2] Leberman (1976) stated that, in the Ligonier Valley, this species is migrating by the second half of Jul. At PISP they have never been recorded in fall. Migration continues through Aug and to the second week of Sep, and stragglers remain until the third week of Oct.

History: The species was first described by Wilson from a male shot somewhere in southeastern Pennsylvania (Wilson 1808–1814; Poole, unpbl. ms.); the first female known to science was not collected until 1825, when one was collected along the Schuylkill not far from Philadelphia on 1 Aug. During the latter part of the nineteenth century and the beginning of the twentieth century, Cerulean Warblers were considered very rare transients and breeding residents in the southwestern counties (Harlow 1913). Sutton (1928b) considered them to be common breeders at Pymatuning Swamp, where they were not found during the BBA project.[3] Todd (1940) considered them to be "one of the most characteristic breeding birds" of Greene Co. and mapped points north to Crawford Co. and east to Bedford Co. where they occurred. Poole (1964) listed this species as a rare transient over most of the state but said they were a "locally common nesting summer resident" west of the Allegheny Mountains and in the lower Susquehanna and Delaware valleys. Ceruleans were not well documented in the Ridge and Valley or in the Glaciated Northeast regions before the BBA project, and even then were probably overlooked because of inaccessible terrain and unfamiliarity with the song.

Comments: The Cerulean Warbler was identified as a high conservation priority in Pennsylvania.[4]

[1] Hall, G.A. 1981. Appalachian region. AB 35:824.
[2] P. Schwalbe and G. Schwalbe, pers. comm.
[3] Ickes, R. 1992. Cerulean Warbler *(Dendroica cerulea)*. BBA (Brauning 1992):328–329.
[4] K. Rosenberg, pers. comm.

Black-and-white Warbler *Mniotilta varia*

General status: Black-and-white Warblers are known by their habit of creeping along branches and up and down the trunks of trees in nuthatch fashion. They breed in Canada from Yukon, southern Mackenzie, across southern Canada to Newfoundland, and south across most of the eastern U.S. except along the Gulf coast east to South Carolina and Florida. Black-and-white Warblers winter from Florida south through the Bahamas and West Indies to Central and South America. They have been recorded many times in early winter in the eastern U.S. north to southern Canada (AOU 1998). Black-and-white Warblers are fairly common regular migrants and breed over most of Pennsylvania. Only one winter record is known.

Habitat: They are found in woodlands and woodland edges during migration. In the breeding season they inhabit a variety of mature deciduous or mixed forest settings, especially along hillsides and ravines with dense undergrowth.

Seasonal status and distribution

Spring: Black-and-white Warblers are among the earliest warblers to appear in the state and are most easily seen during this season. Migrants sometimes return to the state in the fourth week of Mar. Two exceptionally early birds were found on 9 Mar 1987 at Easton in Northampton Co.[1] Their typical migration period is from the second or third week of Apr to the third week of May, with peak migration usually during the first and second weeks of May. Stragglers remain to the second week of Jun.

Breeding: Black-and-white Warblers are locally rare to fairly common in various forest types statewide, except in the Glaciated Northwest section where they are largely absent. They are local and rare in most of the Piedmont, in the southern Ridge and Valley, and south of Pittsburgh and are most common on BBS routes in the Glaciated Northeast and the Poconos. They are among the more widely distributed and common of Pennsylvania's wood-warblers but are area-sensitive and less likely to be found in woodlots smaller than 500 acres (Robbins et al. 1989). Summer records have come from every county. The long-term BBS trend indicates a population decline of 2.9% per year. The nest, made of assorted plant material, is built on the ground and is well concealed from above. Most eggs are found from 19 May to late Jun. Todd (1940) said that about 20 May was the time to look for nests with eggs.

Fall: The first southbound birds are detected as early as the third week of Jul. Their typical migration time is a broad period from the fourth week of Jul to the first week of Oct, and late migrants continue until the third week of Oct. The greatest number of migrants is found from the first to the third week of Sep. There are several records of stragglers through Dec and at least one to the first week of Jan.[2]

Winter: A Black-and-white Warbler remained through the winter of 1997 at Washington Boro in Lancaster Co.[3]

History: Harlow's (1913) description of the Black-and-white Warbler's breeding distribution is very similar to recent patterns; he said it breeds in practically every county but is rather irregularly distributed and in many sections is decidedly rare. Historically they were documented in the Pymatuning area as rare breeders (Sutton 1928b).

Comments: A Black-and-white Warbler × Blackburnian Warbler was netted at PNR in 1980.[4]

[1] Boyle, W.J., Jr., R.O. Paxton, and D.A. Cutler. 1987. Hudson-Delaware region. AB 41:410.
[2] Hall, G.A. 1970. Appalachian region. AB 24:504.
[3] Boyle, W.J., Jr., R.O. Paxton, and D.A. Cutler. 1997. Hudson-Delaware region. NASFN 51:733.
[4] Leberman, R.C. 1981. Bird-banding at Powdermill, 1980. Research report no. 41. Powdermill Nature Reserve, Carnegie Museum of Natural History, Pittsburgh.

American Redstart *Setaphaga ruticilla*

General status: Amercian Redstarts breed across southern Canada south to Oregon, Wyoming, Oklahoma, Texas, and most of the eastern U.S. except along the Gulf coast and most of Florida. They winter from central Florida south through the Bahamas, West Indies, Mexico, and Central and South America. American Redstarts are fairly common regular migrants and are breeding residents over most of Pennsylvania.

Habitat: They inhabit second-growth woodlands, mature forest edges, lowland woodlands with heavy understory, and wooded riparian zones.

Seasonal status and distribution

Spring: Migrants may appear as early as the third week of Apr. Their typical migration period is from the fourth week of Apr or the first week of May to the first week of Jun. Peak migration occurs during the second and third weeks of May.

Breeding: American Redstarts are breeding residents in a variety of forested settings, but they are 20 times more common on BBS routes in the High Plateau than in the Piedmont (see Map 27). They are locally common in most of the northern and western regions of the state, are generally uncommon in the Piedmont and Ridge and Valley regions, and are absent from predominantly agricultural areas.[1] The BBS shows no significant population trend since 1966. The nest is a compact cup of plant fibers and spider webs. Placement varies, but it is often 20 feet or higher in the fork of a deciduous tree. Most nests with eggs have been found from 20 May to 21 Jun.

Fall: Because this species breeds so widely throughout most of the state, the beginning of migration is difficult to detect. An increase in the number of birds during the third or fourth week of Jul indicates that migration is under way. Migration continues to the third week of Oct. The greatest number of birds is recorded from the first to the third week of Sep. Stragglers have remained to the first week of Jan.

History: Mixed and sometimes contradictory accounts of the redstart's status can be found throughout the early literature. Harlow (1913), reflecting his broad experience, reported a status similar to current patterns. Others have described them as uncommon to common, depending on local conditions of forests and composition of tree species. Bent (1953) reported a rather early American Redstart on 3 Apr at Berwyn in Chester Co.

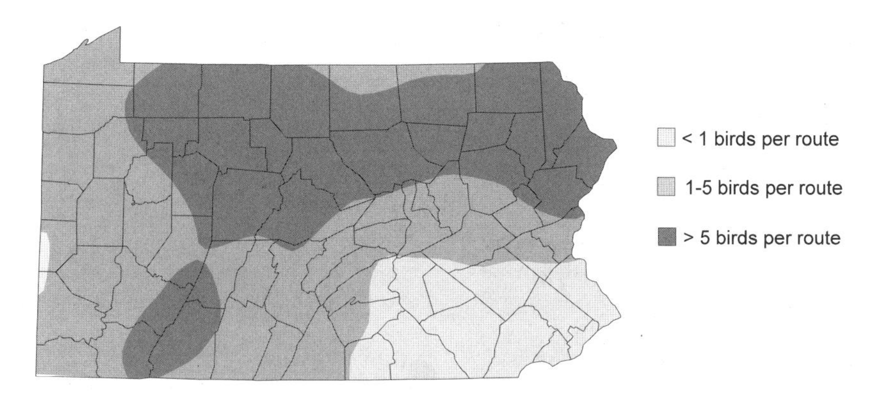

27. American Redstart relative abundance, based on BBS routes, 1985–1994

[1] Mulvihill, R.S. 1992. American Redstart *(Setaphaga ruticilla)*. BBA (Brauning 1992):332–333.

Prothonotary Warbler ***Protonotaria citrea***

General status: Prothonotary Warblers breed from the eastern Great Lakes region west to Minnesota, Kansas, and Texas and east across the southern states and north along the Atlantic coastal plain to New York. They winter from Mexico to Central and South America. Prothonotary Warblers are rarely reported away from breeding sites during migration in Pennsylvania. They are scarce local breeders at scattered sites almost statewide.

Habitat: They inhabit large swamps and wet woodlands bordering lakes and beaver ponds as well as river bottomlands. They almost always are found in wooded settings around standing water.

Seasonal status and distribution

Spring: Prothonotary Warblers are most frequently reported in this season. Early arrivals may appear the third week of Apr. Their usual migration period is from the last week of Apr to the fourth week of May. Most migrants are observed during the second and third weeks of May. Stragglers remain to the first week of Jun.

Breeding: Prothonotary Warblers are rare and local breeders in the Piedmont and Glaciated Northwest, and they are accidental elsewhere. Single pairs are regular at Washington Boro in Lancaster Co., along Gut Road on the Susquehanna River near Conowego Heights in York Co., and in SGL 169 near Newville, Cumberland Co. Prothonotary Warblers are rare breeders at several points in the Pymatuning Swamp and the Conneaut Marsh of Crawford Co. During a period of high lake levels at PISP in the mid- to late 1980s, perhaps as many as nine pairs were nesting. They left when lake levels receded to near normal in the early 1990s.[1] During the BBA project, Prothonotary Warblers were also confirmed breeding in southern Potter Co. and in the Delaware Water Gap in Monroe Co.[2] A pair was observed nest building at New Hope, Bucks Co., in 1996[3] and dummy nests were built at Island Park on the Lehigh River in Northampton Co. in 1994[4] and previously along the Octoraro Creek in Lancaster Co. A variety of sites, primarily in the Piedmont, have supported at least single birds in summer: Buffalo Creek, Union Co.; Codorus Creek, York Co.; Ridley Creek, Delaware Co.; and on several occasions the Unami Creek, Montgomery Co. Prothonotaries place their moss nest in an old woodpecker hole, a natural cavity, holes in old fence posts, and nest boxes placed over water. The nests discovered in the state contained eggs in Jun.

Fall: Very little is known about the fall passage of Prothonotary Warblers through Pennsylvania. Most records are from areas where they breed, and few records exist after young have fledged in midsummer. These records suggest that most birds have left the state by the second week of Sep. Stragglers have been recorded to the second week of Oct.

History: Prothonotary Warblers were reported infrequently in southeastern Pennsylvania during the late 1800s. The first indication of breeding, an adult feeding young, was reported in 1924 in southern Lancaster Co. (Beck 1924). Todd (1940) reported a family group in 1938 on the south shore of Lake LeBoeuf, near Waterford, Erie Co., the first evidence of nesting in western Pennsylvania. Other early breeding records came from Crawford Co. in 1951[5] and in the Wissahickon in 1954.[6] Numerous spring records were made widely across the state in the late 1940s and 1950s and in years following. Poole (unpbl. ms.) listed eight counties where nesting had been confirmed before 1960: Adams, Berks, Crawford, Erie, Lancaster, Montgomery, Union, and York. During the BBA project, it was confirmed in Crawford, Erie, Lancaster, Monroe, and Potter cos.

Comments: The OTC lists the Prothonotary Warbler as Candidate-Rare.

[1] G.M. McWilliams, pers. obs.

[2] Leberman, R.C. 1992. Prothonotary Warbler *(Protonotaria citrea)*. BBA (Brauning 1992): 334–335.

[3] Kitson, K. 1996. Local notes: Bucks County. PB 10:97–98.

[4] Haas, F.C., and B.M. Haas, eds. 1994. Rare and unusual bird reports. PB 8:163–165.

[5] Harrison, H.H. 1961. Status of the Prothonotary Warbler *(Protonotaria citrea)* as a breeding bird in northwestern Pennsylvania. Sandpiper 4:3–6.

[6] Ross, C.C. 1954. The breeding birds of the Wissahickon. Cassinia 46:2–6.

Worm-eating Warbler ***Helmitheros vermivorus***

General status: Worm-eating Warblers breed from Massachusetts and New York west across Ohio to Kansas, south to Louisiana and Mississippi, and along the eastern edge of the Appalachians. They winter in Mexico and Central America. Worm-eating Warblers are rare regular migrants away from breeding sites in Pennsylvania. They breed primarily east of the Allegheny Front.

Habitat: Worm-eating Warblers usually stay within a few feet of the ground, except when singing on territory from high perches. They usually are found on woodland hillsides with dense ground cover, including rhododendron or laurel thickets. Worm-eating Warblers will also use wet lowland areas with dense vegetation. They are among the species most sensitive to forest fragmentation.

Seasonal status and distribution

Spring: Worm-eating Warblers are most often seen in spring in Pennsylvania. Their usual migration time is from the fourth week of Apr or the first week of May to the third week of May. Most birds are recorded during the first and second weeks of May. Stragglers remain to the fourth week of May.

Breeding: Worm-eating Warblers are uncommon to locally fairly common breeding residents from Lycoming Co. south through the Ridge and Valley, in the Blue Ridge, and locally in large woodlands of the Piedmont. They are generally much less common west of the Allegheny Front and are absent in summer from the Glaciated Northwest. Worm-eating Warblers are regular but rare breeders in scattered woodlands in the Southwest, including the western slope of the Allegheny Highlands and Laurel and Chestnut ridges. They are accidental in the western and northern-tier counties.[1] Variable populations on BBS routes indicate no trend since 1966. Todd (1940) described a typical nest placed on the ground at the base of a sapling, built on a foundation of leaves, and lined with moss and grass. Most egg sets have been found from 21 May to 19 Jun.

Fall: This species is not reported as often in the fall, especially after Aug, as in the spring. Worm-eating Warblers become more difficult to find after singing has stopped. In the Susquehanna River valley in Lycoming Co., evidence of migration begins to occur during the fourth week of Jul or the first week of Aug.[2] Banding records suggest that migration is well under way by the second or third week of Aug and continues to the third week of Sep. Stragglers have been recorded until the second week of Oct.

History: Todd (1940) said that, although a "Carolinian" (a species whose breeding range is typically in the southeastern U.S.) species, it appeared to be unaffected by elevation, often occurring high along ridge slopes. In contrast to recent patterns, he considered it to be as common in the wooded southwestern counties (from Greene Co. to Beaver Co.) as in the Ridge and Valley. Warren (1890) and Todd (1940) considered Worm-eating Warblers plentiful in Clarion Co., where the BBA project provided just one "possible" breeding record.[1] A very early Worm-eating Warbler, seen on 28 Mar 1939 at Museum Park in Reading, Berks Co., was observed by Poole (unpbl. ms.).

[1] Santner, S. 1992. Worm-eating Warbler *(Helmitheros vermivorus)*. BBA (Brauning 1992):336–337.
[2] P. Schwalbe and G. Schwalbe, pers. comm.

Swainson's Warbler ***Limnothlypis swainsonii***

General status: Swainson's Warblers breed in the southeastern U.S., reaching their northernmost regular breeding range in West Virginia, southern Ohio, and Maryland. They winter primarily in Cuba, the Cayman Islands, Jamaica, Puerto Rico, and Mexico. They have only recently wandered into Pennsylvania. Swainson's Warblers have been recorded in Pennsylvania during 1975, 1978, 1982, and 1985–1998. All but one record have been in the spring or summer. Like Worm-eating Warblers, they stay hidden in dense thickets and are secretive; even when a singing bird is located, it can be very hard to observe.

Habitat: In Pennsylvania, Swainson's Warblers prefer habitats with rhododendron and hemlock in rich, damp, shaded woods, often in ravines and hollows.

Seasonal status and distribution

Spring and summer: Observations range from 9 May to 14 Jul. Most sightings have been during the second and third weeks of May, when they probably accompany major warbler movements and overshoot their normal breeding range. In western Pennsylvania they have been reported in Allegheny, Erie, Fayette, Indiana, Lawrence, and Westmoreland cos. In the eastern part of the state they have been reported in Dauphin, Delaware, Lancaster, Lehigh, Northampton, and York cos. Most records are from Fayette and Westmoreland cos., where there is an abundance of their preferred habitat. Swainson's Warblers were reported on territory in Jun in Pennsylvania on at least seven occasions. Swainson's Warbler was confirmed in Pennsylvania by D. Krueger on 17 May, when single territorial males were present at the Bear Run Nature Reserve in southern Fayette Co. until early Jun 1975;[1] again from 2 Jun through Aug 1989;[1] and through the summers after 22 May 1990 and 12 May 1991.[2] A bird was present near Delmont in Westmoreland Co. from 9 May to 2 Jun 1998.[3] Two other Jun records in the northeast were far north of any known breeding range: on 4 Jun 1978 (a singing male in a laurel thicket) in Northampton Co.[4] and 17 Jun 1985 in Pike Co. (P. B. Street and Wiltraut 1986). None of these birds was confirmed breeding.

Fall: There are two fall records. One was found at Baer Rocks in Lehigh Co. on 11 Sep 1982,[5] and one was reported from Delaware Co. at Ridley Creek State Park on 14 Aug 1997.[6]

History: Swainson's Warblers were considered hypothetical by Poole (1964), who noted a single record in Philadelphia on 15 May 1954.[7]

Comments: With more intense fieldwork in suitable habitat, especially in the Allegheny Mountain section, nesting may be confirmed in the future.

[1] Krueger, D. 1989. Swainson's Warbler *(Limnothylpis swainsonii)* in western Pennsylvania—Rare vagrant or breeding species? PB 3:86–89.
[2] Krueger, D.E., and R.S. Mulvihill. 1992. Swainson's Warbler *(Limnothylpis swainsonii)*. BBA (Brauning 1992):338–339.
[3] J. Hoffmann, pers. comm.
[4] R. Wiltraut, pers. comm.
[5] Paxton, R.O., W.J. Boyle Jr., and D.A. Cutler. 1983. Hudson-Delaware region. AB 37:164.

[6] Haas, F.C., and B.M. Haas, eds. 1997. Birds of note—July through September 1997. PB 11:149.
[7] Arnett, J.H. 1954. General notes. Cassinia 41:83.

Ovenbird ***Seiurus aurocapillus***

General status: Ovenbirds breed across Canada from eastern British Columbia to Newfoundland and south locally to Colorado and the eastern U.S. except in the Southeast. They winter primarily in Mexico, the West Indies south to Central and South America, and casually north to the Great Lakes region and New England. Ovenbirds are fairly common to common migrants and breeding residents across Pennsylvania. They are accidental in winter.

Habitat: Ovenbirds inhabit a variety of woodland types at all elevations found in Pennsylvania.

Seasonal status and distribution

Spring: Migrants are more frequently observed during this season than in fall. The earliest migrants may appear the third week of Apr, but Ovenbirds typically begin to arrive the fourth week of Apr or the first week of May. Peak migration occurs during the second and third weeks of May. The widespread breeding range of this warbler makes it difficult to determine the end of migration, but by the fourth week of May there is a noticeable decline in the number of birds.

Breeding: Ovenbirds are fairly common statewide, absent only from extensive areas of agricultural and urban development and common in most extensively forested areas. They are more likely to occur in larger wooded units than in small woodlots. Both the density of resident males and success in nesting are positively related to the area of their woodlot or proximity to extensive woodlands,[1] but they are probably less area-sensitive than most other woodland warblers (Robbins et al. 1989). Ovenbirds are found widely in the Piedmont and nest regularly in every Pennsylvania county. They were most common in the extensively forested central and eastern mountains in the BBS (see Map 28). The BBS shows a steady population increase of 2.6% per year since 1966. This warbler is named for the ovenlike dome of vegetation it builds over its nest, which is placed on the ground with leaves and grass. Most nests with eggs have been found from 22 May to 26 Jun.

Fall: The typical migration period is from the first week of Aug to the third or

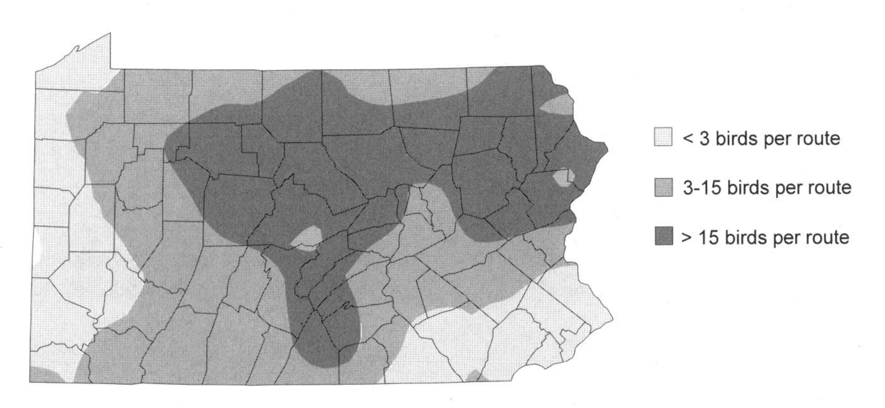

28. Ovenbird relative abundance, based on BBS routes, 1985–1994

fourth week of Oct. The greatest number of birds is recorded any time during the second through the fourth week of Sep. Stragglers have remained through fall into the first week of Jan but rarely through winter.

Winter: Winter records of Ovenbirds are mainly from the southern half of the state. They have been known to survive into spring, but few winter records have been made after Jan. Many sightings are at bird feeding stations. Ovenbirds have been recorded in winter in Berks, Chester, Huntingdon, Luzerne, Northampton, Philadelphia, Warren, Washington, and Westmoreland cos.

History: The historical status of the Ovenbird does not differ notably from recent patterns. Harlow (1913) described the species as generally abundant and widespread as a breeding bird. He considered it "perhaps the most generally and well-distributed warbler nesting in the state," a title that clearly now goes to the Common Yellowthroat in terms of distribution and abundance. Kunkle (1951) reported a rather early spring arrival on 6 Apr in Lewisburg in Union Co.

[1] Porneluzi P., J.C. Bednarz, L.J. Goodrich, N. Zawada, and J. Hoover. 1993. Reproductive performance of territorial Ovenbirds occupying forest fragments and a contiguous forest in Pennsylvania. Conservation Biology 7:618–622.

Northern Waterthrush ***Seiurus noveboracensis***

General status: Northern Waterthrushes breed from Alaska across Canada and south in the West to Washington, Idaho, and Montana and in the East to West Virginia. They winter from southern Florida, the Bahama Islands, and the West Indies to Mexico and Central and South America. Northern Waterthrushes are uncommon to fairly common regular migrants in Pennsylvania, and they breed locally across the northern half of the state and south through the Allegheny Mountain section. There is one historical winter record.

Habitat: Northern Waterthrushes inhabit bogs, beaver dams, swamps, streamside thickets, and poorly drained areas, often those containing hemlock, rhododendron, willow, and alder. They are most often found above 1200 feet elevation during the breeding season.

Seasonal status and distribution

Spring: More Northern Waterthrushes are found during spring migration than in the fall. Earliest spring arrivals have appeared in the first week of Apr. Their typical migration period is from the third or fourth week of Apr to the fourth week of May. Peak migration occurs during the first and second weeks of May. Nonbreeding stragglers remain until the first week of Jun.

Breeding: Northern Waterthrushes are uncommon, local breeders across northern Pennsylvania and rare south through the Allegheny Mountains in Cambria and Somerset cos. to the Maryland line. They are more frequently encountered and may be locally fairly common in extensive wetlands of the Glaciated Northeast (notably in Susquehanna Co.), the Poconos, North Mountain of Luzerne, Sullivan, and Wyoming cos., and the Glaciated Northwest. Scattered pairs are present in the Ridge and Valley and the western edge of the High Plateau, in Butler and Clearfield cos. They are probably underreported because of the inaccessibility of their habitat in some areas. Breeding birds are occasionally found in suitable habitat below 1000 feet elevation. The nest is a grass cup placed on the ground concealed in roots or in upturned roots of a fallen tree near standing water (H. H. Harrison 1975). Late migrants regularly occur and often sing until late May or early Jun in wet habitat well south of established breeding grounds, potentially confusing the nesting distribution. They were reported on only 19 BBS routes, which is insufficient to identify a population trend. Nests have been found with eggs after 27 May, and young were found in Jun.

Fall: Northern Waterthrushes are uncommon to rare at most sites in the fall. Like many other species of warblers that favor dense ground cover, they become difficult to find after singing has stopped. The first migrants have been detected at banding stations as early as the fourth week of Jul. Their usual migration period is from the second or third week of Aug to the fourth week of Sep. The greatest number of birds is recorded during the second or third week of Sep. Stragglers have been recorded until the third week of Dec.

History: Many early ornithologists failed to separate the waterthrush species[1] well after they were distinguished in 1808 (AOU 1983). Baird (1845) mistakenly listed the Northern Waterthrush as "Abundant, Summer," a clear reference to the Louisiana Waterthrush, in Cumberland Co. This error was repeated by J. Thomas (1876) for Bucks Co. and by Pennock (1886) for Chester Co. Warren (1890) clarified the status of the waterthrushes partly by identifying the Northern (then "Water Thrush") as a northerly breeding bird, but mistakenly listed rapidly flowing streams as its habitat. It was left to Sutton (1928a) to distinguish the habitats accurately, stating that the Northern "seems to prefer more quiet, even stagnant, water" and considered it "abundant" in the Pymatuning Swamp (Sutton 1928b). Loss of forested wetland habitat across the northern counties affected this species' population. One was reported on 22 Jan 1928 at Doylestown in Bucks Co. and was observed feeding along Mill Creek the following week.[2]

Comments: During migration Northern Waterthrushes and Louisiana Waterthrushes may be found together and in habitat not typical of their respective breeding haunts. During this time, when they may not be singing, they are frequently misidentified.

[1] Eaton, S.W. 1995. Northern Waterthrush *(Seiurus noveboracensis)*. BNA (A. Poole and Gill 1995), no. 182.

[2] MacReynolds, G. 1927–1928. General notes. Cassinia 27:37.

Louisiana Waterthrush ***Seiurus motacilla***

General status: Louisiana Waterthrushes breed in the U.S. from Maine, Indiana, and Minnesota south to Texas and east away from the Gulf coast across Georgia to North Carolina. They winter in the West Indies, Mexico, and Central and South America. Louisiana Waterthrushes are among the earliest warblers to arrive in the spring and the first to leave in the fall. They are primarily known as breeding residents and are rarely reported in migration away from known nesting sites. They are more widely distributed in the state than are the similar-looking Northern Waterthrushes.

Habitat: Louisiana Waterthrushes are found along flowing streams in forested watersheds. They are sensitive to water quality and forest area and generally are not found at polluted sites or where the woodland has been fragmented. Small, headwater streams with scattered pools through wooded ravines are diagnostic for this species in Pennsylvania.

Seasonal status and distribution

Spring: The first arrivals appear on their breeding grounds the third week of Mar. They typically arrive the first or second week of Apr or about a week later at high elevations and at some northern sites. No noticeable peak migration is evident. Their wide breeding distribution in the state makes it difficult to determine when migration has ended.

Breeding: Uncommon to fairly common along streams statewide, Louisiana Waterthrushes most frequently occur in the southern mountain sections, including slopes of the Ridge and Valley and Allegheny Mountains. They also are found in stream valleys in the High Plateau and Poconos into the higher elevations. They are poorly documented on BBS routes in part because of their specialized riparian habitat and because their nesting season is earlier than that of most warblers. Strongly area-sensitive, they are less likely to occur in forested areas of fewer than 800 acres (Robbins et al. 1989). Because of their dependence on macroinvertebrates for food and sensitivity to water quality, Louisiana Waterthrushes are being studied in Pennsylvania as a riparian forest habitat indicator.[1] The nest is a bulky cup of leaves and grass, placed in a cavity along a stream bank or in an upturned tree root near a stream. Most nests with eggs have been reported from 7 May to 3 Jun, but active nests have been found as early as late Apr.

Fall: Very little is known about the southward movements of Louisiana Waterthrushes in Pennsylvania. They become difficult to find after breeding is completed and singing has stopped. Louisiana Waterthrushes are very early migrants and among the first warblers to leave the state. Banding records suggest that migration may well be under way by the third or fourth week of Jul and ends by the second or third week of Aug, with stragglers to the second week of Sep. A bird on 10 Oct 1996 at Ski Round Top in York Co. was late.[2]

History: Early confusion between Louisiana and Northern waterthrushes leaves some question concerning their historical status. Harlow (1913) considered Louisiana Waterthrushes to be absent from some northern-tier counties (Elk, Forest, and Warren), and Sutton (1928a) identified them as resident in southern and central Pennsylvania. Dwight (1892) distinguished the two species of waterthrush but did not consider Louisiana Waterthrushes to be breeders in the mountains of the Northeast (Bradford and Sullivan cos.) or in the Cresson area (the Alleghenies west of Altoona). Correspondents of Warren (1890) in northern counties attributed waterthrush observations along "cold mountain streams" to the *S. noveboracensis*, probably erroneously. Northern Waterthrushes were noted to occur sympatrically with the Louisiana Waterthrushes in Crawford Co. (Warren 1890; Todd 1940). If those records were in fact of Louisiana Waterthrushes, recent records across northern Pennsylvania reflect the recovery of their range rather than an expansion as suggested by Poole (1964). Extensive timber harvests and degraded waterways may have depressed populations of Louisiana Waterthrushes early in the 1900s, but their range had expanded to nearly the current pattern by the time Poole (1964) published his summary of bird life.

Comments: Records of Louisiana Waterthrushes later than the first week of Sep are questionable and should be carefully documented.

[1] Prosser, D.J., and R.P. Brooks. 1998. A verified habitat suitability index for the Louisiana Waterthrush. J. Field Ornithology 69:288–298.

[2] Haas, F.C., and B.M. Haas, eds. 1997. Birds of note—October though December 1996. PB 10:221.

Kentucky Warbler *Oporornis formosus*

General status: Kentucky Warblers breed in the eastern U.S. from Iowa east to New Jersey south to Texas and east away from the Gulf coast across Georgia to South Carolina. They winter from Mexico to South America. Kentucky Warblers are found mainly across the southern half of Pennsylvania and rarely and locally in the northern half. Their breeding range has been expanding northward in Pennsylvania. These woodland warblers are rarely reported away from breeding sites.

Habitat: They are usually found in dense understory vegetation of shaded bottomland forests, ravines, swamps, and floodplains. Kentucky Warblers may be found in laurel at higher elevations.

Seasonal status and distribution

Spring: Kentucky Warblers may arrive on their territories by the fourth week of Apr, but their typical migration period is from the first week of May to the fourth week of May. During heavy northward passages of warblers, a few Kentucky Warblers may accompany migrant flocks and overfly their normal breeding range.

Breeding: Kentucky Warblers are fairly common within the Southwest region and are uncommon in the Piedmont. Since the 1960s the breeding range has expanded north through Butler, Armstrong, and Indiana cos., and rarely to Jefferson, Warren, Potter, Wyoming, and Lackawanna cos. by 1990.[1] In 1997, they had reached Warren Co.[2] Kentucky Warblers are spottily distributed through the southern Ridge and Valley and east through the Lehigh Valley. They are regular in small woodlots west of the Alleghenies but apparently are restricted to more extensive woodlands in the Piedmont. The irregular records on summer BBS routes do not indicate a population trend. The nest is made of leaves and bark strips, lined with grass or rootlets, and placed on the ground at the foot of a shrub or sapling. Early twentieth century oologists collected numerous nests, with most from 24 May to 19 Jun. The median date of J. Norris's 138 egg sets from the late 1800s in Chester Co. was 2 Jun.

Fall: Very little is known about the southward movement of Kentucky Warblers in Pennsylvania. They become difficult to find after breeding has been completed and singing has stopped. Banding records suggest that this species is southbound by the third or fourth week of Jul. Most birds have left the state by the second week of Sep, with stragglers to the third week of Oct. A very late one was found freshly dead on 27 Dec 1980 at Pittsburgh.[3]

History: Harlow (1913) described Kentucky Warblers "in certain [southern] sections as one of the most abundant of our woodland birds," and Warren (1890) said they were "almost as numerous as the Maryland Yellowthroat" in southeastern Pennsylvania. The large number of eggs collected by several nineteenth-century oologists reflects this regional abundance. They were common in the southwestern part of the state (Jacobs 1893) and occasionally ranged north. Their narrow southern range was reflected in their absence from the turn-of-century list of Westmoreland Co. birds (Townsend 1883). Todd (1940) showed a northern expansion to southern Indiana and Butler cos. in the first half of the twentieth century. Poole (1964) reported a breeding distribution similar to Todd's.

[1] Master, T.L. 1992. Kentucky Warbler *(Oporornis formosus)*. BBA (Brauning 1992):346–347.

[2] Grisez, T. 1997. Local notes: Warren County. PB 11:101.

[3] Hall, G.A. 1981. Appalachian region. AB 35:300.

Connecticut Warbler *Oporornis agilis*

General status: Connecticut Warblers are not common anywhere within their limited breeding range; through central Canada from Alberta to south-central Quebec and south to Minnesota, Wisconsin, and Michigan. Very little is known of their wintering range, with only a few documented records from northeastern Colombia and Venezuela south to Amazonia and central Brazil (AOU 1998). Connecticut Warblers are secretive and usually stay within heavy cover. Banding records suggest that they are more common than sight records indicate. Connecticut Warblers are rare regular migrants in Pennsylvania, with sightings mainly in the fall.

Habitat: They are usually found on or within a few feet of the ground in dense thickets, bogs, or weedy fields.

Seasonal status and distribution

Spring: Connecticut Warblers are rare and irregular at this season and are among of the last warblers to pass through the state. They have been recorded as early as the first week of May. Most records of birds are during the third and fourth weeks of May. Stragglers have been recorded to the second week of Jun. A late bird was banded on 23 Jun 1987 in Lackawanna Co.[1]

Fall: Most Pennsylvania reports of this species are made during the fall at banding stations. Early migrants may appear in the state the fourth week of Aug. Their typical migration period is from the second week of Sep to the first week of Oct. Most birds are banded and observed during the third and fourth weeks of Sep. Stragglers remain to the third week of Oct. Daily banding totals at the PNR banding station have reached five or six birds (Leberman 1976), with season banding totals reaching 25 birds.[2] However, field observers rarely observe more than four or five in a season at these banding stations.

History: This species has always been listed as a rare transient in historical ornithological literature. As are modern ornithologists, many authors, such as Todd (1940), were skeptical of any record before the third week of May. However, Poole (unpbl. ms.) cited early records by "experienced observers" from 25 Apr to 8 May.

Comments: The very similar Mourning Warbler is occasionally misidentified as a Connecticut Warbler, especially in fall, when immature Mourning Warblers show partial eye rings.

[1] Klebauskas, G. 1987. Lackawanna County. PB 1:58.

[2] Hall, G.A. 1967. Appalachian region. AFN 21:35.

Mourning Warbler *Oporornis philadelphia*

General status: As are Connecticut Warblers, Mourning Warblers are difficult to observe because of their preference for skulking in dense thickets. They are most easily located by their slurred song. Mourning Warblers breed in Canada from British Columbia to Newfoundland south across the Great Lakes region through New England, Pennsylvania, and in the Appalachians to West Virginia and Virginia. They winter from Central to South America. Mourning Warblers have a restricted breeding range in Pennsylvania, primarily in the Glaciated Northwest and the western portion of the High Plateau. They are uncommon to rare regular migrants. Most migrants are observed west of the Allegheny Front.

Habitat: Mourning Warblers are usually found on or within a few feet of the ground in thickets and brushy areas in ravines, in clearcuts, or in areas of fallen trees, often near streams, in swamps, or in bogs.

Seasonal status and distribution

Spring: Mourning Warblers are more often observed in spring than in fall. They are among the last warblers to arrive in the state. The earliest arrivals usually appear by the first week of May, but their typical migration time is from the second week of May to the first week of Jun. The greatest numbers are recorded during the third and fourth weeks of May. A few birds away from breeding sites may linger to the second week of Jun.

Breeding: Mourning Warblers are uncommon to locally fairly common from Erie, eastern Crawford, and northern Venango cos. east to Potter Co. East to Wyoming and Wayne cos. they are rare and irregular, and south of the core area to Mercer, Clearfield, and Centre cos. they are accidental in summer. The breeding sites generally are found above 1500 feet elevation in the Glaciated Northwest and higher in the east. Despite their secretive nature in migration, Mourning Warblers are not shy during the breeding season and sing conspicuously in late May and Jun. The bulky nest is placed on or near the ground in dense grass, saplings, or briars and closely resembles that of the Common Yellowthroat. Coinciding with their late migration, egg sets are most likely found from 1 to 24 Jun. No trend in their population is apparent.

Fall: Mourning Warblers become difficult to find after breeding has been completed and singing has stopped. They are rarely reported anywhere in the state in fall migration. Migration is first detected during the second or third week of Aug. One found on 27 Jul 1974 at Villanova in Delaware Co. was far south of any known breeding site.[1] Migration continues during Sep, with stragglers to the fourth week of Oct. One was present at Jonestown in Columbia Co. until 9 Nov 1996.[2]

History: The breeding range of Mourning Warblers has retracted somewhat from that reported at the beginning of the twentieth century in the northeastern and southern mountains, but it has expanded west into Erie and Crawford cos. since the 1960s (Poole 1964). Stone (1900) considered them common in northern mountain regions of Sullivan and Wyoming cos., where they have subsequently been rare. Todd (1940) listed scattered breeding records south along the Allegheny Mountains to the Maryland line, but the "area of comparative abundance" he listed is very similar to the center of the breeding distribution today, Warren to McKean cos., south through Forest and Cameron cos. Poole (1964) showed a cluster of confirmed breeding records in the Poconos, where they have not been regular since early in the twentieth century (P. B. Street and Wiltraut 1996).

[1] Scott, F.R., and D.A. Cutler. 1974. Middle Atlantic Coast region. AB 28:888.

[2] Haas, F.C., and B.M. Haas, eds. 1997. Birds of note—October through December. 1997. PB 10:221.

Common Yellowthroat *Geothlypis trichas*

General status: Common Yellowthroats breed from Alaska across Canada south throughout the U.S. Their winter range in the U.S. extends primarily from California south across the southern states and north to Maryland. A few winter annually in the southern Great Lakes (AOU 1998). Common Yellowthroats are fairly common to common regular migrants and breeding residents throughout Pennsylvania. They are probably the most common and widespread of the breeding warblers in Pennsylvania. Yellowthroats occasionally winter in the state.

Habitat: They can be found in a variety of habitats, including marshes, swamps, bogs, wet meadows, and woodland edges, overgrown fields, fence rows, and brushy roadsides.

Seasonal status and distribution

Spring: These birds may appear as early as the third week of Apr, but they generally arrive the fourth week of Apr or the first week of May. The migration peaks during the second and third weeks of May. The end of migration is difficult to discern because of the widespread breeding residents.

Breeding: Common Yellowthroats are common statewide except in the Piedmont, where they are only slightly less common. Yellowthroats are ubiquitous, the sixth most widely distributed nesting bird (found in 96% of BBA blocks) and one of the most abundant species on BBS routes (average 16.5 birds per route). They are the most common and widely distributed of our breeding warblers. Even in suburban areas yellowthroats will occur in small pockets of shrubby growth. On BBS routes they are most common in northern and western areas, less common in the western Piedmont (see Map 29). The yellowthroat's population is stable according to BBS data. The bulky nest is loosely made of leaves and grass and is placed on or close to the ground. Most nests with eggs are found from 20 May to 20 Jun. Harlow (1918) found the average date of first sets to be 28 May and that of second sets to be 5 Jul.

Fall: The first migrants are detected by at least the third week of Aug. Their typical migration period is from the first week of Sep to the third or fourth week of Oct. Peak migration occurs during the second and third weeks of Sep. Stragglers have remained to the first week of Jan.

Winter: Most winter records come from Tinicum.[1] They have also been recorded at least once in Lancaster Co., where one was found on 9 Feb 1991.[2] Yellowthroats have been recorded in Centre Co. in Feb of 1983[3] and from 6 to 20 Feb 1988 in the State College area.[4] One was present at Ligonier in Westmoreland Co. until 10 Jan 1997.[5] At the Lake Erie Shore they have been recorded at PISP in Erie Co. to 23 Jan 1993.[6]

History: With a few exceptions, early accounts listed this species as common or abundant.

Comments: A Common Yellowthroat banded on 1 Oct 1972 at PNR was trapped and killed near Vereda Nueva, Havana, Cuba, on 6 Jan 1973.[7]

1 J.C. Miller, pers. comm.
2 E. Witmer, pers. comm.
3 J. Peplinski and B. Peplinski, pers. comm.
4 Hall, G.A. 1988. Appalachian region. AB 42:266.
5 Leberman, R.C. 1997. Local notes: Westmoreland County. PB 11:34.
6 Hall, G.A. 1993. Appalachian region. AB 47:258.
7 Leberman, R.C., and M.H. Clench. 1974. Birdbanding at Powdermill, 1973. Research report

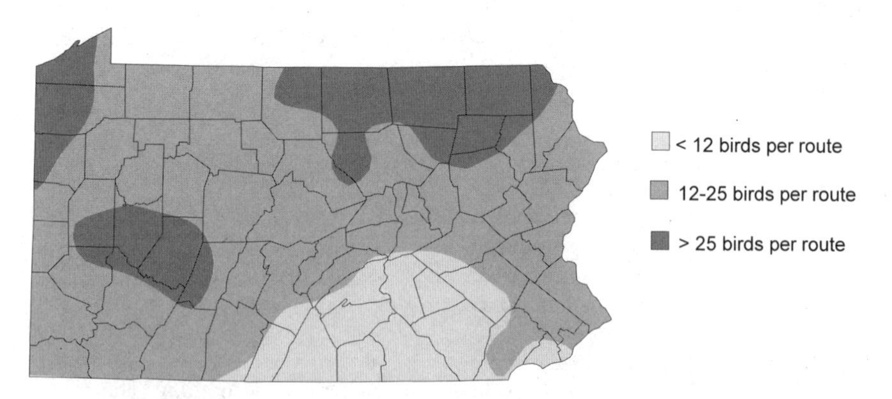

29. Common Yellowthroat relative abundance, based on BBS routes, 1985–1994

no. 33. Powdermill Nature Reserve, Carnegie Museum of Natural History, Pittsburgh.

Hooded Warbler ***Wilsonia citrina***

General status: Hooded Warblers breed in the eastern U.S. from Massachusetts west to Minnesota and south to Texas and northern Florida. They winter primarily along the Gulf-Carribean slope of Central America to Panama. In Pennsylvania, Hooded Warblers are uncommon to rare regular migrants. They breed over most of the state but most commonly in the Southwest and Glaciated Northwest regions.

Habitat: Hooded Warblers are rather secretive, generally remaining close to the ground. They are found in forest understory, usually in mature deciduous forests with thick undergrowth such as stands of laurel and rhododendron, or along hillsides with scattered second-growth trees.

Seasonal status and distribution

Spring: Hooded Warblers are observed mainly during the spring and early summer. The earliest migrants arrive the third week of Apr; they generally begin to appear on their breeding grounds during the fourth week of Apr or the first week of May. The greatest number of migrants occurs during the second and third weeks of May. Because most birds are observed on or near their breeding grounds, it is difficult to discern the last of the northbound birds.

Breeding: Hooded Warblers are widespread and fairly common west of the Allegheny Front from McKean Co. south through the highlands of Somerset Co. In many wooded areas they are locally fairly common. They are more locally distributed and generally uncommon in the Susquehanna and Delaware drainages. They occur in forested valleys and lower slopes locally in the Ridge and Valley and are rare in the High Plateau east of Potter Co. Locations of higher densities in the east include the Blue Ridge section, the valleys of Stony and Clark's creeks (Dauphin Co.), and the Bald Eagle and Rothrock state forests in Lycoming, Centre, and Union cos. They regularly nest in every county except possibly Philadelphia. The sharp contrast between the eastern and the western distribution apparently is not related to elevation or forest cover. Although birds of southern affinities, Hooded Warblers occur regardless of elevation, unlike the pattern of many southern species, which avoid the mountains. There has been a strongly positive trend (6.9% per year) on BBS routes since 1966. The nest is a neat cup of leaves and other plant material woven compactly with spiders' webs, often placed 1–3 feet from the ground in small branches of shrubs or briars. Most nests with eggs are found from 29 May to 27 Jun.

Fall: Migration is first detected by the first week of Aug. The birds' habit of staying low in the dense understory makes them difficult to observe in the fall. Banding records suggest that peak migration occurs from the fourth week of Aug to the second week of Sep. Migration has usually ended by the second week of Oct. Stragglers have been observed to the second week of Nov.

History: Although Hooded Warblers historically have been reported across the state, distribution patterns around the beginning of the twentieth century were somewhat different from those of today. Harlow (1913) considered Hooded Warblers to be regular in Pike Co. and very scarce in Greene Co., the opposite of recent patterns. Few or no records in Somerset, Washington, Indiana, and Armstrong cos. existed 50 years ago (Todd 1940), where they are now common. Changes may be related to conditions in the forest understory but may also reflect a general range expansion.

Wilson's Warbler ***Wilsonia pusilla***

General status: Wilson's Warblers breed from Alaska across central Canada to Labrador and Newfoundland, south through California and the Rockies to New Mexico and to New York in the East. They winter from southern Texas, Louisiana, and Florida south through Central America and casually north to Ontario, Connecticut, and Nova Scotia (AOU 1998). Wilson's Warblers are uncommon to fairly common regular migrants in Pennsylvania. They are accidental in winter in Pennsylvania.

Habitat: Thickets, shrubs, second-growth woodlands, and woodland edges are the preferred habitats of Wilson's Warblers.

Seasonal status and distribution

Spring: They rarely arrive anywhere in the state before the first week of May. Their typical migration period is from the second week of May to the first week of Jun. Peak migration occurs during the third and fourth weeks of May. Stragglers remain to the second week of Jun.

Fall: Earliest southbound birds have been reported the fourth week of Jul. Their typical fall migration period is from the third week of Aug to the second or third week of Oct. Most migrants pass through Pennsylvania during the second and third weeks of Sep. Stragglers have been recorded to the second week of Jan.

Winter: There are two winter records of Wilson's Warblers in Pennsylvania. One was at a feeding station until 4 Feb 1957 at Lititz in Lancaster Co. (Poole, unpbl. ms.); another was found and photographed from 25 to 28 Feb 1993 along Alexander Spring Road in Cumberland Co.[1]

History: Most authors considered Wilson's Warblers to be fairly common migrants, but Baird (1845) listed the species as abundant. Todd (1940) described them as fairly common, but "common only at odd intervals." Poole (1964) believed they were fairly common but somewhat irregular.

[1] Haas, F.C., and B.M. Haas, eds. 1993. Rare and unusual bird reports. PB 7:18.

Canada Warbler ***Wilsonia canadensis***

General status: Canada Warblers breed in Canada from British Columbia across southern Canada to Nova Scotia south through the Great Lakes region, Pennsylvania, and the Appalachians to Georgia. They winter in South America. Canada Warblers are uncommon to fairly common regular migrants in Pennsylvania. They breed at higher elevations across the northern half of the state and locally south in the mountains, notably in the Allegheny Mountain section.

Habitat: Canada Warblers inhabit thickets, woodlands, and woodland edges during migration. In the breeding season they occur above 1000 feet in thick understory of extensive damp woodlands, often along streams, in swamps, or in bogs, especially those with hemlock, mountain laurel, or rhododendron.

Seasonal status and distribution

Spring: They rarely arrive before the first week of May. Their usual migration period is from the second week of May to the first week of Jun. Peak migration occurs during the third and fourth weeks of May.

Breeding: Canada Warblers are uncommon to locally fairly common breeders in extensively forested areas of northern Pennsylvania and south in the Allegheny Mountains to Maryland. They are most widespread and common in the Pocono Plateau and the Allegheny National Forest but may be locally fairly common on slopes thick with shrubs or laurel, or in dense shrub swamps, through the High Plateau. They are locally uncommon in extensively forested higher slopes of the Ridge and Valley, notably in Bald Eagle and Rothrock state forests, and are rare along Blue Mountain south to Stoney Creek, Dauphin Co.[1] In 1997, 6 to 10 males were found in the Blue Ridge singing along streamside rhododendron thickets in Adams,[2] Cumberland,[3] and Franklin[4] cos. State BBS routes indicate a substantial population decline of 4.8% per year since 1966, although it apparently has stabilized since 1980. The nest is a bulky structure, often hidden within the tangled upturned roots of a fallen tree or in a cavity on the ground at the edge of a bank or in a rotting tree stump. Most nests with eggs have been found from 27 May to 14 Jun.

Fall: Away from breeding areas, their typical migration period is from the second or third week of Aug to the first week of Oct. Most migrants are recorded from the fourth week of Aug to the second week of Sep. Stragglers have remained to the fourth week of Nov.

History: Historical accounts indicated that Canada Warblers once were found south of Pennsylvania's High Plateau, but Warren's (1890) report from Dr. Triechler of breeding in Lancaster Co. is doubtful. Todd (1940) listed breeding locations in Armstrong and Mercer cos., well away from localities in which they have recently been reported. Harlow (1918) stated, "nowhere have I found it more abundant than in northern Huntingdon and southern Centre counties," an area in which they currently are spottily distributed. Otherwise their occurrence through the High Plateau is relatively unchanged from historical distributions. The recent records in the Blue Ridge are from areas where they were historically "common to abundant" (Baird 1845; Frey 1943), but they were not reported there during the BBA project.[1]

[1] Mulvihill, R.S. 1992. Canada Warbler *(Wilsonia canadensis)*. BBA (Brauning 1992):354–355.

[2] Kennel, E., and A. Kennel. 1997. Local notes: Adams County. PB 11:91.

[3] Hoffman, D. 1997. Local notes: Cumberland County. PB 11:94.

[4] Henise, D., and R. Henise. 1997. Local notes: Franklin County. PB 11:96.

Yellow-breasted Chat *Icteria virens*

General status: Yellow-breasted Chats are the largest members of the warbler family. They breed across the U.S. except in central and southern Florida, and they winter from southern Texas, Louisiana, and southern Florida south through Central America. Chats are casual in early winter, north in the East to the Great Lakes region and New England (AOU 1998). In Pennsylvania they are uncommon to fairly common regular migrants. Most birds are observed during spring and summer when they are actively singing. They breed most commonly in the Southwest and more locally in the lower elevations of the Ridge and Valley and in the Piedmont. They appear to have steadily declined as a breeding species in Pennsylvania since at least the mid-1960s. Chats are accidental in winter in the southeastern corner of the state.

Habitat: They are habitat-specific, preferring shrubby streamsides, extensive thickets, hedgerows, overgrown pastures, fields, and power-line cuts. They are sensitive to habitat alterations and will leave if thickets are overgrown with trees.

Seasonal status and distribution

Spring: The earliest migrants have appeared by the fourth week of Apr. Their typical migration period is from the first to the fourth week of May. The greatest numbers are observed on or near breeding sites during the second and third weeks of May. Stragglers remain to the first week of Jun.

Breeding: Chats are rare to locally fairly common across southern Pennsylvania and rare elsewhere in the state. They are most common in the Southwest region as far north as Indiana and Mercer cos., and in the southern Ridge and Valley north to Centre Co. Chats are rare and irregular in the northern half of the state and in Cambria and Somerset cos., where previously they had been more common. In the northern-tier counties they are most likely to be found along major river bottoms but may be found casually in large brushy thickets at any elevation. Summer records have come from all counties, except possibly Wayne. Chats tend to be less common south and east of the Ridge and Valley. State BBS data documented a sharp decline of 4.9% per year since 1966. The nest is a bulky structure (more than 4 inches outside diameter) of leaves and grass, usually placed 2–4 feet from the ground in a shrub or briar patch. Nests with eggs are found from 14 May to 14 Jun, with occasional nests (probably second clutches) until early Jul.

Fall: They become difficult to find after singing has stopped. Their secretive habits make the time of departure and peak migration difficult to discern. Southbound movement has been detected through Aug, with fewer birds in Sep and Oct. Stragglers have remained to the first week of Jan.

Winter: Nearly all winter records of chats have been from Tinicum,[1] where they are usually not found later than the third or fourth week of Jan. One remained at a feeder at Ephrata in Lancaster Co. in Jan 1974.[2]

History: Chats responded to the abandonment of farmland, expanded beyond their pre-Colonial southern Pennsylvania range, and became a rare breeder north into the mountains. They were even considered by Todd (1904) to be a "rare summer resident" in Erie Co. Around Pymatuning they had been regular after the 1920s (Sutton 1928b) and had increased to the point of being considered common there by 1960 (Poole, unpbl. ms.). Poole (unpbl. ms.) indicated that they had bred during the twentieth century in all of the northeastern counties, except possibly Bradford, but that they had retreated from these areas except along the North Branch Susquehanna River. One visiting a suet feeder on 17 Feb 1950 at Wyndmoor in Montgomery Co. represents a historical winter record.[3]

[1] J.C. Miller, pers. comm.

[2] E. Witmer, pers. comm.

[3] Imsick, R.C. 1951. General notes. Cassinia 38:38.

Family Thraupidae: Tanagers

The tanager family, Thraupidae, is one of the most colorful groups of birds in the world, and tanagers are sought by both birders and photographers. Most of the hundreds of species in this family are found in the tropics; only three are found in Pennsylvania. Two species, Scarlet and Summer tanagers, breed in the state, and one, Western Tanager, is a vagrant from the West. They can be very difficult to locate in the tree canopy despite the bright colors of the adult males.

Summer Tanager *Piranga rubra*

General status: Summer Tanagers breed from California across the southern states to Florida, north to Iowa, and east to New Jersey. They winter from Mexico south to South America. In Pennsylvania, Summer Tanagers are rare and irregular to casual vagrants, primarily in spring, away from the southwestern corner of the state, where they are regular breeders. Historically, however, they have nested in the southeastern corner of the state. Except for Warren Co., they have not been recorded in the High Plateau north of Centre Co.

Habitat: Summer Tanagers prefer dry upland woodlands, especially those with oak.

Seasonal status and distribution

Spring: Most birds during this season are single males that have overflown their normal breeding range. These birds may appear any time from the second week of May to the first or second week of Jun. Summer residents may arrive on their breeding grounds as early as the fourth week of Apr, but most appear the first or second week of May.

Breeding: Summer Tanagers are rare but regular breeding residents in Greene Co. They were also confirmed nesting in recent years in Beaver and Washington cos.[1] Summer sightings, usually of individual birds seen for just a day, have been reported in Berks, Fulton, Lancaster, and York cos. Breeding birds should be searched for in dry forests in these southern counties. Far north of its usual range, a male at Dingmans Ferry in Pike Co. summered in 1984 and 1985 (P. B. Street and Wiltraut 1986). The nest is not easily distinguishable from that of Scarlet Tanagers; the walls of the cup-shaped nest may be so thin that the eggs can be seen through the bottom. Summer Tanagers lay up to five, but usually three or four, light blue or greenish blue eggs that are spotted or speckled with brownish red or purplish red.

Fall: Very little is known about the movements of this species after the breeding season. Apparently the only records away from their breeding range in the Southwest after Aug are in the Piedmont. Probably most birds have left the state by the fourth week of Sep. Stragglers have been recorded until the second week of Dec.

History: Most specimens and sight records before 1900 were during May and Sep in the southeastern counties. Todd (1940) cited only three records in Allegheny and Mercer cos. Confirmed breeding events were limited to two nineteenth-century instances in Montgomery Co.[2] Summer Tanagers were oddly absent from Sutton's (1928a) summary of Pennsylvania's birds. Even until 1960, most records came from the southeastern counties (Poole 1964). The first record in Greene Co. was in 1957 (Bell 1994)

[1] Ickes, R. 1992. Summer Tanager *(Piranga rubra)*. BBA (Brauning 1992):358–359.

[2] Dearden, R.R., Jr. 1887. A day's tour in the woods of Fort Washington, Pennslyvania. Oologist 4:87–88.

Scarlet Tanager *Piranga olivacea*

General status: Scarlet Tanagers breed in southern Canada from Manitoba to Nova Scotia south to Oklahoma, the northern Gulf states, and South Carolina. They winter from Panama to South America. Scarlet Tanagers are uncommon to fairly common regular migrants in Pennsylvania, and they breed over most of the state.

Habitat: Scarlet Tanagers are found in almost any mature woodlands, including pine stands.

Seasonal status and distribution

Spring: Sightings of Scarlet Tanagers are mainly in spring, when the brightly colored males are most easily observed before trees have completely leafed out. The earliest migrants appear the third week of Apr. Their typical migration period is from the fourth week of Apr or the first week of May to the fourth week of May. Peak migration usually occurs during the second or third week of May. There are records, although none recently, of 50 to 100 tanagers at a site being forced to the ground by bad weather (Leberman 1976).

Breeding: Scarlet Tanagers are breeding residents of woodlands in every Pennsylvania county. They are fairly common over much of the state and uncommon in the Coastal Plain and Piedmont. The BBS suggests that Scarlet Tanagers are most common through the central and southern mountains and least common in extensive agricultural and urban areas of the Piedmont (see Map 30). They are less area-sensitive than many forest birds and are likely to be found in woodlots as small as 30 acres (Robbins et al. 1989). The nest, loosely constructed but not distinctive, is placed at various heights, often 20–40 feet above the ground (Todd 1940). Eggs are most often found from 19 May to 11 Jun. The clutch size is the same as that of the Summer Tanager, and egg color is similar, but markings are not as bold. The population is stable according to BBS data.

Fall: Smaller numbers are observed in fall than in spring because the birds tend to be silent and the young are cryptically colored. The beginning of fall migration is difficult to discern because both young and adults wander widely after the nesting season. Southbound Scarlet Tanagers become conspicuous by the third or fourth week of Aug, with the number of birds remaining high through Sep. Migration is usually over by the third week of Oct. Stragglers have been recorded to the third week of Nov.

History: The status and distribution of Scarlet Tanagers have changed little throughout recorded history in Pennsylvania. They have been recorded as early as the second week of Apr in York Co. (Bent 1958). A bird visiting a feeder for two weeks until 28 Dec 1955, near Lewisburg in Union Co.,[1] provided an unusually late record.

Comments: An analysis of their range and relative abundance has identified Pennsylvania as one of the most important states for the Scarlet Tanager (Rosenberg and Wells 1995). Approximately 13% of its total nesting population is within our boundaries.

[1] Ross, H. 1956. Lewisburg CBC. AFN 10:103.

Western Tanager ***Piranga ludoviciana***

General status: Western Tanagers breed from Alaska, British Columbia, and Mackenzie south in the U.S. west of the Rocky Mountains. They winter from coastal southern California south through Central America. Western Tanagers occasionally stray to the east and have been reported in Pennsylvania at least 15 times in the past 36 years.

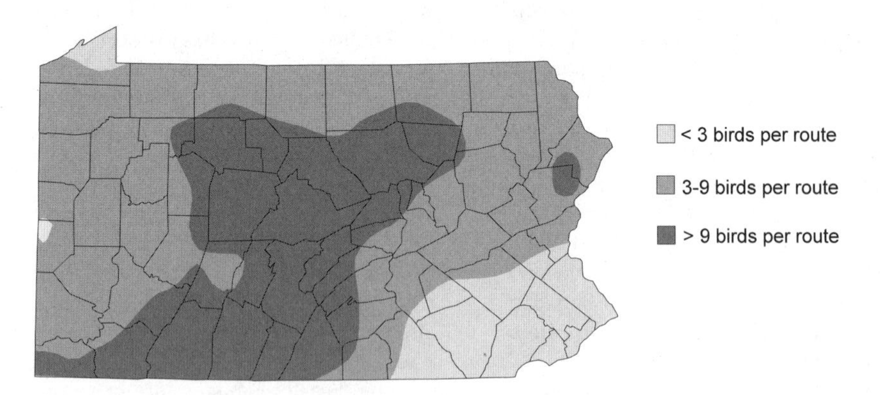

30. Scarlet Tanager relative abundance, based on BBS routes, 1985–1994

Western Tanager at Lake Nockamixon, Bucks County, December 1997. (Photo: Alan Brady)

Nearly all records have been reported during 1958–1960, 1964, 1971–1972, 1974, 1977, 1980–1981, 1989–1990, 1995, and 1997.

Seasonal status and distribution

Fall through spring: The majority of sightings are of single birds, with reports from 13 Aug to 28 May. They have been reported from the counties of Allegheny, Bucks, Butler, Chester, Crawford, Fayette, Indiana, Lackawanna, Lebanon, Montgomery, Perry, and Westmoreland. All but three recent reports are of single-day sightings. A male in breeding plumage was seen by many from 7 Dec 1959 to early Jan 1960 and was photographed at West Chester in Chester Co.;[1] a female or first-winter male was found in Whitpain Township, Montgomery Co., from 30 Dec 1989 until well into Jan 1990.[2] An immature or female, found on the Upper Bucks Co. CBC on 20 Dec 1997, remained to 29 Dec 1997 and was photographed and videotaped.[3]

History: The only Western Tanager before 1959 was one identified as a female coming to a feeding station near Scranton in Lackawanna Co. from 22 Feb to 9 Mar 1952 (Poole, unpbl. ms.).

Comments: Western Tanagers have been confused with Baltimore Orioles, especially in winter when orioles occasionally occur at bird feeding stations.

[1] Scott, F.R., and D.A. Cutler. 1960. Middle Atlantic Coast region. AFN 14:298.
[2] Boyle, W.J., Jr., R.O. Paxton, and D.A. Cutler. 1990. Hudson-Delaware region. AB 44:244.
[3] Kitson, K. 1997. Local notes: Bucks County. PB 11:235.

Family Emberizidae: New World Sparrows

The sparrow family, Emberizidae, makes up another large group, including about 29 species reported in Pennsylvania. Their bill is short and conical, with the cutting edge of the lower mandible angling abruptly downward near the base. They spend much of their time searching for seeds on or near the ground. Some of the grassland species are secretive and are rarely found away from their breeding grounds; they are difficult to locate after singing has stopped. Others are so abundant during migration that they can hardly go unnoticed on lawns and in fields. Most species nest on or near the ground, using various grasses or plant fibers for nesting material. They usually lay from three to five eggs that are white, blue, green, or buff and are spotted with brown, black, or reddish brown throughout or on the large end. This group of birds is quite hardy; many species are frequent visitors to bird feeding stations and remain to winter in areas of the state that have little or no snow cover.

Green-tailed Towhee *Pipilo chlorurus*

General status: Green-tailed Towhees breed in the western U.S. from Washington and Montana, south to Arizona, New Mexico, and western Texas, and east to Colorado. They winter from southern California, east to central Texas, and south through Mexico. Green-tailed Towhees are casual over most of eastern North America. This species has been recorded in Pennsylvania twice.

Seasonal status and distribution

Fall through spring: One was discovered at a feeder at Honey Hollow Environmental Center in Bucks Co. on 23 Dec 1987. It was not seen again until 6 Jan 1988 but then became a regular feeder visitor at the same site until 2 May 1988. It was seen and photographed by many observers.[1] Another Green-tailed Towhee was found at L. Rhoads's feeding station near Green Lane Reservoir in Montgomery Co. on 27 Nov 1994 and remained to 1 Mar 1995. It was documented with a photograph.[2,3]

[1] McNaught, B. 1988. Green-tailed Towhee in Bucks County, first Pennsylvania Record. PB 2:7.
[2] Freed, G. 1995. Notes from the field: Montgomery County. PB 8:225.
[3] Freed, G. 1995. Local notes: Montgomery County. PB 9:34.

Spotted Towhee *Pipilo maculatus*

General status: Until recently the Spotted Towhee was considered a subspecies group of the Eastern (then Rufous-sided) Towhee. Spotted Towhees breed in Canada across southern Canada from British Columbia to Saskatchewan and south through California, southern Nevada, and Arizona. They winter from southern British Columbia, south to Utah and Nevada, and east to Iowa and Texas. Spotted Towhees are casual in the eastern U.S. There was apparently no valid record of Spotted Towhees before 1966 in Pennsylvania. Since then there have been at least three records in the state.

Seasonal status and distribution

Fall through spring: An immature male was collected at the banding station at PNR in Westmoreland Co. on 12 Nov 1966, and an immature female was collected on 16 Oct 1975 (Leberman 1976). A female Spotted Towhee was discovered and photographed by G. M. McWilliams at a bird feeding station on PISP in Erie Co. on 3 Jan 1988. This bird was joined later by another female on 16 Jan, with at least one of them remaining to 1 Apr 1988.[1,2]

Comments: This recent addition to the avifauna of Pennsylvania may be more widespread and common than records indicate. Until the recent split there was probably little effort made by birders to differentiate these subspecies groups. Observers should carefully examine all towhees, especially in fall and winter, when most western vagrants of other species are reported.

[1] McWilliams, G.M. 1988. Erie County. PB 2: 29–30.
[2] McWilliams, G.M. 1988. County reports—April through June 1988: Erie County. PB 2:68.

Eastern Towhee *Pipilo erythrophthalmus*

General status: Previously known as the Red-eyed, or Rufous-sided, Towhee, it was recently renamed Eastern Towhee when the subspecies group, Spotted Towhee *(P. maculatus)*, became recognized as a distinct species. They breed in southern Canada from Saskatchewan to Quebec, south to Louisiana and Florida, and west to Colorado, Kansas, and Arkansas. Eastern Towhees winter from Nebraska east to Massachusetts and south to Texas and Florida. They are common regular migrants and breeding residents throughout Pennsylvania. They are winter residents, most commonly across the southern half of the state.

Habitat: Eastern Towhees are found on or near the ground in thickets (even in suburban yards), hedgerows, woodland edges, brushy fields, clearcuts, parks, and dense understory of woodlands with an open canopy.

Seasonal status and distribution

Spring: Birds that have not wintered in the state may arrive as early as the second week of Mar. Their typical migration period is from the fourth week of Mar or the first week of Apr to the first or second week of May. Migration usually peaks during the second and third weeks of Apr.

Breeding: Eastern Towhees are fairly common statewide and are locally abundant. Although nesting in every county, they are not evenly distributed. Eastern Towhees are essentially a woodland edge species, fairly common to locally common on BBS routes in the Southwest region, in the Ridge and Valley (see Map 31), and in northern clearcuts. They are least common, but still widespread, in urbanized and extensive agricultural areas. This species showed a significant long-term population decline (3.7% per year) since 1966 on state BBS routes until the mid-1980s, when the trend leveled off. The nest is a bulky structure of leaves and bark strips, placed in a depression on the ground under vegetation or up to a few feet in a shrub or thicket. The nesting season extends from late Apr until Aug. Most egg sets have been found from 5 May to 11 Jul, representing two broods. First nests are more likely to be on the ground than in a shrub or tree (Todd 1940).

Fall: The abundance of local birds makes it difficult to determine the beginning of migration, but an increase in the number of birds is usually detected by the second or third week of Sep. Migration peaks during the first and second weeks of Oct. Most migrants have left the state by the first or second week of Nov. A few remain through the fall and winter.

Winter: Wintering birds usually are found in the Piedmont, the Ridge and Valley, and the Southwest, where they may be uncommon to rare. In the Glaciated Northeast and Glaciated Northwest they are rare and local during harsh winters with heavy snow cover and, if present, are most often found at bird feeding stations. In the central and

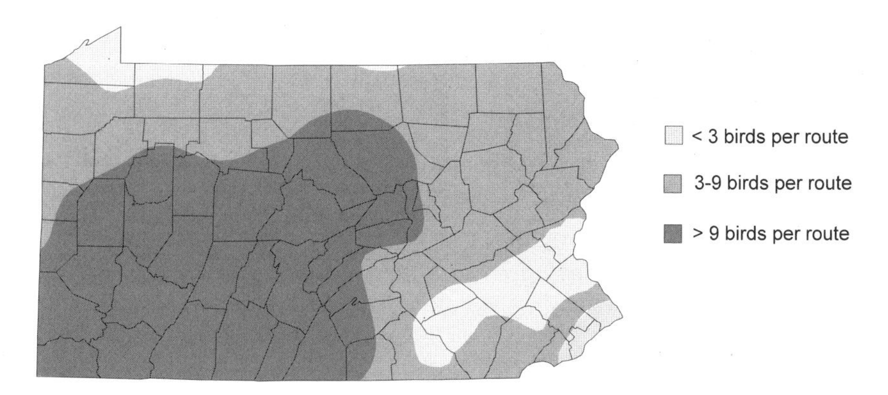

31. Eastern Towhee relative abundance, based on BBS routes, 1985–1994

northern portion of the High Plateau, they are casual to accidental in winter and have been reported only from Potter and Warren cos.

History: Although towhees are generally considered common, some late-nineteenth-century reports from northern forests suggested that they were rare. Todd (1893) said that only a few were found in second-growth forests in Clearfield Co., and Dwight (1892) found only one on North Mountain (Wyoming Co.). Harlow (1913) suggested that the loss of "big timber" contributed to a gradual northward expansion of the "Chewink." The southern bias of early authors is apparent in Turnbull (1869) and Warren's (1890) references to the "Chewink" as "common or abundant." Whether expanding in range or simply exploiting new habitat, towhees were common summer residents in open woodlands statewide by the early twentieth century (Sutton 1928a).

Bachman's Sparrow *Aimophila aestivalis*

General status: Currently Bachman's Sparrows breed from south-central Missouri east to North Carolina and south to eastern Oklahoma, eastern Texas, and central Florida. They winter within their southern breeding range from Texas, along the Gulf states, and north along the Atlantic coast to North Carolina. Bachman's Sparrows formerly bred in the southwestern corner of Pennsylvania, but the species has been extirpated from the state since about the mid-1930s. No record of this species has been accepted since then.

History: W. W. Cooke stated that this species apparently was extending its range and that it became locally common in southern Virginia and even invaded northern Ohio and western Pennsylvania.[1] Todd (1940) added this species to the Pennsylvania bird list in 1910, when he collected an adult male a mile north of Beaver in Beaver Co. Apparently the earliest arrival date recorded was 15 Apr, when two singing males were reported 4 miles south of Waynesburg in Greene Co. (Todd 1940). Todd stated that the first confirmed breeding of this species was around 1909 or 1910, when a nest containing one egg was found in a hillside field adjoining an oak copse close to Waynesburg in Greene Co. He found at least three males during the summer of 1913 near Kirby and in the vicinity of Brock and Rosedale in Greene Co. On 10 May 1916 Todd saw a pair north of Waynesburg in a grove of white oaks, and on 20 May he returned and collected the nest and its contents of five eggs. Todd concluded from his experience with this species that they preferred "sterile fields and open oak groves." Bachman's Sparrows again were found on 12 May 1924 at Patton's Point in Beaver Co., and a bird believed to be a stray was heard and seen in Sewickley Township in Allegheny Co. from 3 to 17 Jun 1928.[2] Apparently the last known nest in Pennsylvania was discovered on 6 Jul 1937 about 2 miles east of Washington in Washington Co., where Sutton observed a brood of young with their parents (Todd 1940). There has been no confirmed sighting of Bachman's Sparrow in the state since 1937.

[1] Cooke, W.W. 1914. The migration of North American sparrows. Bird-Lore 16:176.
[2] Christy, B.H. 1929. Notes. Cardinal 2:130.

American Tree Sparrow *Spizella arborea*

General status: Tree Sparrows breed from Alaska across northern Canada to Newfoundland and south to James Bay. They winter across southern Canada south to North Carolina and west across northern Texas to Nevada and eastern Washington. Most years they are fairly common in Pennsylvania.

They arrive in mid-fall and remain to mid-spring, but the number of wintering birds varies from year to year. Birds are typically found more frequently in late fall, early winter, and early spring than at other times of the year, reflecting the passage through the state of birds to and from southern wintering grounds.

Habitat: Tree sparrows inhabit weedy or brushy fields, marshes, woodland edges, and thickets. They frequently visit bird feeding stations, especially when there is heavy snow cover.

Seasonal status and distribution

Spring: The number of wintering birds remains high through Mar, especially if snow remains on the ground. There is often an influx of sparrows during the second and third weeks of Mar. By the second week of Apr most tree sparrows have left the state, but a few may remain into the third or fourth week of Apr at northern sites and in the High Plateau. Individuals have lingered to the third week of May.

Fall: Tree Sparrows may arrive as early as the fourth week of Sep in northern counties, but most do not appear before the fourth week of Oct. They are frequently found in mixed flocks of migrant sparrows when they first arrive, inhabiting edges of roadways and open fields. In some years, the greatest number of birds is recorded from Dec to the first or second week of Jan.

Winter: By the time winter has arrived loose flocks of a few to more than 50 birds are found. They are widespread during this season and occupy almost every area of suitable habitat. Even during years when American Tree Sparrows have not moved into the state in large numbers, a few can usually be found at rural bird feeding stations. Leberman (1976) wrote that banding records at PNR indicated that this species "shows strong attachment to its wintering grounds, individual Tree Sparrows appear year after year at Powdermill."

History: There has been little change in the status and distribution of these northern visitors throughout history. Most authors, including Turnbull (1869), Todd (1940), and Poole (1964), listed them as common transients or visitors.

Comments: Data from Pennsylvania CBCs suggest a decline of American Tree Sparrows in Pennsylvania.[1]

[1] Hess, P. 1989. American Tree Sparrows *(Spizella arborea)* in Pennsylvania (Part 1), Speculations on a 15-year decline, 1973–1987. PB 3:90–93.

Chipping Sparrow ***Spizella passerina***

General status: Chipping Sparrows breed from Alaska across Canada and most of the U.S. They winter across the southern third of the U.S. and north to California in the West and along the coast to southern New England in the East. Chipping Sparrows are fairly common to common regular migrants and breeding residents throughout Pennsylvania. They are equally at home around towns and along dirt roads in extensively forested regions. They are irregular winter visitors or residents.

Habitat: Chipping Sparrows are found in a wide variety of habitats including open woodlands, woodland edges, pastures, orchards, parks, roadways, lawns, and gardens. During the breeding season they prefer sites where evergreens or various types of dense, ornamental trees or shrubs for nesting are present, but they also are found along roads not overshadowed by forest canopy, small forest openings, and larger clearcuts even in extensively forested areas. In winter Chipping Sparrows may be found at bird feeding stations where suitable cover is nearby.

Seasonal status and distribution

Spring: The earliest migrants have appeared by the second week of Mar. Their typical migration period is from about the third week of Apr to the fourth week of May. Peak migration occurs during the fourth week of Apr or the first week of May.

Breeding: Chipping Sparrows' affinity for scattered shrubs and trees, particularly evergreens, makes for a natural association with suburban and rural residences. They are scarce in the heavily urbanized Coastal Plain and in continuous forested areas. A small, steady decline of 0.9% per year has been noted on state BBS routes since 1966. The nest is a neat cup of grass and pine needles, placed in a shrub, usually less than 10 feet from the ground. Most nests with eggs have been reported from 10 May to 22 Jul, although an early nest was found on 22 Apr (Todd 1940).

Fall: The large breeding population in the state makes it difficult to determine the beginning of migration, but by the second and third weeks of Sep an increase in the number of birds is evident. The greatest number of migrants is recorded from the first to the third week of Oct. By the second or third week of Nov most have left the state. Stragglers may remain to the first week of Jan or later if the weather remains mild.

Winter: There are scattered records of Chipping Sparrows at this season, mostly from the Coastal Plain and Piedmont. Few birds found in winter are found past the third or fourth week of Jan. Those discovered during the fourth week of Feb may be early migrants rather than overwintering birds. Chipping Sparrows are usually found only during mild winters with little snow cover. However, a few individuals have survived through the winter at bird feeding stations as far north as Erie Co.

History: Harlow's (1913) assessment still fits today: "generally distributed thru [*sic*] the state and nesting in every county, irrespective of fauna and life zones." Poole (unpbl. ms.) stated that there were few winter records before 1950 and cited only one record (within this book's definition of winter), a bird found in Reading on 17 Jan 1924.

Comments: Many of the late Dec records are attributable in part to the increase in CBCs within the state, bringing more people into the field at this time. The recent popularity in feeder watching probably contributes to the increase in winter sightings. Some sightings of Chipping Sparrows in late fall and throughout winter may be misidentified American Tree Sparrows.

Clay-colored Sparrow *Spizella pallida*

General status: Clay-colored Sparrows breed in central Canada south in the U.S. to Wyoming and east across the northern states through the Great Lakes region. They winter from southern Texas south through Mexico. They probably are more common in Pennsylvania than records indicate, especially in fall, when they closely resemble immature Chipping Sparrows. They are rare, regular migrants, with sightings from all regions of the state. This species has been recognized as a regular migrant only since about the late 1970s. There were few records before that time. During migration Clay-colored Sparrows are almost always found associated with Chipping Sparrows. Though breeding has not been confirmed in the state by 1998, territorial males have been found in suitable breeding habitat on numerous instances. Clay-colored Sparrows are accidental in winter.

Habitat: During migration Clay-colored Sparrows are observed in brushy fields, on gravel parking lots and roadways at PISP, on lawns, and occasionally at bird feeding stations. Territorial males have been found during the breeding season in weedy or brushy fields and once in an evergreen nursery.

Seasonal status and distribution

Spring: Most Clay-colored Sparrows have been seen during spring. Spring records are reported as early as the second week of Mar and continue to the fourth week of May, with some territorial males remaining through Jun or even later. Most records are of single birds observed from the first to the third week of May.

Summer: Clay-colored Sparrows are accidental summer visitors in scattered western counties but a nest has not yet been discovered in Pennsylvania. Males on territory (for at least a week during late May and Jun) have been reported at YCSP in Indiana Co. in 1985;[1] at Piney Township in Clarion Co. in Jul 1998, possibly with young;[2] along Limber Road, near Stoneboro in Mercer Co. in 1991 and 1992;[3,4] and near West Springfield in Erie Co. in 1984 for 10 days in Jul (Stull et al. 1985). Clay-colored Sparrows became well established in northern New York State by the 1980s (Andrle and Carroll 1988) and expanded eastward into southern Ontario over the past 50 years (Cadman et al. 1987). They are likely to be confirmed breeding in the near future and may become established in Pennsylvania, although surveywide BBS routes show a declining population. Young conifer plantations such as Christmas tree farms may provide a long-term habitat. The nest closely resembles that of the Chipping Sparrow.

Fall: There are undoubtedly more birds in fall than records indicate, but their similarity to basic-plumage Chipping Sparrows makes them easily overlooked. Fall records are from the first week of Sep to the first week of Nov. At least two records exist of birds that lingered into late fall but apparently did not winter: one near Media in Delaware Co. from about 25 Dec 1985 to 4 Jan 1986,[5] and one at a feeder at Charlton in Clinton Co. to 29 Dec 1984.[6]

Winter: They have been reported during the winter period near Kennett Square in Chester Co.[7] and at Wrightsville in York Co.[8] A Clay-colored Sparrow successfully wintered in 1992, at a feeder in Erie, Erie Co.[9]

History: There are supposedly only four reliable historical records of this species in Pennsylvania, all from the Philadelphia area. One was reported by J. H. Austin at Cobb's Creek Park in Philadelphia on 13 Oct 1945;[10] one was along Wissahickon Creek, also in Philadelphia, on 17 Feb 1946 (Poole, unpbl. ms.); and one was at Chestnut Hill on 3 Mar 1946 (Poole, unpbl. ms.). A singing Clay-colored Sparrow was found at Springton Dam near Media in Delaware Co. on 6 May 1959.[11]

[1] Ickes, R. 1992. Clay-colored Sparrow (*Spizella pallida*). BBA (Brauning 1992):376–377.
[2] D.W. Brauning, pers. obs.
[3] McCay, M. 1991. County reports—July through September 1991: Mercer County. PB 5:136.
[4] McCay, M. 1992. County reports—April through June 1992: Mercer County. PB 6:86.
[5] Boyle, W.J., Jr., R.O. Paxton, and D.A. Cutler. 1986. Hudson-Delaware region. AB 40:264.
[6] Hall, G.A. 1985. Appalachian region. AB 39:168.
[7] P. Hurlock, pers. comm.
[8] A. Spiese, pers. comm.
[9] Hall, G.A. 1992. Appalachian region. AB 46:423.
[10] Austin, J.H. 1945. General notes. Cassinia 35:40.
[11] P. Schwalbe and G. Schwalbe, pers. comm.

Field Sparrow *Spizella pusilla*

General status: Field Sparrows breed across the eastern U.S. west to Montana, Colorado, and Texas south along the Gulf coast to northern Florida. They winter primarily from Kansas east to Massachusetts and south to Texas and Florida. Though widespread and fairly common in Pennsylvania both as a migrant and as a breeding species, the Field Sparrow is not as well known as its close relative the Chipping Sparrow. During migration they frequently flock with Chipping Sparrows. Field Sparrows are winter residents over most of the state.

Habitat: Field Sparrows are found in lawns and grassy fields and brushy edges during migration. During the breeding season they prefer overgrown fields and pastures, especially those grown up in brush and small trees to just a few feet in height. They are also found in clearcuts and occasionally in young Christmas tree plantings. In winter Field Sparrows are found primarily outside the forested mountains and along river valleys. They visit bird feeding stations, especially in areas with snow cover.

Seasonal status and distribution

Spring: Migration may begin as early as the first or second week of Mar. Their usual migration time is from the third or fourth week of Mar to the third or fourth week of May. The greatest number of birds usually is recorded during the second or third week of Apr. If cold weather persists into Apr, peak migration may not occur until as late as the first week of May, especially in northwestern Pennsylvania.

Breeding: Field Sparrows are most common in fragmented environments of the southern Ridge and Valley, Glaciated Northeast, and across the western counties (see Map 32). They are widely distributed (found in 87% of BBA blocks) but are spotty in uniformly forested and urbanized regions.[1] Field Sparrows have somewhat more specialized habitat requirements than their ubiquitous congener, the Chipping Sparrow. Their population trend is typical of many early-succession species, declining at 4% per year on BBS routes statewide since 1966. There have been occasional population fluctuations evidenced by the low number of birds recorded on BBS routes between 1977 and 1979. The nesting period is almost identical to that of the Chipping Sparrow, with most egg sets from 11 May to 24 Jul, involving two clutches. The nest is a fine cup of dry grass. Earlier nests are usually placed on the ground at the base of a shrub, and later nests are up to several feet from the ground in weeds or a small shrub.

Fall: The beginning of migration is not well defined in early fall, especially from Aug through about the third week of Sep, when there is some evidence of

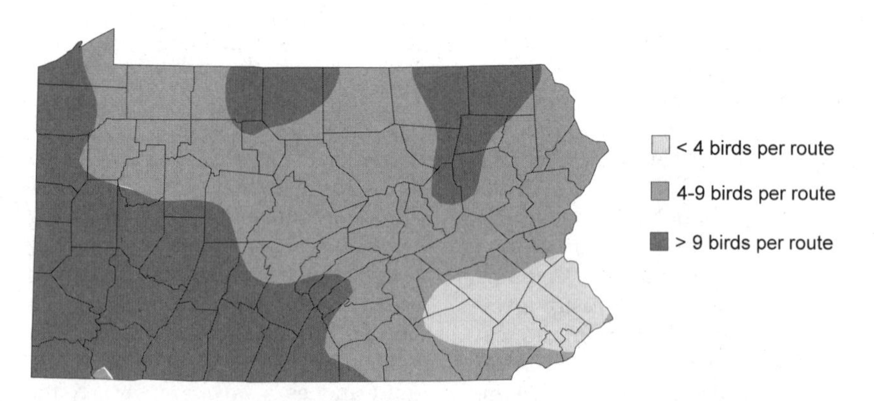

32. Field Sparrow relative abundance, based on BBS routes, 1985–1994

local wandering. Migration may begin by the fourth week of Sep or the first week of Oct. The greatest passage of migrants is recorded during the second and third weeks of Oct. Migration rapidly dwindles by the fourth week of Oct or the first week of Nov, with non-wintering stragglers remaining to the first or second week of Jan. Field Sparrows may be locally common in late fall in some years in the southeastern corner of the state. On the Glenolden CBC in Delaware Co. on 31 Dec 1966, 325 birds were counted, an unusually high total.[2]

Winter: Most wintering Field Sparrows are found in the Coastal Plain, the Piedmont, and the river valleys of the Ridge and Valley, where they may be locally common. Field Sparrows are usually more difficult to find in other areas of the state, where singles or just a few birds are typically found. In the northern-tier counties, south in the High Plateau to the Allegheny Mountain section, and in the Glaciated Northwest, they have been reported in winter from the counties of Bradford, Crawford, Erie, Tioga, and Warren. They are frequently found in mixed flocks with other wintering sparrows.

History: Harlow (1913) considered them abundant, except in the northern counties. Todd (1940) thought that Field Sparrows reached their greatest abundance during the era of agricultural decline, during the early twentieth century. In northern areas that were still heavily forested, Field Sparrows were scarce (e.g., Stone 1891).

[1] Leberman, R.C. 1992. Field Sparrow *(Spizella pusilla)*. BBA (Brauning 1992):378–379.
[2] Rigby, E.H. 1967. Glenolden CBC. AFN 21:152.

Vesper Sparrow *Pooecetes gramineus*

General status: Vesper Sparrows' breeding range extends across southern Canada and through, primarily, the northern half of the U.S. They winter mostly across the southern half of the U.S., north along the Pacific coast to central California, and on the east coast to Connecticut. In Pennsylvania this species depends on extensive open farm fields and grasslands for nesting. Once considered common, they are now listed as uncommon to rare regular migrants. Vesper Sparrows frequently associate with other species of sparrow during migration. They breed widely in the state, with the greatest concentrations now occurring in reclaimed strip mines. Their winter status is poorly known, but they are considered rare and local and at most sites are probably not recorded every year.

Habitat: During migration and in winter Vesper Sparrows are found in bare farm fields, short grassy fields, and along roadways. During the breeding season they prefer extensive agricultural land and extensive grassy fields with bare patches of soil, such as reclaimed strip mines, provided there are widely scattered trees, shrubs, or utility lines for use as singing perches.

Seasonal status and distribution

Spring: Migrant Vesper Sparrows may appear as early as the fourth week of Feb. Their typical migration period is from the third week of Mar or the first week of Apr to the third week of May. The greatest number of migrants may occur from the first to the third week of Apr.

Breeding: Vesper Sparrows apparently benefited from the rapid clearing of forests around the beginning of the twentieth century. Since about the 1960s or 1970s this species has rapidly declined, in part as a result of reforestation, sprawling urbanization, and changing farming practices. Vesper Sparrows are uncommon to locally fairly common nearly statewide in summer. They are notably absent from five counties in the state's southeastern corner (Bucks, Chester, Delaware, Montgomery, and Philadelphia), although they are regular in Lancaster Co. and northern Berks Co. west through the Piedmont. They are uncommon in agricultural areas of the northern tier and probably achieve their greatest abundance in strip mines reclaimed to grass, notably from Allegheny Co. north to Clarion Co.[1] Vesper Sparrows are among the few birds that breed in cornfields,[2] although nesting success there is unknown. The decline of 6.1% per year on state BBS routes is more precipitous than that of any other grassland species except the Henslow's Sparrow. Vesper Sparrows were the most commonly reported of four grassland sparrows in the early years of the BBS in Pennsylvania, but they had slipped behind Grasshopper and Savannah sparrows by the late 1980s (BBS data). The nest is made of grass in a small depression on the ground. Most eggs have been found from 3 May to 19 Jun, although clutches have been found in Aug and some birds are on territory in mid-Apr.

Fall: Fall migration is not well known, but there is a noticeable increase in the number of birds by the first week of Sep. Migration continues to at least the fourth week of Oct, with birds becoming difficult to find after about the second or third week of Nov. Stragglers have been recorded to at least the fourth week of Dec, especially on CBCs.

Winter: There are widely scattered winter records of Vesper Sparrows from the Piedmont and Ridge and Valley; they also have been reported in winter in Allegheny and Westmoreland cos. Their scarcity and patchy distribution may be due in part to lack of birding activity at this season in some of the extensive areas of agriculture, especially in the Piedmont.

History: Numerous nineteenth-century bird lists refer to Vesper Sparrows as "the most abundant and generally distributed of the sparrows" breeding statewide (e.g., Turnbull 1869; Warren 1890). This comment included Susquehanna Co. (Cope 1898), where the species was scarce during the BBA project.[3] Harlow (1913) considered Vesper Sparrows to be a "characteristic species of the open country and uplands" and called them abundant. The abundance and subsequent decline of Vesper Sparrows roughly correspond to dramatic changes in agricultural acreage and practices throughout the twentieth century. Vesper Sparrows were still considered "common" by Todd (1940) and Poole (1964).

[1] Brauning, D.W. 1996. Grassland breeding bird survey. Annual report. Pennsylvania Game Commission, Harrisburg.

[2] Rodenhouse, N.L., and L.B. Best. 1983. Breeding ecology of Vesper Sparrows in corn and soybean fields. American Midland Naturalist 110:265–275.

[3] Santner, S. 1992. Vesper Sparrow *(Pooecetes gramineus)*. BBA (Brauning 1992):380–381.

Lark Sparrow *Chondestes grammacus*

General status: Lark Sparrows breed in southwestern Canada south across the western U.S. and east of the Mississippi Valley locally and irregularly to Wisconsin, Ohio, North Carolina, and Alabama. They winter in California, Arizona, Texas, and along the Gulf coast. In the East, Lark Sparrows are casual in winter along the Atlantic coast to New York. This sparrow historically bred in Pennsylvania but is now only a casual vagrant. Records are from all seasons, but most are during the migration period, when they are usually associated with mixed flocks of sparrows. Nearly all sightings are of single birds.

Habitat: Lark Sparrows are usually found in grassy fields and at bird feeding stations.

Seasonal status and distribution

Spring: Six recent spring records of Lark Sparrows have been reported from the counties of Allegheny, Bucks (twice), Dauphin, Erie, and Lancaster, with dates ranging from 27 Mar to 19 May.

Summer: A Lark Sparrow on Robin Hill Road north of Scandia, Warren Co., on 22 Jun 1976 is the only recent summer sighting.[1]

Fall: Most records are from this season. Lark Sparrows have been reported at least 14 times, with scattered dates ranging from 14 Aug to about 9 Jan.

Winter: There are two winter records. One bird was at Manheim in Lancaster Co. on 24 Jan 1976 (Morrin et al. 1991), and one was at Carlisle in Cumberland Co. from 13 to 21 Jan 1980.[2]

History: Lark Sparrows apparently were first found in the state during the late 1800s and probably peaked as a breeding species before 1919. It is likely that the replacement of forests with agriculture increased their eastward expansion into Pennsylvania until they reached the forested regions of the Allegheny Mountains, which prevented any major expansion beyond that point. Todd's (1940) report that they usually arrived in late Apr and remained to Sep was based on reports from S. S. Dickey in Greene Co. Lark Sparrows were first reported nesting in Pennsylvania when a nest with a set of five eggs was found in Greene Co. (Jacobs 1893) and a nest with eggs was found near Bridgeville in Allegheny Co. on 3 Jun 1897 (Todd 1940). In both instances the adults were not seen, but Todd (1940) felt that there was little chance for error as the eggs of this species are unmistakable. Suspected breeding continued to at least 1919 in Allegheny Co. Todd (1940) discovered a pair about 4 miles south of Beaver in Beaver Co. on 11 May 1902 and observed a female collecting nesting material. The easternmost confirmed nesting, and probably the last in Pennsylvania, was at the Penn State Nature Camp in Huntingdon Co., where a nest containing one fledgling and three eggs was discovered on 27 Jun 1931.[3] Since this last confirmed nesting, the species was infrequently reported until the 1960s. Other records, probably pertaining to migrants, have been reported from Crawford, Perry, and Philadelphia cos. (Poole, unpbl. ms.).

[1] T. Grisez, pers. comm.
[2] Hall, G.A. 1980. Appalachian region. AB 34:274.
[3] Wood, M. 1932. Eastern Lark Sparrow breeding in Central Pennsylvania. Auk 49:98.

Lark Bunting *Calamospiza melanocorys*

General Status: Lark Buntings breed in the plains and prairies from south-central Canada south through the central U.S. They winter from Nevada and Arizona east to Kansas and Texas. All but one record have been since 1961, with the most recent accepted record in 1985.

Seasonal status and distribution

Spring: In 1963 a Lark Bunting was banded by J. K. Gabler at Chambersburg in Franklin Co. on 16 Mar.[1] An adult male reported by W. B. Hicks at Churchville in Bucks Co. was present from 23 Apr to 1 May 1965 and was photographed.[2] A Lark Bunting was discovered by P. Brown, J. Ginader, and B. Ginader at Pymatuning in Crawford Co. on 30 May 1976,[3] and one was photographed by M. Fowles and E. Fowles at Delmont in Westmoreland Co. on 25 May 1980.[4]

Fall: An immature male was found by J. Dougherty at East Falls in Philadelphia on 30 Aug 1969.[5] Lark Buntings have been reported twice by W. Reid during this season in Wilkes-Barre, Luzerne Co.: one on 30 Oct 1968[6] and an immature or female on 21 Sep 1985 in almost the same spot.[7] A male was identified by R. K. Bell near Clarksville in Greene Co. on 9 Aug 1972.[8]

Winter: One record falls within this period: a bird reported by J. P. McGrath on 13 Dec 1961 at Graterford in Montgomery Co.[9] successfully wintered and was last seen on 26 Apr 1962.[10] It was photographed and represents the first documented record of Lark Bunting in Pennsylvania.

History: Todd (1940) included the Lark Bunting in his book *Birds of Western Pennsylvania*, on the basis of an account from F. L. Homer, who stated that a male was found and heard singing in Clarksboro in Mercer Co. on 9 Jun 1896. Todd believed that this report could not have been a misidentification and believed the account to be "entirely credible."

[1] Hall, G.A. 1963. Appalachian region. AFN 17:326.
[2] Scott, F.R., and D.A. Cutler. 1965. Middle Atlantic Coast region. AFN 19:460.
[3] Hall, G.A. 1976. Appalachian region. AB 30:843.
[4] Hall, G.A. 1980. Appalachian region. AB 34:778.
[5] Scott, F.R., and D.A. Cutler. 1970. Middle Atlantic Coast region. AFN 24:30.
[6] Boyajian, N.R. 1969. Hudson St. Lawrence region. AFN 23:28.
[7] Paxton, R.O., W.J. Boyle Jr., and D.A. Cutler. 1986. Hudson-Delaware region. AB 40:90.
[8] Hall, G.A. 1972. Appalachian region. AB 26:860.
[9] Scott, F.R., and D.A. Cutler. 1962. Middle Atlantic Coast region. AFN 16:319.
[10] Scott, F.R., and D.A. Cutler. 1962. Middle Atlantic Coast region. AFN 16:396.

Savannah Sparrow ***Passerculus sandwichensis***

General status: Savannah Sparrows breed from Alaska across most of Canada and south primarily across the northern half of the U.S. They winter from central California, east across the southern states and north to Kansas, Missouri, Kentucky, and Nova Scotia along the Atlantic coast. Savannah Sparrows are uncommon to fairly common regular migrants. During migration they frequently associate with mixed flocks of sparrows. They breed over most of the state but most abundantly across the western and northern portions. In winter, most records are from the southeastern corner of the state.

Habitat: Savannah Sparrows are found primarily in grassy areas, such as hayfields, strip mines reclaimed in grass, meadows, pasture, and to a less extent cultivated fields. During migration they may appear on large open lawns, in brushy fields, or in other open settings.

Seasonal status and distribution

Spring: The first arrivals may appear as early as the first week of Mar. Their typical migration period is from the third or fourth week of Mar to the third week of May. Numbers gradually build through Apr, with most recorded from the second to the fourth week of Apr.

Breeding: Savannah Sparrows are fairly common to locally common in northern and western counties and south in the mountains through Somerset Co. They are uncommon to rare elsewhere. They are uncommon in the Ridge and Valley, rare in the eastern Piedmont, and regular in agricultural areas across the northern portion of the Piedmont, notably in the Cumberland Valley. Pennsylvania is near the southeastern edge of their distribution, and they are generally less common in the southern half of the state. The state's BBS shows a population decline of 4% per year. The nest is made of grass in a depression on the ground under a tuft of grass. Nest dates range from mid-May through Jul.

Fall: The beginning of migration is difficult to discern, but an increase in birds is noted by the second or third week of Sep. The greatest number of birds is found from the fourth week of Sep to the second week of Oct. Migration ends about the first week of Nov, with stragglers remaining to the first or second week of Jan. A few Savannah Sparrows remain to winter. In some years the number of birds may remain high well into the end of Dec, especially in the Tinicum area of the Coastal Plain.

Winter: Most winter records are from the Coastal Plain and the southern portion of the Piedmont, where they are listed by local authorities as uncommon to rare. They have also been found locally in winter in the Ridge and Valley in the counties of Centre, Lycoming, Montour, and Wyoming. West of the Allegheny Front they have only been reported in winter from Allegheny, Butler, Lawrence, and Mercer cos. In the Glaciated Northeast, the High Plateau, the snowbelt area of the Glaciated Northwest, and the Lake Erie Shore no winter records are known.

History: The historical distribution and abundance of Savannah Sparrows in Pennsylvania have shifted along with the balance between forests and fields. They benefited from increased farm acreage that followed extensive logging around the beginning of the twentieth century and more recently from the reclamation of strip mines into grass. However, the regrowth of Pennsylvania's forests, decline of mining, and fewer farmlands have resulted in declines for this and other grassland birds. The Savannah Sparrow was considered local and restricted to relatively few Pennsylvania locations by early ornithologists (Warren 1890; Harlow 1913). Warren's (1890) comment that "I have never observed this species in spring, later than April 25" is corroborated by the notable scarcity of egg sets from early collections. This situation changed quickly. Sutton (1928b) called Savannah Sparrows "fairly common" in Crawford Co., and Todd's (1940) map reflected a widespread distribution, although he considered them "unaccountably rare" in the southwestern counties and generally absent from the southern Ridge and Valley. A range expansion was most likely in progress by Todd's time and continued subsequently. Their breeding distribution eventually reached Tinicum, Philadelphia Co. (J. C. Miller and Price 1959), where Savannah Sparrows have not been reported in recent years.

Grasshopper Sparrow *Ammodramus savannarum*

General status: Grasshopper Sparrows breed in widely scattered populations along the southern border of Canada and south across most of the U.S. They winter from central California, east across the southwestern U.S. through the Gulf states, and north to North Carolina. They breed widely in the state but are more common in the southern half. Fields are occasionally abandoned by birds after a few years' occupation, often for unknown reasons. These sparrows have benefited from the reclamation of strip mines to grasslands. During spring they are occasionally found with mixed flocks of sparrows away from their breeding habitat. Their inconspicuous behavior makes them difficult to observe during migration, especially in fall, when they are no longer singing. Grasshopper Sparrows are accidental in winter.

Habitat: They inhabit meadows and grassy fields, including hayfields and pastures, but especially reclaimed strip mines and recently abandoned farm fields.

Seasonal status and distribution

Spring: In spring Grasshopper Sparrows are observed only as uncommon migrants. The earliest migrants have been recorded the third week of Mar. Their usual migration time is from the third or fourth week of Apr to the first or second week of May.

Breeding: Grasshopper Sparrows are fairly common to locally common in the western Piedmont, Ridge and Valley, and Southwest; they are uncommon in similar habitats in the Glaciated Northeast. They are rare and local in the eastern Piedmont, where suburbanization has reduced agricultural habitats, and are uncommon in Erie and Crawford cos., strongholds of Savannah Sparrows. They were reported in nearly the same number of BBA blocks as the Savannah Sparrow,[1] but as Poole (unpbl. ms.) suggested, they have a more southerly and easterly distribution. They nevertheless occur fairly commonly in the High Plateau from Somerset Co. northeast to Bradford Co. Nests are placed in depressions in the ground, are lined with dry grass, and may be found from mid-May to early Aug. Their nests are difficult to find, but nearly all reports of fledglings and adults carrying food during the BBA were after 4 Jul.[1] BBS data show a population decline in Pennsylvania of 5% per year.

Fall: Very little is known about the status of this species during fall migration. They become difficult to find when singing has stopped. Birds are occasionally found through Sep and Oct, when they are captured in mist nets during banding operations. Stragglers have been recorded to the second week of Jan, with several wintering attempts.

Winter: Fall stragglers have attempted to winter at least four times since 1956. One stayed at State College in Centre Co. until 15 Jan 1956 (Wood 1958). On 18 Feb 1958 one was found dead at Wrightsville in Bucks Co. (Poole, unpbl. ms.). A Grasshopper Sparrow was found at Wyomissing in Berks Co. on 9 Feb 1962,[2] and another was at a feeder from the first snow in Dec 1962 to at least 2 Feb 1963 at Ridgeway in Elk Co.[3]

History: Historical records are similar to recent patterns. The scarcity of Grasshopper Sparrows in the High Plateau from Clinton Co. northwest to Warren Co., noted by Todd (1940), was reflected by the BBA project. Poole (unpbl. ms.) cited early-twentieth-century records from DuBois (Clearfield Co.) to Cooksburg (southern Forest Co.), which is the northwestern edge of current regular occurrence. Owing to habitat loss, nesting has not been documented recently where it had occurred in the Philadelphia area, such as at Tinicum (J. C. Miller 1970).

[1] Santner, S. 1992. Grasshopper Sparrow *(Ammodramus savannarum)*. BBA (Brauning 1992): 384–385.

[2] Scott, F.R., and D.A. Cutler. 1962. Middle Atlantic Coast region. AFN 16:319.

[3] Hall, G.A. 1963. Appalachian region. AFN 17:326.

Henslow's Sparrow ***Ammodramus henslowii***

General status: Henslow's Sparrows breed from South Dakota east across southern Ontario to New York and south to Kansas east to North Carolina. The breeding range in the northwestern and eastern portions has decreased in recent years (AOU 1998). They winter from South Carolina south to Florida and west across the Gulf states to Arkansas and Texas. Henslow's Sparrows are very sensitive to habitat changes and have been on the decline in Pennsylvania. Very little is known about the migratory status of this species. They are local breeders from the Glaciated Northwest south through the southwestern third of the state and in the Glaciated Northeast. Isolated pockets of breeding birds may be found in the High Plateau and Ridge and Valley. The number of breeding birds within an area may fluctuate from year to year.

Habitat: Henslow's Sparrows prefer extensive grasslands, with some peren-

Henslow's Sparrow at West Nicholson, Wyoming County, July 1997. (Photo: Rick Wiltraut)

nial forbs used as perches. They may be found in meadows, uncut hayfields, and abandoned farm fields but most regularly and commonly are in strip mines reclaimed in grass. They prefer generally open landscapes larger than 80 acres and avoid fields that were mowed the previous year.

Seasonal status and distribution

Spring: Henslow's Sparrows are essentially unknown away from breeding habitat during this season. They have been recorded as early as the second week of Apr. Most do not arrive before the fourth week of Apr or the first week of May.

Breeding: Henslow's Sparrows are uncommon and regular, but may be locally fairly common, in extensive grasslands of the Southwest and Glaciated Northwest regions. They are uncommon but local and somewhat irregular in the Glaciated Northeast and are accidental elsewhere east and south of the Allegheny Front. They are notorious for abandoning agricultural fields from one year to the next as a result of mowing, succession, or imperceptible changes in cover. The ephemeral nature of their habitat makes determining population trends difficult, but Henslow's Sparrows have shown a precipitous decline of 10% per year on BBS routes in Pennsylvania since 1966. Despite long-term declines, the BBA project documented a much more widespread distribution than previously thought in northeastern hayfields and reclaimed strip mines in the northwestern portion of the state.[1] They are most abundant and most regular in reclaimed strip mines in Armstrong, southern Clarion, Indiana, and Venango cos., where densities may reach one territorial male per 2 acres. The nest is built on the ground at the base of dense grasses. Most nests are found in Jun and Jul.

Fall: These secretive sparrows are very difficult to locate and are rarely reported once singing has stopped. Banding records suggest that migration may begin by late Aug, with records in Sep, and may continue until the third week of Oct at PNR (Leberman 1976). One was banded at PISP on 1 Oct 1997,[2] and another at PNR on 15 Oct 1998.[3]

History: Valid Pennsylvania breeding records were first reported in 1913 in Huntingdon Co. (Harlow 1913). A small breeding group was located in Crawford Co. in 1922 by Sutton (1928b), including a nest with young. In subsequent decades, Henslow's Sparrows were reported breeding in widely scattered locations across the state, although in only a few places were they found for more than a few years. In the years right after the formation of Lake Arthur, in 1969 at MSP they were numerous, but their habitat grew to brush and they did not return.[4] Todd (1940) stated that "the extremely local distribution and general inconspicuousness of this sparrow have created the impression that it is rare." Poole (unpbl. ms.) reported widely scattered summer records throughout the eastern half of the state in at least 18 counties.

Comments: Henslow's Sparrows were listed by the PGC in 1979 as a State Threatened species. Subsequent surveys identified a much larger population than had previously been suspected, and they were taken off the Threatened list in 1992.[5]

[1] Reid, W. 1992. Henslow's Sparrow *(Ammodramus henslowii).* BBA (Brauning 1992):386–387.
[2] R.F. Leberman, pers. comm.
[3] R.C. Leberman, pers. comm.
[4] G.A. Hall, pers. comm.
[5] Brauning, D.W., M.C. Brittingham, D.A. Gross, R.C. Leberman, T.L. Master, and R.S. Mulvihill. 1994. Pennsylvania breeding birds of special concern: A listing rationale and status update. Journal of the Pennsylvania Academy of Science 68:3–28.

Le Conte's Sparrow ***Ammodramus leconteii***

General status: Le Conte's Sparrows breed across southern Canada from Alberta and east to Quebec, and from Montana east to Michigan. They winter from Missouri and Texas east to South Carolina and Florida. There are presently four accepted records of Le Conte's Sparrow in Pennsylvania. No record exists from before 1959.

Seasonal status and distribution

Spring: In 1992 one was found by L. Lewis and was photographed at Struble Lake in Chester Co. on 16 Apr and remained to 26 Apr for the first accepted eastern Pennsylvania record.[1]

Fall: One was banded and photographed by R. C. Leberman and B. Shaw at PNR on 20 Oct 1991 for the first accepted record in the state.[2] One was photographed by E. Witmer at Bainbridge in Lancaster on 24 Oct 1993.[3,4] In 1994 one was discovered at Green Lane Reservoir in Montgomery Co. on 3 Nov.[5]

Comments: There have been other reports of Le Conte's Sparrows in Pennsylvania. The first report was on 23 Aug 1959 at Saeger Hill in Crawford Co.

Le Conte's Sparrow at Struble Lake, Chester County, April 1992. (Photo: Rick Wiltraut)

(Poole 1964). Le Conte's Sparrows were not reported again until 1974, when one was found on PISP in Erie Co. on 20 Oct.[6] In 1994 a Le Conte's Sparrow was discovered at Struble Lake on 23 Apr.[7] In Lawrence Co. one seen from Wilmington Township on 16 Apr 1995 is currently under review.[8] The most recent sighting of a bird found on 25 Oct 1997 in Williams Township in Northampton Co. is also currently under review.[9] Observers should use caution in identifying this species because some birds reported have proved to be Grasshopper Sparrows. The status of Le Conte's Sparrow is likely to change as observers continue to learn how to identify them and where to look for them.

[1] Lewis, L., and R. Wiltraut. 1992. County reports—April through June 1992: Chester County. PB 6:75.
[2] Leberman, R.C. 1992. First verifiable Pennsylvania record of Le Conte's Sparrow. PB 5: 157–158.
[3] Haas, F.C., and B.M. Haas, eds. 1994. Rare and unusual bird reports. PB 7:163.
[4] Pulcinella, N. 1994. Rare bird reports. PB 8: 17–18.
[5] Freed, G. 1995. Notes from the field: Montgomery County. PB 8:225.
[6] Hall, G.A. 1975. Appalachian region. AB 29:57.
[7] Blust, B. 1994. Notes from the field: Chester County. PB 8:100.
[8] Pulcinella, N. 1995. Rare bird reports. PB 9: 28–29.
[9] Haas, F.C., and B.M. Haas, eds. 1998. Birds of note—October through December 1997. PB 11:232

Nelson's Sharp-tailed Sparrow
Ammodramus nelsoni

General status: Formerly considered to be one of several subspecies of Sharp-tailed Sparrow, the Nelson's Sharp-tailed Sparrow became recognized as a separate species in 1995 (AOU 1998). They breed in central Canada from Mackenzie south into the U.S. to North Dakota, in Ontario and Quebec near Hudson Bay and around James Bay, and in southern Quebec and along the Atlantic coast of Quebec south to Nova Scotia and Maine. They winter along the mid- to south Atlantic coast to Florida and around the Gulf coast. Most sharp-tailed sparrows identified in Pennsylvania are Nelson's. Their shy and secretive behavior and the inaccessibility of their preferred habitat have made them difficult to find. Aggressive fieldwork has shown that this species is more common in the state during fall migration than was previously believed, especially on islands in the lower Susquehanna River. They are currently listed as rare but regular migrants. However, they have recently been found to be uncommon to fairly common in grassy islands of the lower Susquehanna River.

Habitat: Nelson's Sharp-tailed Sparrows prefer marshes, wet meadows, and swamps. They favor areas with sedges or marsh grass that is less than 3 or 4 feet high.

Seasonal status and distribution

Spring: Very few modern records of sharp-tailed sparrows have been reported during spring migration. Although the records are likely to have been of Nelson's Sharp-tailed Sparrows, details are lacking that would totally eliminate Saltmarsh Sharp-tailed Sparrows. In 1975 one was reported on 20 Apr near Shippensburg, near the Cumberland-Franklin co. line,[1] and a rather late bird was reported on 10 Jun 1980 from Montandon Marsh in Northumberland Co.[2]

Fall: Most recent records are from the Coastal Plain and Piedmont, primarily from the lower Susquehanna River on grassy river islands. They arrive during the third week of Sep and peak during the first three weeks of Oct, usually during or after cold fronts. At PISP in Erie Co. during a period of several years in the late 1970s and early 1980s, they were consistently found only during the third and fourth weeks of Sep.[3] On the lower Susquehanna River, where they have recently been found every year since the early 1990s, they occur in Oct. As many as 16 Nelson's Sharp-tailed Sparrows were recorded on grassy islands in the lower Susquehanna River at Columbia in Lancaster Co. on 19 Oct 1963.[4] More recently, more than 13 were recorded on islets in the Susquehanna River near Bainbridge in Lancaster Co. on 6 and 20 Oct 1991,[5] and 12 were at this same location on 8 Oct 1995.[6] In other areas east of the Allegheny Front they have been recorded in the counties of Centre, Chester, Cumberland, Delaware, Lebanon, Luzerne, Montgomery, Northampton, Philadelphia, and York. In other areas west of the Allegheny Front, they have been recorded only in the counties of Crawford and Westmoreland.

History: It is difficult to determine whether all historical sight records of birds found in the state definitely were of this species. However, most authorities considered sight records of Pennsylvania sharp-tailed sparrows to most likely pertain to Nelson's. They were first detected in western Pennsylvania in the fall of 1893 at Erie by S. E. Bacon; he found them to be a rather common transient on PISP (Todd 1940). Bacon apparently collected some of these birds and identified them as Nelson's Sparrow. Todd (1940) mentioned records referred to as Nelson's Sparrow during the spring on PISP. One was collected on 24 May 1900, and another on 27 May 1904. They were not reported again until 1932, when one was observed on 12 May. Apparently this species was not recorded again at PISP until 1957, when one was collected on 22 Sep.[3] Elsewhere in western Pennsylvania they have been recorded only from Warren Co., where a bird was collected on 27 May 1904, and in Crawford Co. in the spring of 1956 or 1957 (Poole, unpbl. ms.). In eastern Pennsylvania, apparently the first record was of one collected from Chester Co. on 13 May 1891 (Poole, unpbl. ms.). Several reports from both spring and fall are from various sites in Berks, Centre, Dauphin, Montgomery, and Lancaster cos. Poole (unpbl. ms.) cited a rather early record of one on 5 Apr 1936 in Harrisburg.

[1] Hall, G.A. 1975. Appalachian region. AB 29:854.
[2] Hall, G.A. 1980. Appalachian region. AB 34:896.
[3] G.M. McWilliams, pers. recs.
[4] Scott, F.R., and D.A. Cutler. 1963. Middle Atlantic Coast region. AFN 18:23.
[5] Paxton, R.O., W.J. Boyle Jr., and D.A. Cutler. 1992. Hudson-Delaware region. AB 46:73.
[6] Carl, B. 1996. Local notes: Lancaster County. PB 9:215.

Saltmarsh Sharp-tailed Sparrow
Ammodramus caudacutus

General status: This species, formerly considered to be one of several subspecies of the Sharp-tailed Sparrow, was recognized as a separate species in 1995 (AOU 1998). Saltmarsh Sharp-tailed Sparrows breed along the Atlantic coast from Maine south to North Carolina, and they winter along the coast from New York to the eastern coast of Florida. The status of this species in Pennsylvania is uncertain, but there is at least one accepted modern record.

Seasonal status and distribution

Fall: One was collected on 3 Oct 1972 at PNR. It was originally identified as a female of the subspecies *altera*, now known as Nelson's Sharp-tailed Sparrow *(A. nelsoni)*. The bird was later reexamined by Dr. Kenneth Parkes and proved to be *A. c. caudacutus*, the Saltmarsh Sharp-tailed Sparrow.[1]

History: There is at least one specimen of this species, collected by S. Wright at Conshohocken in Montgomery Co. on 27 May 1892 (Poole, unpbl. ms.). It is now at the ANSP.

Comments: Now that Saltmarsh Sharp-tailed Sparrows have gained species status, birders are likely to pursue this bird even more. Birds captured for banding (especially those from the Tinicum area and the lower Susquehanna) should be carefully examined for this coastal species. This sparrow is not likely to ever be more than an occasional rare vagrant in Pennsylvania, as its entire life is spent along the Atlantic coast.

[1] Parkes, K.C. 1992. The subspecies of the Sharp-tailed Sparrow and the reidentification of a western Pennsylvania specimen. PB 6:13–14.

Seaside Sparrow ***Ammodramus maritimus***

General status: This nondescript species of saltwater marshes breeds along the Atlantic coast from New Hampshire south to northeastern Florida and along the Gulf coast from western Florida west to southeastern Texas. They winter along the Atlantic coast from Massachusetts south and along the Gulf coast. There are presently two accepted modern records of this species in Pennsylvania.

Seasonal status and distribution

Spring: One was found singing by B. Silfies at the Monocacy Nature Center in Bethlehem City, Northampton Co. on 28–29 Apr 1982.[1] A photograph of this bird was deposited in the files of PORC. On 14 May 1988 one was found by J. Book and T. Garner on the Conejohela Flats on the Susquehanna River in Lancaster Co.[2]

History: Poole (unpbl. ms.) mentioned two records. He wrote that a specimen was in the Reading Public Museum collection labeled "Fritz's Island, April 30, 1887—L. W. Mengel." However, he continued to write that the bird was in "worn late summer plumage" and that he suspected that "there has been some error." He also mentioned in his manuscript that one was seen at Lake Ontelaunee in Berks Co. after a severe storm on 28 Oct 1936 by A. P. Deeter and M. E. Deeter.

Comments: The most recent record, a bird discovered in Williams Township in Northampton Co. on 18–19 Oct 1997, is currently under review.[3]

[1] Boyle, W.J., Jr., R.O. Paxton, and D.A. Cutler. 1982. Hudson-Delaware region. AB 36:836.
[2] Witmer, E. 1988. County reports—April through June 1988: Lancaster County. PB 2:71.
[3] Haas, F.C., and B.M. Haas, eds. 1998. Birds of note—October through December 1997. PB 11:232.

Fox Sparrow ***Passerella iliaca***

General status: Fox Sparrows breed from Alaska across western and central Canada to Newfoundland and south through the mountains of the western U.S. They winter primarily from coastal British Columbia, south through California, across the southern states to northern Florida, and north to Missouri and east to Massachusetts. Fox Sparrows are uncommon to fairly common regular migrants across Pennsylvania, but the number of birds varies from year to year. They are accidental in summer. Fox Sparrows are winter residents, primarily in the Piedmont and Southwest regions.

Habitat: They are usually found on or near the ground in thickets, hedgerows, woodland undergrowth, and edges. In winter, most are observed at bird feeding stations that have sufficient cover nearby.

Seasonal status and distribution

Spring: Migrants may appear as early as the first spring thaw during the third or fourth week of Feb. The majority arrive the first or second week of Mar in southern locations and usually later at more northern locations and at higher elevations. Migration usually ends by the fourth week of Apr. At southern locations peak migration occurs during the fourth week of Mar or the first week of Apr. At northern locations and at higher elevations peak migration may not occur until the second or third week of Apr. Stragglers remain to the second week of May.

Summer: The two records within this period are unusual because the closest breeding area of this species is in Maritime Canada (Erskines 1992). A bird was observed on 16 Aug 1965 in Allegheny Co.,[1] and one was seen in the pine barrens near State College in Centre Co. on 1 Aug 1982.[2] These dates were too early to consider the birds fall migrants because Fox Sparrows do not normally appear in the state until the first week of Oct.

Fall: Fox Sparrows have been recorded as early as the second week of Sep. Their normal migration period is from the first week of Oct to the fourth week of Nov or the first week of Dec. The greatest number of migrants is observed from the fourth week of Oct to the second week of Nov. Many stragglers remain to at least the first or second week of Jan or through the winter. In some years, the number of birds may remain high well into Dec; 248 were counted on the Glenolden CBC on 20 Dec 1958 in Delaware Co.[3]

Winter: Records in this season are mainly during mild winters with little or no snow cover. Most birds have left the state before the third week of Jan. They are uncommon to rare regular winter residents in the Piedmont and Southwest. Across the northern-tier counties and in the High Plateau they are casual visitors or residents, with reports only from the counties of Elk, Erie, Somerset, Tioga, and Westmoreland.

[1] Hall, G.A. 1966. Appalachian region. AFN 20:45.
[2] Hall, G.A. 1983. Appalachian region. AB 37:182.
[3] Rigby, E.H. 1959. Glenolden CBC. AFN 13:119.

Song Sparrow *Melospiza melodia*

General status: Song Sparrows breed from Alaska across the southern half of Canada south in the U.S. to Arizona, Kansas, Arkansas, Tennessee, Georgia, and South Carolina. They winter along the Pacific coast from southern Alaska and Columbia south and east across the entire U.S. Song Sparrows are among the most widely distributed and common birds in Pennsylvania in all seasons. They are found year-round across most of the state but in winter are rare or absent in the higher elevations of the Ridge and Valley, High Plateau, and the snowbelt of the Glaciated Northwest. They are common regular migrants and breed in all regions of the state.

Habitat: Song Sparrows inhabit overgrown fields, brushy areas, thickets, hedges, abandoned pastures, fence rows, second-growth woodlands, and woodland edges.

Seasonal status and distribution

Spring: An increase in the number of birds is detected soon after the first thaw in the third or fourth week of Feb. In areas where few are known to winter, numbers increase usually during the first or second week of Mar or later during prolonged snow cover. Peak migration is from the first to the third week of Mar. The number of birds decreases through Apr, signaling the end of migration.

Breeding: Song Sparrows are common to abundant breeding birds statewide. They were the third most frequently reported species in the BBA[1] and are the sixth most abundant on BBS routes. The

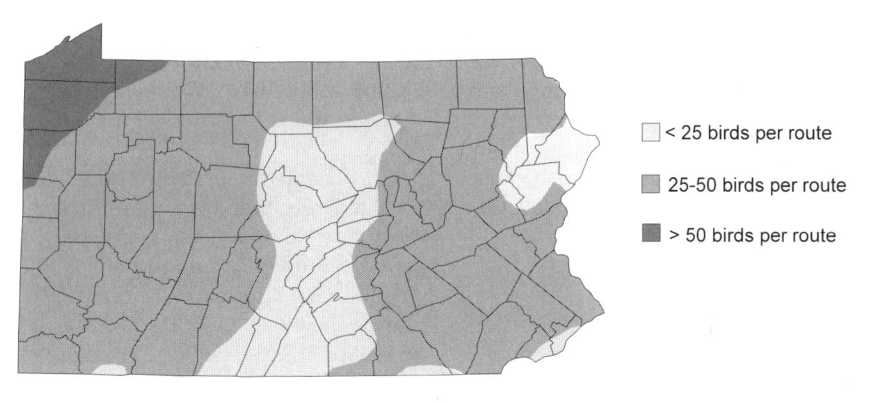

33. Song Sparrow relative abundance, based on BBS routes, 1985–1994

abundance shown on BBS routes (see Map 33) does not appear to be related to any ecological feature. Nests are placed on the ground under a shrub, often on a bank, or in a dense shrub or thicket up to several feet above the ground. The range of dates in the large number of egg sets reflects a long breeding season of two or three clutches; most clutches were found from 27 Apr to 15 Jul, but extremes were 21 Apr and 10 Aug. Later nests are less likely than early nests to be placed on the ground. State BBS data show a small decline in the population (1.6% per year).

Fall: The abundance of breeding birds makes it difficult to determine the beginning of fall migration. By the first week of Oct an increase in the number of birds is apparent. The greatest number of Song Sparrows is recorded during the second and third weeks of Oct. Their numbers decrease through Nov and Dec as cold weather approaches.

Winter: Even though they have been recorded in every county of the state during this season, many birds withdraw from northern counties and from high elevations in the south. They become increasingly rare or even absent at some sites in winter, such as on the mountain ridges of the Ridge and Valley, across the High Plateau, and in the Glaciated Northeast. They also become rare in the snowbelt area of the Glaciated Northwest and along the Lake Erie Shore. At northern sites and in the higher elevations in winter most are usually found at bird feeding stations with suitable habitat or in marshes.

History: Song Sparrows always have been considered common in Pennsylvania.

[1] Master, T.L. 1992. Song Sparrow *(Melospiza melodia)*. BBA (Brauning 1992):388–389.

Lincoln's Sparrow ***Melospiza lincolnii***

General status: Lincoln's Sparrows breed from Alaska across Canada and south through the mountainous region of the western U.S. and Minnesota east to Massachusetts. They winter along the Pacific coast from British Columbia south to California and east across the southern half of the U.S. Unlike most species of passerines, which sing enthusiastically on their northward migration, Lincoln's Sparrows rarely sing while in Pennsylvania and are often difficult to observe as they silently pass through the state. They are uncommon regular migrants over most of the state and are accidental in summer and winter.

Habitat: Lincoln's Sparrows are usually found on or near the ground in thickets, woodland edges, swamps, and wet bottomlands.

Seasonal status and distribution

Spring: The earliest arrivals have appeared in the third week of Apr. Their typical migration period is from the first or second week of May to the beginning of the fourth week of May. Most birds pass through the state rather quickly, with peak migration during the second and third weeks of May. More Lincoln's Sparrows are recorded during spring passage at PISP than at any other single site in the state; daily highs there have reached 30 or more birds. Stragglers may remain to the first week of Jun. One adult, not in breeding condition, was banded on 13 Jun 1998 near Coudersport in Potter Co.[1]

Summer: One record that falls within this period seems too early to be a migrant, but this species is not expected to breed in Pennsylvania because the closest known breeding area is in northern New York. A Lincoln's Sparrow was found singing on 24 Jul 1988 at 2300 feet elevation at Rickett's Glen State Park in Luzerne Co.[2]

Fall: Most records of Lincoln's Sparrows are during this season. However, they are not frequently observed. Birds have been reported as early as the first week of Aug, but their usual migration time is from the first or second week of Sep to the third or fourth week of Oct. Banding records suggest that peak migration occurs during the first and second weeks of Oct. At PNR daily banding highs have reached 15 or more birds. Stragglers have been recorded to the first or second week of Jan.

Winter: Apparently the only recent winter record was of a bird at a feeding station at PNR in Westmoreland Co. through the winter of 1965 (Leberman 1976).

History: The status and distribution have changed very little throughout history. They have always been considered by most authors to be rare in both spring and fall. One bird successfully survived during the winter of 1930 and remained to 19 Apr at Jeffersonville in Montgomery Co.[3]

Comments: A Lincoln's Sparrow banded at 2:25 P.M. at PNR on 15 May 1968 was recovered on 17 May 1968 at about 8:00 A.M. at Lee, Massachusetts, 350 miles away.[4]

[1] D. Hauber, pers. comm.

[2] Paxton, R.O., W.J. Boyle Jr., and D.A. Cutler. 1988. Hudson-Delaware region. AB 42:1277.

[3] Middleton, R.J. 1929–1930. General notes. Cassinia 28:38.

[4] Clench, M.H. 1968. A remarkable recovery of a banded Lincoln's Sparrow. Educational release no. 78. Powdermill Nature Reserve, Carnegie Museum of Natural History, Pittsburgh.

Swamp Sparrow *Melospiza georgiana*

General status: Swamp Sparrows breed in Canada from Mackenzie southeast through central Quebec and Newfoundland, south to Nebraska, and east to Delaware. They winter from Nebraska east through the Great Lakes region to Massachusetts, south to Florida, and west across the Gulf States to southern California. Swamp Sparrows are fairly common to common regular migrants in Pennsylvania. They breed in wetlands over most of the state but most commonly in the glaciated regions. Swamp Sparrows are winter residents primarily in the Coastal Plain and Piedmont.

Habitat: They inhabit swamps, marshes, bogs, fens, wet grassy meadows, as well as roadside ditches or almost any wet area with sedges, reeds, or marsh grasses. During migration they are occasionally observed in unusual places, often far from water, such as along fence rows, on dry hillsides, and even feeding nuthatch-like on tree trunks. In winter they prefer marshy lowlands with some open water.

Seasonal status and distribution

Spring: Migration is difficult to discern in areas where they winter in the southern portions of the state. An increase in their numbers the first or second week of Mar suggests northbound movement. At higher elevations or at northern sites, their arrival may be delayed by ice and snow cover and they may not appear until the third or fourth week of Mar or later. The greatest numbers are found from the second week of Apr to the first week of May. Migration usually ends by the second or third week of May.

Breeding: Swamp Sparrows are rare to locally common breeding birds in a variety of wetland types statewide. They are widely distributed in the Northeast and Northwest Glaciated corners of the state, where nearly half of the BBA blocks in those areas had recorded them.[1] Swamp Sparrows are likely to occur wherever emergent or shrub-scrub wetlands are found,[2] even in sites less than an acre in extent. They are rare in most south-central counties and have disappeared from Washington and Greene cos., where little wetland habitat occurs. No significant population trend has been detected because of their patchy distribution, but populations seem stable where habitat persists. The nest is a sturdy cup of grass, placed in a tussock of grass, old cattail stalks, or goldenrod, usually a foot or less above the surface. Most Pennsylvania egg sets were found in the Delaware River and Crawford Co. marshes. Swamp Sparrow nests with eggs were found from 8 May to 16 Jun, with Harlow's (1918) average on 28 May.

Fall: The number of birds probably remains high through fall migration, but after singing has stopped they become more secretive and usually remain hidden in vegetation unless provoked into view. The beginning of fall migration is difficult to discern, but banding records suggest that it begins by the third week of Sep, with most birds recorded through Oct. The number of birds falls through Nov and Dec, especially as water freezes and snow covers the ground. Stragglers remain to at least the first or second week of Jan. An unusually high count of 510 birds was reported on the Glenolden CBC in Delaware Co. on 20 Dec 1958.[3]

Winter: The number of Swamp Sparrows wintering in the state varies considerably from year to year depending on availability of open water. They are widespread across the southern half of the state at low elevations during mild winters, but during severe winters with heavy snow cover they become local and are found primarily in the Coastal Plain and the southern portion of the Piedmont. Across the northern portion of the state they are very rare and local even during mild winters. Here they have been reported only from the counties of Bradford, Erie, Susquehanna, and Tioga.

History: Historical descriptions were similar to that given above. Todd (1940) said they were "noted in virtually all the smaller swamps in the north western part of the state, north of the terminal moraine." Disappearing wetlands have reduced their distribution in places, including Greene Co., where Jacobs (1893) reported a few nests.

[1] Leberman, R.C. 1992. Swamp Sparrow *(Melospiza georgiana)*. BBA (Brauning 1992):390–391.

[2] Greenberg, R. 1988. Water as a habitat cue for breeding Swamp and Song sparrows. Condor 90:420–427.
[3] Rigby, E.H. 1959. Glenolden CBC. AFN 13:119.

White-throated Sparrow ***Zonotrichia albicollis***

General status: White-throated Sparrows breed from Yukon and Mackenzie east across Canada to Newfoundland, south to North Dakota and east to New Jersey. They winter from New York west to Kansas and south from Florida west across the southern states to California. In Pennsylvania they are best known as common to abundant regular migrants and winter residents. They also breed across the northern portion of the High Plateau and in the Glaciated Northeast.

Habitat: During migration White-throated Sparrows are found in thickets, hedgerows, brushy fields, woodlands, and woodland edges, as well as in suburban areas of towns and cities. During the breeding season they are usually around shrubby wetlands, bogs, swamps, and beaver dams in northern hardwood and hemlock forests. In winter they can be found in habitat similar to that used during migration, but they are more confined to dense thickets near water or bird feeding stations.

Seasonal status and distribution

Spring: The beginning of spring migration is difficult to detect in many places in the state, especially in the Coastal Plain and Piedmont, but a noticeable increase in numbers is usually observed there beginning the second or third week of Mar. Elsewhere in the state spring migrants are not detected until about the third or fourth week of Mar. The peak number of White-throated Sparrows over most of the state may be recorded from the third or fourth week of Apr to the second week of May but three weeks earlier in the Coastal Plain and Piedmont. Daily highs may reach more than a hundred birds at many sites. Migration usually ends by the fourth week of May. Stragglers away from their normal breeding range occasionally remain into summer.

Breeding: White-throated Sparrows are fairly common, regular breeding residents in shrubby wetlands, notably in the Poconos and at North Mountain (Sullivan and Wyoming cos.), but they also may occur in nonwetland shrubs in some locations in the Glaciated Northeast. They are uncommon and irregular in forested areas across the northern tier west to Warren Co. but are rarer to the west. White-throated Sparrows are rare Jun visitors south of this range, but they are not considered breeders far outside the Glaciated Northeast and High Plateau. Regular breeding sites at the periphery of the regular summer range include Black Moshannon State Park (Centre Co.) and White-throat Swamp (Warren Co.). Both their range and population have expanded in recent years. However, in some areas, such as the Pocono section, overbrowsing by deer has reduced their population.[1] Few nests have been discovered, and only hints of timing are available; eggs are expected in late May and Jun. The nest is placed on or near the ground under thick vegetation.

Fall: The first southbound White-throated Sparrows are detected as early as the first week of Sep. Most birds begin to pass through the state the fourth week of Sep with peak migration from the first to the third week of Oct. The number of birds may remain high until the first or second week of Nov, after which numbers gradually fall through Dec. In the southeasternmost corner of the state, some CBC tallies have reached the thousands. For example, on the Glenolden CBC 2926 White-throated Sparrows were recorded in 1957.[2]

Winter: Most winter in valleys and lowlands, especially in the Coastal Plain and Piedmont. After the first or second week of Jan, when the weather typically turns abruptly cold and blustery, they become less common, especially at higher elevations and in the snowbelt area of the Glaciated Northwest. In these regions they are most frequently found close to human habitation or near bird feeding stations. The amount of snow cover usually determines the number of individuals that winter, especially in the mountains and in the Glaciated Northwest.

History: Warren (1890) did not list White-throated Sparrows as summer residents, but males on territory were reported in Pennsylvania in 1893 (Stone 1900). A small "colony" was noted in 1904 at Pocono Lake, where the state's first nest was discovered two years later (Harlow 1913). Further study of the North Mountain documented nesting populations in Wyoming Co. Summer records reported across the northern counties (Todd 1940) match the recent distribution, but at lower numbers. Widely scattered records in summer caused Poole (unpbl. ms.) to describe the breeding distribution as "problematic."

[1] P.B. Street, pers. comm.

[2] Rigby, E.H. 1958. Glenolden CBC. AFN 12:108.

Harris's Sparrow ***Zonotrichia querula***

General status: Casual vagrants to Pennsylvania, Harris's Sparrows breed in the stunted boreal forests of Canada from Mackenzie south to northwestern Ontario. They winter primarily in the central U.S. from Nebraska and Iowa south to Texas. Harris's Sparrows are casual away from their breeding and wintering range throughout North America. Most of the records from Pennsylvania have been from the Piedmont, especially at bird feeding stations in spring and fall. Birds have remained into winter on at least five occasions. All sightings have been since the late 1940s, and all were of single birds.

Seasonal status and distribution

Spring: In addition to those that also wintered, they have appeared at various times in spring from 13 Mar to 11 May. Of the five records west and north of the Allegheny Front, two were in spring: one found by J. Nicholson and C. W. Schuck near Meadville in Crawford Co. from 29 Apr to 8 May 1957 was the first modern record for Pennsylvania.[1] One was reported at a feeder at Butler in Butler Co. from 20 Apr to

Harris's Sparrow at Marion, Franklin County, January 1990. (Photo: Franklin C. Haas)

8 May 1979.[2] East of the Allegheny Front the only spring records away from the Piedmont were of one on 6 May 1984 at SGL 252 in Union Co. (Schweinsberg 1988) and a bird that wintered and remained to 2 May 1996 at Mt. Union in Huntingdon Co.[3]

Fall: Harris's Sparrows have appeared at various times during the fall, with dates ranging from 14 Oct to 29 Dec. Away from the Piedmont they have been recorded once each in Centre, Franklin, and Philadelphia cos. West of the Allegheny Front they have been recorded once in Allegheny and Mercer cos. and twice in Erie Co.

Winter: At least five records of Harris's Sparrows are from during the winter period, with all but one from the Piedmont. In 1967 one was at Limekiln in Berks Co. from 10 to 15 Feb,[4] and one was seen for several days around 18 Mar 1969 in the same area (Uhrich 1997). In 1974 one was at Audubon in Montgomery Co. from 4 Feb to 23 Apr.[5,6] A Harris's Sparrow was at Palmyra in Lebanon Co. from 25 Jan to 15 Feb 1980,[7] and one was present at Marion in Franklin Co. from 16 Dec 1989 through Feb 1990.[8] One was at Cochranville in Chester Co. from 12 Feb into Mar 1993.[9] The only winter record outside of the Piedmont is of one that survived the winter at a feeder at Mt. Union in Huntingdon Co. from Jan through spring in 1996.[10]

History: There is a report of Harris's Sparrow from before 1957: one was in Lancaster on 2–3 Feb 1948 (Morrin et al. 1991).

[1] Hall, G.A. 1957. Appalachian region. AFN 11:345.

[2] Hall, G.A. 1979. Appalachian region. AB 33:772.

[3] Grove, G. 1996. Local notes: Huntingdon County. PB 10:101.

[4] Scott, F.R., and D.A. Cutler. 1967. Middle Atlantic Coast region. AFN 21:404.

[5] Scott, F.R., and D.A. Cutler. 1974. Middle Atlantic Coast region. AB 28:625.

[6] Scott, F.R., and D.A. Cutler. 1974. Middle Atlantic Coast region. AB 28:788.

[7] Richards, K.E., R.O. Paxton, and D.A. Cutler. 1980. Hudson-Delaware region. AB 34:260.

[8] Hall, G.A. 1990. Appalachian region. AB 44:269.

[9] Haas, F.C., and B.M. Haas, eds. 1993. Rare and unusual bird reports. PB 7:18.

[10] Grove, G. 1996. Local notes: Huntingdon County. PB 10:19.

White-crowned Sparrow ***Zonotrichia leucophrys***

General status: White-crowned Sparrows breed from Alaska across northern Canada to Newfoundland and south in the coastal areas and mountains of the western U.S. They winter from British Columbia south through the western states and Mexico and from Texas north to Nebraska and east to New Jersey and North Carolina. White-crowned Sparrows are uncommon to fairly common regular migrants throughout Pennsylvania. They are irregular winter residents over most of the state at low elevations, and most are found in the southeast.

Habitat: This sparrow prefers areas that are more open than those preferred by their close relative, the White-throated Sparrow. White-crowned Sparrows inhabit brushy areas along fences and woodland edges, weedy fields, and patches of bushes in fields (especially in open areas with multiflora rose), lawns, and roadways.

Seasonal status and distribution

Spring: Migrants begin to appear as single individuals and small groups in mid-Apr. They continue to trickle through until May, when larger flocks pass through rather quickly. Numbers usually peak during the first or second week of May. At PISP during peak migration, a single-day count of over 1000 White-crowned Sparrows is not unusual. High counts in other parts of the state rarely exceed a few dozen. Most are gone by the third week of May. Some birds linger to late May and very rarely to the second week of Jun.

Fall: The first southbound White-crowned Sparrows begin to arrive in mid-Sep, rarely late Aug or early Sep. Numbers gradually increase until they peak in early to mid-Oct, and then they rapidly decline by late Oct. Most migrants have departed by early Nov. Stragglers remain to at least the first or second week of Jan or through the winter.

Winter: In winter, White-crowned Sparrows are not expected in mountainous areas or at high elevations. However, one successfully wintered in 1989 at St. Mary's in Elk Co.[1] East of the Allegheny Front they are uncommon in winter in the Coastal Plain and Piedmont and rare in the Ridge and Valley. White-crowned Sparrows are usually found as single birds or in small numbers in winter. In the Southwest they are found in valleys, but are rare. White-crowned Sparrows are very rarely encountered in winter in the Glaciated Northeast, in the Glaciated Northwest, and at Lake Erie Shore.

History: Todd found transient White-crowned Sparrows to be fairly common in spring and more irregular in the fall (Todd 1940). Poole (unpbl. ms.) believed the reverse to be true in Berks Co. Although White-crowned Sparrows are relatively uncommon to rare in winter, they were considered to be even rarer before 1960 (Poole, unpbl. ms.).

Comments: Even today, White-crowned Sparrows seem to be more common in spring than in fall in the western portion of the state, whereas they seem to be more common in fall than in spring in the eastern portion.

[1] Hall, G.A. 1989. Appalachian region. AB 43:314.

Golden-crowned Sparrow *Zonotrichia atricapilla*

General status: This distinctive sparrow of the far West breeds from western Alaska south through British Columbia and winters along the Pacific coast from southern Alaska south through California. They are casual in eastern North America. Golden-crowned Sparrows have been reported in Pennsylvania four times, but only twice in recent history.

Seasonal status and distribution

Spring: A Golden-crowned Sparrow was observed at a feeder by B. Blust and other birders on 29–30 Apr 1979 at Paoli in Chester Co.[1]

Fall: An adult male was observed by J. Devlin at Tinicum on 12 Nov 1961.[2]

History: The first record for the state was of one captured in a banding trap by A. Conway and photographed at Easton in Northampton Co. on 3 Jun 1952. It was kept in a cage for many to see and apparently was released on 6 Jun 1952 (Poole 1964; Morris et al. 1984). One appeared during a storm at the feeder of F. Coffin on 2 Dec 1952 at Scranton in Lackawanna Co. (Poole 1964).

[1] Paxton, R.O., P.W. Smith, and D.A. Cutler. 1979. Hudson-Delaware region. AB 33:758.

[2] Scott, F.R., and D.A. Cutler. 1962. Middle Atlantic Coast region. AFN 16:18.

Dark-eyed Junco *Junco hyemalis*

General status: Dark-eyed Juncos breed from Alaska across Canada south through the west, east across the Great Lakes to Pennsylvania, and south through the Applachians. They winter across southern Canada and the entire U.S. Dark-eyed Juncos are abundant migrants and winter residents throughout Pennsylvania. Breeders are found in most of the mountainous portions of the state.

Habitat: During migration and winter, juncos are found in a broad array of habitats, including woodland undergrowth and edges, thickets, roadsides, suburban yards, and bird feeding stations in towns and cities. During the breeding season, they inhabit a variety of woodlands, especially at higher elevations.

Seasonal status and distribution

Spring: Winter residents mask the beginning of spring migration, but a noticeable increase of birds is usually observed by the third or fourth week of Mar or the first week of Apr. Peak migration occurs from the first to the third week of Apr. Daily highs of hundreds of birds may be counted at most sites. Migration usually ends by the second or third week of May.

Breeding: Juncos are fairly common to common in the forests of the northern Ridge and Valley, Glaciated Northeast, and High Plateau locally south through the Allegheny Mountains to the state line. They are rare and widely distributed in extensively forested mountains through the southern Ridge and Valley and in the northern portions of the Glaciated Northwest. The highest densities are in the High Plateau, where ravines and rock outcrops provide abundant breeding sites. No significant long-term population trends are apparent on BBS routes, but recent observations in new locations suggest an expanding breeding range. For example, the first confirmed breeding record in Lebanon Co. was in 1997.[1] The nest is a compact cup of grass, roots, and other fibers, placed on the ground, often in a vegetated road cut or stream bank or on a dirt ledge in upturned tree roots. Nests are rarely built off of the ground but may be in trees or shrubs, or in odd sites such as under the eaves of a cabin. Most nests with eggs have been found 8 May to 14 Jul, providing ample time for two clutches.

Fall: The beginning of fall migration is difficult to determine because resident populations may still be present when transient birds arrive. A noticeable increase of juncos occurs by the third or fourth week of Sep. Migration peaks from the third week of Oct to the second week of Nov. Migrants may continue to move well into Dec.

Winter: There probably is not a site in the state where juncos have not been found during this season. However, they may be scarce or even absent at high elevations and in heavily forested areas, especially when there is substantial snow cover. Juncos may be common at bird feeding stations.

History: The breeding population apparently is more common recently than when it was described in some historical accounts. Sutton (1928a) considered them rare and local in the higher mountains during summer, but Harlow's (1913) thesis describes them as widespread and sometimes abundant. Recent patterns definitely reflect an expansion of the breeding distribution, particularly into the Ridge and Valley, compared with that described as recently as the 1960s (Poole 1964).

Comments: Numerous subspecies have been described in North America, but most are not safely distinguishable in the field. The most easily identified subspecies is the Oregon Junco *(J. h. oreganus)*, formerly considered a separate species. Only the adult males are readily distinguishable in the field from other subspecies that may occur in the state. They are rare but are reported annually in the state, primarily at bird feeding stations during migration and in winter. There are fewer historical than recent records of Oregon Junco, probably because there were fewer observers and bird feeding stations rather than because of a recent eastern movement of the subspecies. Dark-eyed Juncos have been known to hybridize with White-throated Sparrows. One hybrid was collected at Haverford (Poole 1964), and another was photographed at a feeding station at Jersey Shore in

Lycoming Co.[2] A Dark-eyed Junco banded on 12 Nov 1966 at PNR was recovered on 26 Sep 1967 at Island Lake, Manitoba.[3]

[1] Miller, R. 1997. Local notes: Lebanon County. PB 11:97.
[2] P. Schwalbe and G. Schwalbe, pers. comm.
[3] Leberman, R.C. 1968. Bird-banding at Powdermill, 1967. Research report no. 22. Powdermill Nature Reserve, Carnegie Museum of Natural History, Pittsburgh.

Lapland Longspur ***Calcarius lapponicus***

General status: Lapland Longspurs breed in North America from western and northern Alaska, east across the Canadian Arctic coast and islands to Mackenzie and northern Quebec. They winter across southern Canada and the northern U.S., south through California, and from Oregon southeast to northern Texas and northeast to Maryland. The number of Lapland Longspurs observed in Pennsylvania may vary from year to year, but they are usually uncommon to rare regular migrants and winter visitors or residents. They are occasionally observed in large flocks, especially in fall. Lapland Longspurs often associate with Snow Buntings, Horned Larks, and American Pipits during migration and in winter.

Habitat: Lapland Longspurs usually are found in large open fields, either plowed or with short grass or corn stubble, and also on lakeshores. They are also observed along roadways and in fields with freshly spread manure during winters with heavy snow cover.

Seasonal status and distribution

Spring: Many spring birds probably winter in Pennsylvania, so few go farther south. A flock of 100 birds seen near Waterford in Erie Co. on 21 Mar 1976 (Stull et al. 1985) was likely northbound migrants. Most birds leave the state by the fourth week of Mar. Stragglers have been recorded to the third week of May.

Fall: Lapland Longspurs are more frequently observed in fall than in spring. The earliest arrivals may appear the third week of Sep, but the first week of Oct is more typical. Peak migration is usually recorded from the third week of Oct to the third week of Nov at the Lake Erie Shore. Away from this region, most longspurs are not reported until after the first or second week of Dec. Most sightings are of individuals or small flocks, but flocks containing 30 or more have been recorded.

Winter: The majority of Lapland Longspur sightings are in Jan and Feb in large fields of the lowland valleys, especially in the Piedmont. They have been reported infrequently in winter from the High Plateau counties of Cambria, Clearfield, and Warren. Longspurs are nearly absent in the mostly forested

Lapland Longspur at Klecknersville, Northampton County, February 1994. (Photo: Rick Wiltraut)

High Plateau. They have been known to visit bird feeding stations that adjoin their preferred habitats.

History: Some early authorities thought that this species appeared only in severe winters or during severe snowstorms, but this assumption was later proved to be untrue. Most records were from the large valleys of the Piedmont and along the Lake Erie Shore. They were considered to be fairly common and regular in the region of Lake Erie (Warren 1890) and were believed to be rare and irregular elsewhere. Some sizable flocks were reported from traditional areas, such as the flock containing 50 birds in a field along the Ohio River at Sewickley on 27 Jan 1896 (Todd 1940).

Comments: A report of five birds discovered on 14–16 Aug 1978 at Ulster in Bradford Co. was exceptionally early, but Hall stated that "the observer has sent a careful description which matches in all details."[1]

[1] Hall, G.A. 1979. Appalachian region. AB 33:178.

Snow Bunting ***Plectrophenax nivalis***

General status: Snow Buntings breed from northern Alaska east across the Canadian Arctic coast and islands to Mackenzie and Labrador. They winter from Alaska across southern Canada and the northern half of the U.S. Snow Buntings are regular migrants and winter residents over most of Pennsylvania, but they are scarce or absent in heavily forested areas. The number of birds observed in the state varies considerably from year to year; flock sizes may range from a few birds to several thousand.

Habitat: Snow Buntings inhabit large open fields with short grass, plowed fields, corn stubble and lakeshores. They often are found in fields with freshly spread manure and along roadways during winters with heavy snow cover. During fall migration, they are regularly found on the open beaches of PISP and migrating over hawk-watch sites along mountain ridges.

Seasonal status and distribution

Spring: Snow Buntings may appear any time between the third or fourth week of Feb and the second or third week of Mar. During some years, flocks in the hundreds or thousands have been recorded in the Glaciated Northwest and the Lake Erie Shore during this period, especially in Feb, perhaps indicating a northward influx of migrating birds. Elsewhere across the state, notable counts include 3000 in a sorghum field in northern Berks Co. in late Feb 1978[1] and 2000 at Harvey's Lake in Luzerne Co. on 15 Feb.[2] Some of these flocks are probably winter residents concentrated at food sources. Snow Buntings leave the state by the second or third week of Apr. Stragglers have appeared for a day or so during the first and second weeks of May.

Fall: Earliest arrivals have been recorded the fourth week of Sep. Birds regularly appear with the passage of the first cold front in the last week of Oct or the first week of Nov, especially along the Lake Erie Shore. Southern movements of birds usually correspond with the passage of cold fronts during Nov and Dec and occasionally to the first or second week of Jan.

Winter: As winter progresses Snow Buntings become quite nomadic in their search for food, wandering widely from field to field. They have even been known to visit feeding stations.

History: Their general status has changed little since the beginning of the twentieth century. They were just as variable in their occurrence then as they are now. Todd (1940) mentioned that he occasionally found flocks of Snow Buntings containing several thousand birds in western Pennsylvania, but flock sizes mostly contained far fewer.

Comments: Broun (1949) noted the unusual behavior of single Snow Buntings landing on exposed rocks as they passed the lookouts at Hawk Mountain during fall migration. This behavior is occasionally observed today. Snow Buntings have also been observed resting briefly in trees and on utility wires. It is likely that this species will continue to find fewer places to rest and winter in the state as development encroaches on farmland.

[1] Buckley, P.A., R.O. Paxton, and D.A. Cutler. 1976. Hudson-Delaware region. AB 30:702.
[2] Paxton, R.O., P.A. Buckley, and D.A. Cutler. 1978. Hudson-Delaware region. AB 32:331.

Family Cardinalidae: Cardinals, Grosbeaks, Buntings, and Dickcissels

Eight species in this family are found in Pennsylvania. Five breed in the state, and the other three are rare visitors. This group includes colorful small to medium-sized birds with a heavy finchlike bill. Many of the species, such as the Blue Grosbeak and Indigo and Painted buntings, are characterized by striking sexual dimorphism. Their diet consists of various fruits and seeds, and they may visit bird feeding stations. Most build nests of plant fibers, plant stems, twigs, leaves, rootlets, bark strips, and hair placed in vine tangles, shrubs, or small trees. Three to four white, pale blue, or green eggs are laid. Northern Cardinals and Rose-breasted Grosbeaks lay eggs that are spotted or lightly speckled with brown or reddish brown.

Northern Cardinal *Cardinalis cardinalis*

General status: Cardinals are resident in Canada from southern Ontario and Quebec south across the eastern U.S. to Florida and west to the Dakotas in the north and across Texas to Arizona in the south. Formerly considered strictly limited to southern counties, Northern Cardinals have spread across Pennsylvania by at least the 1950s. They are now considered to be common permanent residents over most of the state.

Habitat: Cardinals are found in woodland edges, thickets, hedgerows, brushy swamps, brush piles, pine plantings, and vine tangles. They may be locally common in suburban and urban areas where such habitats are found, especially near bird feeding stations.

Seasonal status and distribution

Breeding: Cardinals are common permanent residents in western Pennsylvania and the Piedmont and uncommon in the extensively forested regions of the High Plateau (see Map 34), where they

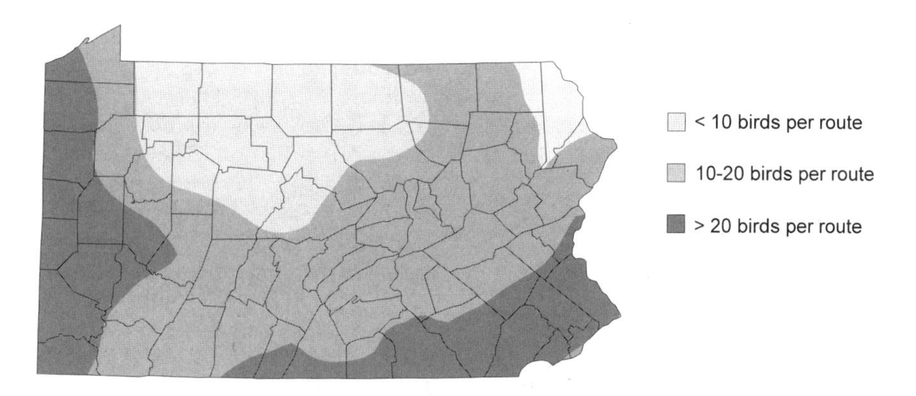

34. Northern Cardinal relative abundance, based on BBS routes, 1985–1994

are generally found in the valleys. Over the rest of the state they are fairly common. Cardinals are reported on all BBS routes in the state, with a long-term average of 13 birds per route. Their long nesting period reflects at least two broods. Egg sets range from 10 Apr to 8 Aug; most eggs are found from 22 Apr to 4 Jul. Nests are bulky and loosely built, are usually placed 2–6 feet from the ground, and are concealed in shrubs or tangled vines. The BBS shows no significant population changes. The cardinal's substantial range expansion occurred before the 1960s.

Winter: During this season they may become concentrated around bird feeding stations or natural sources of food where suitable cover is available nearby.

History: Northern Cardinals were restricted to the Carolinian zone (area consisting of birds that typically breed in the southeastern U.S.) of the southeastern and southwestern counties in the nineteenth century. Baird (1845) considered them "very rare" around Carlisle (Cumberland Co.). They had expanded north into the Ridge and Valley (Harlow 1913) and to Crawford Co. (Sutton 1928b) by the early twentieth century. By 1960 cardinals had "extended their range over practically the entire state" (Poole 1964).

Comments: Their increased activity under the cover of dawn or dusk, especially in winter when there is a complete covering of snow, may be a protective behavior against diurnal predators.

Rose-breasted Grosbeak ***Pheucticus ludovicianus***

General status: Rose-breasted Grosbeaks breed in Canada from British Coulmbia and Mackenzie east to Newfoundland and south through the Dakotas to Nebraska, east to Delaware and in the Appalachians south to Georgia. They winter from Mexico south to South America and occasionally within their breeding range. In Pennsylvania they are fairly common to common regular migrants. They breed over most of the state. They are less common and more local as breeders in the western half of the Ridge and Valley and in the Piedmont than in other areas of the state. Numbers of birds may fluctuate from year to year. Rose-breasted Grosbeaks are accidental in winter.

Habitat: They prefer second-growth deciduous or mixed woodlands or woodlands with nearby openings such as along swamps, utility cuts, clearcuts, or roadways. They may be found in parks or in woodlots in suburban areas.

Seasonal status and distribution

Spring: The earliest arrivals may appear by the third week of Apr. Their usual migration time is the fourth week of Apr or the first week of May to the fourth week of May. Peak migration usually occurs during the second and third weeks of May. They are most easily observed early in migration before trees and shrubs have leafed out. During prolonged cold periods migrating Rose-breasted Grosbeaks may concentrate at bird feeding stations. An example occurred at a feeding station on 7 May 1984 in Bucks Co., where an unusually high concentration of 38 birds was counted.[1]

Breeding: Nesting statewide, Rose-breasted Grosbeaks are most common in the High Plateau. They are fairly common across the northern counties south through the Alleghenies, uncommon in the northern Ridge and Valley areas and the Southwest region, and irregular and rare in southern counties, particularly in the lower Susquehanna River drainage (see Map 35). They nested in Delaware Co., at Ridley Creek State Park, in 1995[2] and in Adams Co. in 1997,[3] the only counties

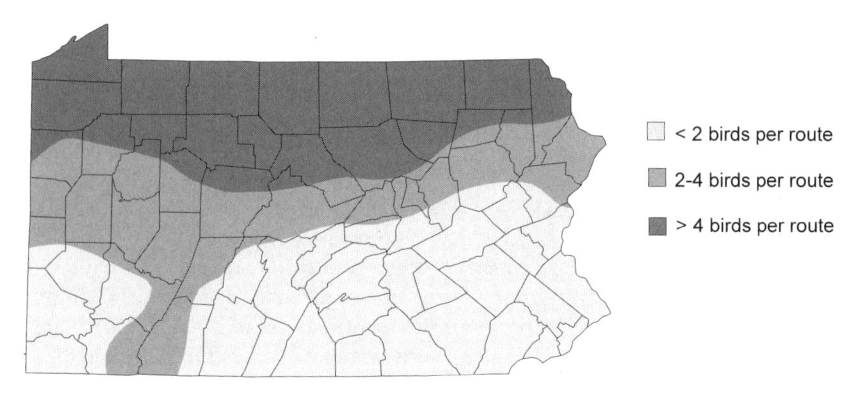

35. Rose-breasted Grosbeak relative abundance, based on BBS routes, 1985–1994

lacking solid breeding reports during the BBA project.[4] Their population and distribution are notoriously variable, as reflected in an increase on the BBS through the 1980s and a decline during the 1990s. The nest is a flimsy cup, loosely constructed and usually placed in the fork of a small tree 10–15 feet from the ground (H. H. Harrison 1975). Most nests with eggs were found from 15 May to 10 Jun. Three to six pale gray, green, or blue eggs spotted or blotched with brown or purple (often wreathed or capped at the large end) are laid.

Fall: Usually fewer birds are seen during fall migration than during spring migration, although banding records indicate that more are passing through than are observed. They become more difficult to observe after singing has stopped in Jul. Some movement is detected as early as the third or fourth week of Jul, but most southbound birds are probably not moving until the third or fourth week of Aug. The greatest number of migrants is recorded during the second and third weeks of Sep. Most migrants have left the state by the second week of Oct. Stragglers have been recorded as late as the second week of Jan. A few have been recorded well into winter.

Winter: They have been recorded in winter at least five times. Apparently the first bird to be recorded during this period in Pennsylvania was a female visiting a feeder at Waterford in Erie Co. from 7 to 9 Feb 1967.[5] A Rose-breasted Grosbeak successfully overwintered at Laverock in Montgomery Co. during the winter of 1976,[6] and still another bird was at Safe Harbor in Lancaster Co. in mid-Feb 1975.[7] One wintered at a feeder to at least early Feb 1988 at Waterford in Erie Co.[8] At Mosertown in Crawford Co. one was found on 28 Feb 1989.[9] An immature male wintered at a feeder near Drummond in Elk Co. to at least mid-Feb 1993.[10]

History: The changeable status of Rose-breasted Grosbeaks is suggested by early reports: B. Everhart called them a "rather common summer resident" in Chester and Delaware cos. in the 1860s, but Warren (1890) called them rare in that area. At the turn of the twentieth century some observers considered these grosbeaks to be abundant summer residents in Clarion, Franklin, and Venango cos. (Harlow 1913), where they probably responded to the growth of fruit-producing shrubs after the extensive timbering of the last century. Todd (1940) noted that the Rose-breasted Grosbeak was "subject to shifting fluctuations in its numbers" and noted that it did not nest in Greene Co., where they were later considered fairly common.[11]

Comments: All out-of-season Rose-breasted Grosbeaks, especially those in late fall and winter, should be carefully examined for the western species, Black-headed Grosbeak, which has also been recorded in late fall and winter. Females of the two species closely resemble one another and may be difficult to separate in the field. A Rose-breasted Grosbeak banded on 22 May 1971 at PNR was shot on 7 Feb 1972, near Urrao, Dept. Antioquia, Colombia.[12]

[1] Boyle, W.J., Jr., R.O. Paxton, and D.A. Cutler. 1984. Hudson-Delaware region. AB 38:889.
[2] Guarente, A. 1995. Local notes: Delaware County. PB 9:92.
[3] Kennell, E., and A. Kennell 1997. Local notes: Adams County. PB 11:91.
[4] Leberman, R.C. 1992. Rose-breasted Grosbeak *(Pheucticus ludovicianus)*. BBA (Brauning 1992): 362–363.
[5] Hall, G.A. 1967. Appalachian region. AFN 21:461.
[6] Buckley, P.A., R.O. Paxton, and D.A. Cutler. 1976. Hudson-Delaware region. AB 30:701.
[7] E. Witmer, pers. comm.
[8] Hall, G.A. 1988. Appalachian region. AB 42:266.
[9] J. Barker, pers. comm.
[10] L. Christenson, pers. comm.
[11] R. Bell, pers. comm.
[12] Leberman, R.C., and M.H. Clench. 1973. Birdbanding at Powdermill, 1972. Research report no. 31. Powdermill Nature Reserve, Carnegie Museum of Natural History, Pittsburgh.

Black-headed Grosbeak *Pheucticus melanocephalus*

General status: This species breeds in the West from British Columbia south to New Mexico and east to South Dakota, Kansas, and Texas. Black-headed Grosbeaks winter primarily in Mexico. They have strayed into Pennsylvania on at least 12 occasions. All but one of the records are recent. Most have been at bird feeding stations during 1960–1962, 1969–1970, 1972, 1974, 1976, 1979, 1987, and 1995–1996.

Seasonal status and distribution

Fall through spring: All sightings are of single birds that have remained for several days or weeks; several have been documented with photographs. Observations range from at least 12 Nov to 10 May. Two birds successfully wintered in the state: one at Falls in Wyoming Co. until at least 24 Apr 1974[1] and one adult at Wexford in Allegheny Co. until 7 May 1996.[2] They have been recorded in Allegheny, Berks, Chester, Clinton, Crawford, Delaware, Philadelphia, Schuylkill, Wyoming, and York cos.

History: There was only one record in Pennsylvania before 1960. In Upper Merion Township in Montgomery Co. a female seen from 15 to 25 Nov 1952 was identified by D. A. Cutler and H. Cutler.[3]

Comments: Caution is advised in separating the female Black-headed Grosbeak from the female Rose-breasted Grosbeak.

[1] Kane R.P., and P.A. Buckley. 1974. Hudson St. Lawrence region. AB 28:783.
[2] Haas, F.C., and B.M. Haas, eds. 1996. Birds of note—April through June 1996. PB 10:91.
[3] Cutler, D.A., and H. Cutler. 1953. Black-headed Grosbeak in Montgomery Co., Pennsylvania. Cassinia 1951–1952:27.

Blue Grosbeak *Guiraca caerulea*

General status: Blue Grosbeaks breed primarily across the southern two-thirds of the U.S. They winter from Mexico to Central America. Blue Grosbeaks formerly bred only in the southeastern and south-central portion of the state, but they have been rapidly expanding north and west and recently have been confirmed nesting in the southwestern portion of the state. They are rare and irregular vagrants outside of their breeding range.

Habitat: Blue Grosbeaks are found in open grassy fields interspersed with brush or scattered trees and along woodland edges and brushy fence rows.

Seasonal status and distribution

Spring: Most migrants away from their normal breeding range have either overflown their breeding territory or perhaps are pioneering males in search of new breeding sites. They may appear any time from the first week of May to the second or third week of Jun. Birds have been recorded outside their normal range as far north as Wyoming Co. in the northeastern, Centre Co. in the central, and Crawford Co. in the northwestern portions of the state. Birds may first arrive on their breeding grounds as early as the fourth week of Apr, but most arrive the first week of May. An exceptional report for the Allegheny Mountains was of three birds in Somerset Co. on 20 May 1976.[1]

Breeding: Since the 1960s the Blue Grosbeak's range has expanded into Fulton Co. and across the Piedmont, where they have become regular, although still rare to uncommon. They may be found in any county in the Piedmont and are well established in southern Lancaster Co., around the Philadelphia International Airport, at Tinicum in southeastern Delaware Co., and in the valleys of southern Fulton Co. This grosbeak is casual to accidental north into the Ridge and Valley. They successfully nested as far north as Northampton Co. in 1996, where two singing males were present in 1997.[2] The first recent summer record in western Pennsylvania was in 1992, when a pair was observed in Westmoreland Co.[3] Nesting was confirmed west of the Allegheny Front for the first time in 1995 at Imperial, Allegheny Co.[4] and again in that same area in 1996 and 1997. The nest is a neat cup placed low in a shrub. Three to five white unmarked eggs are laid.

Fall: Very little is known about the movements of Blue Grosbeaks during this season. Most birds probably leave the state by the second or third week of Sep, but they have been reported as late as the first week of Oct. An unusual location and date were reported when one bird was seen at State College, Centre Co., on 18–19 Nov 1961.[5]

History: The Blue Grosbeak is not a recent addition to the list of birds of southeastern Pennsylvania. Baird (1845) listed it as a rare nester in the Carlisle (Cumberland Co.) area. Blue Grosbeaks were considered very rare and irregular summer residents across the southern counties until the end of the 1800s (Warren 1890). The persistence of spring and summer records suggested that breeding probably occurred irregularly through the early twentieth century, although Poole (1964) reported no nests or eggs until about 1960. A remarkably extralimital summer record from western Pennsylvania was of a pair at Sugar Lake, Crawford Co., on 10 Jul 1959.[6] A very early Blue Grosbeak appeared at a feeding station in Philadelphia on 12 Mar 1952.[7]

[1] Hall, G.A. 1976. Appalachian region. AB 30:843.

[2] R.Wiltraut, pers. comm.

[3] Leberman, R.C., and R.S. Mulvihill. 1992. County reports—April through June 1992: Westmoreland County. PB 6:93.

[4] Floyd, T. 1995. Local notes: Allegheny County. 1995. PB 9:89.

[5] Hall, G.A. 1962. Appalachian region. AFN 16:33.

[6] Leberman, R.C. 1959. Field notes. Sandpiper 2:14.

[7] Miller, J.C. 1953. General notes. Cassinia 1951–1952:34.

Lazuli Bunting *Passerina amoena*

General status: Lazuli Buntings breed across southern Canada from British Columbia east to Saskatchewan, south to Oklahoma, and west to California. They winter from southern Arizona to Mexico. One record of this bunting has been accepted in Pennsylvania.

Seasonal status and distribution

Winter through spring: An adult male Lazuli Bunting identified at a feeder by R. E. Cook and colleagues from 1 Jan to 16 Mar 1975 near Elverson in Chester Co. was documented with a photograph.[1]

[1] Scott, F.R., and D.A. Cutler. 1975. Hudson-Delaware region. AB 29:675.

Indigo Bunting ***Passerina cyanea***

General status: Indigo Buntings breed primarily across southern Canada from Saskatchewan east to Nova Scotia, south through the entire eastern half of the U.S., and west to New Mexico. They winter in the West Indies, Mexico, and Central America. Indigo Buntings are fairly common regular migrants, and they are breeding residents throughout Pennsylvania. They are accidental in winter.

Habitat: Indigo Buntings are found in woodland edges, brushy fields, thickets, and second-growth woodlands with clearings, such as along power line corridors and roadsides. They frequently sing from high perches such as exposed treetops and utility wires.

Seasonal status and distribution

Spring: Birds are usually observed only on their breeding grounds, so migration in progress is most easily detected at PISP, where they do not breed. Indigo Buntings have been recorded as early as the second week of Apr. Their typical migration period is from the fourth week of Apr or the first week of May to the fourth week of May. At PISP a noticeable increase in birds occurs during the second week of May, suggesting a peak migration time. Elsewhere, groups of migrants may concentrate at feeding stations during cold snaps in late Apr and May.

Breeding: Indigo Buntings are among the most frequently detected roadside birds in rural areas during the breeding season. They are fairly common to common breeding residents statewide in every county. Highest numbers were found on BBS routes in the southern mountains, and generally lower numbers were found in eastern counties (see Map 36). The BBS shows a stable population statewide. The nest is a tight cup on a base of leaves, usually placed 1–3

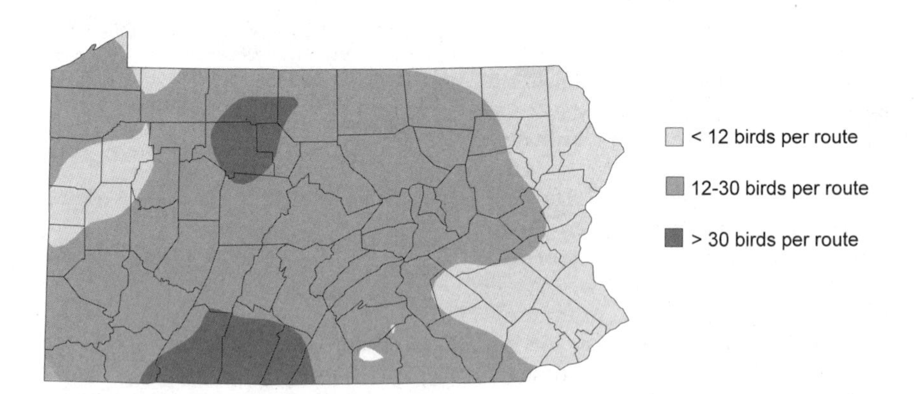

36. Indigo Bunting relative abundance, based on BBS routes, 1985–1994

feet from the ground in the crotch of a sapling. Most nests with eggs have been found from 24 May to 9 Jul, but late nests, probably second clutches, have been found until Aug 3. Usually three or four unmarked white eggs are laid. Unlike most species of birds, which stop singing after mid-Jul, Indigo Buntings continue to sing throughout the summer until at least late Aug.

Fall: Their wide distribution and abundance in the state make the beginning of migration difficult to detect by birders. Banding records suggest that migration is well under way by the third or fourth week of Aug. Peak migration occurs during the third and fourth weeks of Sep. Migration continues to the second or third week of Oct, when large concentrations, containing 50 or more birds, may be found in weedy fields and brushy habitats.[1] Stragglers have been recorded to the end of Dec.

Winter: Three records of Indigo Buntings fall within this season. One visited a feeding station from 8 Dec 1977 to 14 Jan 1978 in Drumore Township in Lancaster Co. (Morrin et al. 1991); an adult male was at a feeder at Devon in Chester Co. from 14 Jan into Apr 1998;[2] and an immature male was banded at PNR in Westmoreland Co. on 2 Feb 1995.[3]

History: They were frequently listed as common or abundant in historical accounts, except in areas of extensive forest. A very early Indigo Bunting was found on 7 Apr 1909 at Westtown in Chester Co. (Conway 1940), and an immature bird was captured and banded at a feeder in Hatboro in Montgomery Co. This bird was captured on 20 Dec 1955 and remained to mid-Mar 1956.[4]

[1] F.C. Haas and B.M. Haas, pers. comm.

[2] P.B. Street, pers. comm.

[3] Leberman, R.C. 1995. Local notes: Westmoreland County. PB 9:28.

[4] Cutler, D.A. 1957. Indigo Bunting at Hatboro, Pa. in winter. Cassinia 42:24.

Painted Bunting *Passerina ciris*

General status: Painted Buntings breed from New Mexico east to Alabama and north to Missouri and along the Atlantic coast in North and South Carolina. They winter from southern Florida to Central America. There are at least nine recent and four historical records of this species in Pennsylvania, most of which are from the southeastern corner.

Seasonal status and distribution

Fall through spring: The first recent record was of a male found by E. L. Poole and colleagues along Wyomissing Creek near Reading on 13 May 1961.[1] A male was discovered at Pennsbury Manor in Bucks Co. on 15 Jun 1966 (Poole, unpbl. ms.), and a male was at a feeder on 12 Dec 1966 and remained through the winter but was found dead on 10 Apr 1967 at West Bethlehem in Northampton Co.[2,3] Uhrich (1997) cited a male in Muhlenberg Township, Berks Co., from Oct 1971 to mid-Apr 1972. Painted Buntings were not recorded again until 1983, when a male visited a feeder at Trappe in Montgomery Co., from 5 Jan to 10 Feb.[4] In 1993 three different birds were reported. One visited feeders at Broomall in Delaware Co. from 11 Feb to 31 Mar;[5] an adult male was at a feeder at Fairview in Erie Co. from 20 to 27 Apr and was photographed;[6] and another adult male was at West Chester in Chester Co. from 10 to 12 May and also was photographed.[7] A female or immature was at Monroeville in Allegheny Co. on 7 Oct 1995.[8] Two birds were recorded in 1996: one male at Point Phillip in Northampton Co. on 27 and 28 Apr was photographed,[9] and a second-year bird was banded at PNR in Westmoreland Co. on 5 May.[10]

History: Todd (1940) cited a record of

a male Painted Bunting that was collected in 1883 or 1884 near Donohoe, west of Latrobe in Westmoreland Co., now believed to be an escaped caged bird.[11] One appeared at Wyncote in Montgomery Co. with Indigo Buntings on 13–14 May 1916 (Poole, unpbl. ms.); an adult male was found near Mercersburg in Franklin Co. on 16 May 1921 (Poole 1964); and one was found at the mouth of Codorus Creek in York Co. on 27 May 1946 (Poole 1964).

Comments: This species has been known to be kept in captivity, so historical records, such as the Donohoe report, are suspect of being escapees. On the basis of the condition of the plumage and timing, the records listed here are believed to have been of birds of wild origin.

[1] Scott, F.R., and D.A. Cutler. 1961. Middle Atlantic Coast region. AFN 15:399.
[2] Scott, F.R., and D.A. Cutler. 1967. Middle Atlantic Coast region. AFN 21:404.
[3] Scott, F.R., and D.A. Cutler. 1967. Middle Atlantic Coast region. AFN 21:494.
[4] Boyle, W.J., Jr., R.O. Paxton, and D.A. Cutler. 1983. Hudson-Delaware region. AB 37:287.
[5] N. Pulcinella, pers. comm.
[6] Haas, F.C., and B.M. Haas, eds. 1993. Rare and unusual bird reports. PB 7:60.
[7] Blust, B. 1993. Chester Co. report. PB 7:60.
[8] Pulcinella, N. 1997. Eighth report of the Pennsylvania Ornithological Records Committee, June 1997. PB 11:130.
[9] Wiltraut, R. 1996. Northampton Co. report. PB 10:103.
[10] Leberman, R.C. 1996. Westmoreland Co. report. PB 10:106.
[11] K. Parkes, pers. comm.

Dickcissel *Spiza americana*

General status: Dickcissels breed primarily in Canada from Saskatchewan south to Texas and east to Georgia and the Appalachian region. They winter primarily from Mexico to South America with small numbers occasionally remaining within their breeding range. Dickcissels are irregular rare visitors in all seasons in Pennsylvania. Most records are during migration and in winter. The breeding status of this species is unpredictable, and the number of individuals may vary considerably from year to year. During invasion years, Dickcissels may appear in widely scattered locations, especially in recently reclaimed strip mines.

Habitat: During migration Dickcissels may be found in grassy fields, but most often they are seen at bird feeding stations. During the breeding season, they inhabit large grassy fields such as hayfields or strip mines recently reclaimed in grass. In winter Dickcissels are found at bird feeding stations near shrubs, thickets, or hedgerows.

Dickcissel at Volant, Lawrence County, July 1994. (Photo: Walter Shaffer)

Seasonal status and distribution

Spring, fall, and winter: They may appear almost any time, anywhere in the state during these seasons, but most records are of individuals at bird feeders in spring and winter. Dickcissels are occasionally found during migration among mixed flocks of sparrows. Rarely are more than one or two birds reported in the state in any given year during these seasons, even after an invasion of birds during the breeding season. They seem to disappear soon after they have nested or when they have been forced out by crop harvesting.

Breeding: Dickcissels have been reported widely across western and southern Pennsylvania. Summer records have come from more than 23 counties since 1960. They were listed as extirpated until a pair nested in a reclaimed strip mine in Clarion Co. in 1983 (Genoways and Brenner 1985). Eastern invasions of Dickcissels may be a result of drought conditions in the Midwestern states. The most dramatic irruption occurred in 1988, when Dickcissels were recorded in 10 counties.[1] A lesser invasion was noted in 1996, with summer records in nine southern counties. Most locations are occupied for only one year. However, Dickcissels have been reported on territory in six of nine years between 1988 and 1996 in the Cumberland Valley of Franklin Co. and have occurred almost annually during the 1990s in neighboring Cumberland Co. and southern Adams Co. The nest is a bulky cup of grass concealed on the ground. From three to five pale blue unmarked eggs are laid. Confirmed breeding records are scant but have come from Adams, Clarion, Franklin, Somerset, Westmoreland, and York cos.[2]

History: In sharp contrast to recent patterns, Dickcissels were considered locally common to abundant in nineteenth-century accounts of southeastern and southwestern Pennsylvania. For example, they were simply listed as "plentiful" from May to Sep by Turnbull (1869). Dickcissels suddenly and inexplicably became rare about 1880, although nesting was confirmed for a few years later.[3] Agriculture, which had originally created habitat, underwent dramatic changes that became unfavorable for this and other grassland birds. After an invasion in 1928 they briefly reoccupied some of their historical range but were only occasionally reported in summer from then until the 1988 incursion. The scant historical records from western Pennsylvania were thoroughly documented in Mulvihill's detailed article[1] and in his BBA account.[2]

Comments: The PGC listed the Dickcissel as threatened in 1999. Female or immature Dickcissels closely resemble female House Sparrows. A few are probably overlooked when they are feeding together.

[1] Mulvihill, R.S. 1988. The occurrence of Dickcissels *(Spiza americana)* in western Pennsylvania during the 1988 nesting season—Its possible bearing on the species' unusual history in eastern North America. PB 2:83–87.

[2] Mulvihill, R.S. 1992. Dickcissel *(Spiza americana)*. BBA (Brauning 1992):370–371.

[3] Rhoads, S.N. 1903. Exit the Dickcissel—A remarkable case of local extinction. Cassinia 7: 17–28.

Family Icteridae: Blackbirds and Orioles

Twelve species from the family Icteridae have been recorded in the state, and all but four rare vagrants breed here. They all are medium-sized birds. Most have a long, sharply pointed, black or fairly long conical bill. The group includes an interesting range of species, from brightly colored orioles, to dark grackles and blackbirds, and melodic meadowlarks and Bobolinks. The blackbirds are known for gathering in huge flocks, seen in migration or at winter roosts. They have benefited from the farmers' plantings of grains and are considered to be agricultural pests as well as unwanted guests in suburban areas. Even though they are federally protected, they have been shot and poisoned in attempts to reduce their populations. Cowbirds parasitize the nests of many species of songbirds. There seems to be a species of blackbird to fill every niche in the state except the heavily forested areas. Meadowlarks and Bobolinks are birds of open fields, where they build their nests on the ground. In recent years, they have suffered from loss of habitat because of changes in farming practices and because of human encroachment. Orioles are the most colorful and perhaps the best songsters of this group and spend most of their time in the tree canopy. The two species of oriole found in the state also breed here. Baltimore Oriole, the better known of the two breeding orioles in the state, makes a unique nest that most people recognize.

Bobolink Dolichonyx oryzivorus

General status: Bobolinks breed across southern Canada and most of the northern half of the U.S. They winter in South America. Bobolinks are common to abundant regular migrants over most of Pennsylvania. They usually do not mix with other flocks of migrating blackbirds. They breed across much of the state but are common and widely distributed primarily in the state's Glaciated corners, in the Allegheny Mountain section, and in the Southwest. Bobolinks have benefited from the conversion of forests into farmland and from mining reclamation practices that have created the open grassy fields that they use during the breeding season.

Habitat: Bobolinks are found in large open grassy fields, meadows, or marshes during migration. During the breeding season they inhabit large open grassy fields, especially hayfields with tall lush grass or strip mines reclaimed in grass.

Seasonal status and distribution

Spring: The earliest arrivals have appeared by the third week of Apr. Their typical migration period is from the fourth week of Apr or the first week of May to the first week of Jun. Migration peaks during the second and third weeks of May. Migration is conspicuous along the Lake Erie Shore in spring. During southerly winds, flocks containing from 25 to 100 birds continually pass over PISP during peak migration.[1] Stragglers may remain to the first week of Jun.

Breeding: Bobolinks have a nearly statewide breeding distribution, ranging from rare and irregular through much of the south to common in the northeastern and western counties, including the mountains to Somerset Co. They are spottily distributed through most of the Piedmont and Ridge and Valley but are regular at MCWMA and in hayfields of the northeastern portion of the Piedmont. The BBS shows no population trend in Pennsylvania, in contrast to the declines noted for most other grassland species. The nest is a loose cup of grass hidden in the vegetation on the ground. Eggs have rarely been found, but most are collected during a brief period from 14 May to 15 Jun. Usually five to six buff or reddish brown eggs, heavily spotted or blotched with dark brown, are laid. They apparently have only one brood. Bobolinks collect in flocks and leave nesting areas in Aug.

Fall: The beginning of migration is difficult to discern in areas where they breed, but in the Coastal Plain, where they do not breed, migration is clearly defined. In the marshes of Tinicum, where large flocks gather, fall migrants may be first detected as early as the fourth week of Jul. Their usual migration period is from the third week of Aug to the first week of Oct. In the Coastal Plain the greatest number of migrants is found from the fourth week of Aug to the third week of Sep. Flocks of Bobolinks numbering in the thousands can be seen daily flying past the mouth of Darby Creek in Delaware and Philadelphia cos. in the fall.[2] Elsewhere in the state the flights are less spectacular, but occasionally concentrations may number into the hundreds. Peak migration usually occurs during the fourth week of Aug and the first week of Sep in the northern portion of the state and west of the Allegheny Front. Stragglers may remain until the third week of Oct. There are two later-than-usual sightings of Bobolinks: one was at Washington Boro in Lancaster Co. on 27 Nov 1994,[3] and an unusually late bird was found on a CBC in Delaware Co. on 20 Dec 1986.[4]

History: Bobolinks have had a long history of conflict with humans. Libhart (1869) stated that they were "eagerly sought by the gunner, and highly prized by the epicure." Identification features took a different twist with Warren's (1890) detailed description of the headless and pickled carcass, a description that lacked only features of flavor. The most extreme persecution came in the southern states, where Bobolinks decimated rice fields. Warren (1890) said, "to prevent total destruction of the crop . . . thousands of men and boys . . . are employed, hundreds of thousands of pounds of gunpowder are burned, and millions of birds are killed." A lesson might be learned from this experience; he went on to say, "still the numbers of birds invading the rice fields each year seems in no way diminished." Bobolinks were widely distributed across Pennsylvania at the end of the 1800s in much the same pattern as is seen recently. They bred in a few scattered locations of southeastern Pennsylvania, including the Philadelphia area, but were "far more numerous in counties of the western and northern parts of the state" (Warren 1890). Significant changes have occurred over time. Poole (unpbl. ms.) stated that they were "most common in northern Pike County," where habitat has now reverted back to forest. Large numbers formerly passed through the Tinicum marshes. An estimated 25,000 were seen there on 1 Sep 1945 (Poole, unpbl. ms.); reduced habitat contributed to making recent counts an order of magnitude smaller.

[1] G.M. McWilliams, pers. obs.
[2] N. Pulcinella, pers. comm.
[3] Heller, J. 1995. Notes from the field: Lancaster County. PB 8:225.
[4] Haas, F.C. 1987. Glenolden CBC. AB 41:778.

Red-winged Blackbird ***Agelaius phoeniceus***

General status: Red-winged Blackbirds breed from Alaska across the southern half of Canada and south throughout the U.S., where they also winter. These abundant migrants are regularly found in flocks of thousands during spring and fall passage and during the winter in the Southeast. Red-winged Blackbirds nest statewide. They winter over most of the state except in the higher elevations of the upper and central High Plateau. Red-winged Blackbirds frequently associate with other species of blackbirds both during migration and in winter.

Habitat: They can be seen almost anywhere during migration, especially in areas of extensive agriculture. During the breeding season they are found in a wide variety of open habitats, including grassy or brushy fields, but they are most abundant in shrub-swamps and marshes with emergent vegetation. In winter, they roost in woodlots or rest in scattered trees but feed at grain fields, especially where there is freshly spread manure, as well as in marshes and at bird feeding stations.

Seasonal status and distribution

Spring: About the third or fourth week of Feb, flocks containing several thousand birds begin to migrate north across Pennsylvania. During mild winters some birds may start moving north as early as the last week of Jan. The greatest number of birds is usually associated with warm fronts, primarily in Mar and Apr. Large flocks, consisting of mostly females and first spring males, may continue to be seen migrating along the Lake Erie Shore until the first or second week of May, when most birds are on territory across the rest of the state.

Breeding: Red-winged Blackbirds are common to abundant summer residents statewide. They were not among the top 10 most frequently reported birds in the BBA project[1] because of their absence from extensively forested areas of the High Plateau. But Red-winged Blackbirds are reported on almost all BBS routes at an average of more than 60 birds per BBS route, the fourth highest species in Pennsylvania in terms of abundance. They are most abundant in northern and western agricultural and strip-mine areas (see Map 37). Although Red-winged Blackbirds are still widespread and common, BBS data indicate a population decline of 2.8% per year since 1966, in part as a result of the loss of agricultural and wetland habitats but also because of reproductive failure associated with modern agricultural practices. The nest is a solid cup woven to stems on or close to the ground in upland sites or several feet or higher above water in a wetland shrub or cattail. Most egg sets have been found from 10 May to 18 Jun; pairs may produce two broods each year. Three or four eggs are laid in a clutch. Eggs are pale blue and are marked mostly on the large end with spots, blotches, or scrawls in black, brown, and purple.

Fall: Red-winged Blackbirds begin to gather into flocks in Aug, when farmers' fields of grain have matured. Flocks increase with the advent of cold north winds. Large migrant flocks include grackles and starlings often. By the second or third week of Nov most Red-winged Blackbirds have departed from all regions west and north of the Piedmont. In the Coastal Plain and Piedmont, flock sizes increase through

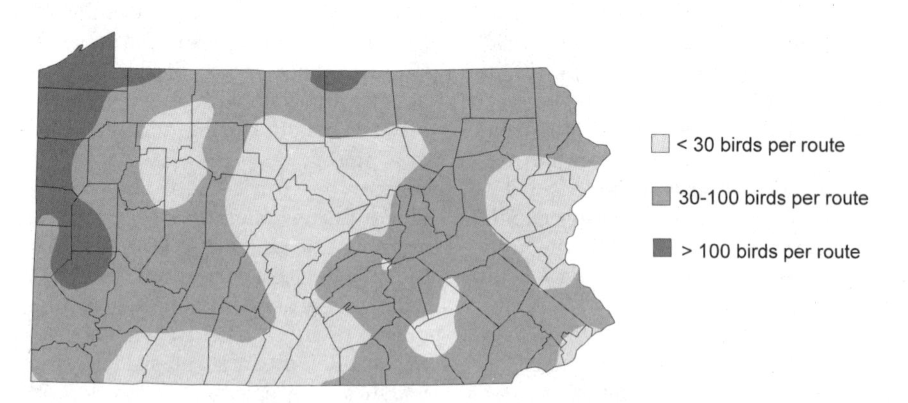

37. Red-winged Blackbird relative abundance, based on BBS routes, 1985–1994

Dec and early Jan. Concentrations have numbered in the tens of thousands in these regions, with a high count of 50,000 birds recorded in southern Lancaster Co. on 2 Jan 1978 (Morrin et al. 1991). The number of birds may remain high through mild winters, but most birds may leave the state after severe winter storms.

Winter: In the Coastal Plain and Piedmont they are uncommon in some years and abundant in others. In these regions the number of birds usually begins to decrease after the second week of Jan and continues to fall until about the third week of Feb. Then the numbers rapidly increase as northbound birds enter the state. Fall and spring migration may overlap during mild winters, making it impossible to separate migrating birds from wintering residents. Away from the Coastal Plain and Piedmont they are uncommon to rare during most winters and usually frequent bird feeding stations. Red-winged Blackbirds may be absent in the state during severe winters, especially in extensively forested areas or at high elevations.

History: Considered common or abundant in Pennsylvania by most early authors (e.g., Baird 1845; Turnbull 1869), Red-winged Blackbirds have used open agricultural habitats since Pennsylvania was settled. They were historically considered most common in emergent wetlands (e.g., cattails) and wet meadows (Sutton 1928b). The exploitation of pastures and old fields for nesting greatly expanded their population and distribution.[1] They were persecuted as agricultural pests (Warren 1890) and hunted for food (Todd 1904) until protected by the Migratory Bird Act in 1913. They were probably most abundant in Pennsylvania during the early 1900s, when agricultural acreage was most extensive. Before the clearing of forests, agricultural habitats were more restricted, and breeding birds were probably confined to nesting in swamps, bogs, or marshy edges of lakes and streams.

[1] Gross, D.A. 1992. Red-winged Blackbird *(Agelaius phoeniceus)*. BBA (Brauning 1992):398–399.

Eastern Meadowlark ***Sturnella magna***

General status: Eastern Meadowlarks breed in southern Canada in Ontario and Quebec, across the eastern U.S., and southwest from Minnesota to Arizona. They winter from New England south, except in the Appalachians, across Ohio, Indiana, Illinois, and through the remainder of their breeding range. Eastern Meadowlarks are fairly common regular migrants over most of Pennsylvania. Despite their abundance, very little is known about their movements during the migration periods. They breed over most of the state except in heavily forested areas or regions of urban development. This species depends on open grasslands for nesting, but with reforestation and development they have rapidly declined as a breeding species since the 1960s. Eastern Meadowlarks are winter residents primarily in the southern half of the state.

Habitat: They inhabit open grasslands and hayfields, roadways, and lawns and are seen or heard overhead during migration. During the breeding season they prefer open grasslands such as hayfields and pastures or reclaimed strip mines. In winter Eastern Meadowlarks are found primarily in fields with manured crop stubble.

Seasonal status and distribution

Spring: Migrant meadowlarks may first appear soon after the first spring thaw during the last week of Feb or the first week of Mar, especially in the Coastal Plain, Piedmont, and Southwest. An unusually high count for such an early date was the 87 birds seen on 25 Feb 1973 at Bath in Northampton Co.[1] At higher elevations and in the northwestern corner of the state, migrants usually do not appear until the second or third week of Mar, especially during extended periods of heavy snowfall. No peak migration is evident, but flocks of 10–20 or more Eastern Meadowlarks periodically appear well into Apr. Migration usually ends by the first or second week of May.

Breeding: Eastern Meadowlarks are unevenly distributed statewide. They are absent from forested north-central areas, rare in fields of increasingly suburbanized eastern counties, and common in northern and western agricultural fields and reclaimed strip mines (see Map 38). On BBS routes they have declined at 5.8% per year since 1966, sufficient to result in an 80% decline in the population in the past 30 years. Yet, they are still more widely distributed than any of the strictly grassland sparrows and were reported on about 75% of BBS routes during the 1990s, but with only one or two individuals on many of those routes. The nest is made of grass, placed on the ground in a small depression, and arched over with grass. The nesting season is long. Most eggs have been found between 8 May and 22 Jul, but early clutches have been reported in late Apr. Usually three to five eggs are laid. They are white and spotted with browns and purples, especially at the large end.

Fall: The beginning of migration is poorly defined, and migrants are not detected until Oct when there is a definite increase in the number of birds. By the first or second week of Nov little migration is noted. Some large concentrations have been found well into Dec; for example, 87 were at Bath in Northampton Co. on 25 Dec 1974.[1] Some are found well into Dec and early Jan, with a few remaining to winter.

Winter: Singles or small flocks are usually found on strips of freshly spread manure in the Piedmont and Southwest, as well as in the lowland river valleys of the Ridge and Valley and the Glaciated Northeast. Eastern Meadowlarks may be locally common in some winters. In the High Plateau there is apparently no winter record north of the Allegheny Mountain section. In the Glaciated Northwest they are casual in winter.

History: Historical lists support the evidence that as habitat was made available, meadowlarks expanded as common birds. Baird (1845) called them "Abundant," and Harlow (1913) stated

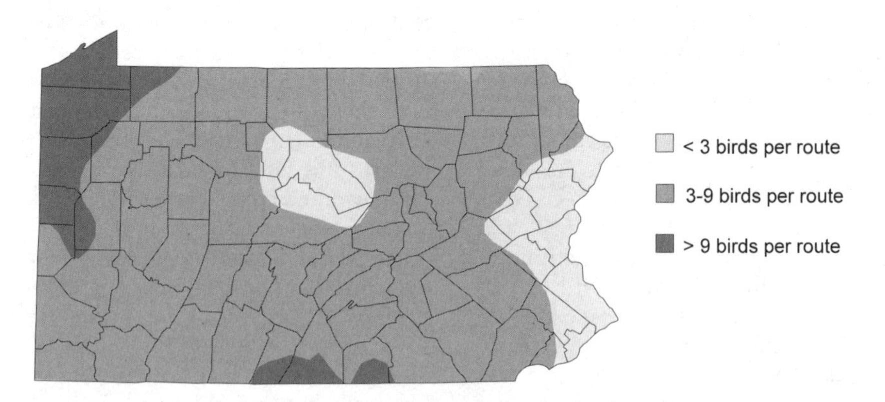

38. Eastern Meadowlark relative abundance, based on BBS routes, 1985–1994

that "it breeds in every county." Understandably, they were not recorded on many local lists from forested settings.

[1] R. Wiltraut, pers. comm.

Western Meadowlark ***Sturnella neglecta***

General status: They are almost indistinguishable from the Eastern Meadowlark in appearance and are best identified by their song. The breeding range of the Western Meadowlark extends across southern Canada from British Columbia to southern Ontario, south to Texas, and west to the Pacific coast. They winter over all but the northern edge of their breeding range and east across the Gulf coast to Mississippi. In Pennsylvania they have been reported about 12 times, always during the spring or summer, when they are in full song. Even though some birds have been on territory, no definite evidence of nesting has been found. Most records have been in the years since 1978.

Seasonal status and distribution

Spring to summer: Breeding has never been definitively confirmed in Pennsylvania, although Western Meadowlarks have been seen carrying food (presumably for young) on at least three occasions. Most birds were seen for only two to four weeks and were not observed in subsequent years. One was found in Berks Co. in 1972 (Uhrich 1997). A singing bird was tape recorded during a BBS on 29 Jun and remained to at least 19 Jul 1978 near Cranesville in Erie Co. (Stull et al. 1985). Also in 1978, one was singing at Brooklyn in Susquehanna Co. from 2 to 7 Jun.[1] One was observed near Evans City in Butler Co. through May of 1985 and again in 1986 and was tape recorded. This bird was believed to have been mated with an Eastern Meadowlark, and both birds were seen carrying food.[2] A Western Meadowlark was found in the State College area in Centre Co. and was apparently heard singing in at least three different locations within a mile of each other on 5–14 May 1987.[2] In Juniata Co., near the village of Center, one was found on 21 Jun[3] and remained to 30 Sep 1997, and the song was recorded on audiotape.[4] A Western Meadowlark was again heard at this location on 26 Jun 1998.[5]

History: This species was first recorded in western Pennsylvania by H. H. Elliot on 9 Jun 1935 south of Hartstown in Crawford Co., when he saw and heard one singing in a meadow by a roadside.[6] In eastern Pennsylvania the first record was a bird reported by G. H. Stuart from 17 to 27 Jul 1942 at Villanova in Delaware Co.[7] Western Meadowlarks were not reported again until 1953, when one was at Johnsville in Bucks Co. from 4 to 11 Jun;[8] and possibly the same bird was found again at the same location in 1955 (Poole, unpbl. ms.). In 1956 one was at Hatboro in Montgomery Co. not far from the Johnsville sighting on 19 May,[9] and another was at University Park in Centre Co. from 10 Apr to 1 May (Wood 1958). Parkes (1956) included this species in the list of birds of the Pittsburgh region, with sight records in Mar and Sep.

Comments: Undoubtedly some Western Meadowlarks pass silently through the state during migration without being detected. Observers should listen for the call note, which is quite different from the Eastern Meadowlark's.

[1] Smith, P.M., R.O. Paxton, and D.A. Cutler. 1978. Hudson-Delaware region. AB 32:1146.

[2] Schwalbe, P.W. 1992. Western Meadowlark *(Sturnella neglecta)*. BBA (Brauning 1992):402–403.

[3] Hoffman, D. 1997. Local notes: Juniata County. PB 11:97.

[4] Haas, F.C., and B.M. Haas, eds. 1998. Birds of note—October through December 1997. PB 11:232.

[5] Troyer, A. 1998. Local notes: Juniata County. PB 12:76.

6 Elliott, H.H. 1936. Western Meadowlark in Crawford County. Cardinal 4:72.
7 Stewart. 1942. General notes. Cassinia 32:48.
8 Reynard, G.B. 1953. General notes. Cassinia 39:22.
9 Brady, A. 1957. General notes. Cassinia 42:27.

Yellow-headed Blackbird *Xanthocephalus xanthocephalus*

General status: The Yellow-headed Blackbird breeds in southern Canada from British Columbia east to southern Ontario, across the western U.S. south to northern Texas, and northeast through Missouri to northwestern Ohio. They winter primarily from Washington south through Arizona to Texas. Yellow-headed Blackbirds are rare irregular visitors during migration in Pennsylvania. Most birds observed are single individuals, but occasionally groups of fewer than five birds are seen, often mixed with other species of blackbirds. Most Yellow-headed Blackbirds are reported from the Piedmont, where the largest concentrations of blackbirds are found. They are accidental in summer and winter.

Habitat: Yellow-headed Blackbirds are usually observed flying overhead with other species of blackbirds, feeding in grain fields or at bird feeding stations, or at roosts in marshes.

Seasonal status and distribution

Spring: Spring birds are usually in large flocks of blackbirds that pass through the state in the third or fourth week of Feb, but many of the records during this season are from Apr through the third week of May. Stragglers have remained to the first week of Jun.

Summer: All four records of Yellow-headed Blackbirds within this period are from western Pennsylvania. The three reports of birds in Jun records are likely to have been spring stragglers, but they may have been range-extension pioneers. One was at Indiana in Indiana Co. on 13 Jun 1961,[1] and an adult male was in suitable breeding habitat at Siegel Marsh in Erie Co. on 11 Jun 1994.[2] At Smith's Marsh in Crawford Co., one was found on 29 Jun 1997.[3] A male was found at PISP in Erie Co. on 14 Jul 1992.[4]

Fall: Yellow-headed Blackbirds have been reported as early as the first week of Aug, but they are usually not reported until blackbird migration is well under way, during the third or fourth week of Aug. Scattered records come from Sep through Dec and into the second week of Jan.

Winter: Most Yellow-headed Blackbirds seen in winter are individuals at feeding station that have remained just into this period. The only ones reported af-

Yellow-headed Blackbird at Hershey, Dauphin County, April 1990. (Photo: Randy C. Miller)

ter the second week of Jan have been at Tinicum and in the area of Bucks and Lancaster cos. There have been reports of at least three winter sightings without specific dates in the Tinicum area.[5] In 1969 one was at Tinicum to 21 Jan.[6] The single record from Bucks Co. was one found in Levittown on 1 Feb 1976.[7] Sightings were reported from Jan in 1978, 1979, and 1980 in southern Lancaster Co. and in the winter of 1993 near Quarryville in Lancaster Co.[8]

History: Since Yellow-headed Blackbirds were first discovered in the state, perhaps as early as the mid-1850s, they were probably recorded fewer than 10 times before 1960. During that hundred-year period, Yellow-headed Blackbirds were reported from the counties of Allegheny, Chester, Crawford, Montgomery, and Washington. It was not until after the 1950s that they became reported at intervals of a few years to almost annually. Perhaps the first report in the state was of a young male that was shot near Philadelphia in Aug 1851 and was placed in the collection of ANSP (Turnbull 1869). Because there was no specific location, there is no way of knowing whether the bird was collected in Pennsylvania. Apparently the first satisfactory record from eastern Pennsylvania was of one observed at Ardmore in Montgomery Co. on 19 Dec 1948.[9] Todd (1940) cited the first positive western Pennsylvania record of a Yellow-headed Blackbird as one shot by D. A. Atkinson from a flock of Red-winged Blackbirds at Wilkinsburg in Allegheny Co. on 26 Apr 1895.

[1] Hall, G.A. 1961. Appalachian region. AFN 15:469.

[2] G.M. McWilliams, pers. recs.

[3] Haas, F.C., and B.M. Haas, eds. 1997. Birds of note—April through June 1997. PB 11:90.

[4] Haas, F.C., and B.M. Haas, eds. 1992. Rare and unusual bird reports. PB 6:130.

[5] J.C. Miller and N. Pulcinella, pers. comm.

[6] E. Fingerhood, pers. comm.

[7] R. French, pers. comm.

[8] E. Witmer, pers. comm.

[9] Mirick, H.D. 1951. Yellow-headed Blackbird at Ardmore, Pa. Cassinia 38:36.

Rusty Blackbird ***Euphagus carolinus***

General status: Rusty Blackbirds breed from Alaska across most of Canada and south to New York. They winter over most of the eastern U.S. west to Montana and south to New Mexico. Their distribution during migration is quite varied over Pennsylvania, from common to abundant regular migrants in the Glaciated Northwest, fairly common to common regular migrants at the Lake Erie Shore and Southwest, to uncommon but regular elsewhere in the state. They mix with other species of blackbirds primarily when feeding in fields, when in migration, or at roosts. Rusty Blackbirds winter in small numbers in the state.

Habitat: Rusty Blackbirds prefer wet habitats such as swamps, beaver dams, slow-moving streams bordered by brush, marshes, and grain fields, especially along the edge of flooded fields.

Seasonal status and distribution

Spring: Away from areas where they winter, migrants have been reported as early as the third or fourth week of Feb. By the third week of Mar the number of birds has increased significantly. Peak migration occurs from the fourth week of Mar to the second or third week of Apr. The greatest number of birds is found in the Glaciated Northwest, especially at Conneaut Marsh and the wooded swamps of the Pymatuning area, where daily counts regularly reach the thousands. Most Rusty Blackbirds have passed through the state by the first week of May. Stragglers may remain to the fourth week of May.

Fall: Early fall migrants may appear as early as the second week of Sep. Their typical migration time is from the first week of Oct to the second week of Nov. The peak migration period is quite short; the greatest number of birds passes through the state from the second to the fourth week of Oct. Stragglers remain until the first or second week of Jan. A few may remain to winter.

Winter: The number of birds varies from year to year and from location to location, but the only area in the state where they seem to be regular winter residents is around Tinicum in the Coastal Plain. They are irregular winter residents in the southern counties of the Piedmont. Rusty Blackbirds are casual winter residents or visitors elsewhere in the state except in the High Plateau north of the Allegheny Mountain section, where there is but one record. One was at Irvine in Warren Co. on 31 Jan 1970.[1] Many sightings away from the southeastern portion of the state are in Jan at bird feeding stations.

History: Most authors believed that the species was rather unpredictable in terms of both the number of birds passing through the state and the frequency of occurrence. Todd (1940) believed they were more or less uncommon and irregular everywhere in western Pennsylvania except at Pymatuning Swamp, where he found Rusty Blackbirds in considerable numbers in both spring and fall. Poole (unpbl. ms.) listed this species as rather irregular, especially in the fall, when they were absent for several years in a row but rather common in other years. More observers in the field and easier access into their preferred habitats have probably contributed to the increased sightings since the 1950s and 1960s.

[1] T. Grisez, pers. comm.

Brewer's Blackbird ***Euphagus cyanocephalus***

General status: Brewer's Blackbirds breed in southern Canada from British Columbia east to Ontario and across the northern states from Michigan west and south to California and New Mexico. The breeding range has recently expanded along its eastern border and they migrate regularly east to the Appalachians (AOU 1998). They winter from British Columbia south, and east across the southern half of the U.S. Though they are probably irregular if not regular visitors in Pennsylvania, Brewer's Blackbirds are currently listed in Pennsylvania as casual. Despite the numerous reports, very few birds have been documented with convincing written descriptions or photographs. Therefore, little information is available on which to substantiate the definite status of this species in Pennsylvania. Brewer's Blackbirds have been reported from all regions of the state during migration, primarily in the fall. All winter sightings are from the southern portion of the Piedmont. Most sightings are of single birds or small groups of fewer than five, but there have been some unsubstantiated reports of large flocks. Reports have continued to increase since the 1950s.

Habitat: Most reports of Brewer's Blackbirds are from farmyards, particularly near barns, in grain fields associating with other species of blackbirds, and at bird feeding stations.

Seasonal status and distribution

Spring: Very little is known about the timing of passage through the state. They usually accompany migrant flocks of blackbirds, with sightings in Mar, Apr, and to the third week of May. Some spring sightings may be of birds that have wintered in the state.

Fall: Like spring migration, very little is known of their fall passage through the state except that they usually accompany large flocks of other species of blackbird. The earliest reports of Brewer's Blackbirds have come from the first week of Sep, but most birds are reported during the largest blackbird migrations in Oct and Nov or when many observers are in the field during CBCs in Dec. A few have remained into the winter period.

Winter: Many reports have been as late as the first and second weeks of Jan. East of the Allegheny Front they have been reported after mid-Jan in the counties of Chester, Delaware, Lancaster, Lehigh, Northampton, and Philadelphia. No record later than mid-Jan is known west of the Allegheny Front.

History: Brewer's Blackbirds were first reported in Pennsylvania as recently as 1953, when a flock of 120 was seen by P. B. Street at Exton in Chester Co. on 19 Dec 1953, and a single bird was found at the same place on 16 Jan 1954.[1] Also in 1954, they were reported at Kempton in Berks Co. and at Tinicum in Delaware and Philadelphia cos. (Poole, unpbl. ms.) A flock of 200 was reported at New Bloomfield in Perry Co. on 25 Oct 1958 (Poole, unpbl. ms.). The first report west of the Allegheny Front was on 4 Apr 1957 at Clarksville in Greene Co.[2] Because no birds were collected or photographed, Poole (1964) listed the species as hypothetical even though he reported having seen several in the spring of 1958 in Berks Co.

Comments: This blackbird probably occurs more frequently in the state than records indicate. However, with their close resemblance to Rusty Blackbirds and Common Grackles, many probably pass through the state undetected.

[1] Cutler, D.A. 1955. General notes. Cassinia 40:35.

[2] Van Cleve, B. 1957. Some records of special interest. ASWPB 22:11.

Common Grackle *Quiscalus quiscula*

General status: Common Grackles are among the most abundant blackbirds as well as one of the most disliked, especially by farmers when flocks of thousands descend upon their grain fields during migration. They are also condemned by people living in suburbs or cities, where the birds roost above houses, lawns, and pools. Attempts to control populations by spraying and shooting, especially at winter roosts, have failed. Common Grackles breed in Canada from British Columbia and Mackenzie east to Newfoundland and south across the eastern U.S. and west to the Rockies. They winter throughout the eastern U.S., except in the northernmost states, west to the Dakotas and south to New Mexico. Common Grackles are abundant regular migrants, and they breed statewide. In Pennsylvania they winter primarily in the Piedmont. They frequently mix with other species of blackbirds during migration and at winter roosts.

Habitat: A wide variety of habitats is used by Common Grackles, especially near human habitation. During migration and in winter they are found in grain fields, roosting or briefly resting in trees or flying overhead. They are also found on lawns or golf courses in spring and fall. During the breeding season, Common Grackles can be found in almost any open areas with trees, especially where conifers are found.

Seasonal status and distribution

Spring: Away from wintering sites, migrants begin to appear after the first warm spring thaw as early as the third week of Feb. In most years they arrive the fourth week of Feb or the first week of Mar. At higher elevations and across the northern-tier counties, migration may be delayed until the second week of Mar or later during years of prolonged severe weather. Birds increase in number in most years by the second or third week of Mar and reach their maximum abundance by the first or second week of Apr. The large number of birds during the breeding season makes it difficult to determine the end of the migration period.

Breeding: Common Grackles are abundant breeding birds statewide in open and fragmented habitats. They penetrate far into forested regions along larger rivers and streams and around residential areas. They are most abundant in agricultural and suburban areas, such as in the Piedmont and western counties north of Pittsburgh, and least common in the extensively forested northern counties (see Map 39). Steady declines of 3.2% per year on BBS routes since 1971 have placed the grackle behind the starling and robin in order of statewide abundance. Their nest is a bulky structure of grass, sometimes plastered with mud. Grackles nest colonially, often more than 25 feet above the ground, usually in evergreen trees near human residences but occasionally in deciduous trees or in other diverse settings. They nest early; most eggs have been observed from 18 Apr to 28 May. Three to six pale greenish white eggs with dark brown and purple spots, blotches, or scrawls are laid.

Fall: A general influx of grackles in late Aug or early Sep signals the beginning of fall migration. The maximum number of birds, flocks containing tens of thousands, is reached between the second week of Sep and the fourth week of Oct. At some sites, such as southern Lancaster Co., there may be concentrations of hundreds of thousands in some years, such as the 218,000 recorded on the Southern Lancaster Co. CBC in 1990 (Morrin et al. 1991). By the first or second week of Nov, Common Grackles have become rather scarce west of the Allegheny Front, and most have left this portion of the state by the first week of Dec.

Winter: In the Piedmont, where most of the state's population winters, they may be abundant in some years, but in other years they may be uncommon. Concentrations of hundreds of thousands, perhaps exceeding a million birds, have been reported in some winters. In the Coastal Plain, Ridge and Valley, and Southwest, Common Grackles are uncommon to rare regular winter residents. In the Glaciated Northeast and Glaciated Northwest, as well as at the Lake Erie Shore, they are casual to accidental winter visitors or residents. Away from the Coastal Plain and Piedmont, most winter birds are observed at bird feeding stations.

History: This adaptable species responded rapidly to human-modified landscapes. They were described as common by the earliest authors. The

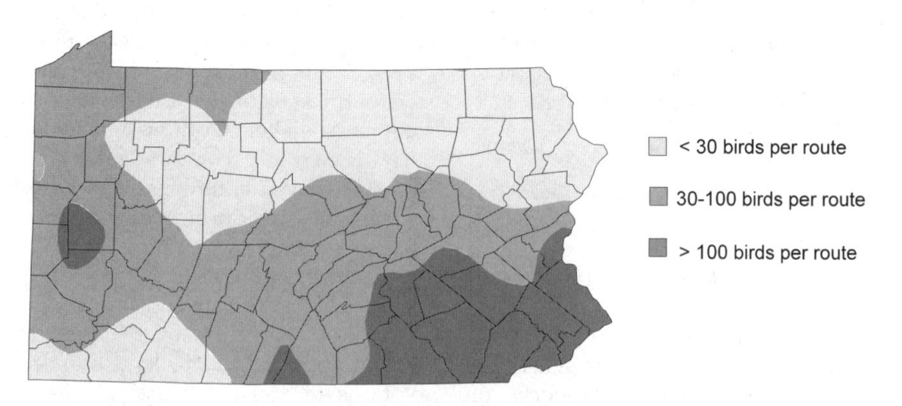

39. Common Grackle relative abundance, based on BBS routes, 1985–1994

presettlement status is unclear, but as Sutton (1928a) stated, "It is a bird of the town, not the wilds." They have traditionally gathered in large flocks in fall and winter, joining with European Starlings, Red-winged Blackbirds, and Brown-headed Cowbirds. Mixed-species flocks have totaled up to 3 million birds (Poole, unpbl. ms.).

[Boat-tailed Grackle] ***Quiscalus major***

General status: Boat-tailed Grackles are residents along the Atlantic coast from New York south and along the Gulf coast. They have been reported in Pennsylvania at least three times. Unfortunately, the reports have not been substantiated with photographs or described sufficiently to eliminate the similar Great-tailed Grackle *(Quiscalus mexicanus)*, which has also been expanding its range but has not been reported in Pennsylvania. The Boat-tailed Grackle is listed as hypothetical in Pennsylvania.

Comments: One was at Tinicum on 13 Dec 1957 (Poole 1964), and another was reported there on 10 Oct 1985.[1] A Boat-tailed Grackle was reported on the Lower Bucks Co. CBC on 17 Dec 1977.[2] More recently was a report of a male near Longwood in Chester Co. on 18 Mar 1995.[3] It is interesting to note that all of the above records fall within the time frame of the Great-tailed Grackle's recent northward expansion.

[1] Paxton, R.O., W.J. Boyle Jr., and D.A. Cutler. 1986. Hudson-Delaware region. AB 40:90.

[2] Herzog, A. 1978. Lower Bucks Co. CBC. AB 32:555.

[3] PORC files.

Brown-headed Cowbird ***Molothrus ater***

General status: The Brown-headed Cowbird is notorious for its breeding strategy, known as nest parasitism, of laying its eggs in the nests of other species of birds. They breed across the southern half of Canada and throughout the U.S. Brown-headed Cowbirds winter along the Pacific coast states, south across the southwestern U.S. and throughout the eastern U.S. north to Nova Scotia. Cowbirds are abundant regular migrants and breed throughout the state. They are birds of open grassland habitat but have become adaptable, infiltrating even heavily forested areas of the High Plateau in pursuit of hosts. In the state they winter primarily in the Piedmont. Cowbirds frequently mix with other species of blackbirds during migration and at winter roosts.

Habitat: During migration and in winter cowbirds are found in grain fields, on lawns, on golf courses, in parks, resting in trees, or flying overhead. In the breeding season they use a wide variety of habitats that are primarily associated with open areas such as farmlands, woodland clearings and edges, and suburban areas of towns and cities. They may range for miles from open areas during the summer into wooded settings as they search for nests in which to place their eggs. In winter they are found in grain fields, often associating with other species of blackbirds, and at bird feeding stations.

Seasonal status and distribution

Spring: Migration may be detected away from major wintering sites soon after the first spring thaw in the third or fourth week of Feb. In most years, cowbirds begin to appear a little later than Red-winged Blackbirds and Common Grackles, often not until the first or second week of Mar. Peak migration usually occurs from the fourth week of Mar to about the third week of Apr. Flock sizes during this period may number in the hundreds or thousands of birds, but they usually do not equal the huge numbers of Red-winged Blackbirds and grackles that pass through the state.

Breeding: Brown-headed Cowbirds are fairly common to common summer residents statewide. They are most common in open agricultural regions and and least common in extensive forested areas (see Map 40). No nest is built, and many passerine species have been parasitized. Warblers and sparrows are frequent hosts in Pennsylvania. Like many grassland-associated species, cowbirds are experiencing significant population declines on BBS routes (4.1% per year). Most nests containing eggs of this species have been observed between 1 May and 25 Jun, although some have been found in Jul. One egg is laid each day, often in different nests (H. H. Harrison 1975). The eggs are grayish white and spotted with brown, often heaviest on one end. Native birds nesting in Jul and Aug are less likely to be parasitized than are birds that nest earlier, providing an advantage for double-brooded and late-nesting species.

Fall: Their southward movements are poorly defined because the breeding population in the state is so large and widespread. The number of cowbirds has noticeably increased by the first or second week of Sep, with the majority passing through the state during Oct. In some years the number of birds may remain high through at least the second week of Jan. Flock sizes are comparable to those in spring.

Winter: Numbers of cowbirds and of wintering sites may vary from year to year. The greatest numbers of birds winter in the Piedmont, especially in the southern portion of the Piedmont. Nearly one half million birds were recorded roosting on Pitney Road in Lancaster Co. in 1978 (Morrin et al. 1991), and as many as a million may have gathered at a winter roost near the Hanover Reservoir in York Co. during the winter of 1959 (Poole, unpbl. ms.). Away from the Piedmont in winter they are most regular and numerous in the Coastal Plain, the lowland valleys of the Ridge and Valley, and the Southwest. In the Glaciated Northeast, the Glaciated Northwest, and through the High Plateau, cowbirds are irregular winter residents, and when present, their abundance varies from rare to common. Away from the Coastal Plain and Piedmont, most wintering birds are found at bird feeding stations.

History: It is generally assumed that cowbirds were very rare or not found in eastern North America before European settlement but had rapidly exploited new agricultural habitats in the

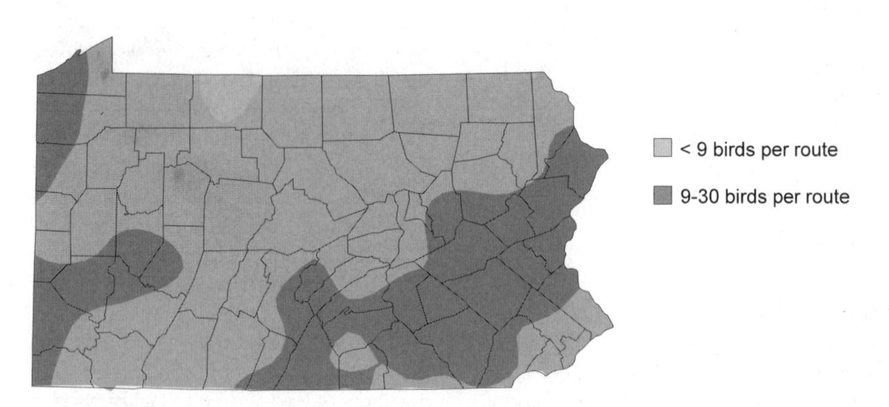

40. Brown-headed Cowbird relative abundance, based on BBS routes, 1985–1994

early 1800s.[1] By the mid-nineteenth century, they were listed as abundant or common in Pennsylvania (Baird 1845; Turnbull 1869). Populations undoubtedly increased to a peak by the end of the 1800s, during the state's agricultural heyday. In Beaver Co. Todd (1940) found "So large a proportion of the nests of small birds . . . invaded by the Cowbird that I have often wondered how these birds contrive to hold their own."

[1] Lowther, P.E. 1993. Brown-headed Cowbird *(Molothrus ater)*. BNA (A. Poole and Gill, 1993), no. 47.

Orchard Oriole ***Icterus spurius***

General status: Orchard Orioles breed from Montana, southern Ontario, and Vermont south to Florida, Louisiana, and Mexico. They winter from Mexico south to South America. Orchard Orioles may spend only about three months out of a year in Pennsylvania, less time than most bird species that nest here. They are uncommon to rare regular migrants, much less common in the northern half of the state at any season than in the southern areas. Orchard Orioles breed in scattered locations throughout the state but most commonly in the Piedmont and the southwestern corner.

Habitat: They inhabit woodland edges, open woodlots, parks, unsprayed orchards, and roadsides, especially where there are scattered mature deciduous trees, near water.

Seasonal status and distribution

Spring: Birds may begin to arrive as early as the third week of Apr. Movement is usually detected from the last week of Apr to the fourth week of May. There does not appear to be a peak migration period over most of the state, but the greatest number of birds reported in the Southwest region is seen in the last few days of Apr or the first week of May.[1]

Breeding: Orchard Orioles are uncommon and local nesters across the southern third of the state and widespread but rare and irregular to the north. They are most common in the Piedmont and southwest of Pittsburgh and are often restricted to major river valleys in the north. They were reported in just 16% of BBA blocks[2] and on 30% of BBS routes, but they have been reported in all counties during the summer. The nest is a hanging basket similar to the Baltimore Oriole's, but only 3 inches deep; it is built 10–20 feet from the ground in a tree or shrub. Most of the relatively few nests with eggs reported have been observed between 24 May and 16 Jun. Usually four or five eggs are laid. They are pale bluish white and spotted, blotched, or scrawled with brown, gray, and purple. Orchard Orioles become secretive in Jul and depart early. Populations are not adequate to determine a trend on BBS routes.

Fall: Orchard Orioles migrate south very early. As soon as they stop singing and nesting has been completed, they become very difficult to locate and are rarely reported after the fourth week of Jul. Most birds recorded at PNR are found between the last week of Jul and the third week of Aug.[1] There are few reports from the third week of Aug to the third week of Sep. Orchard Orioles have been reported only once recently after the third week of Sep, when one was recorded in Nov of 1975 at Sheffield in Warren Co.[3]

History: They were described as "abundant" in southeastern counties (Harlow 1913) and "common" in orchards (Warren 1890). Harlow (1913) reported that Orchard Orioles were expanding into northern Pennsylvania, citing summer records north to Pike and Erie cos. This expansion was short-lived. Todd (1940) stated that the species seemed to be in decline throughout western Pennsylvania. Although Poole (1964) said that it was "becoming scarce everywhere," his distribution map is not significantly different from that reported by the BBA.[2] Bent reports them as leaving as late as 19 Oct (Bent 1958).

[1] R.C. Leberman, pers. comm.
[2] Ickes, R. 1992. Orchard Oriole *(Icterus spurius)*. BBA (Brauning 1992):408–409.
[3] Hall, G.A. 1976. Appalachian region. AB 30:70.

Baltimore Oriole ***Icterus galbula***

General status: Their purselike nest may be as well known as the colorful birds themselves. Baltimore Orioles breed in southern Canada from British Columbia to Nova Scotia south throughout the eastern U.S., except in the southeastern corner, west to Texas and along the western edge of the Great Plains. They winter form Mexico to South America. A few regularly winter along the Atlantic states to Virginia and are casual north to the Great Lakes region and New England (AOU 1998). Baltimore Orioles are fairly common to common regular migrants, and they breed throughout Pennsylvania. They are casual winter visitors or residents, primarily in the southern half of the state.

Habitat: During migration Baltimore Orioles are found in open woodlands and woodland edges or wherever there are tall trees in parks, in backyards, in farmlands, and along roadways. During the breeding season they prefer sites with mature trees in suburban areas and are found along rural roads, streams, ponds, and lakes. In winter Baltimore Orioles are usually found at bird feeding stations with brushy cover.

Seasonal status and distribution

Spring: More birds are observed in spring than during any other time of year. Their typical migration period is from the third or fourth week of Apr to the fourth week of May. Peak migration is during the second and third weeks of May.

Breeding: Fairly common statewide, Baltimore Orioles are most common on the BBS routes in the Glaciated Northeast and are least common in the extensively forested north-central counties (see Map 41). They are absent from uniformly forested regions away from rivers and lakes. The nest, a distinctive hanging basket suspended from small branches high in a deciduous tree, is woven of various materials sometimes including string or yarn. They often use nylon fishing line, and some nests have been nearly completely made of fishing line.[1] Nests are conspicuous after

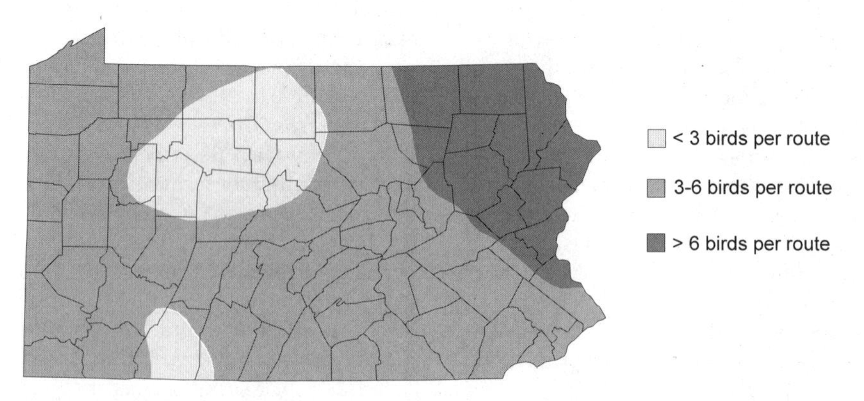

41. Baltimore Oriole relative abundance, based on BBS routes, 1985–1994

leaf-fall. Typical of the egg dates of Neotropical migrants, most eggs are reported during a narrow season from 19 May to 19 Jun. Four or five grayish white eggs, spotted, blotched, or scrawled with brown and black, are laid. For unknown reasons, numbers on the BBS began to decline in the mid-1980s in Pennsylvania and throughout the East, but they have recovered much of those losses in the 1990s.

Fall: Baltimore Orioles become inconspicuous and difficult to find after singing has stopped in Jul. Like Orchard Orioles, they are early migrants, but unlike Orchard Orioles, some birds linger well into the fall. No peak migration period is evident, but the heaviest passage is probably during late Jul and early Aug, as is evidenced by the remarkable concentration of about 75 birds feeding in a peach orchard at Solebury in Bucks Co. in late Jul 1979.[2] Most migrants probably leave the state by the first week of Oct. Stragglers have been observed to the first or second week of Jan. A few have been known to remain through the winter.

Winter: Occasionally, Baltimore Orioles remain in winter, when they are found primarily at bird feeding stations in the Coastal Plain and Piedmont. In these regions they have been recorded in winter in the counties of Berks, Bucks, Delaware, Franklin, Montgomery, Northampton, and Philadelphia. In the Ridge and Valley they have been recorded in winter in the valleys of Centre, Clinton, Huntingdon, Luzerne, Lycoming, Union, and Wyoming cos. In the Southwest, they have been recorded in Allegheny and Indiana cos. Baltimore Orioles are accidental in winter elsewhere in the state. They have been recorded at least once in the High Plateau, where one successfully wintered at PNR in 1974 (Leberman 1976). In the Glaciated Northwest one was in Crawford Co. from 5 to 12 Feb 1967.[3] They have not been recorded in winter at the Lake Erie Shore and in the Glaciated Northeast. Most sightings are of single birds, but up to seven wintered in 1969 at Pennypack Park in Philadelphia Co.[4]

History: Most bird lists of the nineteenth century included Baltimore Orioles. On the basis of habitat and reports such as that of Todd (1940), Baltimore Orioles have been less common in the north-central counties but may have been relatively common anywhere habitat was suitable. Poole (unpbl. ms.) believed that these orioles experience definite population fluctuations that made comparisons of historical accounts difficult. He referred to at least three population fluctuations before 1960. Earlier authors such as Warren, Todd, and Poole reported no winter records later than the third week of Jan.

Comments: As of 1995 this oriole regained its former name of Baltimore Oriole, after being lumped for years with its western cousin, Bullock's Oriole, under the name Northern (AOU 1998). Immature or female Baltimore Orioles have occasionally been misidentified as Bullock's Orioles, especially in late fall or winter.

[1] R. Wiltraut, pers. comm.

[2] Richards, K.E., R.O. Paxton, and D.A. Cutler. 1979. Hudson-Delaware region. AB 33:851.

[3] Hall, G.A. 1967. Appalachian region. AB 21:420.

[4] Scott, F.R., and D.A. Cutler. 1969. Middle Atlantic Coast region. AB 23:578.

Bullock's Oriole *Icterus bullockii*

General status: Bullock's Orioles breed in southern Canada from British Columbia to Saskatchewan and south along the Pacific coast and east to the Dakotas, Kansas, and Texas. They winter primarily in Mexico and Central America. A few winter along the Gulf coast region to Georgia and Florida. They are casual in northeastern North America. Strays are occasionally reported in Pennsylvania. All those reported here have been immature or female orioles at bird feeding stations. There is one accepted record in Pennsylvania.

Seasonal status and distribution

Winter: A female Bullock's Oriole observed by N. Reifsnyder at his feeder from 8 to 18 Jan 1994 near Blue Marsh Lake in Berks Co. was documented with a photograph.[1,2]

Comments: Female orioles were reported as Bullock's in 1980 and 1982: the first bird was observed from 7 to 13 Apr 1980 at Doylestown in Bucks Co.,[3] and the second one was found at Dallas in Luzerne Co. on 27 Dec 1982.[4] Detailed field notes should accompany good photographs to document a female Bullock's Oriole, particularly if found in late fall or in winter. Baltimore and Bullock's orioles occasionally hybridize, and the plumage colors of females are variable and may be similar between the two species. All sightings should be thoroughly documented with good photographs.

[1] Pulcinella, N. 1996. Seventh report of the Pennsylvania Ornithological Records Committee (P.O.R.C.) 1996. PB 10:49.

[2] Keller, R. 1994. Notes from the field: Berks County. PB 8:36.

[3] Paxton, R.O., W.J. Boyle Jr., D.A. Cutler, and K.C. Richards. 1980. Hudson-Delaware region. AB 34:761.

[4] Boyle, W.J., Jr., R.O. Paxton, and D.A. Cutler. 1983. Hudson-Delaware region. AB 37:287.

Family Fringillidae: Finches

These small to medium-sized arboreal birds have a conical bill adapted for extracting seeds. They also eat insects and fruit or pick at salt along roadways in winter. Unlike that of sparrows and buntings, the lower mandible of finches does not turn abruptly down at the base. Most have a relatively short tail that is moderately to deeply forked. They are frequently observed in flocks, flying noisily overhead or extracting seeds from birches, alders, conifers, or other trees. While contentedly feeding in a treetop they may suddenly burst into flight, only to return to a nearby or distant tree to continue actively feeding. They have been likened to parrots in that they typically hang upside down on a seed pod or cone to remove the seeds. Most species readily come to bird feeding stations in winter, preferring sunflower, safflower, or thistle seed. They are usually erratic in their movements, often traveling great distances in search for food. Eleven species have been recorded in Pennsylvania, and six of those have been known to nest there. Several northern species periodically move south in huge numbers from their boreal homes in Canada and the northern U.S. After winter invasions of northern finches, those that are not normally regular breeders in the state (Red Crossbill and Pine Siskin) may sometimes remain to nest. They build their nest of sticks, grasses, rootlets, and strips of bark and line it with fine grasses or hair. Nests of most species are placed in conifers from only a few feet to over 50 feet above the ground. They are usually solitary nesters, but some species will nest in small loose colonies. There are usually two to six white, blue, or bluish green eggs, with some light spotting.

Brambling *Fringilla montifringilla*

General status: The Brambling is a Eurasian species that breeds from northern Scandinavia to Russia and Siberia and winters as far west as the British Isles. They migrate regularly through the Aleutian Islands and Pribilofs and are casual across Canada and much of the northern and western U.S. This species has been recorded once in Pennsylvania.

Seasonal status and distribution

Winter to spring: A male Brambling reported by P. Hess was first discovered on 2 Feb 1978 visiting a feeder in Hampton Township Allegheny Co.[1] and remained there until 1 Apr 1978.[2] Throughout its stay, the Brambling was observed by many birders.

Comments: This bird was unbanded, and its plumage was in fresh condition. Because it was found during an invasion year with concurrent sightings from Canada and Alaska, the Brambling has been accepted as a wild bird and is included on the list of birds of Pennsylvania.[3]

[1] Hall, G.A. 1978. Appalachian region. AB 32:352.
[2] Hall, G.A. 1978. Appalachian Region. AB 32: 1007.
[3] PORC files.

Pine Grosbeak ***Pinicola enucleator***

General status: In North America this irruptive and highly irregular finch breeds in the boreal forests of Alaska southeast across central Canada to Nova Scotia and south in the mountains of the West and to Maine in the East. They winter in the southern portion of their breeding range and casually or sporadically south to Texas, Missouri, Kentucky, and the Carolinas. Pine Grosbeaks have been observed throughout Pennsylvania, but most birds are reported from the mountainous regions. They are often rather tame and allow close approach. As with most northern finches, irruptions south of their normal breeding range are determined by the abundance of their food supply rather than the onset of severe winter weather. There may be many years between invasions when not a single bird is observed in the state.

Habitat: Pine Grosbeaks are most often found around fruit-bearing trees and shrubs and conifers with seed cones. They also occur in woodland edges as well as suburban areas of towns and cities.

Seasonal status and distribution

Fall through spring: Birds arriving during or after a cold front between the third week of Oct and the second week of Nov usually signal the beginning of an invasion year. Pine Grosbeaks may appear any time through the winter during these invasions. There have been only three major invasions (a large number of birds at some sites) since 1960: during 1961–1962, 1968–1969, and 1981–1982. Smaller invasions (a small number of birds at a few scattered sites) have occurred every few years from 1960 to 1989, with at least a few birds reported between invasion years. However, since 1990 they have been

Pine Grosbeak at Hickory Run State Park, Carbon County, December 1985. (Photo: Franklin C. Haas)

irregular, with no invasion and only a few birds reported. Spring sightings are usually of birds that have overwintered, but there have been occasions when flocks suddenly appeared in Mar or Apr, suggesting returning northbound birds. During years when birds are present their numbers vary from individuals or small groups of 10 or fewer to flock sizes reaching into the hundreds. During the invasion winters of 1981 and 1982, as many as 300 were estimated in Carbon Co. At Beltzville State Park 100–120 were observed at one time feeding on multiflora rose, with many singing males remaining into Feb.[1] At HMS in Berks Co. a high count of 208 was reported on 21 Nov 1968.[2] The last birds are usually seen in Apr, with stragglers to the first week of May.

History: Warren (1890) wrote of a large invasion in the northeastern part of the state during the winter of 1890. Todd (1940) stated that between the years 1875 and 1939 the greatest invasion was during the season of 1906–1907. Poole (unpbl. ms.) listed the years of 1929–1930, 1951–1952, 1954–1955, and 1957–1958 for extensive invasions of Pine Grosbeaks in the eastern portion of the state. As in recent history, most historical grosbeaks were found in mountainous areas of the state.

[1] Paxton, R.O., W.J. Boyle Jr., and D.A. Cutler. 1982. Hudson-Delaware region. AB 36:280.
[2] Meritt, J.K. 1970. Field notes. Cassinia 52:28.

Purple Finch ***Carpodacus purpureus***

General status: Purple Finches breed across southern Canada from British Columbia east to Newfoundland and south in the West along the Pacific coast and in the East from North Dakota through the Great Lakes region to West Virginia, Maryland, and Pennsylvania. They winter throughout their breeding range in the U.S. and south in the East to Texas and Florida. Purple Finch populations vary in Pennsylvania; they are quite common in some years and rare in other years. They are regular migrants and winter residents statewide. They breed primarily across the northern half of Pennsylvania and south through the Allegheny Mountain section.

Habitat: Purple Finches use a variety of habitats, especially during migration. They are found in woodlands and woodland edges, swamps, towns and cities, and at bird feeding stations. During the breeding season Purple Finches are in the same habitats used during migration but are more confined to areas with at least a few pine or spruce trees, in which they place their nest. In winter they are found mostly at bird feeding stations where there is adequate cover.

Seasonal status and distribution

Spring: Their typical migration period is from about the third or fourth week of Mar, away from areas where they winter, until the second or third week of May. The greatest number of birds is recorded from the third week of Apr to the first week of May.

Breeding: Purple Finches are uncommon breeding residents in the Glaciated Northeast and in the mountains through the High Plateau. They occur irregularly as rare breeders in the Ridge and Valley and Southwest and are accidental in the Piedmont. Breeding records are not necessarily restricted to highlands. The BBA project presented a distribution that far exceeded that of previous reports, particularly west of the Alleghenies, including a confirmed record in Greene Co.[1] But populations appear to have been declining since 1987 (Uhrich 1997), and competition with House Finches and House Sparrows has been suggested to be detrimental to Purple Finches. Long-term BBS trends are not significant. The nest is a shallow cup placed at varying heights on a horizontal branch of a conifer, occasionally including planted and ornamental pines and spruces. Two broods have been reported.

Fall: The beginning of fall migration is poorly known in the state, but banding records suggest some movement beginning the fourth week of Aug. Migration becomes more detectable by the second or third week of Sep, with the greatest number of Purple Finches usually observed during the second or third week of Oct. Usually fewer birds are reported at this season than in spring, but during heavy flight years, flocks may contain up to 300 birds (Leberman 1976). The number of birds declines through Nov and Dec. Some CBCs in the Piedmont occasionally report hundreds of Purple Finches. These large numbers have not been equaled in recent years.

Winter: They have been recorded in every county in winter, but the number of birds and frequency of occurrence vary from winter to winter. Most birds withdraw from the highest elevations into valleys, where they concentrate at bird feeding stations, especially in the southern half of the state.

History: Todd's (1940) extensive work showed breeding records across the northern mountains and at Pymatuning but none in the Southwest region or the Ridge and Valley. He said that "even in the northern counties this bird is only sparingly distributed, and Tamarack Swamp, Clinton County, is the only locality where I have ever found it at all common in summer." Another traditional stronghold of this species was in the Poconos, where Carter (1917) called it "numerous." Numbers have rebounded during the twentieth century with the reforestation of many areas and planting of conifers.

Comments: There has been some belief that the established House Finches may compete with Purple Finches. This competition is suggested by BBS data that indicate a decline of Purple Finches since the spread of House Finches across the state. House Sparrows may also compete with Purple Finches. House Sparrows displaced wintering Purple Finches at bird feeding stations in Virginia.[2]

[1] Master, T. 1992. Purple Finch *(Carpodacus purpureus)*. BBA (Brauning 1992):412–413.

[2] Shedd, D.H. 1990. Aggressive interactions in wintering House Finches and Purple Finches. WB 102:174–178.

House Finch *Carpodacus mexicanus*

General status: The introduced and established population of House Finches from the western U.S. are now residents in southern Canada from Manitoba east to Nova Scotia and south to Texas and Georgia and west to North Dakota and Kansas. These finches rapidly spread across Pennsylvania soon after their introduction and are now common year-round residents throughout Pennsylvania. They are familiar and frequent visitors at bird feeding stations, where they often outnumber other species of birds.

Habitat: House Finches are usually associated with human habitation where shrubs, thickets, or hedgerows are present for cover and nesting. They are occasionally found in forests away from human structures.

Seasonal status and distribution

Breeding: House Finches are common to locally abundant statewide. They are most abundant in the suburbanized habitats of southeastern Pennsylvania (see Map 42) but achieve high numbers around towns and cities across the state. The BBS reported an increase of 15% per year since 1966, but this does not adequately reflect the phenomenal expansion this species has experienced. A 30% decline was detected in Penn-

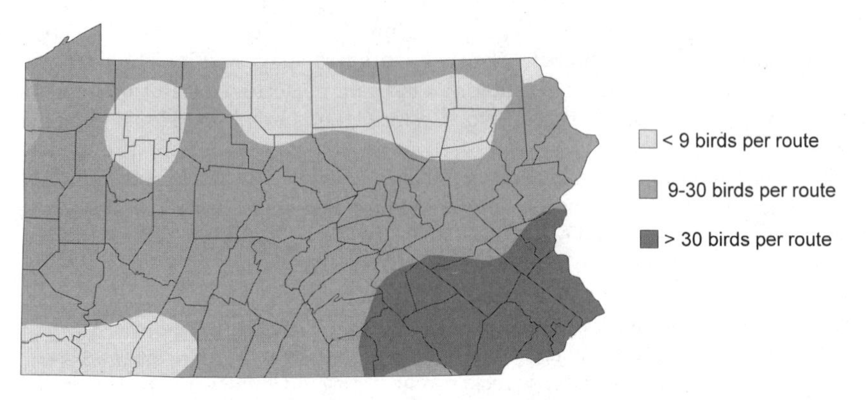

42. House Finch relative abundance, based on BBS routes, 1985–1994

sylvania on the BBS in 1996, probably a direct result of a conjunctivitis (eye disease) epidemic. Nests are usually placed in ornamental evergreens and occasionally in hanging flowerpots, nest boxes, or building crevices. Most eggs dates come from the Cornell Nest Record program and span from May to Jul.

History: House Finches were introduced from the western U.S. in the late 1930s to pet dealers in New York City,[1] and in 1940 some were released.[1,2] They rapidly spread south through New Jersey and then west into southeastern Pennsylvania in the mid-1950s (Long 1981). The first Pennsylvania record of House Finch was obtained on 1 Dec 1955, when one was trapped and photographed by W. Middleton at Norristown in Montgomery Co.[3] Winter visitors were first found in the southeastern counties in 1955 (Poole 1964), and probable nesting was first reported in 1962 in Montgomery Co.[4] Nesting was confirmed by two pairs in Berks Co. in 1969 (Uhrich 1997) and in Philadelphia Co. in 1971.[5] They had spread to Centre Co. by 1969 and were nesting in Erie Co. by 1977 (Wood 1979).

[1] Woods, R.S. 1968. House Finch. Pages 290–314 *in* Life histories of North American cardinals, grosbeaks, buntings, finches, sparrows, and allies: Part 1, ed. A.C. Bent and O.L. Austin Jr. U.S. National Museum Bulletin No. 237. Washington, D.C.

[2] Elliot, J.J., and R. Arbib. 1953. Origin and status of the House Finch in the eastern U.S. Auk 70: 31–37.

[3] Potter, J.K. 1956. Middle Atlantic Coast region. AFN 10:241.

[4] Middleton, R.J. 1963. Probable breeding of the House Finch in Pennsylvania. Cassinia 47:40.

[5] Pepper, W.E. 1972. Nesting House Finch in Pennsylvania. Cassinia 53:46.

Red Crossbill *Loxia curvirostra*

General status: In North America, Red Crossbills breed in Alaska and Canada from British Columbia east across southern Ontario and Quebec to Newfoundland. In the U.S., they breed primarily in the mountains of the West, across the northern-tier to the New England states south to New York, and locally in the Appalachians to North Carolina and Tennessee. They winter throughout their breeding range and wander irregularly and sporadically over most of the U.S. They move south of their normal range during years of low cone production. Reports of at least one or two birds are made somewhere in Pennsylvania every year. However, during invasion years there may be scattered flocks throughout the state. Most birds are usually observed in the mountainous regions. The two crossbill species have been observed together, especially during 1997–1998, but they typically do not mix. Red Crossbills have occasionally bred in Pennsylvania, especially after invasion years.

Habitat: Red Crossbills most often are found in conifers, usually preferring the cones of white, pitch, scots, and red pine; but they will take seeds from hemlock, spruce, American larch, and occasionally deciduous trees. They are often detected by voice as they are flying overhead or from the cracking sound made by their extracting seeds from cones in the top of a conifer. Red Crossbills occasionally visit bird feeding stations.

Seasonal status and distribution

Fall through spring: They may appear anywhere in the state during any month of the year, but during invasion years they often appear during or after a cold front in the third or fourth week of Oct. Since 1960 notable invasion years have been 1960–1961, 1967, 1969–1973, 1980–1982, 1984–1985, 1987–1989, 1991, and 1997–1998. Flock sizes have varied from a few birds to as many as 150 or occasionally more. Most flocks reported are of fewer than 35–40 birds. However, the most spectacular irruption of this species in recorded history in Pennsylvania occurred in 1997–1998, when Red Crossbills were widespread across the state. At Cook Forest State Park in Clarion Co., small flocks began to be seen in Dec, and their numbers continued to build until well over an estimated 1000 birds were present by Jan.[1] The numbers remained high through Feb before falling off in Mar. Smaller flocks were seen throughout the state until at least late Mar 1998. Large numbers of White-winged Crossbills were frequently observed with them. During some years Red Crossbills appear in great numbers during Oct and Nov, are absent for part of the winter, and then make sporadic appearances during the winter and spring. Most have left the state by the fourth week of Apr, with stragglers remaining to the third week of May and occasionally later.

Breeding: Breeding records of Red Crossbills in the state are very scarce but could potentially occur in conifers statewide. After invasion years, crossbills occasionally have nested, although evidence typically has involved only dependent young. Crossbills may nest at any time of year but are most likely to nest in spring and early summer. They may wander widely from breeding areas, so observations of family groups during the summer do not necessarily indicate local nesting. Recent nesting observations have been confined in the mountainous regions. Adults were observed feeding young at Leonard Harrison State Park in Tioga Co. in 1980.[2] A territorial bird on Mt. Davis in Somerset Co. in 1986 was the most likely report of nesting during the BBA project.[2] Nest building was observed in Wyoming Co. in 1993.[3] The unprecedented invasion during the winter of 1998 failed to generate any evidence of nesting activity.

History: Red Crossbills have been noted in Pennsylvania since the 1700s (Barton 1799) and have always been considered irregular, with occasional notable invasion years. These include 1887–1888, 1904–1907, 1922–1923, and 1940–1941 (Todd 1940; Poole, unpbl. ms.). Most flocks reported had fewer than 35 birds (Poole, unpbl. ms.), but during the invasion of 1887–1888 as many as 500 birds were seen together in Warren Co. (Todd 1940). They were reported more often and in greater numbers than White-winged Crossbills. Nineteenth-century records assumed that crossbills nested in our mountains but rarely provided specific evidence. The irregular nesting behavior of crossbills was noted by Gentry (1876–1877), but some authors, notably Warren (1890), failed to distinguish nonbreeding summer records from breeding populations. The best historical breeding evidence came from widely scattered locations, including expected sites in the state's mountains (Poole 1964) but also sites in the pine barrens of Chester Co.[4] No regularly occurring breeding population was ever documented.

Comments: Perhaps as many as nine subspecies of Red Crossbill are found in North America (AOU 1998). Several of these subspecies may occur in Pennsylvania.

[1] M. Leahy, pers. comm.

[2] Fingerhood, E.D. 1992. Red Crossbill *(Loxia curvirostra)*. BBA (Brauning 1992):437–438.

[3] D.A. Gross, pers. comm.; D.W. Brauning, pers. obs.

[4] Pennock, C.J. 1912. Crossbills in Chester Co., Pennsylvania, in summer. Auk 29:245–246.

White-winged Crossbill ***Loxia leucoptera***

General status: In North America White-winged Crossbills breed in Alaska, across Canada, south locally in the mountains of the West, east across the northern states, and to New England south to New York. They winter throughout their breeding range and wander irregularly and sporadically south to Arkansas, Kentucky, and North Carolina in the East. They move south of their normal range during years of low cone production. At least one or two birds are found somewhere in Pennsylvania almost every year. Scattered flocks are found throughout the state during invasion years. Most birds are usually observed in the mountainous regions of the state from fall through spring. White-winged Crossbills are accidental in summer. Except for the winter of 1998, they were reported more frequently and in greater numbers before 1950.

Habitat: White-winged Crossbills are found in conifers, usually preferring the softer cones of hemlock, spruce, and American larch, but they also will take seeds from pines with stiff cones. They are occasionally found feeding on the seeds of deciduous trees. White-winged Crossbills are often discovered by voice as they are flying overhead or by the cracking sound they make when extracting seeds from cones. They are occasionally found at bird feeding stations.

Seasonal status and distribution

Fall through spring: Like other northern finches, their irruptions often first reach Pennsylvania beginning with cold fronts during late Oct through Nov. A rather early record was at HMS in Schuylkill-Berks co. on 22 Sep 1997.[1] Since 1960 notable invasions have been during 1963–1966, 1969–1970, 1971–1972, 1977–1978, 1981–1982, 1985–1986, 1989–1990, and 1997–1998. Most flocks during invasion years consist of fewer than 35–40 birds, but counts of hundreds of birds have been observed. At PISP on 14 Nov 1981 a flock consisting of about 450 birds was observed crossing Lake Erie from Canada.[2] This flock signaled the beginning of the 1981–1982 invasion. Another large invasion of White-winged Crossbills was observed in Bradford Co. on 27 Nov 1989, when an estimated 800 birds were observed at SGL 36.[3] Probably one of the largest invasions ever recorded in Pennsylvania occurred across most of the state in 1997–1998, when over 1000 birds were present through the winter at Cook Forest State Park in Clarion Co.[4] They were frequently observed mixed with Red Crossbills. Once a flock has found a tract of coniferous trees, they usually remain through the winter or until the seeds from all the cones have been consumed. Most birds leave the state by the end of Mar or the first half of Apr, but some have remained to the second week of Jun.

Summer: Though less likely to occur in the summer than Red Crossbills, they have been recorded at least four times since 1960. One was with a group of House Finches at Chestnut Hill in Philadelphia Co. on 30 Jul 1972,[5] and in 1977, an immature male was seen at a feeder at HMS on 15–16 Aug.[6] A female White-winged Crossbill visited a feeder at Lincolnville in Crawford Co. from 10 to 18 Jul 1985;[7] and 15 birds were counted at Elmhurst in Lackawanna Co. on 1 Aug 1989.[8]

History: Both Poole (unpbl. ms.) and Todd (1940) wrote that this species was more irregular and less frequent in Pennsylvania than the Red Crossbill. Apparently the first recorded invasion in western Pennsylvania was in 1874–1875 (Todd 1904). Warren (1890) wrote that they were common in 1889–1890 in the northeastern part of the state. There were invasions of varying numbers throughout the state during 1906–1907, 1922–1923, 1940–1941, and 1952–1957. Flock sizes were generally about the same as what is reported now, but, on rare occasions, some reached hundreds of birds (Todd 1940).

[1] Haas, F.C., and B.M. Haas, eds. 1997. Birds of note—July through September 1997. PB 11:149.
[2] G.M. McWilliams, pers. obs.
[3] Reid, W. 1989. County reports—October through December 1989: Bradford County. PB 3:138.
[4] M. Leahy, pers. comm.
[5] Scott, F.R., and D.A. Cutler. 1972. Middle Atlantic Coast region. AB 26:844.
[6] Buckley, P.A., R.O. Paxton, and D.A. Cutler. 1978. Hudson-Delaware region. AB 32:188.
[7] Hall, G.A. 1985. Appalachian region. AB 39:914.
[8] Klebauskas, G. 1989. County reports—July through September 1989: Lackawanna County. PB 3:108.

Common Redpoll *Carduelis flammea*

General status: In North America, Common Redpolls breed in Alaska across northern Canada, south to southeastern Quebec, and east to Newfoundland. They winter in the southern part of their breeding range and across southern Canada and the northern U.S. Common Redpolls are irregular or casual in winter south to Arkansas, Alabama, and South Carolina in the East. During major invasion years, they may be locally abundant throughout Pennsylvania, but during many years they can be quite rare and are usually found only in the more northern counties or in the mountain valleys. Irruption years of redpolls are usually more frequent than for other finches such as Pine Grosbeaks and crossbills, but intervals of several years have passed when they were very rare or absent. They frequently associate with Pine Siskins and American Goldfinches.

Habitat: Common Redpolls inhabit open woodlands or clearcuts with stands of birches or American larch, open swamps or wet areas bordered by alders, and grassy fields. They also visit bird feeding stations. Common Redpolls are frequently found in parks or suburban areas of towns and cities where various species of birches are present.

Seasonal status and distribution

Fall through spring: As with other northern finches, the arrival of birds during the last two weeks of Oct or the first couple of weeks in Nov usually indicates the onset of an invasion year. However, there have been years of heavy influx in which birds did not arrive until late Nov or Dec. The greatest number of birds is usually found within the first few weeks of their arrival in the northwestern portion of the state and perhaps eastward across the northern-tier counties. As winter progresses, flock sizes begin to dwindle, and they wander more widely. Toward the end of the winter season, birds have begun to drift south and east, and by Feb or Mar large numbers begin to appear in the valleys and lowlands across the southern half of the state. By the second week of Mar, there is often a sudden influx of birds in the northwestern portion of the state that probably represents returning birds. Flock sizes vary from a few birds to hundreds during invasion years, whereas in noninvasion years there may be as few as one or two birds reported for the entire state. Common Redpolls usually move through Pennsylvania more rapidly on their return trip, and most leave the state by the second week of Apr, with stragglers to the fourth week of Apr.

History: Very little has changed throughout this species' history in Pennsylvania, except that in the old days they were rarely reported before midwinter anywhere in the state. Poole (unpbl. ms.) and Todd (1940) describe invasion years less frequently than they have been reported recently. Todd (1940) cited notable invasions during 1875, 1911–1912, and 1916–1917 and wrote that "for the years between and since, there are only a few scattered records." Poole (unpbl. ms.) wrote of invasions following Todd's accounts, during the years 1946–1947, 1953, and others of "less magnitude" in 1956 and 1958.

Hoary Redpoll ***Carduelis hornemanni***

General status: In North America, Hoary Redpolls breed in western and northern Alaska across northern Canada from Yukon east to Manitoba, northern Quebec and Labrador. They winter in their breeding range south, irregularly, to southern Canada and in the eastern U.S. from Illinois to Maryland. A few Hoary Redpolls sometimes accompany flocks of Common Redpolls, especially during strong southward movements. Since 1960 there have been dozens of reports. Unfortunately only four have been well documented; the accepted records have all been from bird feeding stations between 1994 and 1998. Three of the four records were in 1994 with the most recent Common Redpoll invasion.

Seasonal status and distribution

Winter to spring: The first accepted record in Pennsylvania was of a bird identified by R. Grubb at Perkiomenville in Montgomery Co. on 30 Jan 1994;[1] the second was found and photographed by F. C. Haas and B. M. Haas and was seen by many birders at Churchtown in Lancaster Co. from 1 to 14 Mar 1994.[2] There was also a Hoary Redpoll identified by R. Leberman at Meadville in Crawford Co. on 19 Mar 1994.[3] The latest record was of one observed and well described by a visiting British birder, J. L. Muddeman, at Henningsville in Berks Co. on 18 Feb 1996.[4]

History: Todd (1940) cited two records from western Pennsylvania: one identified by C. S. Beardslee on 9 Jan 1934 at Presque Isle in Erie Co., probably the same bird seen by G. M. Cook at the same location on 28 Jan 1934. Poole (unpbl. ms.) cited one identified by M. Broun at HMS from 18 to 23 Mar 1956 and one by E. L. Poole near Fleetwood in Berks Co. on 26 Feb 1960.

Comments: Separating this species from Common Redpoll requires careful observation and familiarity with aging characteristics and sexing and the various degrees of plumage coloration in Common Redpolls. Many birds reported as Hoary Redpolls have shown more of the characteristics of very pale Common Redpolls. In one instance, a partial albino Common Redpoll was photographed in Crossingville, Crawford Co., in late Mar 1996.[5]

[1] Freed, G. 1994. Notes from the field: Montgomery County. PB 8:39.

[2] Heller, J. 1994. Notes from the field: Lancaster County. PB 8:38.

[3] Pulcinella, N. 1995. Sixth report of the Pennsylvania Ornithological Records Committee (P.O.R.C.) June 1995. PB 9:72.

Hoary Redpoll at Caernarvon Township, Lancaster County, March 1994. (Photo: Franklin C. Haas)

[4] Keller, R. 1996. Local notes: Berks County. PB 10:16.
[5] A. Troyer, pers. comm.

Pine Siskin ***Carduelis pinus***

General status: Pine Siskins breed from Alaska across southern Canada, south through the western U.S., and south to Missouri, Indiana, Ohio, New Jersey, and through the Appalachians to North Carolina and Tennessee. They winter throughout their breeding range and the remainder of the U.S. and Mexico. Like other northern finches, Pine Siskins are irruptive. In some years, they are abundant in Pennsylvania, while in other years quite rare, but winter flocks are found almost annually at least in the northern mountains. The greatest number of birds is found across the mountainous areas of the state. Following the years of greatest abundance some may stay and nest in widely scattered sites throughout the state. They frequently associate with Common Redpolls or American Goldfinches from fall through spring.

Habitat: Siskins are found in mixed or coniferous woodlands, especially those with hemlock, spruce, or American larch. They also are found in open woodlands or clearcuts where there are stands of birches, in wet areas bordered by alders, in grassy fields, and at bird feeding stations. They will use ornamental spruce plantings for nesting.

Seasonal status and distribution

Fall through spring: They usually arrive earlier in the state than other northern finches. Pine Siskins may appear as early as the third week of Sep. Most arrive after the passage of the first cold front during the third or fourth week of Oct. The number of birds varies from year to year, but sizable flocks containing hundreds of birds have been reported from Nov through the winter and into the spring. After some years of heavy movements, flocks of siskins may remain until the second or third week of May. A few breeding or nonbreeding stragglers may remain through the summer.

Breeding: Pine Siskins are rare irregular nesting birds, most often found in the northern mountainous regions, where they may occur in considerable numbers after major irruptions. Siskins are not known to nest regularly anywhere in the state, but suitable habitat exists in Warren and McKean cos. and south in the Laurel Highlands, where many breeding records have been reported. Although nesting could occur anywhere in the state, few nesting records have been obtained in the southern tier, and none in the western-tier counties. Nesting attempts have been observed at

PNR in about half of the years between 1980 and 1998, but colonies are nearly always destroyed by grackles.[1] Breeding has been confirmed most often by reports of fledged young around backyard conifers, but nesting activity is probably overlooked in remote coniferous areas. The breeding status may be confused by winter flocks that linger well into the nesting season. When breeding, flocks break into groups of about a half-dozen birds and nest semicolonially.[2] Nests containing eggs or young have been found from Apr 1 into early May, but nesting dates vary from year to year and nests should be looked for as early as Mar. Nests are placed at various heights from the ground, well out from the trunk of a conifer. The last major invasion occurred in 1988, when most BBA records were reported.[3]

History: The pattern of movements into Pennsylvania has changed little over the years except that invasions may have been less frequent in the past. The first confirmation of nesting in Pennsylvania was in Warren Co. in 1912.[4] Nesting was reported widely across the northern tier in 1925.[5] Historical breeding records came from across the High Plateau.

1 R.C. Leberman, pers. comm.
2 Peterson, J.M.C. 1988. Pine Siskin *(Carduelis pinus)*. Pages 494–495 *in* The Atlas of Breeding Birds in New York State, ed. R.F. Andrle and J.R. Carroll. Cornell University Press, Ithaca, N.Y.
3 Gross, D.A. 1992. Pine Siskin *(Carduelis pinus)*. BBA (Brauning 1992):416–417.
4 Simpson, R.S. 1912. The Pine Siskin. Oologist 29:372–373.
5 Harlow, R.C. 1951. Tribal nesting of the Pine Siskin in Pennsylvania. Cassinia 38:4–9.

[Lesser Goldfinch] *Carduelis psaltria*

General status: This western species is resident in the U.S. from Washington south through California and east to Colorado, New Mexico, and Central Texas. They are casual or accidental in eastern North America in Missouri, Arkansas, Louisiana, Kentucky, North Carolina, Maine, and Ontario (AOU 1998). This species has been reported once in Pennsylvania.

Comments: A Lesser Goldfinch was studied by an experienced birder on 3 Feb 1982 at Meadville, Crawford Co.[1] The bird was found again in nearby Saegertown on 22–25 Mar 1982 in Crawford Co. (Leberman 1988). Because this bird may have been an escapee, it is listed as hypothetical.

1 Hall, G.A. 1982. Appalachian region. AB 36:295.

American Goldfinch *Carduelis tristis*

General status: American Goldfinches breed across southern Canada from British Columbia east to Newfoundland and primarily across the northern two-thirds of the U.S. They winter over most of their breeding range and throughout the remainder of the U.S. American Goldfinches are widespread, found throughout most of the year in all regions of Pennsylvania. They are common to abundant regular migrants, and they breed statewide. The number of birds may vary from year to year in winter. They occasionally associate with Pine Siskins and Common Redpolls.

Habitat: American Goldfinches inhabit pastures, old or abandoned weedy or brushy fields, and open woodlands and woodland edges, as well as marshes, where they extract cattail down for nesting material. In winter they are also found where there are birches, alders, or American larches and at bird feeding stations.

Seasonal status and distribution

Spring: An increase in numbers beginning the last week of Mar or the first week of Apr indicates northbound movement. Their usual migration period is from the second or third week of Apr to the fourth week of May. Peak migration occurs from the fourth week of Apr to the second week of May.

Breeding: Goldfinches may be found commonly statewide except in the central counties, where they are uncommon. They may be found even in our largest cities, but are less common there, and in uniformly forested areas (see Map 43). They initiate nesting later than any other bird in Pennsylvania. Few nests have been found before 5 Jul, and nests with eggs regularly are found through the end of Aug and occasionally into Sep. Nesting is synchronized with thistle seed production. Nests are tight cups containing plant down, placed in the crotch of a shrub or small tree, just a few feet (occasionally more than 15 feet) above the ground. Typical of populations of birds of brushy habitats, goldfinch populations have been declining on the BBS by 3.7% per year since 1966.

Fall: The large number of nesting birds and their wide breeding range make it difficult to determine the beginning of fall migration. The end of the nesting season and the beginning of fall migration may overlap from the middle of Aug to perhaps the first week of Sep. Banding records show a noticeable increase in birds during the third or fourth week of Aug, but most birds are probably not southbound until the third or fourth week of Sep.[1] At northern sites, such as along the Lake Erie Shore, goldfinches become less common by the first week of Oct, with fewer birds recorded through the fall. However, at more-southern sites, such as the Coastal Plain, high numbers may still be found well into Dec or later. At PNR numbers may remain high into Nov (Leberman 1976).

Winter: American Goldfinches winter throughout Pennsylvania, but they are most common in the lowland valleys in the southern half of the state. Their variability in numbers is most notable at higher elevations in the mountains, across the northern-tier counties, and in the snowbelt area of the Glaciated Northwest, where they may be quite rare in some winters and rather common in other winters.

[1] Stewart, R.E., and C.S. Robbins. 1958. Birds of Maryland and the District of Columbia. U.S. Fish and Wildlife Service. North American Fauna No. 62.

Evening Grosbeak ***Coccothraustes vespertinus***

General status: This handsome finch breeds across the boreal forest of southern Canada and through the mountain-

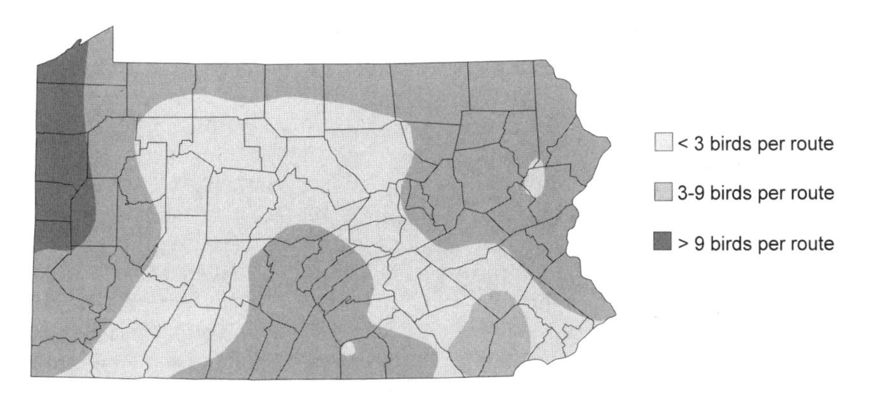

43. American Goldfinch relative abundance, based on BBS routes, 1985–1994

ous regions of the western U.S., and in the eastern U.S. as far south as New York and Massachusetts. They winter throughout their breeding range and south through most of the U.S., except from southern Texas and the southern Gulf states to South Carolina. Evening Grosbeaks are nomadic, wandering widely in search of food. They are regular residents or visitors to Pennsylvania in the mountainous regions from the fall through spring, but the number of birds varies from year to year. Evening Grosbeaks are irregular visitors to the remainder of the state, where they are usually found only during years of heavy movements. There have been infrequent summer records since 1960, primarily from the northern counties. The summer records suggest possible breeding in the state; and in 1994 breeding was confirmed. Evening Grosbeak irruptions into the state have become rarer and less frequent since 1990.

Habitat: They prefer mixed woodlands and coniferous forests, where they feed on a wide variety of buds, fruits, and seeds. Away from bird feeding stations Evening Grosbeaks are more often heard flying overhead than seen perched. In winter they often invade towns and cities, especially where sunflower seeds are offered at feeding stations.

Seasonal status and distribution

Fall through spring: Evening Grosbeaks have been recorded as early as the fourth week of Sep, but generally they first appear during the passage of a cold front from the second week of Oct to the second week of Nov. Flocks numbering anywhere from a few to several hundred birds may be found any time after the second week of Oct through the winter and into early spring. Most Evening Grosbeaks have left the state by the third week of Apr, but often there is a brief influx of northbound birds from the fourth week of Apr to the second week of May. Stragglers may remain to at least the second week of Jun.

Breeding: They have been recorded in Jun, Jul, and Aug in Pennsylvania, but most summer sightings have been during the second half of Aug. Most observations have been from the northern counties including Crawford, Erie, Potter, Warren, and Wyoming. However, a flock of 15 was observed at Holicong in Bucks Co. on the unlikely dates of 24–26 Aug 1962.[1] Also out of its normal range was one at Peach Ridge in Juniata Co. on 1 Jul 1996.[2] On 11 Jul 1994 nesting was confirmed in the state by S. Conant and D. Brauning when two juveniles were photographed being fed by adults at Schmitthenner Lake in Forkston Township in western Wyoming Co.[3] Recently fledged young and families were also observed near Coalbed Swamp and nearby locations in July 1994.[4] Adults were reported feeding young during the BBA project in McKean Co. but were never adequately documented.[5] Wood (1979) suggested that Evening Grosbeaks were "Perhaps a casual breeder in some northern counties," but specific documentation was lacking.

History: Evening Grosbeaks were essentially unknown east of Ohio or Ontario before the winter of 1890, when they invaded the New England states and Pennsylvania. They were recorded in our state primarily from northern counties, but they did manage to penetrate as far south as Allegheny Co. (Todd 1940). Most historical observations were from Feb through Apr, but a specimen was taken at Meadville, Crawford Co., on the unusual date of 21 Jul 1910 (Todd 1940). Occasionally birds lingered until mid-May, and a few summer records were obtained. They were periodically reported after the

1889–1990 invasion during 1910–1911, 1914, 1920, 1922, 1926, 1929–1930, and 1933–1936 (Todd 1940; Poole unpbl. ms.). From 1942 through at least the late 1980s they became more frequent and occurred in greater numbers; since then, both frequency and numbers have declined.

Comments: An unusual phenomenon of an east-west migration by Evening Grosbeaks had been discussed as early as 1892[6] and was documented from banding returns as early as 1928.[7] Banding studies at PNR in Westmoreland Co. have confirmed that of 77 banding recoveries, most birds came from the northeastern population (e.g., New Brunswick) and returned to the northeast along the mountain ridges in spring.[8,9]

[1] Scott, F.R., and D.A. Cutler. 1963. Middle Atlantic Coast region. AFN 17:21.

[2] Haas, F.C., and B.M. Haas, eds. 1996. Birds of note—July through September 1997. PB 10:151.

[3] Conant, S. 1994. First confirmed Evening Grosbeak nest in Pennsylvania. PB 8:133–135.

[4] D.A. Gross, pers. comm.

[5] Brauning, D.W. 1994. Comments on the breeding Evening Grosbeaks in Pennsylvania. PB 8:135.

[6] Butler, A.W. 1892. Some notes concerning the Evening Grosbeak. Auk 9:238–247.

[7] Bulletin Northeastern Bird-Banding Association. 1928. 4:56–59.

[8] Leberman, R.C., and A. Heimerdinger. 1966. Bird-banding at Powdermill, 1965. Research report no. 15. Powdermill Nature Reserve, Carnegie Museum of Natural History, Pittsburgh.

[9] Leberman, R.C., and M.H. Clench. 1969. Bird-banding at Powdermill. Research report no. 26. Powdermill Nature Reserve, Carnegie Museum, p. 12.

Family Passeridae: Old World Sparrows

These Old World birds are similar to New World sparrows, but they have shorter legs and a heavier bill. Two species have been introduced into North America, and one, the House Sparrow, has flourished wherever humans have settled. House Sparrows are aggressive and will readily take over nest sites of other cavity-nesting species.

House Sparrow *Passer domesticus*

General status: House Sparrows were introduced into North America in 1853 (Bent 1958) and became one of the most abundant birds in the U.S. Along with two other naturalized species, the European Starling and the Rock Dove, this species is closely associated with human habitation but is unpopular among many people. In North America they are resident from southern Alaska across Canada from southern Yukon and Mackenzie, east through the southern half of Quebec to Newfoundland and south throughout the U.S. House Sparrows are permanent residents throughout the state. Studies have shown a decline in this species' population (Forbush 1929) since its peak around the beginning of the twentieth century.

Habitat: House Sparrows are found anywhere around human habitation, particularly around farms and especially in urban areas of the largest cities where they are found foraging for food scraps on sidewalks, streets, and around buildings.

Seasonal status and distribution

Breeding: They are less common in extensively forested areas across the north-central mountains (see Map 44) than elsewhere. House Sparrows are most common in the southeast and northwest and are, with starlings and pigeons, among the most prominent of urban residents. House Sparrows nest almost anywhere: in cavities of buildings, on building ledges, in forks of trees, in ivy thickets, in natural tree cavities, in bird houses erected for bluebirds, and even in Cliff Swallow nests or cavities excavated by Bank Swallows. Their nests are a bulky construc-

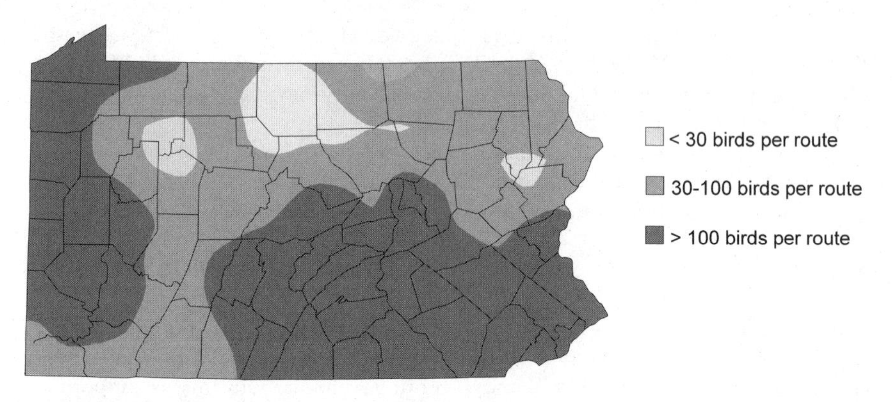

44. House Sparrow relative abundance, based on BBS routes, 1985–1994

tion of grasses, string, paper, cloth, and feathers, packed inside a cavity. When in the fork of a tree, the nest is built into a large untidy dome of the same material listed above, with the entrance hole in the side. Usually five white, pale green, or blue eggs, finely and uniformly spotted with brown and gray, are laid. House Sparrow populations are experiencing a declining trend of 3.0% per year, although they still are among the top five species on BBS routes, with an average of 39 birds per route statewide. The nesting season is long, ranging from mid-Apr through mid-Jul.

History: House Sparrows were introduced into North America multiple times during the mid-nineteenth century and were well established across Pennsylvania by the time of Warren's writing (1890). They were thought to have reached a peak population by the early 1900s, and declined when the automobile replaced the horse as the primary means of transportation.

LITERATURE CITED

Alerich, C.L. 1993. Forest statistics for Pennsylvania—1978 and 1989. Resource Bulletin NE-126. U.S. Department of Agriculture, Northeast Forest Experiment Station, Radnor.

American Ornithologists' Union (AOU). 1957. Check-list of North American birds, 5th ed. Lord Baltimore Press, Baltimore.

American Ornithologists' Union (AOU). 1983. Check-list of North American birds, 6th ed. Allen Press, Lawrence, Kans.

American Ornithologists' Union (AOU). 1998. Check-list of North American birds, 7th ed. Allen Press, Lawrence, Kans.

Andrle, R.F., and J.R. Carroll, eds. 1988. The atlas of breeding birds in New York State. Cornell Univ. Press, Ithaca, N.Y.

Audubon, J.J. 1831–1839. Ornithological biography, vols. 1–5. Adam and Charles Black, Edinburgh.

Audubon, J.J. 1840. (ca. 1840–1844). The birds of America, vols. 1–7. J.B. Chevalier, Philadelphia.

Baird, S.F. 1845. Catalogue of birds found in the neighborhood of Carlisle, Cumberland County, Pennsylvania. Literary Record and Journal of the Linnaean Association of Pennsylvania College 1:249–257.

Barton, B.S. 1799. Fragments of the natural history of Pennsylvania. James and Johnson, Philadelphia.

Beals, E. 1960. Forest bird communities in the Apostle Islands of Wisconsin. Wilson Bulletin 72:156–181.

Beck, H.H. 1924. A chapter on the ornithology of Lancaster County, Pennsylvania, with supplementary notes on the mammals. Pages 1–39 *in* Lancaster County, Pennsylvania—A history. Lewis Historical Publishing Company, New York.

Beck, H.H. 1955. Birds of the Lower Susquehanna Lake Region. Pennsylvania Water and Power Company.

Bell, R.K. 1994. The summer birds of Greene County, Pennsylvania: A 100-year update, 1893–1993. The Redstart 61:119–140.

Bellrose, F.C. 1980. Ducks, geese, and swans of North America. Stackpole Books, Harrisburg.

Bent, A.C. 1938. Life histories of North American birds of prey: Part 2. Smithsonian Institution U.S. National Museum Bulletin No. 174. Washington, D.C.

Bent, A.C. 1950. Life histories of North American wagtails, shrikes, vireos, and their allies. Smithsonian Institution U.S. National Museum Bulletin No. 197. Washington, D.C.

Bent, A.C. 1953. Life histories of North American wood warblers: Part 1. Smithsonian Institution U.S. National Museum Bulletin No. 203. Washington, D.C.

Bent, A.C. 1958. Life histories of North American blackbirds, orioles, tanagers, and allies. Smithsonian Institution U.S. National Museum Bulletin No. 211. Washington, D.C.

Brauning, D.W., ed., 1992. Atlas of breeding birds in Pennsylvania. Univ. of Pittsburgh Press, Pittsburgh.

Broun, M. 1949. Hawks aloft: The story of Hawk Mountain. Dodd, Mead Company, New York.

Buckelew, A.R., Jr., and G.A. Hall. 1994. The West Virginia breeding bird atlas. Univ. Pittsburgh Press, Pittsburgh.

Bull, J., and J. Farrand Jr. 1977. The Audubon Society guide to North American birds—Eastern region. Alfred A. Knopf, New York.

Burleigh, T.D. 1931. Notes on the breeding birds of State College, Centre County, Pennsylvania. Wilson Bulletin 43:37–54.

Burns, F.L. 1919. The ornithology of Chester County, Pennsylvania. Gorham Press, Boston.

Cadman, M.D., P.F.J. Eagles, and F.M. Helleiner. 1987. Atlas of breeding birds of Ontario. Univ. Waterloo Press, Waterloo, Ontario.

Carter, J.D. 1904. Summer birds of Pocono Lake, Monroe County, Pennsylvania. Cassinia 8:29–35.

Carter, J.D. 1917. Summer birds of Pocono. Cassinia 21:6–25.

Cashen, S.T., and M.C. Brittingham. 1998.

Avian use of restored wetlands in Pennsylvania. Final Report to Pennsylvania Game Commission, Harrisburg.

Cassin, J. 1862. Birds of Delaware County. Pennsylvania, ed. G. Smith. Henry B. Ashmead, Philadelphia. Pages 435–439 *in* History of Delaware County.

Chandler, R.J. 1989. The Facts on File field guide to North Atlantic shorebirds. Facts on File, Inc., New York.

Christy, B.H., and G.M. Sutton. 1928. The summer birds of Cook Forest. Cardinal 2:68–75.

Conway, A.F. 1940. Checklist of the birds of Chester County, Pennsylvania. Mimeographed. State Teachers Collection, West Chester.

Cope, F.R., Jr. 1898. The summer birds of Susquehanna County, Pennsylvania. Proceedings of the Academy of Natural Sciences of Philadelphia 50:76–88.

Cope, F.R., Jr. 1902. Observations on the summer birds of parts of Clinton and Potter counties, Pennsylvania. Cassinia 5:8–21.

Davis, W.H., and P. Roca. 1995. Bluebirds and their survival. Univ. Press of Kentucky, Lexington.

Diefenbach, D.R. 1996. Game take and furtaker survey. Annual report. Pennsylvania Game Commission, Harrisburg.

Dunn, J., and K. Garrett. 1997. Warblers. Peterson Field Guide Series. Houghton Mifflin, Boston.

Dunn, J. and K. Jacobs. 1995. Waterfowl population monitoring. Pennsylvania Game Commission Annual Report, Project no. 51004, Harrisburg.

Dwight, J., Jr. 1892. Summer birds of the crest of the Pennsylvania Alleghenies. Auk 9: 129–141.

Ehrlich, P.R., D.S. Dobkin, and D. Wheye, 1988. The birder's handbook: A field guide to the natural history of North American birds. Simon and Schuster, New York.

Erskines, A.J. 1992. Atlas of breeding birds of the Maritime Provinces. Nova Scotia Museum and Nimbus, Halifax, Nova Scotia.

Forbush, E.H. 1929. Birds of Massachusetts and other New England states, vol. 3, Land birds from sparrows to thrushes. Massachusetts Department of Agriculture, Norwood Press, Norwood, Mass.

Frey, E.S. 1943. The centennial check-list of the birds of Cumberland County, Pennsylvania and her borders, 1840–1943. Published privately, Lemoyne, Pa.

Genoways, H.H. and F.J. Brenner, eds. 1985. Chapter 5 *in* Species of special concern in Pennsylvania, ed. F.B. Gill. Carnegie Museum of Natural History. Special Publication No. 11. Carnegie Museum of Natural History, Pittsburgh.

Gentry, T.G. 1876–1877. Life histories of the birds of eastern Pennsylvania, vol. 1. Published privately, Philadelphia.

Gentry, T.G. 1877. Life histories of the birds of eastern Pennsylvania, vol. 2. The Naturalist's Agency, Salem, Mass.

Gill, F.B. 1990. Ornithology. W.H. Freeman, New York.

Godfrey, E.W. 1986. Birds of Canada, rev. ed. National Museum of Natural Sciences; National Museums of Canada, Ottawa, Canada.

Grant, P.J. 1986. Gulls, a guide to identification. Buteo Books, Vermillion, S.D.

Grimm, J.W., and R.H. Yahner. 1986. Status and management of select species of avifauna in Pennsylvania with emphasis on raptors. 1985 final report to Pennsylvania Game Commission, Harrisburg.

Grimm, W.C. 1952. Birds of the Pymatuning region. Pennsylvania Game Commission, Harrisburg.

Grube, G.E. 1959. The breeding birds of Adams County, Pennsylvania. Lock Haven Bullletin 1:45–57.

Gullion, G.W. 1970. Factors influencing Ruffed Grouse populations. Transactions of the North American Wildlife Natural Resource Conference 35:93–105.

Hall, G.A. 1983. West Virginia birds: Distribution and ecology. Special Publication No. 7. Carnegie Museum of Natural History, Pittsburgh.

Haney, J.C., and C.P. Schaadt. 1996. Functional roles of eastern old growth in promoting forest bird diversity. Pages 76–88 *in* Eastern old-growth forests: Perspectives for rediscovery and recovery, ed. M.B. Davis. Island Press, Washington, D.C.

Harlow, R.C. 1906. Summer birds of western Pike County, Pennsylvania. Cassinia 10: 16–25.

Harlow, R.C. 1913. The breeding birds of Pennsylvania. M.S. thesis, Pennsylvania State Univ.

Harlow, R.C. 1918. Notes on the breeding birds of Pennsylvania and New Jersey. Auk 39:399–410.

Harrison, H.H. 1975. A field guide to birds' nests in the United States east of the Mississippi. Houghton Mifflin, Boston.

Harrison, C. 1978. A field guide to the nests, eggs, and nestlings of North American birds. Collins, Cleveland.

Harrison, P. 1983. Seabirds, An identification guide. Houghton Mifflin, Boston.

Hayman, P., J. Marchant, and T. Prater. 1986. Shorebirds, an identification guide. Houghton Mifflin, Boston.

Jacobs, J.W. 1893. Summer birds of Greene County, Pennsylvania. Republican Book and Job Office, Waynesburg, Pa.

Johnsgard, P.A. 1978. Ducks, geese, and swans of the world. Univ. of Nebraska Press, Lincoln.

Jones, A.L., and P.D. Vickery. undated (ca. 1996). Conserving grassland birds: Managing large grasslands including conservation lands, airports, and landfills over 75 acres, for grassland birds. Massachusetts Audubon Society, Lincoln, Mass.

Kalm, P. 1753–1761. Travels in North America, 3 vols., ed. by A.B. Benson, 1937. Dover, New York.

Kaufman, K. 1990. Advanced birding. Peterson Field Guide Series. Houghton Mifflin, Boston.

Keim, T.D. 1905. Summer birds of Port Alleghany [*sic*], McKean County, Pennsylvania. Cassinia 8:36–41.

Kitson, K. 1998. Birds of Bucks County. Bucks County Audubon Society, New Hope, Pa.

Klinger, S., J. Dunn, and J. Hassinger. 1998. Game Birds in trouble. Keystone Conservationist 1:40–45.

Kosack, J. 1995. The Pennsylvania Game Commission, 1895–1995: 100 years of wildlife conservation. Pennsylvania Game Commission, Harrisburg.

Krider, J. 1879. Forty years' notes of a field ornithologist. Privately published, Philadelphia.

Kunkle, D.E. 1951. The birds of the Lewisburg region. Bucknell Ornithology Club, Lewisburg, Pa.

Leberman, R.C. 1976. The birds of the Ligonier Valley. Carnegie Museum of Natural History, Special Publication No. 3. Pittsburgh.

Leberman, R.C. 1988. The birds of western Pennsylvania and adjacent regions. Special Publication No. 13. Carnegie Museum of Natural History, Pittsburgh.

Libhart, J.J. 1869. Birds of Lancaster County, Pennsylvania. Pages 502–516 *in* An authentic history of Lancaster County, Pennsylvania, ed. J.I. Mombert. J.E. Barr and Company, Lancaster, Pa.

Long, J.L. 1981. Introduced birds of the world. Universe Books, New York.

Madge, S., and H. Burn. 1988. Waterfowl: An identification guide to the ducks, geese, and swans of the world. Houghton Mifflin, Boston.

Michner, E. 1863. Insectivorous birds of Chester County, Pennsylvania. Pages 287–307 *in* United States Agriculture Report.

Miller, J.C. 1970. Additional notes on the birds of Tinicum. Cassinia 52:21.

Miller, J.C., and C.E. Price Jr. 1959. Birds of Tinicum. Cassinia 44:3–15.

Miller, R.F. 1946. Breeding birds of the Philadelphia region: Part 3. Cassinia 33:1–23.

Miller, R.F. 1949. The breeding birds of the Philadelphia region: Part 4. Cassinia 37:1–8.

Mombert, J.I., ed. 1869. An authentic history of Lancaster County, Pennsylvania. J. E. Barr and Company, Lancaster, Pa.

Morrin, H.B., ed. committee chairman. 1991. A guide to the birds of Lancaster County, Pennsylvania. Lancaster County Bird Club.

Morris, B.L., R.E. Wiltraut, and F.E. Brock. 1984. Birds of the Lehigh Valley area. Lehigh Valley Audubon Society, Emmaus, Pa.

Muller, E.K., ed. 1989. Pennsylvania's past. Pages 73–115 *in* The atlas of Pennsylvania, ed. D.J. Cuff, W.J. Young, E.K. Muller, W. Zelinsky, and R.F. Aber. Temple Univ. Press, Philadelphia.

Palmer, R.S., ed. 1988. Handbook of North American birds, vol. 4. Diurnal raptors. Yale Univ. Press, New Haven.

Parkes, K.C. 1956. A field list of birds of the Pittsburgh region. Carnegie Museum of Natural History, Pittsburgh.

Pennock, C.J. 1886. The birds of Chester County. Quarterly Report of the Pennsylvania Board of Agriculture 39:78–91.

Peterjohn, B.G. 1989. The birds of Ohio. Indiana Univ. Press, Bloomington.

Poole, A., and F. Gill, eds. 1992–1998. Birds of North america series. The Academy of Natural Sciences, Philadelphia, and the American Ornithologists' Union, Washington, D.C.

Poole, E.L. 1947. A half century of bird life in Berks County, Pennsylvania. Reading Public Museum and Art Gallery Bulletin No. 19. Reading, Pa.

Poole, E.L. ca. 1960. Unpublished manuscript in the department of ornithology. Academy of Natural Sciences, Philadelphia.

Poole, E.L. 1964. Pennsylvania birds, An annotated list. Livingston, Narbeth, Pa.

Price, J., S. Droege, and A. Price. 1995. The summer atlas of North American birds. Academic Press, New York.

Reimann, E.J. 1947. Summer birds of Tamarack Swamp, 1900 and 1947. Cassinia 36:17–24.

Rhoads, S.N. 1899. Notes on some of the rarer birds of western Pennsylvania. Auk 16: 308–313.

Richards, K.C. 1976. Some declining bird species of southeastern Pennsylvania. Cassinia 55:33–36.

Robbins, C.S., and E.A.T. Blom. 1996. Atlas of the breeding birds of Maryland and the District of Columbia. Univ. Pittsburgh Press, Pittsburgh.

Robbins, C.S., D. Bystrak, and P.H. Geissler. 1986. The breeding bird survey: Its first 15 years, 1965–1979. U.S. Dept. of the Interior, Fish and Wildlife Service Resource Publication 157. Washington, D.C.

Robbins, C.S., D.K. Dawson, and B.A. Dowell. 1989. Habitat area requirements of breeding forest birds of the middle Atlantic states. Wildlife Monograph No. 103. The Wildlife Society.

Rosenberg, K.V., and J.V. Wells. 1996. Importance of geographic areas to Neotropical migrant birds in the northeast. Report submitted to U.S. Fish and Wildlife Service, Hadley, Mass.

Sauer, J.R., J.E. Hines, G. Gough, I. Thomas, and B.G. Peterjohn. 1997. The North American breeding bird survey results and analysis. Version 96:4. Patuxent Wildlife Research Center, Laurel, MD. http://www.mbr.nbs.gov/bbs/bbs.html

Sauffer, D.F., and L.B. Best. 1980. Habitat selection by birds of riparian communities: Evaluating effects of habitat alterations. Journal of Wildlife Management 44:1–15.

Schweinsberg, A.R. 1988. Birds of the Central Susquehanna Valley. Privately published.

Stone, W. 1891. The summer birds of Harvey's Lake, Luzerne County, Pennsylvania, with remarks on the faunal position of the region. Proceedings of the Academy of Natural Sciences of Philadelphia 43:431–438.

Stone, W. 1894. Birds of eastern Pennsylvania and New Jersey. Delaware Valley Ornithological Club, Philadelphia.

Stone, W. 1900. The summer birds of the higher parts of Sullivan and Wyoming counties, Pennsylvania. Abstract of Proceedings of the Delaware Valley Ornithological Club 3:20–23.

Stone, W. 1937. Bird studies at Old Cape May. Delaware Valley Ornithology Club, Philadelphia.

Street, J.F. 1915. Nesting birds of Pocono Lake. Cassinia 19:14–23.

Street, J.F. 1921. Summer birds of Adams and Franklin counties, Pennsylvania. Cassinia 24:8–19.

Street, P.B. 1954. Birds of the Pocono Mountains, Pennsylvania. Cassinia 41:3–76.

Street, P.B., and R. Wiltraut. 1986. Birds of the Pocono Mountains, Pennsylvania. Cassinia 61:3–19.

Street, P.B., and R. Wiltrant. 1996. Birds of the Pocono Mountains. Cassinia 66:2–27.

Stull, J., J.A. Stull, and G.M. McWilliams. 1985. Birds of Erie County, Pennsylvania including Presque Isle. Allegheny Press, Elgin, Pa.

Sutton, G.M. 1928a. An introduction to the birds of Pennsylvania. Horace McFarland, Harrisburg.

Sutton, G.M. 1928b. The birds of Pymatuning swamp and Conneaut Lake, Crawford County, Pennsylvania. Annals of the Carnegie Museum 18:19–239.

Thomas, J. 1876. Birds of Bucks County, Pennsylvania. *In* The history of Bucks County, Pennsylvania, ed. W.W.H. Davis. Doylestown, Pa.

Thomas, L.S. 1955. Birds of Bucks County, Pennsylvania. Cassinia 39:3–26.

Thorne, S.G., K.C. Kim, and K.C. Steiner, codirectors, and B.J. Guinness, ed. 1995. A heritage for the 21st century: Conserving Pennsylvania's native biological diversity. Pennsylvania Fish and Boat Commission, Harrisburg.

Tiner, R.W., Jr. 1987. Mid-Atlantic wetlands: A disappearing natural treasure. U.S. Fish and Wildlife Service and U.S. Environmental Protection Agency, Washington, D.C.

Todd, W.E.C. 1893. Summer birds of Indiana and Clearfield counties, Pennsylvania. Auk 10:35–46.

Todd, W.E.C. 1904. The birds of Erie and Presque Isle, Erie County, Pennsylania. [Reprinted from Annals of the Carnegie Museum, Vol. II, 1904.]

Todd, W.E.C. 1940. Birds of western Pennsylvania. Univ. Pittsburgh Press, Pittsburgh.

Townsend, C.H. 1883. Notes on the birds of Westmoreland County, Pennsylvania. Proceedings of the Academy of Natural Sciences of Philadelphia 35:59–68.

Trimble, R. 1940. Changes in the bird life at Pymatuning Lake, Pennsylvania. Annals of the Carnegie Museum 28:83–132.

Turnbull, W.P. 1869. The birds of east Pennsylvania and New Jersey. Henry Grambo and Company, Philadelphia.

Uhrich, W.D., ed. 1997. A century of bird life in Berks County, Pennsylvania. Reading Public Museum, Reading, Pa.

Warren, B.H. 1890. Report on the birds of Pennsylvania. 2nd ed., revised and augmented. State Board of Agriculture, Harrisburg. [The first edition was published in 1888.]

Wilson, A. 1808–1814. American ornithology, 9 vols. Bradford and Inskeep, Philadelphia.

Wilson, A., and C.L. Bonaparte. 1870. American ornithology, 3 vols. Porter and Coates, Philadelphia.

Wood, M. 1958. Birds of the State College region. Pennsylvania Agricultural Experiment Station Bulletin No. 558.

Wood, M. 1979. Birds of Pennsylvania, when and where to find them. Pennsylvania State Univ., University Park.

Wood, M. 1983. Birds of central Pennsylvania, 3rd ed. Pennsylvania State Univ., University Park.

Yahner, R.H., and R.W. Rohrbaugh. 1996. Long-term status and management of Northern Harriers, Short-eared Owls, and associated wildlife species in Pennsylvania; final research report—1996. Pennsylvania Game Commission, Harrisburg.

Yohn, C., and R.H. Yahner. 1995. Effects of size and shape of forest clearcut stands on breeding-bird communities in a northern hardwood forest. Final Report to International Paper Company.

Young, R.T. 1896. Summer birds of the anthracite coal regions of Pennsylvania. Auk 13: 278–285.

Zeleny, L. 1976. The bluebird. Indiana Univ. Press, Bloomington.

Zelinsky, W., ed. 1989. Human patterns. Pages 116–165 *in* The atlas of Pennsylvania, ed. D.J. Cuff, W.J. Young, E.K. Muller, W. Zelinsky, and R.F. Aber. Temple Univ. Press, Philadelphia.

INDEX OF ENGLISH NAMES

INDEX OF SCIENTIFIC NAMES

About the Authors

JERRY MCWILLIAMS has been fascinated with birds and butterflies since childhood. He began collecting butterflies and watching birds in the vicinity of his home in southern Erie County in his teens. He later expanded his travels from Pennsylvania to across the United States and worldwide to places such as Costa Rica, Trinidad, Ecuador, Peru, Kenya, Australia, and Papua New Guinea in pursuit of his obsession. Throughout his travels he has amassed a large collection of bird and butterfly photographs, many of which have been published and several of which have won awards. McWilliams is co-author of *Birds of Erie County* and has been a member of the Pennsylvania Ornithological Records Committee since its inception in 1989.

DAN BRAUNING was born and raised in Philadelphia. An interest in nature and open spaces drew him to the extensive park system of that city, and a fascination with birds was born. In pursuit of that interest, he received a B.S. in biology at Geneva College and went on to Penn State for an M.S. in ecology. After returning to Philadelphia, he began work at the Academy of Natural Sciences and was asked to coordinate the Breeding Bird Atlas project. At the completion of that project in 1990, he was hired as an Ornithologist by the Pennsylvania Game Commission.